Curves and Surfaces

Marco Abate · Francesca Tovena

Curves and Surfaces

Marco Abate
Department of Mathematics
University of Pisa (Italy)

Francesca Tovena
Department of Mathematics
University of Rome Tor Vergata
(Italy)

Translated by:
Daniele A. Gewurz
Department of Mathematics, University of Rome La Sapienza (Italy)

Translated from the original Italian edition:
M. Abate, F. Tovena: Curve e superfici, ©Springer-Verlag Italia 2006

UNITEXT – La Matematica per il 3+2
ISSN print edition: 2038-5722   ISSN electronic edition: 2038-5757

ISBN 978-88-470-1940-9   ISBN 978-88-1941-6(eBook)
DOI 10.1007/978-88-1941-6

Library of Congress Control Number: 2011934365

Springer Milan Dordrecht Heidelberg London New York

9 8 7 6 5 4 3 2 1

Cover design: Beatrice ᗺ, Milano
Cover image: inspired by "Tetroid with 24 Heptagons" by Carlo H. Sequin

Typesetting with LaTeX: PTP-Berlin, Protago TeX-Production GmbH, Germany
(www.ptp-berlin.eu)

Springer-Verlag Italia S.r.l., Via Decembrio 28, I-20137 Milano
Springer is a part of Springer Science+Business Media (www.springer.com)

Preface

This book is the story of a success.

Classical Euclidean geometry has always had a weak point (as Greek geometers were well aware): it was not able to satisfactorily study curves and surfaces, unless they were straight lines or planes. The only significant exception were conic sections. But it was an exception confirming the rule: conic sections were seen as intersections of a cone (a set consisting of the lines through a point forming a constant angle with a given line) with a plane, and so they could be dealt with using the linear geometry of lines and planes. The theory of conic sections is rightly deemed one of the highest points of classical geometry; beyond it, darkness lay. Some special curves, a couple of very unusual surfaces; but of a general theory, not even a trace.

The fact is, classical geometers did not have the language needed to speak about curves or surfaces in general. Euclidean geometry was based axiomatically on points, lines, and planes; everything had to be described in those terms, while curves and surfaces in general do not lend themselves to be studied by means of that vocabulary. All that had to be very frustrating; it suffices to take a look around to see that our world is full of curves and surfaces, whereas lines and planes are just a typically human construction.

Exeunt Greeks end Egyptians, centuries go by, Arabs begin looking around, Italian algebraists break the barrier of third and fourth degree equations, enter Descartes and discovers Cartesian coordinates. We are now at the beginning of 1600s, more than a millennium after the last triumphs of Greek geometry; at last, very flexible tools to describe curves and surfaces are available: they can be seen as images or as vanishing loci of functions given in Cartesian coordinates. The menagerie of special curves and surfaces grows enormously, and it becomes clear that in this historical moment the main theoretical problem is to be able to precisely define (and measure) how curves and surfaces differ from lines and planes. What does it mean that a curve is curved (or that a surface is curved, for that matter)? And how can we measure just *how much* a curve or a surface is curved?

Compared to the previous millennium, just a short wait is enough to get satisfactory answers to these questions. In the second half of 17th century, Newton and Leibniz discovered differential and integral calculus, redefining what mathematics is (and what the world will be, opening the way for the coming of the industrial revolution). Newton's and Leibniz's calculus provides effective tools to study, measure, and predict the behavior of moving objects. The path followed by a point moving on a plane or in space is a curve. The point moves along a straight line if and only if its velocity does not change direction; the more the direction of its velocity changes, the more its path is curved. So it is just natural to measure how much curved is the path by measuring how much the direction of the velocity changes; and differential calculus was born exactly to measure variations. *Voilà*, we have an effective and computable definition of the curvature of a curve: the length of the acceleration vector (assuming, as we may always do, that the point moves along the curve at constant scalar speed).

The development of differential geometry of curves and surfaces in the following two centuries is vast and impetuous. In 18th century many gifted mathematicians already applied successfully to geometry the new calculus techniques, culminating in the great accomplishments of 19th century French school, the so-called *fundamental theorems of local theory of curves and surfaces*, which prove that the newly introduced instruments are (necessary and) sufficient to describe *all* local properties of curves and surfaces.

Well, at least locally. Indeed, as in any good success story, this is just the beginning; we need a *coup de théatre*. The theory we have briefly summarized works very well for curves; not so well for surfaces. Or, more precisely, in the case of surfaces we are still at the surface (indeed) of the topic; the local description given by calculus techniques is not enough to account for all global properties of surfaces. Very roughly speaking, curves, even on a large scale, are essentially straight lines and circles somewhat crumpled in plane and space; on the other hand, describing surfaces simply as crumpled portions of plane, while useful to perform calculations, is unduly restrictive and prevents us from understanding and working with the deepest and most significant properties of surfaces, which go far beyond than just measuring curvature in space.

Probably the person who was most aware of this situation was Gauss, one of the greatest (if not *the* greatest) mathematician of first half of 19th century. With two masterly theorems, Gauss succeeded both in showing how much was still to be discovered about surfaces, and in pointing to the right direction to be followed for further research.

With the first theorem, his *Theorema Egregium*, Gauss showed that while, seen from inside, all curves are equal (an intelligent one-dimensional being from within such a one-dimensional world would not be able to decide whether he lives in a straight line or in a curve), this is far from true for surfaces: there is a kind of curvature (the *Gaussian curvature*, in fact) that can be measured from within the surface, and can differ for different surfaces. In other words, it is possible to decide whether Earth is flat or curved without leaving one's

backyard: it is sufficient to measure (with sufficiently precise instruments...)
the Gaussian curvature of our garden.

The second theorem, called *Gauss-Bonnet theorem* since it was completed
and extended by Pierre Bonnet, one of Gauss's brightest students, disclosed
that by just studying local properties one can fail to realize that deep and
significant phenomena take place at a global level, having geometrical im-
plications at a local level too. For instance, one of the consequences of the
Gauss-Bonnet theorem is that no matter how much we deform in space a
sphere, by stretching it (without breaking) so to *locally* change in an appar-
ently arbitrary way its Gaussian curvature, *the integral of Gaussian curvature
on the whole surface is constant*: every local variation of curvature is necessarily compensated by an opposite variation somewhere else.

Another important topic pointed to the interrelation between local and
global properties: the study of curves on surfaces. One of the fundamental
problems in classical differential geometry (even more so because of its practical importance) is to identify the shortest curve (the *geodesic*) between two
given points on a surface. Even simply proving that when the points are close
to each other there exists a unique geodesic joining them is not completely
trivial; if the two points are far from each other, infinitely many geodesics
could exist, or none at all. This is another kind of phenomenon of a typically
global nature: it cannot be studied with differential calculus techniques only,
which are inherently local.

Once more, to go ahead and study global properties a new language and
new tools were needed. So we arrive at the twentieth century and at the creation and development (by Poincaré and others) of topology, which turns out
to be the perfect environment to study and understand global properties of
surfaces. To give just one example, the description given by Hopf and Rinow
of the surfaces in which all geodesics can be extended for all values of the
time parameter (a property implying, in particular, that each pair of points
is connected by a geodesic) is intrinsically topological.

And this is just the beginning of an even more important story. Starting from seminal insights by Riemann (who, in particular, proved that the
so-called non-Euclidean geometries were just geometries on surfaces different
from the plane and not necessarily embedded in Euclidean space), definitions
and ideas introduced to study two-dimensional surfaces have been extended to
n-dimensional manifolds, the analogous of surfaces in any number of dimensions. Differential geometry of manifolds has turned out to be one of the most
significant areas in contemporary mathematics; its language and its results are
used more or less everywhere (and not just in mathematics: Einstein's general theory of relativity, to mention just one example, could not exist without
differential geometry). And who knows what the future has in store for us ...

This book's goal is to tell the main threads of this story from the viewpoint of contemporary mathematics. Chapter 1 describes the local theory of
curves, from the definition of what a curve is to the fundamental theorem
of the local theory of curves. Chapters 3 and 4 deal with the local theory

of surfaces, from the (modern) definition of a surface to Gauss's *Theorema Egregium*. Chapter 5 studies geodesics on a surface, up to the proof of their existence and local uniqueness.

The remaining chapters (and some of the chapters' supplementary material; see below) are devoted instead to global properties. Chapter 2 discusses some fundamental results in the global theory of plane curves, including the *Jordan curve theorem*, which is a good example of the relatively belated development of the interest in global properties of curves: the statement of the Jordan curve theorem seems obvious until you try to actually prove it — and then it turns out to be surprisingly difficult, and far deeper than expected.

Chapter 6 is devoted to the Gauss-Bonnet theorem, with a complete proof is given, and to some of its innumerable applications. Finally, in Chapter 7, we discuss a few important results about the global theory of surfaces (one of them, the Hartman-Nirenberg theorem, proved only in 1959), mainly focusing on the connections between the sign of Gaussian curvature and the global (topological and differential) structure of surfaces. We shall confine ourselves to curves and surfaces in the plane and in space; but the language and the methods we shall introduce are compatible with those used in differential geometry of manifolds of any dimension, and so this book can be useful as a training field before attacking n-dimensional manifolds.

We have attempted to provide several possible paths for using this book as a textbook. A minimal path consists in just dealing with local theory: Chapters 1, 3, and 4 can be read independently of the remaining ones, and can be used to offer, in a two-month course, a complete route from initial definitions to Gauss's Theorema Egregium; if time permits, the first two sections of Chapter 5 can be added, describing the main properties of geodesics. Such a course is suitable and useful both for Mathematics (of course) and Physics students, from the second year on, and for Engineering and Computer Science students (possibly for a Master degree) who need (or are interested in) more advanced mathematical techniques than those learnt in a first-level degree.

In a one-semester course, on the other hand, it is possible to adequately cover the global theory too, introducing Chapter 2 about curves, the final section of Chapter 5 about vector fields, and, at the professor's discretion, Chapter 6 about the Gauss-Bonnet theorem or Chapter 7 about the classification of closed surfaces with constant Gaussian curvature. In our experience, in a one-semester course given to beginning or intermediate undergraduate Mathematics or Physics students, it is difficult to find time to cover both topics, so the two chapters are completely independent; but in the case of an one-year course, or in an one-semester course for advanced students, it might be possible to do so, possibly even touching upon some of the supplementary material (see below).

Each chapter comes with several guided problems, that is, solved exercises, both computational (one of the nice features of local differential geometry of curves and surfaces is that almost everything can be explicitly computed — well, with the exception of geodesics) and of a more theoretical character, whose goal is to teach you how to effectively use the techniques given in the corresponding chapter. You will also be able to test your skill by solving the large number of exercises we propose, subdivided by topic.

You might have noticed another feature peculiar to this book: we address you directly, dear reader. There is a precise reason for this: we want you to feel actively involved while reading the book. A mathematics textbook, no matter at what level, is a sequence of arguments, exposed one after another with (hopefully) impeccable logic. While reading, we are led by the argument, up to a point where we have no idea why the author followed that path instead of another, and — even worse —we are not able to reconstruct that path on our own. In order to learn mathematics, it is not enough to read it; you must *do* mathematics. The style we adopted in this book is chosen to urge you in this direction; in addition to motivations for each notion we shall introduce, you will often find direct questions trying to stimulate you to an active reading, without accepting anything just because you trust us (and, with any luck, they will also help you to stay awake, should you happen to study in the wee hours of the morning, the same hours we used to write this book...).

This book also has the (vain?) ambition not to be just a textbook, but something more. This is the goal of the supplementary material. There is a wealth of extremely interesting and important results that does not usually find its place in courses (mostly due to lack of time), and that is sometimes hard to find in books. The supplementary material appended to each chapter display a choice (suggested, as is natural, by our taste) of these results. We go from complete proofs of the Jordan curve theorem, also for curves that are no more than continuous, and of the existence of triangulations on surfaces (theorems often invoked and used but very rarely proved), to a detailed exposition of the Hopf-Rinow theorem about geodesics or Bonnet's and Hadamard's theorems about surfaces with Gaussian curvature having a well-defined sign. We shall also prove that all closed surfaces are orientable, and the fundamental theorem of the local theory of surfaces (which, for reasons made clear at the end of Chapter 4, is not usually included in a standard curriculum), and much more. We hope that this extra material will answer questions you might have asked yourself while studying, arouse further you curiosity, and provide motivation and examples for moving towards the study of differential geometry in any dimension. A warning: with just some exceptions, the supplementary material is significantly more complicated than the rest of the book, and to be understood it requires a good deal of participation on your part. A reassurance: nothing of what is presented in the supplementary material is used in the main part of the book. On a first reading, the supplementary results may be safely ignored with no detriment to the comprehension of the rest of the book. Lastly, mainly due to space considerations, the supplementary material includes only a limited number of exercises.

Two words about the necessary prerequisites. As you can imagine, we shall use techniques and ideas from differential and integral calculus in several real variables, and from general topology. The necessary notions from topology are really basic: open sets, continuous functions, connectedness and compactness, and even knowing those in metric spaces is enough. Since these topics are usually covered in any basic Geometry course, and often in Calculus courses too, we did not feel the need to provide specific references for the few topology theorems we shall use. If necessary, you will find all of the above (and much more) in the first four chapters of [11].

Much more care has been put in giving the results in calculus we shall need, both because they typically are much deeper than those in general topology, and to give you statements consistent with what we need. All of them are standard results, covered in any Advanced Calculus course (with the possible exception of Theorem 4.9.1); a good reference text is [5].

On the other hand, we do not require you to have been exposed to algebraic topology (in particular, if you do not have any idea of what we are talking about, you do not have to worry). For this reason, Section 2.1 contains a complete introduction to the theory of the degree for continuous maps of the circle to itself, and Section 7.5 discusses everything we shall need about the theory of covering maps.

Needless to say, this is not the first book about this topic, and it could not have been written ignoring the earlier books. We found especially useful the classical texts by do Carmo [4] and Spivak [22], and the less classical but not less useful ones by Lipschutz [13] and Montiel and Ros [16]; if you will like this book, you might want to have a look at those books as well. On the other hand, good starting points for studying differential geometry in any dimension are the already cited [22], and [10] and [12].

Lastly, the pleasant duty of thanks. This book would never have seen the light of the day, and would certainly have been much worse, without the help, assistance, understanding, and patience of (in alphabetical order) Luigi Ambrosio, Francesca Bonadei, Piermarco Cannarsa, Cinzia Casagrande, Ciro Ciliberto, Michele Grassi, Adele Manzella, and Jasmin Raissy. We are particularly grateful to Daniele A. Gewurz, who flawlessly completed the daunting task of translating our book from Italian to English. A special thanks goes to our students of all these years, who put up with several versions of our lecture notes and have relentlessly pointed out even the smallest error. Finally, a very special thanks to Leonardo, Jacopo, Niccolò, Daniele, Maria Cristina, and Raffaele, who have fearlessly suffered the transformation of their parents in an appendix to the computer keyboard, and that with their smiles remind us that the world still deserved to be lived.

Pisa and Rome, July 2011

Marco Abate
Francesca Tovena

Contents

1

Local theory of curves

Elementary geometry gives a fairly accurate and well-established notion of what is a straight line, whereas is somewhat vague about curves in general. Intuitively, the difference between a straight line and a curve is that the former is, well, straight while the latter is curved. But is it possible to measure how curved a curve is, that is, how far it is from being straight? And what, exactly, is a curve? The main goal of this chapter is to answer these questions. After comparing in the first two sections advantages and disadvantages of several ways of giving a formal definition of a curve, in the third section we shall show how Differential Calculus enables us to accurately measure the curvature of a curve. For curves in space, we shall also measure the torsion of a curve, that is, how far a curve is from being contained in a plane, and we shall show how curvature and torsion completely describe a curve in space. Finally, in the supplementary material, we shall present (in Section 1.4) the local canonical shape of a curve; we shall prove a result (Whitney's Theorem 1.1.7, in Section 1.5) useful to understand what *cannot* be the precise definition of a curve; we shall study (in Section 1.6) a particularly well-behaved type of curves, foreshadowing the definition of regular surface we shall see in Chapter 3; and we shall discuss (in Section 1.7) how to deal with curves in $\mathbb{R}^n$ when $n \geq 4$.

1.1 How to define a curve

What is a curve (in a plane, in space, in $\mathbb{R}^n$)? Since we are in a mathematical textbook, rather than in a book about military history of Prussian light cavalry, the only acceptable answer to such a question is a precise definition, identifying exactly the objects that deserve being called curves and those that do not. In order to get there, we start by compiling a list of objects that we consider without a doubt to be curves, and a list of objects that we consider without a doubt not to be curves; then we try to extract properties possessed by the former objects and not by the latter ones.

Abate M., Tovena F.: Curves and Surfaces.
DOI 10.1007/978-88-470-1941-6_1, © Springer-Verlag Italia 2012

Example 1.1.1. Obviously, we have to start from straight lines. A line in a plane can be described in at least three different ways:

- as the graph of a first degree polynomial: $y = mx + q$ or $x = my + q$;
- as the vanishing locus of a first degree polynomial: $ax + by + c = 0$;
- as the image of a map $f: \mathbb{R} \to \mathbb{R}^2$ having the form $f(t) = (\alpha t + \beta, \gamma t + \delta)$.

A word of caution: in the last two cases, the coefficients of the polynomial (or of the map) are not uniquely determined by the line; different polynomials (or maps) may well describe the same subset of the plane.

Example 1.1.2. If $I \subseteq \mathbb{R}$ is an interval and $f: I \to \mathbb{R}$ is a (at least) continuous function, then its *graph*

$$\Gamma_f = \big\{ (t, f(t)) \mid t \in I \big\} \subset \mathbb{R}^2$$

surely corresponds to our intuitive idea of what a curve should be. Note that we have

$$\Gamma_f = \big\{ (x, y) \in I \times \mathbb{R} \mid y - f(x) = 0 \big\} ,$$

that is a graph can always be described as a vanishing locus too. Moreover, it also is the image of the map $\sigma: I \to \mathbb{R}^2$ given by $\sigma(t) = (t, f(t))$.

Remark 1.1.3. To be pedantic, the graph defined in last example is a graph with respect to the *first* coordinate. A graph with respect to the *second* coordinate is a set of the form $\big\{ (f(t), t) \mid t \in I \big\}$, and has the same right to be considered a curve. Since we obtain one kind of graph from the other just by permuting the coordinates (an operation which geometrically amounts to reflecting with respect to a line), both kinds of graphs are equally suitable, and in what follows dealing with graphs we shall often omit to specify the coordinate we are considering.

Example 1.1.4. A *circle* (or *circumference*) with center $(x_0, y_0) \in \mathbb{R}^2$ and radius $r > 0$ is the curve having equation

$$(x - x_0)^2 + (y - y_0)^2 = r^2 .$$

Note that it is not a graph with respect to either coordinate. However, it can be represented as the image of the map $\sigma: \mathbb{R} \to \mathbb{R}^2$ given by

$$\sigma(t) = (x_0 + r \cos t, y_0 + r \sin t) .$$

Example 1.1.5. Open sets in the plane, closed disks and, more generally, subsets of the plane with non-empty interior do not correspond to the intuitive idea of curve, so they are to be excluded. The set $[0,1] \times [0,1] \setminus \mathbb{Q}^2$, in spite of having an empty interior, does not look like a curve either.

Let us see which clues we can gather from these examples. Confining ourselves to graphs for defining curves is too restrictive, since it would exclude circles, which we certainly want to consider as curves (however, note that circles locally are graphs; we shall come back to this fact later).

The approach via vanishing loci of functions looks more promising. Indeed, all the examples we have seen (lines, graphs, circles) can be described in this way; on the other hand, an open set in the plane or the set $[0,1] \times [0,1] \setminus \mathbb{Q}^2$ cannot be the vanishing locus of a continuous function (why?).

So we are led to consider sets of the form

$$C = \{(x,y) \in \Omega \mid f(x,y) = 0\} \subset \mathbb{R}^2$$

for suitable (at least) continuous functions $f \colon \Omega \to \mathbb{R}$, where $\Omega \subseteq \mathbb{R}^2$ is open.

We must however be careful. Sets of this kind are closed in the open set Ω, and this is just fine. But the other implication hold as well:

Proposition 1.1.6. *Let $\Omega \subseteq \mathbb{R}^n$ be an open set. Then a subset $C \subseteq \Omega$ is closed in Ω if and only if there exists a continuous function $f \colon \Omega \to \mathbb{R}$ such that $C = \{x \in \Omega \mid f(x) = 0\} = f^{-1}(0)$.*

Proof. It is enough to define $f \colon \Omega \to \mathbb{R}$ by setting

$$f(x) = d(x, C) = \inf\{\|x - y\| \mid y \in C\} \, ,$$

where $\| \cdot \|$ is the usual Euclidean norm in $\mathbb{R}^n$. Indeed, f is obviously continuous, and $x \in C$ if and only if $f(x) = 0$ (why?). □

So, using continuous functions we get sets that clearly cannot be considered curves. However, the problem could be caused by the fact that continuous functions are too many and not regular enough; we might have to confine ourselves to smooth functions.

(Un)fortunately this precaution is not enough. In Section 1.5 of the supplementary material to this chapter we shall prove the following:

Theorem 1.1.7 (Whitney). *Let $\Omega \subseteq \mathbb{R}^n$ be an open set. Then a subset $C \subseteq \Omega$ is closed in Ω if and only if there exists a function $f \colon \Omega \to \mathbb{R}$ of class C^∞ such that $C = f^{-1}(0)$.*

In other words, *any* closed subsets is the vanishing locus of a C^∞ function, not just of a continuous function, and the idea of defining the curves as vanishing loci of arbitrary smooth functions has no chance of working.

Let's take a step back and examine again Examples 1.1.1, 1.1.2, and 1.1.4. In all those cases, it is possible to describe the set as the image of a mapping. This corresponds, in a sense, to a dynamic vision of a curve, thought of as a locus described by a continuously (or differentiably) moving point in a plane or in space or, more in general, in $\mathbb{R}^n$. With some provisos we shall give shortly, this idea turns out to be the right one, and leads to the following definition.

Definition 1.1.8. Given $k \in \mathbb{N} \cup \{\infty\}$ and $n \geq 2$, a *parametrized curve* of class C^k in $\mathbb{R}^n$ is a map $\sigma: I \to \mathbb{R}^n$ of class C^k, where $I \subseteq \mathbb{R}$ is an interval. The image $\sigma(I)$ is often called *support* (or *trace*) of the curve; the variable $t \in I$ is the *parameter* of the curve. If $I = [a, b]$ and $\sigma(a) = \sigma(b)$, we shall say that the curve is *closed*.

Remark 1.1.9. If I is not an open interval, and $k \geq 1$, saying that σ is of class C^k in I means that σ can be extended to a C^k function defined in an open interval properly containing I. Moreover, if σ is closed of class C^k, unless stated otherwise we shall always assume that

$$\sigma'(a) = \sigma'(b), \ \sigma''(a) = \sigma''(b), \ \ldots, \ \sigma^{(k)}(a) = \sigma^{(k)}(b) \ .$$

In particular, a closed curve of class C^k can always be extended to a *periodic* map $\hat{\sigma}: \mathbb{R} \to \mathbb{R}^n$ of class C^k.

Example 1.1.10. The graph of a map $f: I \to \mathbb{R}^{n-1}$ of class C^k is the image of the parametrized curve $\sigma: I \to \mathbb{R}^n$ given by $\sigma(t) = \big(t, f(t)\big)$.

Example 1.1.11. Given $v_0, v_1 \in \mathbb{R}^n$ such that $v_1 \neq O$, the parametrized curve $\sigma: \mathbb{R} \to \mathbb{R}^n$ given by $\sigma(t) = v_0 + tv_1$ has as its image the *straight line* through v_0 in the direction v_1.

Example 1.1.12. The two parametrized curves $\sigma_1, \sigma_2: \mathbb{R} \to \mathbb{R}^2$ given by

$$\sigma_1(t) = (x_0 + r\cos t, y_0 + r\sin t) \quad \text{and} \quad \sigma_2(t) = (x_0 + r\cos 2t, y_0 + r\sin 2t)$$

both have as their image the *circle* having center $(x_0, y_0) \in \mathbb{R}^2$ and radius $r > 0$.

Example 1.1.13. The parametrized curve $\sigma: \mathbb{R} \to \mathbb{R}^3$ given by

$$\sigma(t) = (r\cos t, r\sin t, at) \ ,$$

with $r > 0$ and $a \in \mathbb{R}^*$, has as its image the *circular helix* having *radius* r e *pitch* a; see Fig 1.1.(a). The image of the circular helix is contained in the right circular cylinder having equation $x^2 + y^2 = r^2$. Moreover, for each $t \in \mathbb{R}$ the points $\sigma(t)$ and $\sigma(t + 2\pi)$ belong to the same line parallel to the cylinder's axis, and have distance $2\pi|a|$.

Example 1.1.14. The curve $\sigma: \mathbb{R} \to \mathbb{R}^2$ given by $\sigma(t) = (t, |t|)$ is a continuous parametrized curve which is not of class C^1 (but see Exercise 1.11).

All t parametrized curves we have seen so far (with the exception of the circle; we'll come back to it shortly) provide a homeomorphism between their domain and their image. But it is not always so:

Example 1.1.15. The curve $\sigma: \mathbb{R} \to \mathbb{R}^2$ given by $\sigma(t) = (t^3 - 4t, t^2 - 4)$ is a non-injective parametrized curve; see Fig. 1.1.(b).

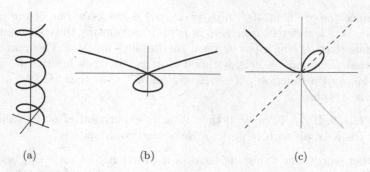

(a)   (b)   (c)

Fig. 1.1. (a) circular helix; (b) non-injective curve; (c) folium of Descartes

Example 1.1.16. The curve $\sigma: (-1, +\infty) \to \mathbb{R}^2$ given by

$$\sigma(t) = \left(\frac{3t}{1+t^3}, \frac{3t^2}{1+t^3} \right)$$

is an injective parametrized curve, but it is not a homeomorphism with its image (why?). The set obtained by taking the image of σ, together with its reflection across the line $x - y$, is the *folium of Descartes*; see Fig. 1.1.(c).

We may also recover some vanishing loci as parametrized curves. Not all of them, by Whitney's Theorem 1.1.7; but we shall be able to work with vanishing loci of functions f having nonzero *gradient* ∇f, thanks to a classical Calculus theorem, the *implicit function theorem* (you can find its proof, for instance, in [5, p. 148]):

Theorem 1.1.17 (Implicit function theorem). *Let Ω be an open subset of $\mathbb{R}^m \times \mathbb{R}^n$, and $F: \Omega \to \mathbb{R}^n$ a map of class C^k, with $k \in \mathbb{N}^* \cup \{\infty\}$. Denote by (x, y) the coordinates in $\mathbb{R}^{m+n}$, where $x \in \mathbb{R}^m$ and $y \in \mathbb{R}^n$. Let $p_0 = (x_0, y_0) \in \Omega$ be such that*

$$F(p_0) = O \qquad and \qquad \det\left(\frac{\partial F_i}{\partial y_j}(p_0) \right)_{i,j=1,\ldots,n} \neq 0 .$$

Then there exist a neighborhood $U \subset \mathbb{R}^{m+n}$ of p_0, a neighborhood $V \subset \mathbb{R}^m$ of x_0 and a map $g: V \to \mathbb{R}^n$ of class C^k such that $U \cap \{p \in \Omega \mid F(p) = O\}$ precisely consists of the points of the form $(x, g(x))$ with $x \in V$.

Using this we may prove that the vanishing locus of a function having nonzero gradient is (at least locally) a graph:

Proposition 1.1.18. *Let $\Omega \subseteq \mathbb{R}^2$ be an open set, and $f: \Omega \to \mathbb{R}$ a function of class C^k, with $k \in \mathbb{N}^* \cup \{\infty\}$. Choose $p_0 \in \Omega$ such that $f(p_0) = 0$ but $\nabla f(p_0) \neq O$. Then there exists a neighborhood U of p_0 such that $U \cap \{p \in \Omega \mid f(p) = 0\}$ is the graph of a function of class C^k.*

Proof. Since the gradient of f in $p_0 = (x_0, y_0)$ is not zero, one of the partial derivatives of f is different from zero in p; up to permuting the coordinates we can assume that $\partial f / \partial y(p_0) \neq 0$. Then the implicit function Theorem 1.1.17 tells us that there exist a neighborhood U of p_0, an open interval $I \subseteq \mathbb{R}$ including x_0, and a function $g: I \to \mathbb{R}$ of class C^k such that $U \cap \{f = 0\}$ is exactly the graph of g. $\qquad\qquad\qquad\qquad\qquad\qquad\qquad\qquad\qquad\qquad\qquad\quad\square$

Remark 1.1.19. If $\partial f / \partial x(p) \neq 0$ then in a neighborhood of p the vanishing locus of f is a graph with respect to the second coordinate.

In other words, the vanishing locus of a function f of class C^1, being locally a graph, is *locally* the support of a parametrized curve near the points where the gradient of f is nonzero.

Example 1.1.20. The gradient of the function $f(x, y) = (x - x_0)^2 + (y - y_0)^2 - r^2$ is zero only in (x_0, y_0), which does not belong to the vanishing locus of f. Accordingly, each point of the circle with center (x_0, y_0) and radius $r > 0$ has a neighborhood which is a graph with respect to one of the coordinates.

Remark 1.1.21. Actually, it can be proved that a subset of $\mathbb{R}^2$ which is locally a graph always is the support of a parametrized curve; see Theorem 1.6.8.

However, the definition of a parametrized curve is not yet completely satisfying. The problem is that it may well happen that two parametrized curves that are different as maps describe what seems to be the same geometric set. An example is given by the two parametrized curves given in Example 1.1.12, both having as their image a circle; the only difference between them is the speed with which they describe the circle. Another, even clearer example (one you have undoubtedly stumbled upon in previous courses) is the straight line: as recalled in Example 1.1.1, the same line can be described as the image of infinitely many distinct parametrized curves, just differing in speed and starting point.

On the other hand, considering just the image of a parametrized curve is not correct either. Two different parametrized curves might well describe the same support in geometrically different ways: for instance, one could be injective whereas the other comes back more than once on sections already described before going on. Or, more simply, two different parametrized curves might describe the same image a different number of times, as is the case when restricting the curves in Example 1.1.12 to intervals of the form $[0, 2k\pi]$.

These considerations suggest to introduce an equivalence relation on the class of parametrized curves, such that two equivalent parametrized curves really describe the same geometric object. The idea of only allowing changes in speed and starting point, but not changes in direction or retracing our steps, is formalized using the notion of diffeomorphism.

Definition 1.1.22. A *diffeomorphism of class* C^k (with $k \in \mathbb{N}^* \cup \{\infty\}$) between two open sets $\Omega, \Omega_1 \subseteq \mathbb{R}^n$ is a homeomorphism $h: \Omega \to \Omega_1$ such that both h and its inverse h^{-1} are of class C^k.

More generally, a *diffeomorphism of class* C^k between two sets A, $A_1 \subseteq \mathbb{R}^n$ is the restriction of a diffeomorphism of class C^k of a neighborhood of A with a neighborhood of A_1 and sending A onto A_1.

Example 1.1.23. For instance, $h(x) = 2x$ is a diffeomorphism of class C^∞ of $\mathbb{R}$ with itself, whereas $g(x) = x^3$, even though it is a homeomorphism of $\mathbb{R}$ with itself, is not a diffeomorphism, not even of class C^1, since the inverse function $g^{-1}(x) = x^{1/3}$ is not of class C^1.

Definition 1.1.24. Two parametrized curves $\sigma \colon I \to \mathbb{R}^n$ and $\tilde{\sigma} \colon \tilde{I} \to \mathbb{R}^n$ of class C^k are *equivalent* if there exists a diffeomorphism $h \colon \tilde{I} \to I$ of class C^k such that $\tilde{\sigma} = \sigma \circ h$; we shall also say that $\tilde{\sigma}$ is a *reparametrization* of σ, and that h is a *parameter change*.

In other words, two equivalent curves only differ in the speed they are traced, while they have the same image, they curve (as we shall see) in the same way, and more generally they have the same geometric properties. So we have finally reached the official definition of what a curve is:

Definition 1.1.25. A *curve* of class C^k in $\mathbb{R}^n$ is an equivalence class of parametrized curves of class C^k in $\mathbb{R}^n$. Each element of the equivalence class is a *parametrization* of the curve. The *support* of a curve is the support of any parametrization of the curve. A *plane curve* is a curve in $\mathbb{R}^2$.

Remark 1.1.26. In what follows we shall almost always use the phrase "let $\sigma \colon I \to \mathbb{R}^n$ be a curve" to say that σ is a particular parametrization of the curve under consideration.

Some curves have a parametrization keeping an especially strong connection with its image, and so they deserve a special name.

Definition 1.1.27. A *Jordan* (or *simple*) *arc* of class C^k in $\mathbb{R}^n$ is a curve admitting a parametrization $\sigma \colon I \to \mathbb{R}^n$ that is a homeomorphism with its image, where $I \subseteq \mathbb{R}$ is an interval. In this case, σ is said to be a *global parametrization* of C. If I is an open (closed) interval, we shall sometimes say that C is an *open* (*closed*) *Jordan arc*.

Definition 1.1.28. A *Jordan curve* of class C^k in $\mathbb{R}^n$ is a closed curve C admitting a parametrization $\sigma \colon [a, b] \to \mathbb{R}^n$ of class C^k, injective both on $[a, b)$ and on $(a, b]$. In particular, the image of C is homeomorphic to a circle (why?). The periodic extension $\hat{\sigma}$ of σ mentioned in Remark 1.1.9 is a *periodic parametrization* of C. Jordan curves are also called *simple curves* (mostly when $n > 2$).

Example 1.1.29. Graphs (Example 1.1.2), lines (Example 1.1.11) and circular helices (Example 1.1.13) are Jordan arcs; the circle (Example 1.1.4) is a Jordan curve.

Example 1.1.30. The *ellipse* $E \subset \mathbb{R}^2$ with *semiaxes* $a, b > 0$ is the vanishing locus of the function $f: \mathbb{R}^2 \to \mathbb{R}$ given by $f(x,y) = (x/a)^2 + (y/b)^2 - 1$, that is,

$$E = \left\{ (x,y) \in \mathbb{R}^2 \;\middle|\; \frac{x^2}{a^2} + \frac{y^2}{b^2} = 1 \right\} .$$

A periodic parametrization of E of class C^∞ is the map $\sigma: \mathbb{R} \to \mathbb{R}^2$ given by $\sigma(t) = (a \cos t, b \sin t)$.

Example 1.1.31. The *hyperbola* $I \subset \mathbb{R}^2$ with *semiaxes* $a, b > 0$ is the vanishing locus of the function $f: \mathbb{R}^2 \to \mathbb{R}$ given by $f(x,y) = (x/a)^2 - (y/b)^2 - 1$, that is,

$$I = \left\{ (x,y) \in \mathbb{R}^2 \;\middle|\; \frac{x^2}{a^2} - \frac{y^2}{b^2} = 1 \right\} .$$

A global parametrization of the component of I contained in the right half-plane is the map $\sigma: \mathbb{R} \to \mathbb{R}^2$ given by $\sigma(t) = (a \cosh t, b \sinh t)$.

In the Definition 1.1.24 of equivalence of parametrized curves we allowed the direction in which the curve is described to be reversed; in other words, we also admitted diffeomorphisms with negative derivative everywhere. As you will see, in some situations it will be important to be able to distinguish the direction in which the curve is traced; so we introduce a slightly finer equivalence relation.

Definition 1.1.32. Two parametrized curves $\sigma: I \to \mathbb{R}^n$ and $\tilde{\sigma}: \tilde{I} \to \mathbb{R}^n$ of class C^k are *equivalent with the same orientation* if there exists a parameter change $h: \tilde{I} \to I$ from $\tilde{\sigma}$ to σ with positive derivative everywhere; they are *equivalent with opposite orientation* if there exists a parameter change $h: \tilde{I} \to I$ from $\tilde{\sigma}$ to σ with negative derivative everywhere (note that the derivative of a diffeomorphism between intervals cannot be zero in any point, so it is either positive everywhere or negative everywhere). An *oriented curve* is then an equivalent class of parametrized curves with the same orientation.

Example 1.1.33. If $\sigma: I \to \mathbb{R}^n$ is a parametrized curve, then the parametrized curve $\sigma^-: -I \to \mathbb{R}^n$ given by $\sigma^-(t) = \sigma(-t)$, where $-I = \{ t \in \mathbb{R} \mid -t \in I \}$, is equivalent to σ but with the opposite orientation.

In general, working with equivalence classes is always a bit tricky; you have to choose a representative element and to check that all obtained results do not depend on that particular representative element. Nevertheless, there is a large class of curves, the *regular curves*, for which it is possible to choose in a canonical way a parametrization that represents the geometry of the curve particularly well: the *arc length parametrization*. The existence of this canonical parametrization permits an effective study of the geometry (and, in particular, of the differential geometry) of curves, confirming a posteriori that this is the right definition.

In the next section we shall introduce this special parametrization; on the other hand, in Section 1.6 of the supplementary material of this chapter we shall discuss another way of definining what a curve is.

1.2 Arc length

This is a book about differential geometry; so our basic idea is to study geometric properties of curves (and surfaces) by using techniques borrowed from Mathematical Analysis, and in particular from Differential Calculus. Accordingly, apart from a few particular situations (such as Section 2.8 of the supplementary material in Chapter 2), we shall always work with curves of class at least C^1, in order to be able to compute derivatives.

The derivative of a parametrization of a curve tells us the speed at which we are describing the image of the curve. The class of curves for which the speed is nowhere zero (so we always know the direction we are going) is, as we shall see, the right class for differential geometry.

Definition 1.2.1. Let $\sigma: I \rightarrow \mathbb{R}^n$ be a parametrized curve of class (at least) C^1. The vector $\sigma'(t)$ is the *tangent vector* to the curve at the point $\sigma(t)$. If $t_0 \in I$ is such that $\sigma'(t_0) \neq O$, then the line through $\sigma(t_0)$ and parallel to $\sigma'(t_0)$ is the *affine tangent line* to the curve at the point $\sigma(t_0)$. Finally, if $\sigma'(t) \neq O$ for all $t \in I$ we shall say that σ is *regular*.

Remark 1.2.2. The notion of a tangent vector depends on the parametrization we have chosen, while the affine tangent line (if any) and the fact of being regular are properties of the curve. Indeed, let $\sigma: I \rightarrow \mathbb{R}^n$ and $\tilde{\sigma}: \tilde{I} \rightarrow \mathbb{R}^n$ be two equivalent parametrized curves of class C^1, and $h: \tilde{I} \rightarrow I$ the parameter change. Then, by computing $\tilde{\sigma} = \sigma \circ h$, we find

$$\tilde{\sigma}'(t) = h'(t)\, \sigma'\big(h(t)\big) \,. \tag{1.1}$$

Since h' is never zero, we see that the *length* of the tangent vector depends on our particular parametrization, but its *direction* does not; so the affine tangent line in $\tilde{\sigma}(t) = \sigma\big(h(t)\big)$ determined by $\tilde{\sigma}$ is the same as that determined by σ. Moreover, $\tilde{\sigma}'$ is never zero if and only if σ' is never zero; so, being regular is a property of the curve, rather than of a particular representative.

Example 1.2.3. Graphs, lines, circles, circular helices, and the curves in Examples 1.1.15 and 1.1.16 are regular curves.

Example 1.2.4. The curve $\sigma: \mathbb{R} \rightarrow \mathbb{R}^2$ given by $\sigma(t) = (t^2, t^3)$ is a non-regular curve whose image cannot be the image of a regular curve; see Fig 1.2 and Exercises 1.4 and 1.10.

As anticipated in the previous section, what makes the theory of curves especially simple to deal with is that every regular curve has a canonical parametrization (unique up to its starting point; see Theorem 1.2.11), strongly

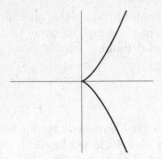

Fig. 1.2. A non-regular curve

related to the geometrical properties common to all parametrizations of the curve. In particular, to study the geometry of a regular curve, we often may confine ourselves to working with the canonical parametrization.

This canonical parametrization basically consists in using as our parameter the length of the curve. So let us start by defining what we mean by length of a curve.

Definition 1.2.5. Let $I = [a,b]$ be an interval. A *partition* $\mathcal{P}$ of I is a $(k+1)$-tuple $(t_0, \ldots, t_k) \in [a,b]^{k+1}$ with $a = t_0 < t_1 < \cdots < t_k = b$. If $\mathcal{P}$ is partition of I, we set

$$\|\mathcal{P}\| = \max_{1 \le j \le k} |t_j - t_{j-1}| .$$

Definition 1.2.6. Given a parametrized curve $\sigma \colon [a,b] \to \mathbb{R}^n$ and a partition $\mathcal{P}$ of $[a,b]$, denote by

$$L(\sigma, \mathcal{P}) = \sum_{j=1}^{k} \|\sigma(t_j) - \sigma(t_{j-1})\|$$

the length of the polygonal closed curve having vertices $\sigma(t_0), \ldots, \sigma(t_k)$. We shall say that σ is *rectifiable* if the limit

$$L(\sigma) = \lim_{\|\mathcal{P}\| \to 0} L(\sigma, \mathcal{P})$$

exists and is finite. This limit is the *length* of σ.

Theorem 1.2.7. *Every parametrized curve* $\sigma \colon [a,b] \to \mathbb{R}^n$ *of class* C^1 *is rectifiable, and we have*

$$L(\sigma) = \int_a^b \|\sigma'(t)\| \, \mathrm{d}t .$$

Proof. Since σ is of class C^1, the integral is finite. So we have to prove that, for each $\varepsilon > 0$ there exists a $\delta > 0$ such that if $\mathcal{P}$ is a partition of $[a,b]$ with $\|\mathcal{P}\| < \delta$ then

$$\left| \int_a^b \|\sigma'(t)\| \, \mathrm{d}t - L(\sigma, \mathcal{P}) \right| < \varepsilon . \tag{1.2}$$

We begin by remarking that, for each partition $\mathcal{P} = (t_0, \ldots, t_k)$ of $[a, b]$ and for each $j = 1, \ldots, k$, we have

$$\|\sigma(t_j) - \sigma(t_{j-1})\| = \left\| \int_{t_{j-1}}^{t_j} \sigma'(t)\,dt \right\| \leq \int_{t_{j-1}}^{t_j} \|\sigma'(t)\|\,dt \; ;$$

so summing over j we find

$$L(\sigma, \mathcal{P}) \leq \int_a^b \|\sigma'(t)\|\,dt \, , \tag{1.3}$$

independently of the partition $\mathcal{P}$.

Now, fix $\varepsilon > 0$; then the uniform continuity of σ' over the compact interval $[a, b]$ provides us with a $\delta > 0$ such that

$$|t - s| < \delta \Longrightarrow \|\sigma'(t) - \sigma'(s)\| < \frac{\varepsilon}{b - a} \tag{1.4}$$

for all $s, t \in [a, b]$. Let $\mathcal{P} = (t_0, \ldots, t_k)$ be a partition of $[a, b]$ with $\|\mathcal{P}\| < \delta$. For all $j = 1, \ldots, k$ and $s \in [t_{j-1}, t_j]$ we have

$$\sigma(t_j) - \sigma(t_{j-1}) = \int_{t_{j-1}}^{t_j} \sigma'(s)\,dt + \int_{t_{j-1}}^{t_j} \left(\sigma'(t) - \sigma'(s) \right) dt$$

$$= (t_j - t_{j-1})\sigma'(s) + \int_{t_{j-1}}^{t_j} \left(\sigma'(t) - \sigma'(s) \right) dt \, .$$

Hence,

$$\|\sigma(t_j) - \sigma(t_{j-1})\| \geq (t_j - t_{j-1})\|\sigma'(s)\| - \int_{t_{j-1}}^{t_j} \|\sigma'(t) - \sigma'(s)\|\,dt$$

$$\geq (t_j - t_{j-1})\|\sigma'(s)\| - \frac{\varepsilon}{b - a}(t_j - t_{j-1}) \, ,$$

where the last step follows from the fact that $s, t \in [t_{j-1}, t_j]$ implies $|t - s| < \delta$, so we may apply (1.4). Dividing by $t_j - t_{j-1}$ we get

$$\frac{\|\sigma(t_j) - \sigma(t_{j-1})\|}{t_j - t_{j-1}} \geq \|\sigma'(s)\| - \frac{\varepsilon}{b - a} \; ;$$

then integrating with respect to s over $[t_{j-1}, t_j]$ it follows that

$$\|\sigma(t_j) - \sigma(t_{j-1})\| \geq \int_{t_{j-1}}^{t_j} \|\sigma'(s)\|\,ds - \frac{\varepsilon}{b - a}(t_j - t_{j-1}) \, .$$

Summing over $j = 1, \ldots, k$ we get

$$L(\sigma, \mathcal{P}) \geq \int_a^b \|\sigma'(s)\|\,ds - \varepsilon \, ,$$

which taken together with (1.3) gives (1.2). $\square$

Corollary 1.2.8. *Length is a geometric property of C^1 curves, and it does not depend on a particular parametrization. In other words, any two equivalent parametrized curves of class C^1 (defined on a compact interval) have the same length.*

Proof. Let $\sigma\colon [a,b] \to \mathbb{R}^n$ and $\tilde{\sigma}\colon [\tilde{a},\tilde{b}] \to \mathbb{R}^n$ be equivalent parametrized curves, and $h\colon [\tilde{a},\tilde{b}] \to [a,b]$ the parameter change. Then (1.1) implies

$$L(\tilde{\sigma}) = \int_{\tilde{a}}^{\tilde{b}} \|\tilde{\sigma}'(t)\|\, \mathrm{d}t = \int_{\tilde{a}}^{\tilde{b}} \|\sigma'(h(t))\|\, |h'(t)|\, \mathrm{d}t = \int_a^b \|\sigma'(\tau)\|\, \mathrm{d}\tau = L(\sigma)\,,$$

thanks to the classical theorem about change of variables in integrals.   □

Remark 1.2.9. Note that the length of a curve does not depend only on its support, since a non-injective parametrization may describe some arc more than once. For instance, the two curves in Example 1.1.12, restricted to $[0, 2\pi]$, have different lengths even though they have the same image.

The time has come for us to define the announced canonical parametrization:

Definition 1.2.10. Let $\sigma\colon I \to \mathbb{R}^n$ be a curve of class C^k (with $k \geq 1$). Having fixed $t_0 \in I$, the *arc length* of σ (measured starting from t_0) is the function $s\colon I \to \mathbb{R}$ of class C^k given by

$$s(t) = \int_{t_0}^t \|\sigma'(\tau)\|\, \mathrm{d}\tau\,.$$

We shall say that σ is *parametrized by arc length* if $\|\sigma'\| \equiv 1$. In other words, σ is parametrized by arc length if and only if its arc length is equal to the parameter t up to a translation, that is $s(t) = t - t_0$.

A curve parametrized by arc length is clearly regular. The fundamental result is that the converse implication is true too:

Theorem 1.2.11. *Every regular oriented curve admits a unique (up to a translation in the parameter) parametrization by arc length. More precisely, let $\sigma\colon I \to \mathbb{R}^n$ be a regular parametrized curve of class C^k. Having fixed $t_0 \in I$, denote by $s\colon I \to \mathbb{R}$ the arc length of σ measured starting from t_0. Then $\tilde{\sigma} = \sigma \circ s^{-1}$ is (up to a translation in the parameter) the unique regular C^k curve parametrized by arc length equivalent to σ and having the same orientation.*

Proof. First of all, $s' = \|\sigma'\|$ is positive everywhere, so $s\colon I \to s(I)$ is a monotonically increasing function of class C^k having inverse of class C^k between the intervals I and $\tilde{I} = s(I)$. So $\tilde{\sigma} = \sigma \circ s^{-1}\colon \tilde{I} \to \mathbb{R}^n$ is a parametrized curve equivalent to σ and having the same orientation. Furthermore,

$$\tilde{\sigma}'(t) = \frac{\sigma'\big(s^{-1}(t)\big)}{\|\sigma'\big(s^{-1}(t)\big)\|}\,,$$

so $\|\tilde{\sigma}'\| \equiv 1$, as required.

To prove uniqueness, let σ_1 be another parametrized curve satisfying the hypotheses. Being equivalent to σ (and so to $\tilde{\sigma}$) with the same orientation, there exists a parameter change h with positive derivative everywhere such that $\sigma_1 = \tilde{\sigma} \circ h$. As both $\tilde{\sigma}$ and σ_1 are parametrized by arc length, (1.1) implies $|h'| \equiv 1$; but $h' > 0$ everywhere, so necessarily $h' \equiv 1$. This means that $h(t) = t + c$ for some $c \in \mathbb{R}$, and thus σ_1 is obtained from $\tilde{\sigma}$ by translating the parameter. $\square$

So, every regular curve admits an essentially unique parametrization by arc length. In some textbooks this parametrization is called the *natural parametrization*.

Remark 1.2.12. In what follows, we shall always use the letter s to denote the arc-length parameter, and the letter t to denote an arbitrary parameter. Moreover, the derivatives with respect to the arc-length parameter will be denoted by a dot (˙), while the derivatives with respect to an arbitrary parameter by a prime ('). For instance, we shall write $\dot{\sigma}$ for $d\sigma/ds$, and σ' for $d\sigma/dt$. The relation between $\dot{\sigma}$ and σ' easily follows from the chain rule:

$$\sigma'(t) = \frac{d\sigma}{dt}(t) = \frac{d\sigma}{ds}(s(t))\frac{ds}{dt}(t) = \|\sigma'(t)\|\,\dot{\sigma}(s(t)) . \tag{1.5}$$

Analogously we have

$$\dot{\sigma}(s) = \frac{1}{\|\sigma'(s^{-1}(s))\|}\,\sigma'(s^{-1}(s)) ,$$

where in last formula the letter s denotes both the parameter and the arc length function. As you will see, using the same letter to represent both concepts will not cause, once you get used to it, any confusion.

Example 1.2.13. Let $\sigma : \mathbb{R} \to \mathbb{R}^n$ be a line parametrized as in Example 1.1.11. Then the arc length of σ starting from 0 is $s(t) = \|v_1\|t$, and thus $s^{-1}(s) = s/\|v_1\|$. In particular, a parametrization of the line by arc length is $\tilde{\sigma}(s) = v_0 + sv_1/\|v_1\|$.

Example 1.2.14. Let $\sigma : [0, 2\pi] \to \mathbb{R}^2$ be the parametrization of the circle with center $(x_0, y_0) \in \mathbb{R}^2$ and radius $r > 0$ given by $\sigma(t) = (x_0 + r\cos t, y_0 + r\sin t)$. Then the arc length of σ starting from 0 is $s(t) = rt$, so $s^{-1}(s) = s/r$. In particular, a parametrization $\tilde{\sigma} : [0, 2\pi r] \to \mathbb{R}^2$ by arc length of the circle is given by $\tilde{\sigma}(s) = (x_0 + r\cos(s/r), y_0 + r\sin(s/r))$.

Example 1.2.15. The circular helix $\sigma : \mathbb{R} \to \mathbb{R}^3$ with radius $r > 0$ and pitch $a \in \mathbb{R}^*$ described in Example 1.1.13 has $\|\sigma'\| \equiv \sqrt{r^2 + a^2}$. So an arc length parametrization is

$$\tilde{\sigma}(s) = \left(r\cos\frac{s}{\sqrt{r^2 + a^2}}, r\sin\frac{s}{\sqrt{r^2 + a^2}}, \frac{as}{\sqrt{r^2 + a^2}} \right) .$$

Example 1.2.16. The *catenary* is the graph of the hyperbolic cosine function; so a parametrization is the curve $\sigma\colon\mathbb{R}\to\mathbb{R}^2$ given by $\sigma(t)=(t,\cosh t)$. It is one of the few curves for which we can explicitly compute the arc length parametrization using elementary functions. Indeed, $\sigma'(t)=(1,\sinh t)$; so

$$s(t)=\int_0^t\sqrt{1+\sinh^2\tau}\,\mathrm{d}\tau=\int_0^t\cosh\tau\,\mathrm{d}\tau=\sinh t$$

and

$$s^{-1}(s)=\operatorname{arc\,sinh}s=\log\bigl(s+\sqrt{1+s^2}\bigr)\,.$$

Now, $\cosh\bigl(\log\bigl(s+\sqrt{1+s^2}\bigr)\bigr)=\sqrt{1+s^2}$, and thus the parametrization of the catenary by arc length is

$$\tilde{\sigma}(s)=\left(\log\bigl(s+\sqrt{1+s^2}\bigr),\sqrt{1+s^2}\right)\,.$$

Example 1.2.17. Let E be an ellipse having semiaxes a, $b>0$, parametrized as in Example 1.1.30, and assume $b>a$. Then

$$s(t)=\int_0^t\sqrt{a^2\sin^2\tau+b^2\cos^2\tau}\,\mathrm{d}\tau=b\int_0^t\sqrt{1-\left(1-\frac{a^2}{b^2}\right)\sin^2\tau}\,\mathrm{d}\tau$$

is an elliptic integral of the second kind, whose inverse is expressed using Jacobi elliptic functions. So, to compute the arc-length parametrization of the ellipse we have to resort to non-elementary functions.

Remark 1.2.18. Theorem 1.2.11 says that every regular curve can be parametrized by arc length, at least in principle. In practice, finding the parametrization by arc length of a particular curve might well be impossible: as we have seen in the previous examples, in order to do so it is necessary to compute the inverse of a function given by an integral. For this reason, from now on we shall use the parametrization by arc length to introduce the geometric quantities (like curvature, for instance) we are interested in, but we shall always explain how to compute those quantities starting from an arbitrary parametrization too.

1.3 Curvature and torsion

In a sense, a straight line is a curve that never changes direction. More precisely, the image of a regular curve is contained in a line if and only if the direction of its tangent vector σ' is constant (see Exercise 1.22). As a result, it is reasonable to suppose that the variation of the direction of the tangent vector could tell us how far a curve is from being a straight line. To get an effective way of measuring this variation (and so the curve's curvature), we shall use the tangent versor.

Definition 1.3.1. Let $\sigma: I \to \mathbb{R}^n$ be a regular curve of class C^k. The *tangent versor* (also called *unit tangent vector*) to σ is the map $\mathbf{t}: I \to \mathbb{R}^n$ of class C^{k-1} given by

$$\mathbf{t} = \frac{\sigma'}{\|\sigma'\|} \; ;$$

we shall also say that the versor $\mathbf{t}(t)$ is tangent to the curve σ at the point $\sigma(t)$.

Remark 1.3.2. Equation (1.1) implies that the tangent vector only depends on the oriented curve, and not on a particular parametrization we might have chosen. In particular, if the curve σ is parametrized by arc length, then

$$\mathbf{t} = \dot{\sigma} = \frac{\mathrm{d}\sigma}{\mathrm{d}s} \; .$$

On the other hand, the tangent versor does depend on the orientation of the curve. If $\mathbf{t}^-$ is the tangent versor to the curve (introduced in Example 1.1.33) σ^- having opposite orientation, then

$$\mathbf{t}^-(t) = -\mathbf{t}(-t) \; ,$$

that is the tangent versor changes sign when the orientation is reversed.

The variations in the direction of the tangent vector can be measured by the variation of the tangent versor, that is, by the derivative of $\mathbf{t}$.

Definition 1.3.3. Let $\sigma: I \to \mathbb{R}^n$ be a regular curve of class C^k (with $k \geq 2$) parametrized by arc length. The *curvature* of σ is the function $\kappa: I \to \mathbb{R}^+$ of class C^{k-2} given by

$$\kappa(s) = \|\dot{\mathbf{t}}(s)\| = \|\ddot{\sigma}(s)\| \; .$$

Clearly, $\kappa(s)$ is the curvature of σ at the point $\sigma(s)$. We shall say that σ is *biregular* if κ is everywhere nonzero. In this case the *radius of curvature* of σ at the point $\sigma(s)$ is $r(s) = 1/\kappa(s)$.

Remark 1.3.4. If $\sigma: I \to \mathbb{R}^n$ is an arbitrary regular parametrized curve, the *curvature* $\kappa(t)$ of σ at the point $\sigma(t)$ is defined by reparametrizing the curve by arc length. If $\sigma_1 = \sigma \circ s^{-1}$ is a parametrization of σ by arc length, and κ_1 is the curvature of σ_1, then we define $\kappa: I \to \mathbb{R}^+$ by setting $\kappa(t) = \kappa_1\big(s(t)\big)$, so the curvature of σ at the point $\sigma(t)$ is equal to the curvature of σ_1 at the point $\sigma_1\big(s(t)\big) = \sigma(t)$.

Example 1.3.5. A line parametrized as in Example 1.2.13 has a constant tangent versor. So the curvature of a straight line is everywhere zero.

Example 1.3.6. Let $\sigma: [0, 2\pi r] \to \mathbb{R}^2$ be the circle with center $(x_0, y_0) \in \mathbb{R}^2$ and radius $r > 0$, parametrized by arc length as in Example 1.2.14. Then

$$\mathbf{t}(s) = \dot{\sigma}(s) = \big(-\sin(s/r), \cos(s/r)\big) \text{ and } \dot{\mathbf{t}}(s) = \frac{1}{r}\big(-\cos(s/r), -\sin(s/r)\big) \; ,$$

so σ has constant curvature $1/r$. This is the reason why the reciprocal of the curvature is called radius of curvature; see also Example 1.4.3.

Example 1.3.7. Let $\sigma \colon \mathbb{R} \to \mathbb{R}^3$ be the circular helix with radius $r > 0$ and pitch $a \in \mathbb{R}^*$, parametrized by arc length as in Example 1.2.15. Then,

$$\mathbf{t}(s) = \left(-\frac{r}{\sqrt{r^2 + a^2}} \sin \frac{s}{\sqrt{r^2 + a^2}}, \frac{r}{\sqrt{r^2 + a^2}} \cos \frac{s}{\sqrt{r^2 + a^2}}, \frac{a}{\sqrt{r^2 + a^2}}\right)$$

and

$$\dot{\mathbf{t}}(s) = -\frac{r}{r^2 + a^2} \left(\cos \frac{s}{\sqrt{r^2 + a^2}}, \sin \frac{s}{\sqrt{r^2 + a^2}}, 0\right) ;$$

so the helix has constant curvature

$$\kappa \equiv \frac{r}{r^2 + a^2} .$$

Example 1.3.8. Let $\sigma \colon \mathbb{R} \to \mathbb{R}^2$ be the catenary, parametrized by arc length as in Example 1.2.16. Then

$$\mathbf{t}(s) = \left(\frac{1}{\sqrt{1 + s^2}}, \frac{s}{\sqrt{1 + s^2}}\right)$$

and

$$\dot{\mathbf{t}}(s) = \left(-\frac{s}{(1 + s^2)^{3/2}}, \frac{1}{(1 + s^2)^{3/2}}\right) ;$$

so the catenary has curvature

$$\kappa(s) = \frac{1}{1 + s^2} .$$

Now, it stands to reason that the direction of the vector $\dot{\mathbf{t}}$ should also contain significant geometric information about the curve, since it gives the direction the curve is following. Moreover, the vector $\dot{\mathbf{t}}$ cannot be just any vector. Indeed, since $\mathbf{t}$ is a versor, we have

$$\langle \mathbf{t}, \mathbf{t} \rangle \equiv 1 ,$$

where $\langle \cdot, \cdot \rangle$ is the canonical scalar product in $\mathbb{R}^n$; hence, after taking the derivative, we get

$$\langle \dot{\mathbf{t}}, \mathbf{t} \rangle \equiv 0 .$$

In other words, $\dot{\mathbf{t}}$ *is orthogonal to* $\mathbf{t}$ *everywhere.*

Definition 1.3.9. Let $\sigma \colon I \to \mathbb{R}^n$ be a biregular curve of class C^k (with $k \geq 2$) parametrized by arc length. The *normal versor* (also called *unit normal vector*) to σ is the map $\mathbf{n} \colon I \to \mathbb{R}^n$ of class C^{k-2} given by

$$\mathbf{n} = \frac{\dot{\mathbf{t}}}{\|\dot{\mathbf{t}}\|} = \frac{\dot{\mathbf{t}}}{\kappa} .$$

The plane through $\sigma(s)$ and parallel to $\mathrm{Span}\big(\mathbf{t}(s), \mathbf{n}(s)\big)$ is the *osculating plane* to the curve at $\sigma(s)$. The *affine normal line* of σ at the point $\sigma(s)$ is the line through $\sigma(s)$ parallel to the normal versor $\mathbf{n}(s)$.

Before going on, we must show how to compute the curvature and the normal versor without resorting to the arc-length, fulfilling the promise we made in Remark 1.2.18:

Proposition 1.3.10. *Let* $\sigma\colon I \to \mathbb{R}^n$ *be any regular parametrized curve. Then the curvature* $\kappa\colon I \to \mathbb{R}^+$ *of* σ *is given by*

$$\kappa = \frac{\sqrt{\|\sigma'\|^2\|\sigma''\|^2 - |\langle \sigma'', \sigma'\rangle|^2}}{\|\sigma'\|^3} . \tag{1.6}$$

In particular, σ *is biregular if and only if* σ' *and* σ'' *are linearly independent everywhere; in this case,*

$$\mathbf{n} = \frac{1}{\sqrt{\|\sigma''\|^2 - \frac{|\langle \sigma'', \sigma'\rangle|^2}{\|\sigma'\|^2}}} \left(\sigma'' - \frac{\langle \sigma'', \sigma'\rangle}{\|\sigma'\|^2}\sigma' \right) . \tag{1.7}$$

Proof. Let $s\colon I \to \mathbb{R}$ be the arc length of σ measured starting from an arbitrary point. Equation (1.5) gives

$$\mathbf{t}\big(s(t)\big) = \frac{\sigma'(t)}{\|\sigma'(t)\|} ;$$

since

$$\frac{\mathrm{d}}{\mathrm{d}t}\mathbf{t}\big(s(t)\big) = \frac{\mathrm{d}\mathbf{t}}{\mathrm{d}s}\big(s(t)\big)\frac{\mathrm{d}s}{\mathrm{d}t}(t) = \|\sigma'(t)\|\dot{\mathbf{t}}\big(s(t)\big) ,$$

we find

$$\dot{\mathbf{t}}\big(s(t)\big) = \frac{1}{\|\sigma'(t)\|}\frac{\mathrm{d}}{\mathrm{d}t}\left(\frac{\sigma'(t)}{\|\sigma'(t)\|} \right)$$

$$= \frac{1}{\|\sigma'(t)\|^2}\left(\sigma''(t) - \frac{\langle \sigma''(t), \sigma'(t)\rangle}{\|\sigma'(t)\|^2}\sigma'(t) \right) ; \tag{1.8}$$

note that $\dot{\mathbf{t}}\big(s(t)\big)$ is a multiple of the component of $\sigma''(t)$ orthogonal to $\sigma'(t)$. Finally,

$$\kappa(t) = \big\|\dot{\mathbf{t}}\big(s(t)\big)\big\| = \frac{1}{\|\sigma'(t)\|^2}\sqrt{\|\sigma''(t)\|^2 - \frac{|\langle \sigma''(t), \sigma'(t)\rangle|^2}{\|\sigma'(t)\|^2}} ,$$

and the proof is complete, as the last claim follows from the Cauchy-Schwarz inequality, and (1.7) follows from (1.8).   □

Let us see how to apply this result in several examples.

Example 1.3.11. Let $\sigma\colon \mathbb{R} \to \mathbb{R}^2$ be the ellipse having semiaxes a, $b > 0$, parametrized as in Example 1.1.30. Then $\sigma'(t) = (-a\sin t, b\cos t)$, and hence $\sigma''(t) = (-a\cos t, -b\sin t)$. Therefore

$$\mathbf{t}(t) = \frac{\sigma'(t)}{\|\sigma'(t)\|} = \frac{1}{\sqrt{a^2\sin^2 t + b^2\cos^2 t}}(-a\sin t, b\cos t)$$

and the curvature of the ellipse is given by

$$\kappa(t) = \frac{ab}{(a^2 \sin^2 t + b^2 \cos^2 t)^{3/2}} \, .$$

Example 1.3.12. The normal versor of a circle with radius $r > 0$ is

$$\mathbf{n}(s) = \left(-\cos(s/r), -\sin(s/r) \right) ;$$

that of a circular helix with radius $r > 0$ and pitch $a \in \mathbb{R}^*$ is

$$\mathbf{n}(s) = \left(-\cos \frac{s}{\sqrt{r^2 + a^2}}, -\sin \frac{s}{\sqrt{r^2 + a^2}}, 0 \right) ;$$

that of the catenary is

$$\mathbf{n}(s) = \left(-\frac{s}{\sqrt{1 + s^2}}, \frac{1}{\sqrt{1 + s^2}} \right) ;$$

and that of the ellipse with semiaxes $a, b > 0$ is

$$\mathbf{n}(t) = \frac{1}{\sqrt{a^2 \sin^2 t + b^2 \cos^2 t}} (-b \cos t, -a \sin t) \, .$$

Example 1.3.13. Let $\sigma \colon I \to \mathbb{R}^n$, given by $\sigma(t) = \left(t, f(t) \right)$, be the graph of a map $f \colon I \to \mathbb{R}^{n-1}$ of class (at least) C^2. Then

$$\mathbf{t} = \frac{1}{\sqrt{1 + \|f'\|^2}} (1, f') \, ,$$

$$\mathbf{n}(t) = \frac{1}{\sqrt{\|f''\|^2 - |\langle f'', f' \rangle|^2 / (1 + \|f'\|^2)}} \left(-\frac{\langle f'', f' \rangle}{1 + \|f'\|^2}, f'' - \frac{\langle f'', f' \rangle}{1 + \|f'\|^2} f' \right) \, .$$

and

$$\kappa = \frac{\sqrt{(1 + \|f'\|^2) \|f''\|^2 - |\langle f', f' \rangle|^2}}{(1 + \|f'\|^2)^{3/2}} \, .$$

In particular, σ is biregular if and only if f'' is never zero (why?).

Remark 1.3.14. To define the normal versor we had to assume the biregularity of the curve. However, if the curve is *plane*, to define a normal versor regularity is enough.

Indeed, if $\sigma \colon I \to \mathbb{R}^2$ is a plane curve of class C^k parametrized by arc length, for all $s \in I$ there exists a unique versor $\tilde{\mathbf{n}}(s)$ that is orthogonal to $\mathbf{t}(s)$ and such that the pair $\{\mathbf{t}(s), \tilde{\mathbf{n}}(s)\}$ has the same orientation as the canonical basis. In coordinates,

$$\mathbf{t}(s) = (a_1, a_2) \quad \Longrightarrow \quad \tilde{\mathbf{n}}(s) = (-a_2, a_1) ;$$

in particular, the map $\tilde{\mathbf{n}} \colon I \to \mathbb{R}^2$ is of class C^{k-1}, just like $\mathbf{t}$. Moreover, since $\dot{\mathbf{t}}(s)$ is orthogonal to $\mathbf{t}(s)$, it has to be a multiple of $\tilde{\mathbf{n}}(s)$; so there exists a function $\tilde{\kappa} \colon I \to \mathbb{R}$ of class C^{k-2} such that we have

$$\dot{\mathbf{t}} = \tilde{\kappa} \tilde{\mathbf{n}} \, . \tag{1.9}$$

Definition 1.3.15. If $\sigma\colon I \to \mathbb{R}^2$ is a regular plane curve of class C^k (with $k \geq 2$) parametrized by arc length, the map $\tilde{\mathbf{n}}\colon I \to \mathbb{R}^2$ of class C^{k-1} just defined is the *oriented normal versor* of σ, while the function $\tilde{\kappa}\colon I \to \mathbb{R}$ of class C^{k-2} is the *oriented curvature* of σ.

Remark 1.3.16. Since, by construction, we have $\det(\mathbf{t}, \tilde{\mathbf{n}}) \equiv 1$, the oriented curvature of a plane curve is given by the formula

$$\tilde{\kappa} = \det(\mathbf{t}, \dot{\mathbf{t}}) . \tag{1.10}$$

To put it simply, this means that, if $\tilde{\kappa} > 0$, then the curve is bending in a counterclockwise direction, while if $\tilde{\kappa} < 0$ then the curve is bending in a clockwise direction. Finally, if $\sigma\colon I \to \mathbb{R}^2$ is an arbitrary parametrized plane curve, then the oriented curvature of σ in the point $\sigma(t)$ is given by (see Problem 1.1)

$$\tilde{\kappa}(t) = \frac{1}{\|\sigma'(t)\|^3} \det\big(\sigma'(t), \sigma''(t)\big) . \tag{1.11}$$

Remark 1.3.17. The oriented curvature $\tilde{\kappa}$ of a plane curve is related to the usual curvature κ by the identity $\kappa = |\tilde{\kappa}|$. In particular, the normal versor introduced in Definition 1.3.9 coincides with the oriented normal versor $\tilde{\mathbf{n}}$ when the oriented curvature is positive, and with its opposite when the oriented curvature is negative.

Example 1.3.18. Example 1.3.6 show that the oriented curvature of the circle with center $(x_0, y_0) \in \mathbb{R}^2$ and radius $r > 0$, parametrized by arc length as in Example 1.2.14, is equal to the constant $1/r$ everywhere. On the other hand, let $\sigma = (\sigma_1, \sigma_2)\colon I \to \mathbb{R}^2$ be a regular curve parametrized by arc length, with constant oriented curvature equal to $1/r \neq 0$. Then the coordinates of σ satisfy the linear system of ordinary differential equations

$$\begin{cases} \ddot{\sigma}_1 = -\frac{1}{r}\dot{\sigma}_2 , \\ \ddot{\sigma}_2 = \frac{1}{r}\dot{\sigma}_1 . \end{cases}$$

Keeping in mind that $\dot{\sigma}_1^2 + \dot{\sigma}_2^2 \equiv 1$, we find that there exists a $s_0 \in \mathbb{R}$ such that

$$\dot{\sigma}(s) = \left(-\sin\frac{s+s_0}{r}, \cos\frac{s+s_0}{r}\right),$$

so the support of σ is contained (why?) in a circle with radius $|r|$. In other words, circles are characterized by having a constant nonzero oriented curvature.

As we shall shortly see (and as the previous example suggests), the oriented curvature completely determines a plane curve in a very precise sense: two plane curves parametrized by arc length having the same oriented curvature only differ by a rigid plane motion (Theorem 1.3.37 and Exercise 1.49).

Space curves, on the other hand, are not completely determined by their curvature. This is to be expected: in space, a curve may bend and also twist, that is leave any given plane. And, if $n > 3$, a curve in $\mathbb{R}^n$ may hypertwist in even more dimensions. For the sake of clarity, *in the rest of this section we shall* (almost) *uniquely consider curves in the space* $\mathbb{R}^3$, postponing the general discussion about curves in $\mathbb{R}^n$ until Section 1.7 of the supplementary material to the present chapter.

If the support of a regular curve is contained in a plane, it is clear (why? see the proof of Proposition 1.3.25) that the osculating plane of the curve is constant. This suggests that it is possible to measure how far a space curve is from being plane by studying the variation of its osculating plane. Since a plane (through the origin of $\mathbb{R}^3$) is completely determined by the direction orthogonal to it, we are led to the following:

Definition 1.3.19. Let $\sigma: I \to \mathbb{R}^3$ be a biregular curve of class C^k. The *binormal versor* (also called *unit binormal vector*) to the curve is the map $\mathbf{b}: I \to \mathbb{R}^3$ of class C^{k-2} given by $\mathbf{b} = \mathbf{t} \wedge \mathbf{n}$, where $\wedge$ denotes the vector product in $\mathbb{R}^3$. The *affine binormal line* of σ at the point $\sigma(s)$ is the line through $\sigma(s)$ parallel to the binormal versor $\mathbf{b}(s)$.

Finally, the triple $\{\mathbf{t}, \mathbf{n}, \mathbf{b}\}$ of $\mathbb{R}^3$-valued functions is the *Frenet frame* of the curve. Sometimes, the maps $\mathbf{t}$, $\mathbf{n}$, $\mathbf{b}: I \to \mathbb{R}^3$ are also called *spherical indicatrices* because their image is contained in the unit sphere of $\mathbb{R}^3$.

So we have associated to each point $\sigma(s)$ of a biregular space curve σ an orthonormal basis $\{\mathbf{t}(s), \mathbf{n}(s), \mathbf{b}(s)\}$ of $\mathbb{R}^3$ having the same orientation as the canonical basis, and varying along the curve (see Fig. 1.3).

Remark 1.3.20. The Frenet frame depends on the orientation of the curve. Indeed, if we denote by $\{\mathbf{t}^-, \mathbf{n}^-, \mathbf{b}^-\}$ the Frenet frame associated with the curve $\sigma^-(s) = \sigma(-s)$ equivalent to σ having opposite orientation, we have

$$\mathbf{t}^-(s) = -\mathbf{t}(-s)\,, \qquad \mathbf{n}^-(s) = \mathbf{n}(-s)\,, \qquad \mathbf{b}^-(s) = -\mathbf{b}(-s)\,.$$

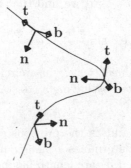

Fig. 1.3. The Frenet frame

On the other hand, since it was defined using a parametrization by arc length, the Frenet frame only depends on the oriented curve, and not on the specific parametrization chosen to compute it.

Example 1.3.21. Let $\sigma: \mathbb{R} \to \mathbb{R}^3$ be the circular helix with radius $r > 0$ and pitch $a \in \mathbb{R}^*$, parametrized by arc length as in Example 1.2.15. Then

$$\mathbf{b}(s) = \left(\frac{a}{\sqrt{r^2 + a^2}} \sin \frac{s}{\sqrt{r^2 + a^2}}, -\frac{a}{\sqrt{r^2 + a^2}} \cos \frac{s}{\sqrt{r^2 + a^2}}, \frac{r}{\sqrt{r^2 + a^2}} \right).$$

Example 1.3.22. If $\sigma: I \to \mathbb{R}^3$ is the graph of a map $f = (f_1, f_2): I \to \mathbb{R}^2$ such that f'' is nowhere zero, then

$$\mathbf{b} = \frac{1}{\sqrt{\|f''\|^2 + |\det(f', f'')|^2}} \left(\det(f', f''), -f_2'', f_1'' \right).$$

Example 1.3.23. If we identify $\mathbb{R}^2$ with the plane $\{z = 0\}$ in $\mathbb{R}^3$, we may consider every plane curve as a space curve. With this convention, it is straightforward (why?) to see that the binormal versor of a biregular curve $\sigma: I \to \mathbb{R}^2$ is everywhere equal to $(0, 0, 1)$ if the oriented curvature of σ is positive, and everywhere equal to $(0, 0, -1)$ if the oriented curvature of σ is negative.

Remark 1.3.24. Keeping in mind Proposition 1.3.10, we immediately find that the binormal versor of an arbitrary biregular parametrized curve $\sigma: I \to \mathbb{R}^3$ is given by

$$\mathbf{b} = \frac{\sigma' \wedge \sigma''}{\|\sigma' \wedge \sigma''\|}. \tag{1.12}$$

In particular, we obtain another formula for the computation of the normal versor of curves in $\mathbb{R}^3$:

$$\mathbf{n} = \mathbf{b} \wedge \mathbf{t} = \frac{(\sigma' \wedge \sigma'') \wedge \sigma'}{\|\sigma' \wedge \sigma''\| \, \|\sigma'\|}.$$

Moreover, formula (1.6) for the computation of the curvature becomes

$$\kappa = \frac{\|\sigma' \wedge \sigma''\|}{\|\sigma'\|^3}. \tag{1.13}$$

The next proposition confirms the correctness of our idea that the variation of the binormal versor measures how far a curve is from being plane:

Proposition 1.3.25. *Let $\sigma: I \to \mathbb{R}^3$ be a biregular curve of class C^k (with $k \geq 2$). Then the image of σ is contained in a plane if and only if the binormal versor is constant.*

Proof. Without loss of generality, we may assume that the curve σ is parametrized by arc length.

If the image of σ is contained in a plane, then there is a plane $H \subset \mathbb{R}^3$ containing the origin such that $\sigma(s) - \sigma(s') \in H$ for all s, $s' \in I$. Dividing by $s - s'$ and taking the limit as $s' \to s$ we immediately find that $\mathbf{t}(s) \in H$ for all $s \in I$. In the same way, it can be shown that $\dot{\mathbf{t}}(s) \in H$ for all $s \in I$, so $\mathbf{n}(s) \in H$ for all $s \in I$. Hence $\mathbf{b}(s)$ must always be one of the two versors orthogonal to H; since it changes continuously, it is constant.

On the other hand, assume the binormal versor is a constant vector $\mathbf{b}_0$; we want to prove that the support of σ is contained in a plane. Now, a plane is determined by one of its points and an orthogonal versor: a point $p \in \mathbb{R}^3$ is in the plane passing through $p_0 \in \mathbb{R}^3$ and orthogonal to the vector $v \in \mathbb{R}^3$ if and only if $\langle p - p_0, v \rangle = 0$. Take $s_0 \in I$; we want to show that the support of σ is contained in the plane through $\sigma(s_0)$ and orthogonal to $\mathbf{b}_0$. This is the same as showing that $\langle \sigma(s), \mathbf{b}_0 \rangle \equiv \langle \sigma(s_0), \mathbf{b}_0 \rangle$, or that the function $s \mapsto \langle \sigma(s), \mathbf{b}_0 \rangle$ is constant. And indeed we have

$$\frac{\mathrm{d}}{\mathrm{d}s} \langle \sigma, \mathbf{b}_0 \rangle = \langle \mathbf{t}, \mathbf{b}_0 \rangle \equiv 0 \,,$$

as $\mathbf{t}$ is always orthogonal to the binormal versor, so the support of σ really is contained in the plane of equation $\langle p - \sigma(s_0), \mathbf{b}_0 \rangle = 0$. $\square$

This result suggests that the derivative of the binormal versor might measure how far a biregular curve is from being plane. Now, $\mathbf{b}$ is a versor; so, taking the derivative of $\langle \mathbf{b}, \mathbf{b} \rangle \equiv 1$ we get $\langle \dot{\mathbf{b}}, \mathbf{b} \rangle \equiv 0$, that is $\dot{\mathbf{b}}$ is always orthogonal to $\mathbf{b}$. On the other hand,

$$\dot{\mathbf{b}} = \dot{\mathbf{t}} \wedge \mathbf{n} + \mathbf{t} \wedge \dot{\mathbf{n}} = \mathbf{t} \wedge \dot{\mathbf{n}} \,,$$

so $\dot{\mathbf{b}}$ is perpendicular to $\mathbf{t}$ too; hence, $\dot{\mathbf{b}}$ has to be a multiple of $\mathbf{n}$.

Definition 1.3.26. Let $\sigma : I \to \mathbb{R}^3$ be a biregular curve of class C^k (with $k \geq 3$) parametrized by arc length. The *torsion* of σ is the function $\tau : I \to \mathbb{R}$ of class C^{k-3} such that $\dot{\mathbf{b}} = -\tau \mathbf{n}$. (*Warning*: in some texts the torsion is defined to be the opposite of the function we have chosen.)

Remark 1.3.27. Proposition 1.3.25 may then be rephrased by saying that *the image of a biregular curve σ is contained in a plane if and only if the torsion of σ is everywhere zero* (but see Exercise 1.30 for an example of what can happen if the curve is not biregular even in a single point). In Section 1.4 of the supplementary material of this chapter we shall give a geometrical interpretation of the sign of the torsion.

Remark 1.3.28. Curvature and torsion do not depend on the orientation of the curve. More precisely, if $\sigma : I \to \mathbb{R}^3$ is a biregular curve parametrized by arc length, and σ^- is the usual curve parametrized by arc length equivalent to σ but with the opposite orientation given by $\sigma^-(s) = \sigma(-s)$, then the curvature κ^- and the torsion τ^- of σ^- are such that

$$\kappa^-(s) = \kappa(-s) \quad \text{and} \quad \tau^-(s) = \tau(-s) \,.$$

Remark 1.3.29. On the other hand, the oriented curvature and the oriented normal versor of a plane curve depend on the orientation of the curve. Indeed, with the notation of the previous remark applied to a plane curve σ, we find

$$\mathbf{t}^-(s) = -\mathbf{t}(-s) , \quad \tilde{\kappa}^-(s) = -\tilde{\kappa}(-s) , \quad \tilde{\mathbf{n}}^-(s) = -\tilde{\mathbf{n}}(-s) .$$

Remark 1.3.30. To find the torsion of a biregular curve $\sigma: I \to \mathbb{R}^3$ with an arbitrary parametrization, first of all note that $\tau = -\langle \dot{\mathbf{b}}, \mathbf{n} \rangle$. Taking the derivative of (1.12), we get

$$\dot{\mathbf{b}} = \frac{d\mathbf{b}}{ds} = \frac{dt}{ds}\frac{d\mathbf{b}}{dt} = \frac{1}{\|\sigma'\|}\left[\frac{\sigma' \wedge \sigma'''}{\|\sigma' \wedge \sigma''\|} - \frac{\langle \sigma' \wedge \sigma'', \sigma' \wedge \sigma''' \rangle}{\|\sigma' \wedge \sigma''\|^3}\sigma' \wedge \sigma'' \right] .$$

Therefore, keeping in mind Equation (1.7), we obtain

$$\tau = -\frac{\langle \sigma' \wedge \sigma''', \sigma'' \rangle}{\|\sigma' \wedge \sigma''\|^2} = \frac{\langle \sigma' \wedge \sigma'', \sigma''' \rangle}{\|\sigma' \wedge \sigma''\|^2} .$$

Example 1.3.31. If $\sigma: I \to \mathbb{R}^3$ is the usual parametrization $\sigma(t) = \big(t, f(t)\big)$ of the graph of a function $f: I \to \mathbb{R}^2$ with f'' nowhere zero, then

$$\tau = \frac{\det(f'', f''')}{\|f''\|^2 + |\det(f', f'')|^2} .$$

Example 1.3.32. Let $\sigma: \mathbb{R} \to \mathbb{R}^3$ be the circular helix with radius $r > 0$ and pitch $a \in \mathbb{R}^*$, parametrized by arc length as in Example 1.2.15. Then, taking the derivative of the binormal versor found in Example 1.3.21 and keeping in mind Example 1.3.12, we find

$$\tau(s) \equiv \frac{a}{r^2 + a^2} .$$

Thus both the curvature and the torsion of the circular helix are constant.

We have computed the derivative of the tangent versor and of the binormal versor; for the sake of completeness, let us compute the derivative of the normal versor too. We get

$$\dot{\mathbf{n}} = \dot{\mathbf{b}} \wedge \mathbf{t} + \mathbf{b} \wedge \dot{\mathbf{t}} = -\tau\mathbf{n} \wedge \mathbf{t} + \mathbf{b} \wedge \kappa\mathbf{n} = -\kappa\mathbf{t} + \tau\mathbf{b} .$$

Definition 1.3.33. Let $\sigma: I \to \mathbb{R}^3$ be a biregular space curve. The three equations

$$\begin{cases} \dot{\mathbf{t}} = \kappa\mathbf{n} , \\ \dot{\mathbf{n}} = -\kappa\mathbf{t} + \tau\mathbf{b} , \\ \dot{\mathbf{b}} = -\tau\mathbf{n} , \end{cases} \tag{1.14}$$

are the *Frenet-Serret formulas* of σ.

Remark 1.3.34. There are Frenet-Serret formulas for plane curves too. Since
$\dot{\tilde{n}}$ is, for the usual reasons, orthogonal to $\tilde{n}$, it has to be a multiple of t. Taking
the derivative of $\langle t, \tilde{n} \rangle \equiv 0$, we find $\langle t, \dot{\tilde{n}} \rangle = -\tilde{\kappa}$. So the *Frenet-Serret formulas
for plane curves* are

$$\begin{cases} \dot{t} = \tilde{\kappa}\tilde{n} \,, \\ \dot{\tilde{n}} = -\tilde{\kappa}t \,. \end{cases}$$

In Section 1.7 of the supplementary material of this chapter we shall see analogous formulas for curves in $\mathbb{R}^n$.

The basic idea of the local theory of space curves is that the curvature
and the torsion completely determine a curve (compare Example 1.3.32 and
Problem 1.7). To convey in precise terms what we mean, we need a definition.

Definition 1.3.35. A *rigid motion* of $\mathbb{R}^n$ is an affine map $\rho: \mathbb{R}^n \to \mathbb{R}^n$ of the
form $\rho(x) = Ax + b$, where $b \in \mathbb{R}^n$ and

$$A \in SO(n) = \{ A \in GL(n, \mathbb{R}) \mid A^T A = I \text{ and } \det A = 1 \} \,.$$

In particular, when $n = 3$ every rigid motion is a rotation about the origin,
followed by a translation.

If a curve is obtained from another through a rigid motion, both curves
have the same curvature and torsion (Exercise 1.26); conversely, the *fundamental theorem of the local theory of curves* states that any two curves with
equal curvature and torsion can always be obtained from one another through
a rigid motion. Frenet-Serret formulas are exactly the tool that will enable us
to prove this result, using the classical Analysis theorem about the existence
and uniqueness of the solutions of a linear system of ordinary differential
equations (see [24, p. 162]):

Theorem 1.3.36. *Given an interval* $I \subseteq \mathbb{R}$, *a point* $t_0 \in I$, *a vector*
$u_0 \in \mathbb{R}^n$, *and two functions* $f: I \to \mathbb{R}^n$ *and* $A: I \to M_{n,n}(\mathbb{R})$ *of class* C^k,
with $k \in \mathbb{N}^* \cup \{\infty\}$, *where* $M_{p,q}(\mathbb{R})$ *denotes the space of* $p \times q$ *real matrices,
there exists a unique solution* $u: I \to \mathbb{R}^n$ *of class* C^{k+1} *to the Cauchy problem*

$$\begin{cases} u' = Au + f \,, \\ u(t_0) = u_0 \,. \end{cases}$$

In particular, the solution of the Cauchy problem for *linear* systems of
ordinary differential equations exists over the whole domain of definition of
the coefficients. This is what we need to prove the *fundamental theorem of the
local theory of curves:*

**Theorem 1.3.37 (Fundamental theorem of the local theory of cur-
ves).** *Given two functions* $\kappa: I \to \mathbb{R}^+$ *and* $\tau: I \to \mathbb{R}$, *with* κ *always positive
and of class* C^{k+1} *and* τ *of class* C^k *(with* $k \in \mathbb{N}^* \cup \{\infty\}$*), there exists a unique
(up to a rigid motion) biregular curve* $\sigma: I \to \mathbb{R}^3$ *of class* C^{k+3} *parametrized
by arc length with curvature* κ *and torsion* τ.

Proof. We prove existence first. Frenet-Serret formulas (1.14) form a linear system of ordinary differential equations in 9 unknowns (the components of $\mathbf{t}$, $\mathbf{n}$, and $\mathbf{b}$); so we can apply Theorem 1.3.36.

Fix a point $s_0 \in I$ and an orthonormal basis $\{\mathbf{t}_0, \mathbf{n}_0, \mathbf{b}_0\}$ with the same orientation as the canonical basis. Theorem 1.3.36 provides us with a unique triple of functions $\mathbf{t}$, $\mathbf{n}$, $\mathbf{b} : I \to \mathbb{R}^3$, with $\mathbf{t}$ of class C^{k+2} and $\mathbf{n}$ and $\mathbf{b}$ of class C^{k+1}, satisfying (1.14) and such that $\mathbf{t}(s_0) = \mathbf{t}_0$, $\mathbf{n}(s_0) = \mathbf{n}_0$, and $\mathbf{b}(s_0) = \mathbf{b}_0$.

We want to prove that the triple $\{\mathbf{t}, \mathbf{n}, \mathbf{b}\}$ we have just found is the Frenet frame of some curve. We show first that being an orthonormal basis in s_0 forces it to be so in every point. From (1.14) we deduce that the functions $\langle \mathbf{t}, \mathbf{t} \rangle$, $\langle \mathbf{t}, \mathbf{n} \rangle$, $\langle \mathbf{t}, \mathbf{b} \rangle$, $\langle \mathbf{n}, \mathbf{n} \rangle$, $\langle \mathbf{n}, \mathbf{b} \rangle$, and $\langle \mathbf{b}, \mathbf{b} \rangle$ satisfy the following system of six linear ordinary differential equations in 6 unknowns

$$
\begin{cases}
\frac{\mathrm{d}}{\mathrm{d}s} \langle \mathbf{t}, \mathbf{t} \rangle = 2\kappa \langle \mathbf{t}, \mathbf{n} \rangle \,, \\
\frac{\mathrm{d}}{\mathrm{d}s} \langle \mathbf{t}, \mathbf{n} \rangle = -\kappa \langle \mathbf{t}, \mathbf{t} \rangle + \tau \langle \mathbf{t}, \mathbf{b} \rangle + \kappa \langle \mathbf{n}, \mathbf{n} \rangle \,, \\
\frac{\mathrm{d}}{\mathrm{d}s} \langle \mathbf{t}, \mathbf{b} \rangle = -\tau \langle \mathbf{t}, \mathbf{n} \rangle + \kappa \langle \mathbf{n}, \mathbf{b} \rangle \,, \\
\frac{\mathrm{d}}{\mathrm{d}s} \langle \mathbf{n}, \mathbf{n} \rangle = -2\kappa \langle \mathbf{t}, \mathbf{n} \rangle + 2\tau \langle \mathbf{n}, \mathbf{b} \rangle \,, \\
\frac{\mathrm{d}}{\mathrm{d}s} \langle \mathbf{n}, \mathbf{b} \rangle = -\kappa \langle \mathbf{t}, \mathbf{b} \rangle - \tau \langle \mathbf{n}, \mathbf{n} \rangle + \tau \langle \mathbf{b}, \mathbf{b} \rangle \,, \\
\frac{\mathrm{d}}{\mathrm{d}s} \langle \mathbf{b}, \mathbf{b} \rangle = -2\tau \langle \mathbf{n}, \mathbf{b} \rangle \,,
\end{cases}
$$

with initial conditions

$$\langle \mathbf{t}, \mathbf{t} \rangle(s_0) = 1 \,, \ \langle \mathbf{t}, \mathbf{n} \rangle(s_0) = 0 \,, \ \langle \mathbf{t}, \mathbf{b} \rangle(s_0) = 0 \,,$$
$$\langle \mathbf{n}, \mathbf{n} \rangle(s_0) = 1 \,, \ \langle \mathbf{n}, \mathbf{b} \rangle(s_0) = 0 \,, \ \langle \mathbf{b}, \mathbf{b} \rangle(s_0) = 1 \,.$$

But it is straightforward to verify that

$$\langle \mathbf{t}, \mathbf{t} \rangle \equiv \langle \mathbf{n}, \mathbf{n} \rangle \equiv \langle \mathbf{b}, \mathbf{b} \rangle \equiv 1 \,, \quad \langle \mathbf{t}, \mathbf{n} \rangle \equiv \langle \mathbf{t}, \mathbf{b} \rangle \equiv \langle \mathbf{n}, \mathbf{b} \rangle \equiv 0 \qquad (1.15)$$

is a solution of the same system of differential equations, satisfying the same initial conditions in s_0. So the functions $\mathbf{t}$, $\mathbf{n}$ and $\mathbf{b}$ have to satisfy equalities (1.15), and the triple $\{\mathbf{t}(s), \mathbf{n}(s), \mathbf{b}(s)\}$ is orthonormal for all $s \in I$. Moreover, it has the same orientation of the canonical basis of $\mathbb{R}^3$ everywhere: indeed, $\langle \mathbf{t} \wedge \mathbf{n}, \mathbf{b} \rangle$ is a continuous function on I with values in $\{+1, -1\}$, whose value is $+1$ in s_0; hence, necessarily, $\langle \mathbf{t} \wedge \mathbf{n}, \mathbf{b} \rangle \equiv +1$, which implies (why?) that $\{\mathbf{t}(s), \mathbf{n}(s), \mathbf{b}(s)\}$ has the same orientation as the canonical basis everywhere.

Finally, define the curve $\sigma : I \to \mathbb{R}^3$ by setting

$$\sigma(s) = \int_{s_0}^{s} \mathbf{t}(t) \, \mathrm{d}t \,.$$

The curve σ is of class C^{k+3} with derivative $\mathbf{t}(s)$, so it is regular, parametrized by arc length, and with tangent versor $\mathbf{t}$. Since the equations (1.14) give $\ddot{\sigma} = \kappa \mathbf{n}$ with $\kappa > 0$ everywhere, we deduce that κ is the curvature and $\mathbf{n}$

the normal versor of σ (in particular, σ is biregular). It follows that $\mathbf{b}$ is the binormal versor and, thanks to (1.14) once more, that τ is the torsion of σ, as required.

Let us now prove uniqueness. Let $\sigma_1 \colon I \to \mathbb{R}^3$ be another biregular curve of class C^{k+3}, parametrized by arc length, with curvature κ and torsion τ. Fix $s_0 \in I$; up to a rigid motion, we may assume that $\sigma(s_0) = \sigma_1(s_0)$, and that σ and σ_1 have the same Frenet frame at s_0. By the uniqueness of the solution of (1.14), it follows that σ and σ_1 have the same Frenet frame at all points of I; in particular, $\dot\sigma \equiv \dot\sigma_1$. But this implies

$$\sigma(s) = \sigma(s_0) + \int_{s_0}^{s} \dot\sigma(t)\,\mathrm{d}t = \sigma_1(s_0) + \int_{s_0}^{s} \dot\sigma_1(t)\,\mathrm{d}t = \sigma_1(s)\,,$$

and $\sigma_1 \equiv \sigma$.   □

Therefore curvature and torsion are all we need to completely describe a curve in space. For this reason, curvature and torsion are sometimes called *intrinsic* or *natural equations* of the curve.

Remark 1.3.38. Exactly in the same way (Exercise 1.49) it is possible to prove the following result: *Given a function $\tilde\kappa \colon I \to \mathbb{R}$ of class C^k, with $k \in \mathbb{N}^* \cup \{\infty\}$, there exists a unique (up to a rigid motion in the plane) regular curve $\sigma \colon I \to \mathbb{R}^2$ of class C^{k+2} parametrized by arc length having oriented curvature $\tilde\kappa$.*

Guided problems

For convenience, we repeat here the Frenet-Serret formulas, and the formulas (given in Remarks 1.3.24 and 1.3.30, and useful to solve the exercises) for the computation of curvature, torsion and Frenet frame of an arbitrarily parametrized biregular space curve:

$$\mathbf{t} = \frac{\sigma'}{\|\sigma'\|}\,, \quad \mathbf{b} = \frac{\sigma' \wedge \sigma''}{\|\sigma' \wedge \sigma''\|}\,, \quad \mathbf{n} = \frac{(\sigma' \wedge \sigma'') \wedge \sigma'}{\|\sigma' \wedge \sigma''\|\,\|\sigma'\|}\,,$$

$$\kappa = \frac{\|\sigma' \wedge \sigma''\|}{\|\sigma'\|^3}\,, \quad \tau = \frac{\langle \sigma' \wedge \sigma'', \sigma'''\rangle}{\|\sigma' \wedge \sigma''\|^2}\,;$$

$$\begin{cases} \dot{\mathbf{t}} = \kappa\mathbf{n}\,, \\ \dot{\mathbf{n}} = -\kappa\mathbf{t} + \tau\mathbf{b}\,, \\ \dot{\mathbf{b}} = -\tau\mathbf{n}\,. \end{cases}$$

Problem 1.1. Let $\sigma \colon I \to \mathbb{R}^2$ be a biregular plane curve, parametrized by an arbitrary parameter t. Show that the oriented curvature of σ is given by

$$\tilde\kappa = \frac{1}{\|\sigma'\|^3}\det(\sigma', \sigma'') = \frac{x'y'' - x''y'}{\left((x')^2 + (y')^2\right)^{3/2}}\,,$$

where x, $y: I \to \mathbb{R}$ are defined by $\sigma(t) = (x(t), y(t))$.

Solution. By formula (1.10), the oriented curvature is given by $\tilde{\kappa} = \det(\mathbf{t}, \dot{\mathbf{t}})$. To complete the proof, it is sufficient to substitute $\mathbf{t} = \sigma'/\|\sigma'\|$ and

$$\dot{\mathbf{t}}(s(t)) = \frac{1}{\|\sigma'(t)\|} \frac{d}{dt}\left(\frac{\sigma'(t)}{\|\sigma'(t)\|}\right)$$
$$= \frac{1}{\|\sigma'(t)\|^2}\left(\sigma''(t) - \frac{\langle\sigma''(t), \sigma'(t)\rangle}{\|\sigma'(t)\|^2}\sigma'(t)\right)$$

in (1.10). Since the determinant is linear and alternating with respect to columns, we get $\tilde{\kappa} = \det(\sigma', \sigma'')/\|\sigma'(t)\|^3$, as desired. $\square$

Problem 1.2. *Let $\sigma: I \to \mathbb{R}^n$ be a regular curve of class C^2 parametrized by arc length. Denote by $\theta(\varepsilon)$ the angle between the versors $\mathbf{t}(s_0)$ and $\mathbf{t}(s_0 + \varepsilon)$, tangent to σ respectively in $\sigma(s_0)$ and in a nearby point $\sigma(s_0 + \varepsilon)$, for $\varepsilon > 0$ small. Show that the curvature $\kappa(s_0)$ of σ in $\sigma(s_0)$ satisfies the equality*

$$\kappa(s_0) = \lim_{\varepsilon \to 0}\left|\frac{\theta(\varepsilon)}{\varepsilon}\right|.$$

Deduce that the curvature κ measures the rate of variation of the direction of the tangent line, with respect to the arc length. [Note: This problem will find a new interpretation in Chapter 2; see Proposition 2.4.2 and its proof.]

Solution. Consider the versors $\mathbf{t}(s_0)$ and $\mathbf{t}(s_0+\varepsilon)$, having as their initial point the origin O; the triangle they determine is isosceles, and the length of the third side is given by $\|\mathbf{t}(s_0 + \varepsilon) - \mathbf{t}(s_0)\|$. The Taylor expansion of the sine function yields

$$\|\mathbf{t}(s_0 + \varepsilon) - \mathbf{t}(s_0)\| = 2\left|\sin(\theta(\varepsilon)/2)\right| = \left|\theta(\varepsilon) + o(\theta(\varepsilon))\right|.$$

Keeping in mind the definition of curvature, we conclude that

$$\kappa(s_0) = \|\dot{\mathbf{t}}(s_0)\| = \lim_{\varepsilon \to 0}\left\|\frac{\mathbf{t}(s_0 + \varepsilon) - \mathbf{t}(s_0)}{\varepsilon}\right\|$$
$$= \lim_{\varepsilon \to 0}\left|\frac{\theta(\varepsilon) + o(\theta(\varepsilon))}{\varepsilon}\right|.$$

As $\lim_{\varepsilon \to 0}\theta(\varepsilon) = 0$, the assertion follows. $\square$

Problem 1.3 (the tractrix). *Let $\sigma: (0, \pi) \to \mathbb{R}^2$ be the plane curve defined by*

$$\sigma(t) = \left(\sin t, \cos t + \log\tan\frac{t}{2}\right);$$

the image of σ is called tractrix *(Fig. 1.4). [Note: This curve will be used in later chapters to define surfaces with important properties; see Example 4.5.23.]*

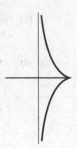

Fig. 1.4. The tractrix

(i) *Prove that σ is a parametrization of class C^∞, regular everywhere except in $t = \pi/2$.*

(ii) *Check that the length of the segment of the tangent line to the tractrix from the point of tangency to the y-axis is always 1.*

(iii) *Determine the arc length of σ starting from $t_0 = \pi/2$.*

(iv) *Compute the curvature of σ where it is defined.*

Solution. (i) Since $\tan(t/2) > 0$ for all $t \in (0, \pi)$, the curve σ is of class C^∞. Moreover,

$$\sigma'(t) = \left(\cos t, \frac{\cos^2 t}{\sin t} \right) \qquad \text{and} \qquad \|\sigma'(t)\| = \frac{|\cos t|}{\sin t} \ ,$$

so $\sigma'(t)$ is zero only for $t = \pi/2$, as desired.

(ii) If $t_0 \neq \pi/2$, the affine tangent line $\eta \colon \mathbb{R} \to \mathbb{R}^2$ to σ at the point $\sigma(t_0)$ is given by

$$\eta(x) = \sigma(t_0) + x\sigma'(t_0) = \left(\sin t_0 + x \cos t_0, \cos t_0 + \log \tan \frac{t_0}{2} + x \frac{\cos^2 t_0}{\sin t_0} \right) \ .$$

The tangent line intersects the y-axis in the point where the first coordinate of η is zero, that is, for $x = -\tan t_0$. So the length we are looking for is

$$\|\eta(-\tan t_0) - \eta(0)\| = \|(-\sin t_0, -\cos t_0)\| = 1 \ ,$$

as stated.

In a sense, this result is true for $t_0 = \pi/2$ too. Indeed, even if the tangent vector to σ tends to O for $t \to \pi/2$, the tangent line to σ at $\sigma(t)$ tends to the x-axis for $t \to \pi/2$, since

$$\lim_{t \to \pi/2-} \frac{\sigma'(t)}{\|\sigma'(t)\|} = (1,0) = -(-1,0) = - \lim_{t \to \pi/2+} \frac{\sigma'(t)}{\|\sigma'(t)\|} \ .$$

So, if we consider the x-axis as the tangent line to the support of the tractrix at the point $\sigma(\pi/2) = (1,0)$, in this case too the segment of the tangent line from the point of the curve to the y-axis has length 1.

(iii) If $t > \pi/2$ we have

$$s(t) = \int_{\pi/2}^{t} \|\sigma'(\tau)\| \, d\tau = -\int_{\pi/2}^{t} \frac{\cos \tau}{\sin \tau} \, d\tau = -\log \sin t \,.$$

Analogously, if $t < \pi/2$ we have

$$s(t) = \int_{\pi/2}^{t} \|\sigma'(\tau)\| \, d\tau = -\int_{t}^{\pi/2} \frac{\cos \tau}{\sin \tau} \, d\tau = \log \sin t \,.$$

In particular,

$$s^{-1}(s) = \begin{cases} \pi - \arcsin e^{-s} \in [\pi/2, \pi) & \text{if } s \in [0, +\infty) \,, \\ \arcsin e^{s} \in (0, \pi/2] & \text{if } s \in (-\infty, 0] \,, \end{cases}$$

and using the formula $\tan \frac{x}{2} = \frac{\sin x}{1+\cos x}$ we see that the reparametrization of σ by arc length is given by

$$\sigma\big(s^{-1}(s)\big) = \begin{cases} \big(e^{-s}, -s - \sqrt{1 - e^{-2s}} - \log\big(1 - \sqrt{1 - e^{-2s}}\big)\big) & \text{if } s > 0 \,, \\ \big(e^{s}, s + \sqrt{1 - e^{2s}} - \log\big(1 + \sqrt{1 - e^{2s}}\big)\big) & \text{if } s < 0 \,. \end{cases}$$

(iv) Using the reparametrization $\sigma_1 = \sigma \circ s^{-1}$ of σ by arc length we have just computed, we find

$$\dot{\sigma}_1(s) = \begin{cases} \left(-e^{-s}, -\dfrac{1 - e^{-2s} - \sqrt{1 - e^{-2s}}}{1 - \sqrt{1 - e^{-2s}}}\right) & \text{if } s > 0 \,, \\[3ex] \left(e^{s}, \dfrac{1 - e^{2s} + \sqrt{1 - e^{2s}}}{1 + \sqrt{1 - e^{2s}}}\right) & \text{if } s < 0 \,, \end{cases}$$

and

$$\ddot{\sigma}_1(s) = \begin{cases} \left(e^{-s}, \dfrac{e^{-2s}}{\sqrt{1 - e^{-2s}}}\right) & \text{if } s > 0 \,, \\[3ex] \left(e^{s}, -\dfrac{e^{2s}}{\sqrt{1 - e^{2s}}}\right) & \text{if } s < 0 \,. \end{cases}$$

So the curvature κ_1 of σ_1 for $s \neq 0$ is given by

$$\kappa(s) = \|\ddot{\sigma}(s)\| = \frac{1}{\sqrt{e^{2|s|} - 1}} \,,$$

and (keeping in mind Remark 1.3.4) the curvature κ of σ for $t \neq \pi/2$ is

$$\kappa(t) = \kappa_1\big(s(t)\big) = |\tan t| \,.$$

As an alternative, we could have computed the curvature of σ by using the formula for curves with an arbitrary parametrization (see next problem and Problem 1.1). $\square$

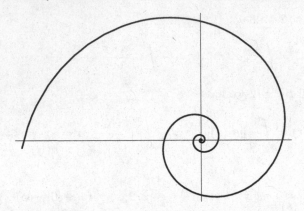

Fig. 1.5. Logarithmic spiral

Problem 1.4 (logarithmic spiral). *Fix two real numbers $a > 0$ and $b < 0$. The* logarithmic spiral *(Fig. 1.5) is the plane curve $\sigma \colon \mathbb{R} \to \mathbb{R}^2$ given by*

$$\sigma(t) = (ae^{bt} \cos t, ae^{bt} \sin t) \, .$$

(i) *Show that the support of the spiral satisfies the equation $r = ae^{b\theta}$, expressed in the polar coordinates (r, θ).*
(ii) *Show that $\sigma(t)$ winds around the origin O tending to it as $t \to \infty$.*
(iii) *Determine the arc length of σ, starting from $t = 0$. Find the arc length in the case $a = 1/2$ and $b = -1$.*
(iv) *Determine the curvature and the torsion of σ, and remark that the curvature is never zero.*

Solution. (i) We have $r^2 = x^2 + y^2 = a^2 e^{2bt}(\cos^2 t + \sin^2 t) = a^2 e^{2bt}$, and the assertion follows because r is always positive.

(ii) First of all, by (i) we have $\|\sigma(t)\| = ae^{bt}$, and thus $\sigma(t) \to O$ as $t \to \infty$, because $b < 0$.

Moreover, t coincides with the argument θ of $\sigma(t)$ up to a multiple of 2π; so the argument of $\sigma(t)$ is periodic of period 2π and assumes all possible values, that is σ winds around the origin.

(iii) Note that the parametrization of σ is of class C^∞. Differentiating, we find

$$\sigma'(t) = ae^{bt}(b \cos t - \sin t, b \sin t + \cos t)$$

and so

$$\|\sigma'(t)\| = ae^{bt} \sqrt{b^2 + 1} \, .$$

We deduce from this that the arc length of σ starting from $t = 0$ is given by:

$$s(t) = \int_0^t \|\sigma'(\tau)\| \, d\tau = a\sqrt{b^2 + 1} \int_0^t e^{b\tau} \, d\tau = a\sqrt{b^2 + 1} \left[\frac{e^{bt} - 1}{b} \right] \, .$$

In the case $a = 1/2$, $b = -1$, the arc length is $s(t) = (1 - e^{-t})/\sqrt{2}$.

(iv) By applying the usual formulas we find

$$
\begin{aligned}
\kappa(t) &= \frac{\|\sigma' \wedge \sigma''\|}{\|\sigma'\|^3} \\
&= \frac{-(b^2 - 1)\sin^2 t + 2b^2 \cos^2 t - (b^2 - 1)\cos^2 t + 2b^2 \sin^2 t}{a\, e^{bt}\, (b^2 + 1)^{3/2}} \\
&= \frac{2b^2 - (b^2 - 1)}{a\, e^{bt}\, (b^2 + 1)^{3/2}} = \frac{b^2 + 1}{a\, e^{bt}\, (b^2 + 1)^{3/2}} = \frac{1}{a\, e^{bt}\, (b^2 + 1)^{1/2}} \, .
\end{aligned}
$$

In particular, the curvature is never zero, and the curve is biregular.

Finally, since the curve σ is plane and biregular, its torsion is defined and is zero everywhere. □

Problem 1.5 (Twisted cubic). *Determine the curvature, the torsion, and the Frenet frame of the curve $\sigma \colon \mathbb{R} \to \mathbb{R}^3$ defined by $\sigma(t) = (t, t^2, t^3)$.*

Solution. Differentiating the expression of σ we find

$$
\sigma'(t) = (1, 2t, 3t^2) \,, \quad \sigma''(t) = (0, 2, 6t) \quad \text{and} \quad \sigma'''(t) = (0, 0, 6) \,.
$$

In particular, σ' is nowhere zero; thus σ is regular and

$$
\mathbf{t}(t) = \frac{1}{\sqrt{1 + 4t^2 + 9t^4}} (1, 2t, 3t^2) \,.
$$

Next,

$$
\sigma'(t) \wedge \sigma''(t) = (6t^2, -6t, 2)
$$

is never zero, so σ' and σ'' are always linearly independent, and σ is biregular. Using the formulas recalled at the beginning of this section we get

$$
\mathbf{b}(t) = \frac{\sigma'(t) \wedge \sigma''(t)}{\|\sigma'(t) \wedge \sigma''(t)\|} = \frac{1}{2\sqrt{1 + 9t^2 + 9t^4}} (6t^2, -6t, 2) \,,
$$

$$
\mathbf{n}(t) = \mathbf{b}(t) \wedge \mathbf{t}(t) = \frac{(-9t^3 - 2t, 1 - 9t^4, 6t^3 + 3t)}{\sqrt{(1 + 4t^2 + 9t^4)(1 + 9t^2 + 9t^4)}} \,,
$$

$$
\kappa(t) = \frac{\|\sigma'(t) \wedge \sigma''(t)\|}{\|\sigma'(t)\|^3} = \frac{2\sqrt{1 + 9t^2 + 9t^4}}{(1 + 4t^2 + 9t^4)^{3/2}} \,,
$$

and

$$
\tau(t) = \left\langle \mathbf{b}(t), \frac{\sigma'''(t)}{\|\sigma'(t) \wedge \sigma''(t)\|} \right\rangle = \frac{3}{1 + 9t^2 + 9t^4} \,.
$$

□

Problem 1.6. *Prove that the curve* $\sigma\colon (0, +\infty) \to \mathbb{R}^3$ *defined by*

$$\sigma(t) = \left(t, \frac{1+t}{t}, \frac{1-t^2}{t} \right)$$

is contained in a plane.

Solution. By noting that

$$\sigma'(t) = \left(1, -\frac{1}{t^2}, -\frac{1}{t^2} - 1 \right) \qquad \text{and} \qquad \sigma''(t) = \left(0, \frac{2}{t^3}, \frac{2}{t^3} \right),$$

we find that the vector product

$$\sigma'(t) \wedge \sigma''(t) = \frac{2}{t^3}\,(1, -1, 1)$$

is nowhere zero, so the curvature $\kappa = \|\sigma' \wedge \sigma''\|/\|\sigma'\|^3$ is nowhere zero. By Remark 1.3.27, we may conclude that σ is a plane curve if and only if the torsion $\tau = \langle \sigma' \wedge \sigma'', \sigma''' \rangle / \|\sigma' \wedge \sigma''\|^2$ is zero everywhere, that is, if and only if $\langle \sigma' \wedge \sigma'', \sigma''' \rangle$ is zero everywhere. But

$$\langle \sigma' \wedge \sigma'', \sigma''' \rangle = \det \begin{pmatrix} 1 & -\frac{1}{t^2} & -\frac{1}{t^2} - 1 \\ 0 & \frac{2}{t^3} & \frac{2}{t^3} \\ 0 & -\frac{6}{t^4} & -\frac{6}{t^4} \end{pmatrix} \equiv 0 \,,$$

and the assertion follows. $\qquad\qquad\qquad\qquad\qquad\qquad\qquad\qquad\qquad\qquad\quad$ $\square$

Problem 1.7. *Let* $\sigma\colon I \to \mathbb{R}^3$ *be a biregular curve parametrized by arc length, having constant curvature* $\kappa_0 > 0$ *and constant torsion* $\tau_0 \in \mathbb{R}$. *Prove that, up to rotations and translations of* $\mathbb{R}^3$, σ *is an arc of a circular helix.*

Solution. If $\tau_0 = 0$, then Proposition 1.3.25 and Example 1.3.18 tell us that σ is an arc of a circle, so it can be considered an arc of the degenerate circular helix with pitch 0.

Assume, on the other hand, $\tau_0 \neq 0$. Then

$$\frac{\mathrm{d}}{\mathrm{d}s}(\tau_0 \mathbf{t} + \kappa_0 \mathbf{b}) = \tau_0 \kappa_0 \mathbf{n} - \kappa_0 \tau_0 \mathbf{n} \equiv O \,;$$

so $\tau_0 \mathbf{t} + \kappa_0 \mathbf{b}$ has to be everywhere equal to a constant vector $\mathbf{v}_0$ having length $\sqrt{\kappa_0^2 + \tau_0^2}$. Up to rotations in $\mathbb{R}^3$ (which do not change the curvature nor the torsion; see Exercise 1.26), we may assume

$$\mathbf{v}_0 = \sqrt{\kappa_0^2 + \tau_0^2}\, \mathbf{e}_3 \quad \Longrightarrow \quad \mathbf{e}_3 \equiv \frac{\tau_0}{\sqrt{\kappa_0^2 + \tau_0^2}} \mathbf{t} + \frac{\kappa_0}{\sqrt{\kappa_0^2 + \tau_0^2}} \mathbf{b} \,,$$

where $\mathbf{e}_3 = (0, 0, 1)$ is the third vector of the canonical basis of $\mathbb{R}^3$. Let then $\sigma_1\colon I \to \mathbb{R}^3$ be defined by

$$\sigma_1(s) = \sigma(s) - \frac{\tau_0 s}{\sqrt{\kappa_0^2 + \tau_0^2}} \mathbf{e}_3$$

(beware: as we shall see shortly, s is *not* the arc length parameter of σ_1). We want to show that σ_1 is the parametrization of an arc of a circle contained in a plane orthogonal to $\mathbf{e}_3$. First of all,

$$\frac{\mathrm{d}}{\mathrm{d}s}\langle\sigma_1,\mathbf{e}_3\rangle = \langle\sigma_1',\mathbf{e}_3\rangle = \langle\mathbf{t},\mathbf{e}_3\rangle - \frac{\tau_0}{\sqrt{\kappa_0^2+\tau_0^2}} \equiv 0\,;$$

hence $\langle\sigma_1,\mathbf{e}_3\rangle$ is constant, and so the support of σ_1 is contained in a plane orthogonal to $\mathbf{e}_3$, as claimed. Moreover,

$$\sigma_1' = \mathbf{t} - \frac{\tau_0}{\sqrt{\kappa_0^2+\tau_0^2}}\mathbf{e}_3 = \frac{\kappa_0^2}{\kappa_0^2+\tau_0^2}\mathbf{t} - \frac{\kappa_0\tau_0}{\kappa_0^2+\tau_0^2}\mathbf{b} \quad\text{and}\quad \sigma_1'' = \kappa_0\mathbf{n}\,;$$

hence

$$\|\sigma_1'\| \equiv \frac{\kappa_0}{\sqrt{\kappa_0^2+\tau_0^2}} \quad\text{and}\quad \sigma_1'\wedge\sigma_1'' = \frac{\kappa_0^3}{\kappa_0^2+\tau_0^2}\mathbf{b} + \frac{\kappa_0^2\tau_0}{\kappa_0^2+\tau_0^2}\mathbf{t}\,.$$

So, using (1.13) we find that the curvature κ_1 of σ_1 is

$$\kappa_1 = \frac{\|\sigma_1'\wedge\sigma_1''\|}{\|\sigma_1'\|^3} \equiv \frac{\kappa_0^2+\tau_0^2}{\kappa_0}\,.$$

Thus σ_1, being a plane curve with constant curvature, by Example 1.3.18 parametrizes an arc of a circle with radius $r = \kappa_0/(\kappa_0^2+\tau_0^2)$ and contained in a plane orthogonal to $\mathbf{e}_3$. Up to a translation in $\mathbb{R}^3$, we may assume that this circle is centered at the origin, and hence σ indeed is a circular helix with radius r and pitch $a = \tau_0/(\kappa_0^2+\tau_0^2)$, as stated. $\square$

Problem 1.8 (Curves on a sphere). *Let $\sigma\colon I \to \mathbb{R}^3$ be a biregular curve parametrized by arc length.*

(i) *Prove that if the support of σ is contained in a sphere with radius $R > 0$ then*

$$\tau^2 + (\dot\kappa/\kappa)^2 \equiv R^2\kappa^2\tau^2\,. \tag{1.16}$$

(ii) *Prove that if $\dot\kappa$ is nowhere zero and σ satisfies (1.16) then the support of σ is contained in a sphere with radius $R > 0$. [Note: You can find further information about curves contained in a sphere in Exercise 1.56.]*

Solution. (i) Up to a translation in $\mathbb{R}^3$ (which does not change curvatures and torsions; see Exercise 1.26), we may assume the center of the sphere to be in the origin. So $\langle\sigma,\sigma\rangle \equiv R^2$; differentiating three times and applying Frenet-Serret formulas we find

$$\langle\mathbf{t},\sigma\rangle \equiv 0, \quad \kappa\langle\mathbf{n},\sigma\rangle + 1 \equiv 0 \quad\text{and}\quad \dot\kappa\langle\mathbf{n},\sigma\rangle + \kappa\tau\langle\mathbf{b},\sigma\rangle \equiv 0\,. \tag{1.17}$$

Now, $\{\mathbf{t},\mathbf{n},\mathbf{b}\}$ is an orthonormal basis; in particular, we may write

$$\sigma = \langle\sigma,\mathbf{t}\rangle\mathbf{t} + \langle\sigma,\mathbf{n}\rangle\mathbf{n} + \langle\sigma,\mathbf{b}\rangle\mathbf{b}\,,$$

so $|\langle\sigma,\mathbf{t}\rangle|^2 + |\langle\sigma,\mathbf{n}\rangle|^2 + |\langle\sigma,\mathbf{b}\rangle|^2 \equiv R^2$, and (1.17) implies (1.16).

(ii) Since $\dot\kappa$ is nowhere zero, by equation (1.16) so is τ; hence we may divide (1.16) by $\tau^2\kappa^2$, obtaining

$$\frac{1}{\kappa^2} + \left(\frac{1}{\tau}\frac{d}{ds}\left(\frac{1}{\kappa}\right)\right)^2 \equiv R^2\,.$$

Differentiating and recalling that $\dot\kappa \neq 0$, we find

$$\frac{\tau}{\kappa} + \frac{d}{ds}\left(\frac{1}{\tau}\frac{d}{ds}\left(\frac{1}{\kappa}\right)\right) \equiv 0\,.$$

Define now $\eta\colon I \to \mathbb{R}^3$ by setting

$$\eta = \sigma + \frac{1}{\kappa}\mathbf{n} + \frac{1}{\tau}\frac{d}{ds}\left(\frac{1}{\kappa}\right)\mathbf{b}\,.$$

Then

$$\frac{d\eta}{ds} = \mathbf{t} + \frac{d}{ds}\left(\frac{1}{\kappa}\right)\mathbf{n} - \mathbf{t} + \frac{\tau}{\kappa}\mathbf{b} + \frac{d}{ds}\left(\frac{1}{\tau}\frac{d}{ds}\left(\frac{1}{\kappa}\right)\right)\mathbf{b} - \frac{d}{ds}\left(\frac{1}{\kappa}\right)\mathbf{n} \equiv O\,,$$

that is the curve η is constant. This means that there exists a point $p \in \mathbb{R}^3$ such that

$$\|\sigma - p\|^2 = \frac{1}{\kappa^2} + \left(\frac{1}{\tau}\frac{d}{ds}\left(\frac{1}{\kappa}\right)\right)^2 \equiv R^2\,;$$

hence, the support of σ is contained in the sphere with radius R and center p.
$\square$

Problem 1.9. *Let $f\colon \mathbb{R}^2 \to \mathbb{R}$ be a C^∞ function; in this problem we shall write $f_x = \partial f/\partial x$, $f_y = \partial f/\partial y$, $f_{xx} = \partial^2 f/\partial x^2$, and so on. Choose a point $p \in f^{-1}(0) = C$, with $f_y(p) \neq 0$, and let $g\colon I \to \mathbb{R}$, with $I \subseteq \mathbb{R}$, a C^∞ function such that $f^{-1}(0)$ is given, in a neighborhood of p, by the graph of g, as in Proposition 1.1.18. Finally, choose $t_0 \in I$ so that $p = (t_0, g(t_0))$.*

(i) *Show that the tangent vector to C in p is parallel to $(-f_y(p), f_x(p))$, and thus the vector $\nabla f(p) = (f_x(p), f_y(p))$ is orthogonal to the tangent vector.*

(ii) *Show that the oriented curvature at p of C is given by*

$$\tilde\kappa = \frac{f_{xx}f_y^2 - 2f_{xy}f_x f_y + f_{yy}f_x^2}{\|\nabla f\|^3}\,,$$

where we have oriented C in such a way that the tangent versor at p is a positive multiple of $(-f_y(p), f_x(p))$.

(iii) *If $f(x,y) = x^4 + y^4 - xy - 1$ and $p = (1,0)$, compute the oriented curvature of C at p.*

Solution. (i) Consider the parametrization $\sigma(t) = (t, g(t))$. The tangent vector is parallel to $\sigma'(t_0) = (1, g'(t_0))$. Since $f(t, g(t)) \equiv 0$, differentiating with respect to t we find that

$$f_x(t, g(t)) + f_y(t, g(t)) g'(t) \equiv 0 ; \tag{1.18}$$

so

$$g'(t_0) = -\frac{f_x(p)}{f_y(p)} ,$$

and the assertion immediately follows.

(ii) By differentiating again we find $\sigma''(t_0) = (0, g''(t_0))$; so using the formula $\tilde{\kappa} = \|\sigma'\|^{-3} \det(\sigma', \sigma'')$ proved in Problem 1.1 we get

$$\tilde{\kappa} = \frac{|f_y(p)|^3 g''(t_0)}{\|\nabla f(p)\|^3} . \tag{1.19}$$

Take now one more derivative of (1.18) and evaluate it at t_0: we find that

$$f_{xx}(p) + f_{xy}(p) g'(t_0) + [f_{yx}(p) + f_{yy}(p) g'(t_0)] g'(t_0) + f_y(p) g''(t_0) \equiv 0 .$$

The parametrization σ is oriented as required if and only if $f_y(p) < 0$. So, extracting $g''(t_0)$ from the previous expression and inserting it in (1.19), we find the formula we were seeking.

(iii) In this case,

$$f_x(p) = 4 , \quad f_y(p) = -1 , \quad f_{xx}(p) = 12 , \quad f_{xy}(p) = -1 \quad f_{yy}(p) = 0 ,$$

and so $\tilde{\kappa} = 4/17^{3/2}$.   □

Exercises

PARAMETRIZATIONS AND CURVES

1.1. Prove that the curve $\sigma\colon \mathbb{R} \to \mathbb{R}^2$ given by $\sigma(t) = (t/(1 + t^4), t/(1 + t^2))$ is an injective regular parametrization, but not a homeomorphism with its image.

1.2. Draw the support of the curve parametrized, in polar coordinates (r, θ), by $\sigma_1(\theta) = (a \cos\theta, \theta)$, for $\theta \in [0, 2\pi]$. Note that the image is contained in a circle, and that it is defined by the equation $r = a \cos\theta$.

1.3. Prove that the relation introduced in Definition 1.1.24 actually is an equivalence relation on the set of parametrizations of class C^k.

1.4. Prove that the curve $\sigma\colon \mathbb{R} \to \mathbb{R}^2$ given by $\sigma(t) = (t^2, t^3)$ is not regular and that no parametrization equivalent to it can be regular.

1.5. Prove that every open interval $I \subseteq \mathbb{R}$ is C^∞-diffeomorphic to $\mathbb{R}$.

1.6. Prove that every interval $I \subseteq \mathbb{R}$ is C^∞-diffeomorphic to one of the following: $[0,1)$, $(0,1)$, or $[0,1]$. In particular, every regular curve admits a parametrization defined in one of these intervals.

1.7. Determine the parametrization $\sigma_1 : (-\pi, \pi) \to \mathbb{R}^3$ equivalent to the parametrization $\sigma : \mathbb{R} \to \mathbb{R}^2$ given by $\sigma(t) = (r \cos t, r \sin t)$ of the circle, obtained by the parameter change $s = \arctan(t/4)$.

1.8. Prove that the two parametrizations σ, $\sigma_1 : [0, 2\pi] \to \mathbb{R}^2$ of class C^∞ of the circle defined by $\sigma(t) = (\cos t, \sin t)$ and $\sigma_1(t) = (\cos 2t, \sin 2t)$ are not equivalent (see Example 1.1.12 and Remark 1.2.9).

1.9. Let $\sigma_1 : [0, 2\pi] \to \mathbb{R}^2$ be defined by

$$\sigma_1(t) = \begin{cases} (\cos t, \sin t) & \text{for } t \in [0, \pi] \,, \\ (-1, 0) & \text{for } t \in [\pi, 2\pi] \,. \end{cases}$$

(i) Show that σ_1 is continuous but not of class C^1.
(ii) Prove that σ_1 is not equivalent to the usual parametrization of the circle $\sigma : [0, 2\pi] \to \mathbb{R}^2$ given by $\sigma(t) = (\cos t, \sin t)$.

1.10. Prove that the support of the curve of the Example 1.2.4 cannot be the image of a regular curve (in particular, it is not a 1-submanifold of $\mathbb{R}^2$; see Section 1.6).

1.11. For all $k \in \mathbb{N}^* \cup \{\infty\}$, find a parametrized curve $\sigma : \mathbb{R} \to \mathbb{R}^2$ of class C^k having as its support the graph of the absolute value function. Show next that no such curve can be regular.

1.12. Let $\sigma : [0, 1] \to \mathbb{R}^2$ given by

$$\sigma(t) = \begin{cases} (-1 + \cos(4\pi t), \sin(4\pi t)) & \text{for } t \in [0, 1/2] \,, \\ (1 + \cos(-4\pi t - \pi), \sin(-4\pi t - \pi)) & \text{for } t \in [1/2, 1] \,; \end{cases}$$

see Fig. 1.6.(a).

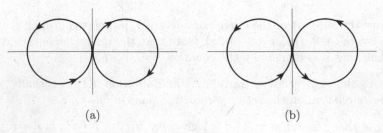

(a)   (b)

Fig. 1.6.

(i)   Show that σ defines a parametrization of class C^1 but not C^2.
(ii) Consider $\sigma_1 : [0,1] \to \mathbb{R}^2$ given by

$$\sigma_1(t) = \begin{cases} \sigma(t) & \text{for } t \in [0, 1/2] , \\ \big(1 + \cos(4\pi t + \pi), \sin(4\pi t + \pi)\big) & \text{for } t \in [1/2, 1] ; \end{cases}$$

see Fig. 1.6.(b). Show that σ and σ_1 are not equivalent, not even as continuous parametrizations.

1.13. The *conchoid of Nicomedes* is the plane curve described, in polar coordinates, by the equation $r = b + a / \cos\theta$, with fixed a, $b \neq 0$, and $\theta \in [-\pi, \pi]$. Draw the support of the conchoid and determine a parametrization in Cartesian coordinates.

1.14. Show, using the parameter change $v = \tan(t/2)$, that the parametrizations $\sigma_1 : [0, \infty) \to \mathbb{R}^3$ and $\sigma_2 : [0, \pi) \to \mathbb{R}^3$ of the circular helix, given by

$$\sigma_1(v) = \left(r \, \frac{1 - v^2}{1 + v^2}, \frac{2rv}{1 + v^2}, 2a \arctan v \right) \quad \text{and} \quad \sigma(t) = (r \cos t, r \sin t, at)$$

respectively, are equivalent.

1.15. Epicycloid. An *epicycloid* is the plane curve described by a point P of a circle C with radius r that rolls without slipping on the outside of a circle C_0 with radius R. Assume that the center of C_0 is the origin, and that the point P starts in $(R, 0)$ and moves counterclockwise. Finally, denote by t the angle between the positive x-axis and the vector OA, joining the origin and the center A of C; see Fig. 1.7.(a).

(i)   Show that the center A of C has coordinates $\big((r + R) \cos t, (r + R) \sin t\big)$.
(ii) Having computed the coordinates of the vector AP, determine a parametrization of the epicycloid.

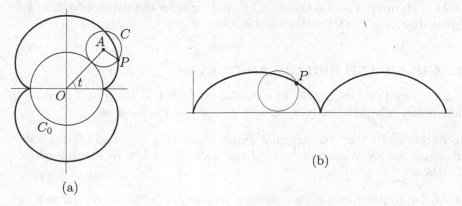

(a)   (b)

Fig. 1.7. (a) epicycloid; (b) cycloid

LENGTH AND RECTIFIABLE CURVES

1.16. Let $\sigma:[a,b] \to \mathbb{R}^n$ be a rectifiable curve. Show that

$$L(\sigma) \geq \|\sigma(b) - \sigma(a)\| \, ,$$

and deduce that a line segment is the shortest curve between two points.

1.17. Let $f:\mathbb{R} \to \mathbb{R}$ be the function given by

$$f(t) = \begin{cases} t\sin(\pi/t) & \text{if } t \neq 0 \, , \\ 0 & \text{if } t = 0 \, . \end{cases}$$

Show that the curve $\sigma:[0,1] \to \mathbb{R}^2$ given by $\sigma(t) = (t, f(t))$ is an injective, non-rectifiable, continuous curve.

1.18. Cycloid. In the xy-plane consider a circle with radius 1 rolling without slipping on the x-axis, as in Fig. 1.7.(b). The path described by a point of the circle is called *cycloid*. Following the motion of such a point P, starting from the origin and up to the moment when it arrives back to the x-axis, we get a regular curve $\sigma:[0,2\pi] \to \mathbb{R}^2$ with the cycloid as its support. Show that σ is defined by $\sigma(t) = (t - \sin t, 1 - \cos t)$, and determine its length.

1.19. Let $\sigma:[a,b] \to \mathbb{R}^3$ be the usual parametrization $\sigma(t) = (t, f(t))$ of the graph of a function $f:[a,b] \to \mathbb{R}$ of class C^1. Prove that the length of σ is

$$L(\sigma) = \int_a^b \sqrt{1 + |f'(t)|^2} \, dt \, .$$

1.20. Prove that if $\sigma:[0,+\infty) \to \mathbb{R}^2$ is the logarithmic spiral parametrized as in Problem 1.4, then the limit $\lim\limits_{t \to +\infty} \int_0^t \|\sigma'(\lambda)\| \, d\lambda$ exists and is finite. In a sense, we may say that the logarithmic spiral has a finite length.

1.21. Determine a parametrization by arc length for the parabola $\sigma:\mathbb{R} \to \mathbb{R}^2$ given by $\sigma(t) = (t, a\,t^2)$ with a fixed $a > 0$.

REGULAR AND BIREGULAR CURVES

1.22. Prove that the support of a regular curve $\sigma:I \to \mathbb{R}^n$ is contained in a line if and only if the tangent versor $\mathbf{t}:I \to \mathbb{R}^n$ of σ is constant.

1.23. Let $\sigma:I \to \mathbb{R}^3$ be a regular curve. Show that $\sigma(t)$ and $\sigma'(t)$ are orthogonal for every value of $t \in I$ if and only if $\|\sigma\|$ is a constant non-zero function.

1.24. Determine which of the following maps $\sigma_i:\mathbb{R} \to \mathbb{R}^3$ are regular and/or biregular curves:

(i) $\sigma_1(t) = (e^{-t}, 2t, t - 1)$;

(ii) $\sigma_2(t) = (2t, (t^2 - 1)^2, 3t^3)$;

(iii) $\sigma_3(t) = (t, 2t, t^3)$.

1.25. Let $\sigma: [-2\pi, 2\pi] \to \mathbb{R}^3$ be the curve given by

$$\sigma(t) = (1 + \cos t, \sin t, 2\sin(t/2)) .$$

Prove that it is a regular curve having as its support the intersection of the sphere of radius 2 centered at the origin with the cylinder having equation $(x - 1)^2 + y^2 = 1$.

CURVATURE AND TORSION

1.26. Let $\sigma: I \to \mathbb{R}^3$ be a biregular curve parametrized by arc length, and $\rho: \mathbb{R}^3 \to \mathbb{R}^3$ a rigid motion. Prove that $\rho \circ \sigma$ is a biregular curve parametrized by arc length with the same curvature and the same torsion as σ.

1.27. Let $\sigma: \mathbb{R} \to \mathbb{R}^3$ be the curve given by $\sigma(t) = (1 + \cos t, 1 - \sin t, \cos 2t)$. Prove that σ is a regular curve, and compute its curvature and its torsion without reparametrizing it by arc length.

1.28. Let $f: U \to \mathbb{R}$ be a function of class C^∞ defined on an open subset U of the plane $\mathbb{R}^2$, and $\sigma: I \to U$ a regular curve such that $f \circ \sigma \equiv 0$. Prove that for every $t \in I$ the tangent vector $\sigma'(t)$ is orthogonal to the gradient of f computed in $\sigma(t)$, and determine the oriented curvature of σ as a function of the derivatives of f.

1.29. Let $\sigma: I \to \mathbb{R}^2$ be a regular plane curve, given in polar coordinates by the equation $r = \rho(\theta)$, that is,

$$\sigma(\theta) = (\rho(\theta) \cos \theta, \rho(\theta) \sin \theta)$$

for some function $\rho: I \to \mathbb{R}^+$ of class C^∞ nowhere zero. Prove that the arc length of σ is given by

$$s(\theta) = \int_{\theta_0}^{\theta} \sqrt{\rho^2 + (\rho')^2} \, d\theta ,$$

and that the oriented curvature of σ is

$$\tilde{\kappa} = \frac{2(\rho')^2 - \rho\rho'' + \rho^2}{(\rho^2 + (\rho')^2)^{3/2}} .$$

1.30. Let $\sigma: \mathbb{R} \to \mathbb{R}^3$ be the map defined by

$$\sigma(t) = \begin{cases} (t, e^{-1/t}, 0) & \text{if } t \geq 0 , \\ (t, 0, e^{1/t}) & \text{if } t \leq 0 . \end{cases}$$

Prove that σ is a regular curve of class C^∞, biregular everywhere except at the origin and for $t = \pm 1/2$, with zero torsion everywhere in $\mathbb{R} \setminus \{0, 1/2, -1/2\}$, but whose support is not contained in a plane. (*Hint:* see the proof of Lemma 1.5.1.)

1.31. Let $\sigma: (0, +\infty) \to \mathbb{R}^3$ be the curve given by $\sigma(t) = (t, 2t, t^4)$. Prove that σ is a regular curve whose support is contained in a plane, and compute the curvature of σ at each point.

1.32. Determine the arc length, the curvature and the torsion of the curve $\sigma: \mathbb{R} \to \mathbb{R}^3$ defined by $\sigma(t) = (a \cosh t, b \sinh t, a t)$. Prove that, if $a = b = 1$, then the curvature is equal to the torsion for every value of the parameter.

1.33. Let $\sigma: \mathbb{R} \to \mathbb{R}^3$ be the mapping defined by

$$\sigma(t) = (2\sqrt{2}\, t - \sin t, 2\sqrt{2} \sin t + t, 3 \cos t) \ .$$

Prove that the curve defined by σ is a circular helix (up to a rigid motion of $\mathbb{R}^3$).

1.34. Consider a plane curve $\sigma: I \to \mathbb{R}^2$ parametrized by arc length. Prove that if the vector $O\sigma(s)$ forms a constant angle θ with the tangent versor $\mathbf{t}(s)$ then σ is a logarithmic spiral (see Problem 1.4).

1.35. Let $\sigma: [a, b] \to \mathbb{R}^3$ be a curve of class at least C^2.

(i)   Show that if the support of σ is contained in a plane through the origin then the vectors σ, σ', and σ'' are linearly dependent.
(ii) Show that if the vectors σ, σ', and σ'' are linearly dependent but the vectors σ, σ' are linearly independent then the support of σ is contained in a plane through the origin.
(iii) Find an example in which σ and σ' are linearly dependent but the support of σ is not contained in a plane through the origin.

FRENET FRAME AND OSCULATING PLANE

1.36. Let $\sigma: \mathbb{R} \to \mathbb{R}^3$ be the curve $\sigma(t) = (e^t, e^{2t}, e^{3t})$. Find the values of $t \in \mathbb{R}$ for which the tangent vector $\sigma'(t)$ is orthogonal to the vector $\mathbf{v} = (1, 2, 3)$.

1.37. Let $\sigma: \mathbb{R} \to \mathbb{R}^3$ be the curve $\sigma(t) = \big((4/5) \cos t, 1 - \sin t, -(3/5) \cos t\big)$. Determine the Frenet frame of σ.

1.38. Let $\sigma: \mathbb{R} \to \mathbb{R}^2$ be the plane curve parametrized by $\sigma(t) = (t, \frac{1}{3}t^3)$. Determine the curvature of σ and study the values of the parameter for which it is zero. Determine the normal versor and the oriented normal versor, wherever they are defined.

1.39. Let $\sigma\colon I \to \mathbb{R}^3$ be a regular curve of class at least C^2. Show that the vector σ'' is parallel to the osculating plane and that its components along the vectors $\mathbf{t}$ and $\mathbf{n}$ are $\|\sigma'\|'$ and $\kappa\|\sigma'\|^2$, respectively. Furthermore, if σ is biregular, show that the osculating plane in $\sigma(t_0)$ is the plane through $\sigma(t_0)$ and parallel to $\sigma'(t_0)$ and $\sigma''(t_0)$. In particular, the equation of the osculating plane is $\langle \sigma'(t_0) \wedge \sigma''(t_0), p - \sigma(t_0) \rangle = 0$.

1.40. Let σ be a curve such that all its affine tangent lines pass through a given point. Show that if σ is regular then its support is contained in a straight line, and find a counterexample with a non-regular σ (considering the tangent lines only in the points of the curve where they are defined).

1.41. Let $\sigma\colon \mathbb{R} \to \mathbb{R}^3$ be a Jordan arc of class C^2, not necessarily regular, such that all its affine tangent lines pass through a given point P. Show that the support of σ is contained in a straight line, and find a counterexample with a Jordan arc not of class C^2 (considering the tangent lines only in the points of the curve where they are defined).

1.42. Let $\sigma\colon [a,b] \to \mathbb{R}^3$ be a biregular curve such that all its affine normal line pass through a given point. Show that the support of σ is contained in a circle.

1.43. Consider the curve $\sigma\colon \mathbb{R} \to \mathbb{R}^3$ given by $\sigma(t) = (t, (1/2)t^2, (1/3)t^3)$. Show that the osculating planes of σ in three different points $\sigma(t_1)$, $\sigma(t_2)$, $\sigma(t_3)$ intersect in a point belonging to the plane generated by the points $\sigma(t_1)$, $\sigma(t_2)$ and $\sigma(t_3)$.

1.44. Show that the binormal vector to a circular helix parametrized as in Example 1.1.13 forms a constant angle with the axis of the cylinder containing the helix.

1.45. Show that the curve $\sigma\colon \mathbb{R} \to \mathbb{R}^3$ parametrized by

$$\sigma(t) = (t + \sqrt{3}\sin t, 2\cos t, \sqrt{3}\,t - \sin t),$$

is a circular helix (up to a rigid motion of $\mathbb{R}^3$).

1.46. Let $\gamma\colon \mathbb{R} \to \mathbb{R}^3$ be the curve parametrized by $t \mapsto \gamma(t) = (t, At^2, Bt^n)$, where $A, B > 0$ are real numbers and $n \geq 1$ is an integer.

(i) Determine a map $\eta\colon \mathbb{R} \to \mathbb{R}^3$ such that $\eta(t)$ is the intersection point between the affine tangent line to γ in $\gamma(t)$ and the plane $z = 0$.
(ii) Find conditions on A, B, n in order for η to be a regular curve.

FRENET-SERRET FORMULAS

1.47. Determine the curvature and the torsion at each point of the curve $\sigma\colon \mathbb{R} \to \mathbb{R}^3$ parametrized by $\sigma(t) = (3t - t^3, 3t^2, 3t + t^3)$.

1.48. Determine the curvature, the torsion, and the Frenet frame of the curve $\sigma\colon \mathbb{R} \to \mathbb{R}^3$ given by $\sigma(t) = \big(a(t - \sin t), a(1 - \cos t), bt\big)$.

1.49. Given a function $\tilde{\kappa}\colon I \to \mathbb{R}$ of class C^k, prove that there exists a unique (up to plane rigid motions) regular curve $\sigma\colon I \to \mathbb{R}^2$ of class C^{k+2} parametrized by arc length having oriented curvature $\tilde{\kappa}$.

1.50. Generalized helices. Let $\sigma\colon I \to \mathbb{R}^3$ be a biregular curve parametrized by arc length. Prove that the following assertions are equivalent:

(i)   there exist two constants $a,\ b \in \mathbb{R}$, not both zero, such that $a\kappa + b\tau \equiv 0$;
(ii) there exists a nonzero versor $\mathbf{v}_0$ such that $\langle \mathbf{t}, \mathbf{v}_0 \rangle$ is constant;
(iii) there exists a plane π such that $\mathbf{n}(s) \in \pi$ for all $s \in I$;
(iv) there exists a nonzero versor $\mathbf{v}_0$ such that $\langle \mathbf{b}, \mathbf{v}_0 \rangle$ is constant;
(v)   there exist $\theta \in (0,\pi) \setminus \{\pi/2\}$ and a biregular plane curve parametrized by arc length $\eta\colon J_\theta \to \mathbb{R}^3$, where $J_\theta = (\sin \theta)I$, such that

$$\sigma(s) = \eta(s \sin \theta) + s \cos \theta \, \mathbf{b}_\eta$$

for all $s \in I$, where $\mathbf{b}_\eta$ is the (constant!) binormal versor of η;
(vi) the curve σ has a parametrization of the form $\sigma(s) = \eta(s) + (s - s_0)\mathbf{v}$, where η is a plane curve parametrized by arc length, and v is a vector orthogonal to the plane containing the support of η.

A curve σ satisfying any of these equivalent conditions is called *(generalized) helix*; see Fig. 1.8. Finally, write the curvature, the torsion and the Frenet frame of σ as functions of the curvature and of the Frenet frame of η.
(*Hint:* if $\tau/\kappa = c$ is constant, set $c = \frac{\cos \alpha}{\sin \alpha}$ and $\mathbf{v}_0(s) = \cos \alpha \, \mathbf{t}(s) + \sin \alpha \, \mathbf{n}(s)$ and prove that $\mathbf{v}_0$ is constant.)

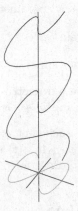

Fig. 1.8. Generalized helix $\sigma(t) = (\cos t, \sin(2t), t)$

1.51. Check for which values of the constants $a, b \in \mathbb{R}$ the curve $\sigma \colon \mathbb{R} \to \mathbb{R}^3$ parametrized by $\sigma(t) = (at, bt^2, t^3)$ is a generalized helix.

1.52. Let $\sigma \colon I \to \mathbb{R}^3$ be a biregular curve parametrized by arc length. Prove that $\kappa \equiv \pm\tau$ if and only if there exists a nonzero versor $\mathbf{v}$ such that $\langle \mathbf{t}, \mathbf{v} \rangle \equiv \langle \mathbf{b}, \mathbf{v} \rangle$. Prove furthermore that, in this case, $\langle \mathbf{t}, \mathbf{v} \rangle$ is constant.

1.53. Let $\sigma \colon \mathbb{R} \to \mathbb{R}^3$ be the curve $\sigma(t) = \big(1 + \cos t, \sin t, 2\sin(t/2)\big)$. Prove that σ is not a plane curve and that its support is contained in the sphere with radius 2 and center at the origin.

1.54. Let $\sigma \colon [a, b] \to \mathbb{R}^3$ be a biregular curve parametrized by arc length. Show that

$$\frac{\mathrm{d}^3\sigma}{\mathrm{d}s^3} = -\kappa^2\mathbf{t} + \dot{\kappa}\mathbf{n} + \kappa\tau\mathbf{b} \,.$$

1.55. Let $\sigma \colon I \to \mathbb{R}^3$ be a biregular curve parametrized by arc length having tangent versor $\mathbf{t}$, and for every $\varepsilon \neq 0$ set $\sigma_\varepsilon = \sigma + \varepsilon\mathbf{t}$. Prove that σ_ε always is a regular curve, and that the normal versor of σ_ε is always orthogonal to the normal versor of σ if the curvature κ of σ is of the form

$$\kappa(s) = c(\mathrm{e}^{2s/\varepsilon} - c^2\varepsilon^2)^{-1/2}$$

for some constant $0 < c < 1/\varepsilon$.

1.56. Let $\sigma \colon I \to \mathbb{R}^3$ be a biregular curve parametrized by arc length, having constant curvature $\kappa_0 > 0$. Prove that the support of σ is contained in a sphere with radius $R > 0$ if and only if $\kappa_0 > 1/R$ and $\tau \equiv 0$.

1.57. Let $\sigma \colon I \to \mathbb{R}^3$ and $\alpha \colon I \to \mathbb{R}^3$ be two different biregular curves parametrized by arc length, having equal affine binormal lines in the points corresponding to the same parameter. Prove that the curves σ and α are plane.

1.58. Determine curvature and torsion of the biregular curve $\sigma \colon (0, +\infty) \to \mathbb{R}^3$ defined by $\sigma(t) = \big(2t, \frac{1+2t}{2t}, \frac{1-4t^2}{2t}\big)$.

1.59. Determine curvature and torsion of the regular curve $\sigma \colon \mathbb{R} \to \mathbb{R}^3$ defined by $\sigma(t) = (\cos t, \sin t, 2\sin \frac{t}{2})$.

1.60. The Darboux vector. Let $\sigma \colon I \to \mathbb{R}$ be a biregular curve parametrized by arc length. The vector $\mathbf{d}(s) = \tau(s)\mathbf{t}(s) + \kappa(s)\mathbf{b}$ is called the Darboux vector at $\sigma(s)$. Show that $\mathbf{d}$ satisfies $\dot{\mathbf{t}} = \mathbf{d} \wedge \mathbf{t}$, $\dot{\mathbf{n}} = \mathbf{d} \wedge \mathbf{n}$, $\dot{\mathbf{b}} = \mathbf{d} \wedge \mathbf{b}$.

1.61. Let $\sigma \colon I \to \mathbb{R}^3$ be a regular curve parametrized by arc length, having nowhere zero curvature κ and torsion τ; let $s_0 \in I$ be fixed. For every $\varepsilon \in \mathbb{R}$ let $\gamma^\varepsilon \colon I \to \mathbb{R}^3$ be the curve given by $\gamma^\varepsilon(t) = \sigma(t) + \varepsilon\,\mathbf{b}(t)$, where $\{\mathbf{t}, \mathbf{n}, \mathbf{b}\}$ is the Frenet frame of σ. Denote by $\mathbf{t}^\varepsilon$, $\mathbf{n}^\varepsilon$ e $\mathbf{b}^\varepsilon$ the tangent, normal and binormal versors of γ^ε, and by κ^ε, τ^ε the curvature and the torsion of γ^ε. Prove that:

(i) γ^ε is always a biregular curve;

(ii) σ is a plane curve if and only if $\mathbf{b}^\varepsilon \equiv \pm\mathbf{b}$;

(iii) σ is a plane curve if and only if $\mathbf{t}^\varepsilon \equiv \mathbf{t}$ if and only if $\mathbf{n}^\varepsilon \equiv \mathbf{n}$.

1.62. Bertrand curves. Two biregular curves σ, $\sigma_1 \colon I \to \mathbb{R}^3$, having normal versors $\mathbf{n}$ and $\mathbf{n}_1$ respectively, are called *Bertrand curves* if they have the same affine normal line at every point. In particular, possibly modifying the orientation, it is always possible to assume that $\mathbf{n} \equiv \mathbf{n}_1$, that is, that the curves have the same normal versors.

(i)   Show that if σ and σ_1 are the parametrizations by arc length of two Bertrand curves then there exists a real-valued function $\alpha \colon I \to \mathbb{R}$ such that $\sigma_1 \equiv \sigma + \alpha\mathbf{n}$.

(ii)   Show that the distance between points corresponding to the same parameter of two Bertrand curves is constant, that is, the function α in (i) is constant.

(iii) Show that the angle between the tangent lines in two corresponding points of two Bertrand curves is constant.

(iv)   Show that if σ and σ_1 are biregular Bertrand curves with never vanishing torsion then there exist constants $a \in \mathbb{R}$, $b \in \mathbb{R}^*$ such that $\kappa + a\tau \equiv b$, where κ and τ are the curvature and the torsion of σ.

(v)   Prove the converse of the previous statement: if σ is a curve having curvature κ and torsion τ, both nowhere zero, such that $\kappa + a\tau \equiv b$ for suitable constants $a \in \mathbb{R}$ and $b \in \mathbb{R}^*$, then there exists another curve σ_1 such that σ and σ_1 are Bertrand curves.

(vi)   Show that if σ is a biregular curve with nowhere zero torsion τ then σ is a circular helix if and only if there exist at least two curves σ_1 and σ_2 such that σ and σ_i are Bertrand curves, for $i = 1$, 2. Show that, in this case, there exist infinitely many curves $\tilde\sigma$ such that σ and $\tilde\sigma$ are Bertrand curves.

(vii) Prove that if two Bertrand curves σ and σ_1 have the same binormal versor then there exists a constant $a > 0$ such that $a(\kappa^2 + \tau^2) = \kappa$.

FUNDAMENTAL THEOREM OF THE LOCAL THEORY OF CURVES

1.63. Find a plane curve, parametrized by arc length $s > 0$, with curvature $\kappa(s) = 1/s$. Do the same with the oriented curvature $\tilde\kappa(s) = 1/s$ instead of the usual curvature.

1.64. Compute the curvature of the catenary $\sigma \colon \mathbb{R} \to \mathbb{R}^2$ parametrized by $\sigma(t) = \big(a\cosh{(t/a)}, t\big)$, where a is a real constant.

1.65. Given $a > 0$, determine a curve having curvature and torsion given respectively by $\kappa(s) = \sqrt{1/2as}$ and $\tau(s) = 0$ for $s > 0$.

1.66. Compute curvature and torsion of the curve $\sigma: \mathbb{R} \to \mathbb{R}^3$ parametrized by $\sigma(t) = e^t(\cos t, \sin t, 3)$.

1.67. Regular curves with nonzero constant torsion. We know from Example 1.3.32 and Problem 1.7 that the circular helix is characterized by having both curvature and torsion constant (and both nonzero, except for the degenerate case with support contained in a plane circle). The aim of this exercise is to study biregular curves having nonzero constant torsion in $\mathbb{R}^3$.

(i) Show that if σ is a biregular curve having constant torsion $\tau \equiv a$ then

$$\sigma(t) = a^{-1} \int_{t_0}^{t} \mathbf{b}(s) \wedge \dot{\mathbf{b}}\, \mathrm{d}s.$$

Moreover, prove that the vectors $\mathbf{b}$, $\dot{\mathbf{b}}$, and $\ddot{\mathbf{b}}$ are linearly independent for all values of the parameter.

(ii) Consider, on the other hand, a function $f: I \to \mathbb{R}^3$ of class at least C^2, having values in the unitary sphere (that is, $\|f\| \equiv 1$), and such that the vectors $f(\lambda)$, $f'(\lambda)$, and $f''(\lambda)$ are linearly independent for all $\lambda \in I$. Consider the curve $\sigma: I \to \mathbb{R}^3$ given by

$$\sigma(t) = a \int_{t_0}^{t} f(\lambda) \wedge f'(\lambda)\, \mathrm{d}\lambda\,,$$

for some nonzero constant a and a fixed value $t_0 \in I$. Show that σ is regular and that it has constant torsion $\tau \equiv a^{-1}$.

EVOLUTE AND INVOLUTE

Definition 1.E.1. Let $\sigma: I \to \mathbb{R}^2$ be a biregular plane curve, parametrized by arc length. The plane curve $\beta: I \to \mathbb{R}^2$ parametrized by

$$\beta(s) = \sigma(s) + \frac{1}{\kappa(s)}\mathbf{n}(s)$$

is the *evolute* of σ. Now let $\sigma: I \to \mathbb{R}^3$ be a regular plane curve, parametrized by arc length. A curve $\tilde{\sigma}: I \to \mathbb{R}^3$ (not necessarily parametrized by arc length) is an *involute* of σ if $\dot{\sigma}(s)$ is parallel to $\tilde{\sigma}(s) - \sigma(s)$ and orthogonal to $\tilde{\sigma}'(s)$ for all $s \in I$.

1.68. Show that the affine normal line to a biregular plane curve $\sigma: I \to \mathbb{R}^2$ at the point $\sigma(s)$ is equal to the affine tangent line of its evolute β at the point $\beta(s)$. In particular, the affine tangent line to the evolute at $\beta(s)$ is orthogonal to the affine tangent line to the original curve at $\sigma(s)$.

1.69. Show that the evolute of the catenary $\sigma(t) = (t, \cosh t)$ is parametrized by $\beta(t) = (t - \sinh t \cosh t, 2 \cosh t)$.

1.70. Find the evolute β of the curve $\sigma(t) = (\cos^3 t, \sin^3 t)$.

1.71. Given $a > 0$ and $b < 0$, find the evolute of the logarithmic spiral parametrized by $\sigma(t) = (ae^{bt} \cos t, ae^{bt} \sin t)$.

1.72. Prove that any biregular plane curve σ is an involute of its evolute. Moreover, prove that any couple of involutes of σ are Bertrand curves (see Exercise 1.62).

1.73. Let $\sigma \colon I \to \mathbb{R}^3$ be a regular curve parametrized by arc length. For $c \in \mathbb{R}$, define $\sigma_c \colon I \to \mathbb{R}^3$ by $\sigma_c(s) = \sigma(s) + (c - s)\dot{\sigma}(s)$ for all $s \in I$.

(i)   Prove that a curve $\tilde{\sigma} \colon I \to \mathbb{R}^3$ is an involute of σ if and only if there exists $c \in \mathbb{R}$ such that $\tilde{\sigma}(s) = \sigma_c(s)$ for all $s \in I$.
(ii) Assume that σ is biregular and prove that the involute σ_c is biregular in $I \backslash \{c\}$. Prove, moreover, that the tangent versor of σ_c in $\sigma_c(s)$ is parallel to the normal versor of σ in $\sigma(s)$ and that, in general, σ_c is not parametrized by arc length.
(iii) Assume that σ is biregular and prove that the curvature of an involute σ_c is given by $\frac{\sqrt{\kappa^2 + \tau^2}}{|(c-s)\kappa|}$, in terms of the curvature κ and the torsion τ of σ.
(iv) Let σ be the circular helix, as in the Example 1.2.15. Prove that each involute σ_c of σ is a plane curve.
(v)   Determine the involute of the catenary $\sigma(t) = (t, \cosh t)$ and of the circle $\sigma_1(t) = (r \cos t, r \sin t)$.

1.74. Let $\sigma \colon I \to \mathbb{R}^3$ be a biregular curve parametrized by arc length and let $\hat{\sigma} = \sigma - \kappa^{-1}\mathbf{n}$, where $\mathbf{n}$ is the normal versor of σ. Prove that if σ is an involute of $\hat{\sigma}$, then σ is a plane curve.

SPHERICAL INDICATRICES

1.75. Let $\sigma \colon I \to \mathbb{R}^3$ be a regular curve of curvature κ. Prove that $\mathbf{t} \colon I \to \mathbb{R}^3$ is regular if and only if σ is biregular, and that the arc length of σ is an arc length for $\mathbf{t}$ as well if and only if $\kappa \equiv 1$.

1.76. Let $\sigma \colon I \to \mathbb{R}^3$ be a biregular curve of curvature κ and torsion τ. Denote by s the arc length of σ, and by s_1 the arc length of the normal curve $\mathbf{n} \colon I \to \mathbb{R}^3$. Prove that $ds_1/ds = \sqrt{\kappa^2 + \tau^2}$.

1.77. Let $\sigma \colon \mathbb{R} \to \mathbb{R}^3$ be the circular helix given by $\sigma(t) = (r \cos t, r \sin t, at)$. Prove that its tangent versor $\mathbf{t} \colon \mathbb{R} \to \mathbb{R}^3$ is a circle with center on the z-axis, and compute its radius of curvature.

1.78. Let $\sigma \colon I \to \mathbb{R}^3$ be a biregular curve. Prove that if the support of its tangent versor $\mathbf{t} \colon I \to \mathbb{R}^3$ is a circle then σ is (up to a rigid motion of $\mathbb{R}^3$) a circular helix.

1.79. Let $\sigma: I \to \mathbb{R}^3$ be a biregular curve. Show that the tangent vector at a point of the tangent curve $\mathbf{t}: I \to \mathbb{R}^3$ to σ is parallel to the affine normal line at the corresponding point of σ.

1.80. Let $\sigma: I \to \mathbb{R}^3$ be a biregular curve having curvature κ and torsion τ. Prove that the curvature κ_1 of the tangent curve $\mathbf{t}: I \to \mathbb{R}^3$ of σ is given by

$$\kappa_1 = \sqrt{1 + \frac{\tau^2}{\kappa^2}} \, .$$

Supplementary material

1.4 The local canonical form

It is well known that the affine tangent line is the straight line that best approximates a curve at a given point. But which is the circle that best approximates a curve at a given point? And which is the plane that best approximates a curve in space?

To answer these (and other) questions, let us begin by proving in detail that the affine tangent line is indeed the line that best approximates a curve. Let $\sigma: I \to \mathbb{R}^n$ be a curve of class (at least) C^1, and $t_0 \in I$. A line approximating σ at t_0 must, at the very least, pass through $\sigma(t_0)$, and so it has a parametrization of the form $\eta(t) = \sigma(t_0) + (t - t_0)\mathbf{v}$ for some $\mathbf{v} \in \mathbb{R}^3$. Then

$$\sigma(t) - \eta(t) = \sigma(t) - \sigma(t_0) - (t - t_0)\mathbf{v} = (t - t_0)\left(\frac{\sigma(t) - \sigma(t_0)}{t - t_0} - \mathbf{v} \right) .$$

The line that best approximates σ is the one for which the difference $\sigma(t) - \eta(t)$ has the fastest rate of approach to 0 when t tends to t_0. Recalling that σ is of class C^1, we find

$$\sigma(t) - \eta(t) = (t - t_0)\big(\sigma'(t_0) - \mathbf{v}\big) + o(t - t_0) \, ,$$

and so the line that best approximates σ is the one for which $\mathbf{v} = \sigma'(t_0)$, that is the affine tangent line.

This argument suggests that to answer analogous questions it might be handy to know the Taylor expansion of a curve. The following proposition express this expansion in a particularly useful way:

Proposition 1.4.1 (Local canonical form). *Let $\sigma: I \to \mathbb{R}^3$ be a biregular curve of class (at least) C^3, parametrized by arc length. Given $s_0 \in I$, denote by $\{\mathbf{t}_0, \mathbf{n}_0, \mathbf{b}_0\}$ the Frenet frame of σ at s_0. Then*

$$\sigma(s) - \sigma(s_0) = \left((s - s_0) - \frac{\kappa^2(s_0)}{6}(s - s_0)^3 \right) \mathbf{t}_0$$

$$+ \left(\frac{\kappa(s_0)}{2}(s - s_0)^2 + \frac{\dot{\kappa}(s_0)}{6}(s - s_0)^3 \right) \mathbf{n}_0 \qquad (1.20)$$

$$+ \frac{\kappa(s_0)\tau(s_0)}{6}(s - s_0)^3 \mathbf{b}_0 + o\big((s - s_0)^3\big) \ .$$

Proof. The usual Taylor expansion of σ about s_0 is

$$\sigma(s) = \sigma(s_0) + \dot{\sigma}(s_0)(s - s_0) + \frac{\ddot{\sigma}(s_0)}{2}(s - s_0)^2$$

$$+ \frac{1}{3!}\frac{\mathrm{d}^3\sigma}{\mathrm{d}s^3}(s_0)(s - s_0)^3 + o\big((s - s_0)^3\big) \ .$$

Recalling that

$$\dot{\sigma}(s_0) = \mathbf{t}_0 \ , \qquad \ddot{\sigma}(s_0) = \kappa(s_0)\mathbf{n}_0 \ ,$$

and

$$\frac{\mathrm{d}^3\sigma}{\mathrm{d}s^3}(s_0) = \frac{\mathrm{d}(\kappa\mathbf{n})}{\mathrm{d}s}(s_0) = \dot{\kappa}(s_0)\mathbf{n}_0 - \kappa^2(s_0)\mathbf{t}_0 + \kappa(s_0)\tau(s_0)\mathbf{b}_0 \ ,$$

we get (1.20). $\qquad\qquad\qquad\qquad\qquad\qquad\qquad\qquad\qquad\qquad\qquad\qquad\square$

We describe now a general procedure useful for answering questions of the kind we have seen at the beginning of this section. Suppose we have a curve $\sigma\colon I \to \mathbb{R}^n$ parametrized by arc length, and a family of maps $F_\lambda\colon \Omega \to \mathbb{R}^k$ depending on a parameter $\lambda \in \mathbb{R}^m$, where $\Omega \subseteq \mathbb{R}^n$ is an open neighborhood of the support of σ. Having set $C_\lambda = \{x \in \Omega \mid F_\lambda(x) = O\}$, we want to find for which value of λ the set C_λ best approximates the curve σ at a point $s_0 \in I$. In order for C_λ to approximate σ at s_0, it has at least to pass through $\sigma(s_0)$; then, we require $F_\lambda\big(\sigma(s_0)\big) = O$. Recalling what we did for the tangent line, it is natural to consider the map $F_\lambda \circ \sigma$, and to say that *the value of λ for which C_λ best approximates σ at s_0 is the one for which the map $s \mapsto F_\lambda\big(\sigma(s)\big)$ has the fastest rate of approach to zero as $s \to s_0$.*

Let's see how to apply this procedure in two examples.

Example 1.4.2. We want to find the plane that best approximates a biregular curve parametrized by arc length $\sigma\colon I \to \mathbb{R}^3$ at the point $s_0 \in I$. The equation of a generic plane through $\sigma(s_0)$ is $F_\mathbf{v}(x) = 0$, where

$$F_\mathbf{v}(x) = \langle x - \sigma(s_0), \mathbf{v}\rangle$$

and $\mathbf{v} \in \mathbb{R}^3$ is a versor orthogonal to the plane. Proposition 1.4.1 gives

$$F_\mathbf{v}\big(\sigma(s)\big) = \langle \mathbf{t}_0, \mathbf{v}\rangle(s - s_0) + \frac{\kappa(s_0)}{2}\langle \mathbf{n}_0, \mathbf{v}\rangle(s - s_0)^2 + o\big((s - s_0)^2\big) \ ,$$

so the plane best approximating the curve at σ_0 is the one for which $\mathbf{v}$ is orthogonal both to $\mathbf{t}_0$ and $\mathbf{n}_0$. But this means $\mathbf{v} = \pm\mathbf{b}_0$, and so we have proved that *the plane best approximating a curve in a point is the osculating plane.*

Example 1.4.3. We want to find the circle that best approximates a biregular curve parametrized by arc length $\sigma\colon I \to \mathbb{R}^3$ at the point $s_0 \in I$. First of all, the circle with radius $r > 0$ and center $p_0 \in H$ contained in the plane H through $\sigma(s_0)$ has equation $F_{p_0,r,\mathbf{v}}(x) = O$, where

$$F_{p_0,r,\mathbf{v}}(x) = \left(\langle x - \sigma(s_0), \mathbf{v}\rangle, \|x - p_0\|^2 - r^2\right),$$

and $\mathbf{v}$ is a versor orthogonal to H. Moreover, p_0 is such that $\langle p_0 - \sigma(s_0), \mathbf{v}\rangle = 0$ and $\|\sigma(s_0) - p_0\|^2 = r^2$, since the point p_0 belongs to H and the circle has to pass through $\sigma(s_0)$.

Clearly, the circle best approximating σ at s_0 has to be contained in the plane best approximating σ at s_0, that is, by the previous example, in the osculating plane. If we denote by $\{\mathbf{t}_0, \mathbf{n}_0, \mathbf{b}_0\}$ the Frenet frame of σ at s_0, we may take $\mathbf{v} = \mathbf{b}_0$ and write $\sigma(s_0) - p_0 = r\cos\theta\,\mathbf{t}_0 + r\sin\theta\,\mathbf{n}_0$ for a suitable $\theta \in \mathbb{R}$. Then

$$F_{p_0,r,\mathbf{v}}\big(\sigma(s)\big) = \left(0, 2r(\cos\theta)(s - s_0) + \big(1 + \kappa(s_0)r\sin\theta\big)(s - s_0)^2\right)$$
$$+o\big((s - s_0)^2\big),$$

so the circle best approximating σ at s_0 (called *osculating circle* of σ at s_0) can be obtained by taking $\theta = -\pi/2$ (that is, $p_0 = \sigma(s_0) + r\mathbf{n}_0$) and $r = 1/\kappa(s_0)$, further justifying the name "radius of curvature" for the inverse of the curvature. In particular, the point

$$p_0 = \sigma(s_0) + \frac{1}{\kappa(s_0)}\mathbf{n}_0$$

is the *center of curvature* of σ at s_0. For a plane curve, the evolute is the locus of the centers of curvature (see Definition 1.E.1).

The local canonical form (1.20) can also be used to understand something more about the shape of a biregular curve σ; for instance, it illustrates the geometrical meaning of torsion.

Definition 1.4.4. Let $H \subset \mathbb{R}^3$ be a plane through $p_0 \in \mathbb{R}^3$ and orthogonal to a versor $\mathbf{v} \in \mathbb{R}^3$. The *positive* (with respect to $\mathbf{v}$) *half-space* bounded by H is the open half-space consisting of the points $p \in \mathbb{R}^3$ such that $\langle p - p_0, \mathbf{v}\rangle > 0$. The other half-space is the *negative half-space* bounded by H.

Definition 1.4.5. Let $\sigma\colon I \to \mathbb{R}^3$ be a be a biregular curve parametrized by arc length, and $\{\mathbf{t}_0, \mathbf{n}_0, \mathbf{b}_0\}$ the Frenet frame of σ at $s_0 \in I$. The plane through $\sigma(s_0)$, parallel to $\mathrm{Span}(\mathbf{n}_0, \mathbf{b}_0)$ is the *normal plane* to σ at $\sigma(s_0)$, while the plane through $\sigma(s_0)$, parallel to $\mathrm{Span}(\mathbf{t}_0, \mathbf{b}_0)$ is the *rectifying plane* of σ at $\sigma(s_0)$.

By (1.20), for s near to s_0, the inner product $\langle\sigma(s) - \sigma(s_0), \mathbf{b}_0\rangle$ has the same sign as $\tau(s_0)(s - s_0)$; so, if $\tau(s_0) > 0$, *the curve σ, when s increases,* moves from the negative (with respect to $\mathbf{b}_0$) half-space bounded by the osculating plane to the positive one, while it moves form the positive half-space to the negative one if $\tau(s_0) < 0$.

The behavior with respect to the other two coordinate planes determined by the Frenet frame, however, is qualitatively the same for all biregular curves. Indeed, by (1.20), for s close to s_0, the inner product $\langle \sigma(s) - \sigma(s_0), \mathbf{t}_0 \rangle$ has the same sign as $s - s_0$, while the inner product $\langle \sigma(s) - \sigma(s_0), \mathbf{n}_0 \rangle$ is always positive; so the curve σ crosses the normal plane always moving from the negative half-space (with respect to $\mathbf{t}_0$) to the positive half-space, while it always stays in the positive half-space (with respect to $\mathbf{n}_0$) determined by the rectifying plane.

Exercises

1.81. Show that a biregular curve is plane if and only if all of its osculating planes pass through the same point. (*Hint*: assume that σ is parametrized by arc length. The equation of the osculating plane in $\sigma(s)$ is $\langle \mathbf{b}(s), \mathbf{x} - \sigma(s) \rangle = 0$. If $\mathbf{x}_0$ belongs to each osculating plane, then $\langle \mathbf{b}(s), \mathbf{x}_0 - \sigma(s) \rangle = 0$. Differentiating, you conclude a part of the proof.)

1.82. Prove that if all normal planes to a biregular curve σ pass through the same point then σ is plane.

1.83. The circular helix. Fix a real non zero number a and consider the circular helix $\sigma_a \colon \mathbb{R} \to \mathbb{R}^3$, having radius r and pitch a, parametrized by arc length as in Example 1.2.15. Consider the map $\beta_a \colon \mathbb{R} \to \mathbb{R}^3$ defined by $\beta_a(s) = \sigma_a(s) + \frac{1}{\kappa(s)} \mathbf{n}(s)$, where $\kappa(s)$ and $\mathbf{n}(s)$ are, respectively, the curvature and the normal versor to σ_a at $\sigma_a(s)$.

(i) Prove that β_a parametrizes a circular helix.
(ii) Determine the value of a and r in order for β_a to be contained in the same cylinder as σ_a.

1.84. Let $\sigma \colon I \to \mathbb{R}^3$ be a regular curve whose support is contained in a sphere with radius r. Show that the curvature of σ is greater than or equal to $1/r$ in every point.

1.85. Given $\delta > 0$, let $\sigma \colon (-\delta, \delta) \to \mathbb{R}^3$ be a curve of class at least C^3, parametrized by arc length, having curvature and torsion positive everywhere. Considering just the lowest degree terms in each component (with respect to the Frenet frame $\mathbf{t}_0, \mathbf{n}_0, \mathbf{b}_0$ in $s = 0$) of the local canonical form of σ at $s = 0$, we get a curve $\sigma_1 \colon (-\delta, \delta) \to \mathbb{R}^3$ parametrized by $\sigma_1(s) = s\, \mathbf{t}_0 + \frac{\kappa(0)}{2} s^2\, \mathbf{n}_0 + \frac{\kappa(0)\tau(0)}{6} s^3\, \mathbf{b}_0$.

(i) Show that the orthogonal projection of σ_1 on the osculating plane of σ at O is a parabola, and compute its curvature.

(ii) Draw the orthogonal projection of σ_1 on the normal plane to σ at O, and show that it does not have a regular parametrization.

(iii) Show that the orthogonal projection of σ_1 on the rectifying plane of σ at O is the graph of a function. Finally, show that this orthogonal projection has a point of inflection at the origin (that is, its curvature is zero at the origin).

1.86. Show that the difference between the length s of a curve $\sigma : [a, b] \to \mathbb{R}^3$ of class C^1 and the length of the line segment from $\sigma(a)$ to $\sigma(b)$ is of the order of s^3 (for sufficiently small $b - a$).

1.87. Consider the helix $\sigma : \mathbb{R} \to \mathbb{R}^3$ parametrized by $\sigma(t) = (\cos t, \sin t, t)$.

(i) Show that the Cartesian equation of the osculating plane of σ at the point corresponding to the parameter $t = \frac{\pi}{2}$ is given by $x + y = \frac{\pi}{2}$.

(ii) Determine the projection of σ on the plane $z = 0$.

(iii) Determine the parametrization of the curve obtained by intersecting the plane $z = 0$ with the affine tangent lines of the helix.

(iv) Show that, for a fixed point P in space, the points of σ for which the osculating plane passes through P lie in a same plane.

1.88. Show that there does not exist a rigid motion $\rho : \mathbb{R}^3 \to \mathbb{R}^3$ such that $\rho \circ \sigma_+ = \sigma_-$, where $\sigma_+, \sigma_- : \mathbb{R} \to \mathbb{R}^3$ are the helices parametrized by $\sigma_\pm(t) = (r \cos t, \sin t, \pm t)$.

1.89. Show that the straight lines (with the usual parametrization) are the only regular curves in $\mathbb{R}^3$ such that the affine tangent lines at all points are parallel to a given line.

1.90. We shall call *inflection* point for a regular curve a point at which the curvature is zero. Show that if $\sigma(s_0)$ is an isolated inflection point of a regular curve parametrized by arc length then the normal versor is defined and is continuous in a neighborhood of s_0.

1.91. Let $\sigma : I \to \mathbb{R}^3$ be a biregular curve. Show that there is no point $P \in \mathbb{R}^3$ such that all binormal lines to σ pass through P.

1.92. Let $C_\lambda = \{(x, y, z) \in \mathbb{R}^3 \mid x + \lambda y - x^3 + y^3 = 0\}$. Determine the value of $\lambda \in \mathbb{R}$ for which C_λ best approximates in the origin the curve implicitly defined by $x + y = x^4 + y^4$.

ORDER OF CONTACT

Definition 1.E.2. Given a curve $\sigma : I \to \mathbb{R}^3$ of class C^m, let $F : \Omega \to \mathbb{R}^k$ be a map of class C^m, where $\Omega \subseteq \mathbb{R}^3$ is an open neighborhood of the support of σ, and assume that there is $s_0 \in I$ such that $F(\sigma(s_0)) = 0$, that is

$\sigma(s_0) \in C = \{x \in \Omega \mid F(x) = 0\}$. We say that F (or C) and σ have *contact of order* $r \leq m$ at $p_0 = \sigma(s_0)$ if and only if

$$f(s_0) = f'(s_0) = f''(s_0) = \cdots = f^{(r-1)}(s_0) = O \quad \text{and} \quad f^{(r)}(s_0) \neq O,$$

where $f = F \circ \sigma$.

1.93. Show that the definition of order of contact of a map F with a curve σ is independent of the parametrization of the curve.

1.94. Given a curve $\sigma: I \to \mathbb{R}^3$ of class C^m, let $F: \Omega \to \mathbb{R}$ be a function of class C^m, where $\Omega \subseteq \mathbb{R}^3$ is an open neighborhood of the support of σ. Assume that $C = F^{-1}(0)$ and the support of σ intersect in $p_0 = \sigma(t_0)$ and in $0 < r - 1 < m - 1$ more distinct points. Show that there exist $t_1, t_2, \ldots, t_{r-1}$ in a neighborhood of t_0 such that

$$f(t_0) = f'(t_1) = f''(t_2) = \ldots = f^{(r-1)}(t_{r-1}) = 0.$$

(*Hint:* use Rolle's theorem.)

1.95. Osculating sphere Let $\sigma: I \to \mathbb{R}^3$ be a biregular curve parametrized by arc length, and let $s_0 \in I$ be such that σ has non zero torsion in $\sigma(s_0)$. Show that there exists a unique sphere with contact order 4 in $\sigma(s_0)$. This sphere is called the *osculating sphere* in $\sigma(s_0)$. Show that the center of this sphere is $\sigma(s_0) + \frac{1}{\kappa(s_0)}\mathbf{n}(s_0) - \frac{\dot{\kappa}(s_0)}{(\kappa(s_0))^2\tau(s_0)}\mathbf{b}(s_0)$.

1.96. Determine the order of contact at the origin between the plane regular curve implicitly defined by $x^4 + y^7 + 3x = 0$ and the plane curve parametrized by $\sigma(t) = (t^2, t)$.

1.97. Determine the order of contact at the origin between the parabola parametrized by $\sigma(t) = (t, t^2)$ and the regular curve implicitly defined by the equation $x^4 + y^7 + 3x = 0$ in a neighborhood of $(0, 0)$.

1.98. Determine the order of contact at $p_0 = (1, 0)$ between the curve parametrized by $\sigma(t) = (\sqrt{1 - 2t^2}, t)$ and the line $x - 1 = 0$.

1.99. Determine the order of contact at $p_0 = (0, 2)$ between the plane curves defined by equations $x^2 + y^2 - 2y = 0$ and $3x^2 + 3xy + 2y^2 - 6x - 6y + 4 = 0$ by considering one of the curves as a parametrized curve and the other as the vanishing locus of a function. Is there any difference if the roles are exchanged?

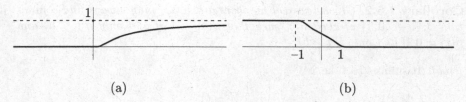

Fig. 1.9. (a) graph of α; (b) graph of β for $a = -1$ and $b = 1$

1.5 Whitney's Theorem

The goal of this section is to give a proof of Whitney's Theorem 1.1.7. Let us start with some preliminary results.

Lemma 1.5.1. *There exists a function* $\alpha: \mathbb{R} \to [0, 1)$ *which is monotonic, of class* C^∞ *and such that* $\alpha(t) = 0$ *if and only if* $t \leq 0$.

Proof. Set

$$\alpha(t) = \begin{cases} e^{-1/t} & \text{if } t > 0, \\ 0 & \text{if } t \leq 0; \end{cases}$$

see Fig. 1.9.(a). Clearly, α takes values in $[0, 1)$, is monotonic, is zero only in $\mathbb{R}^-$, and is of class C^∞ in $\mathbb{R}^*$; we have only to check that it is of class C^∞ in the origin too. To verify this, it suffices to prove that the right and left limits of all derivatives in the origin coincide, that is, that

$$\lim_{t \to 0^+} \alpha^{(n)}(t) = 0$$

for all $n > 0$. Assume we have proved the existence, for all $n \in \mathbb{N}$, of a polynomial p_n of degree $2n$ such that

$$\forall t > 0 \qquad \alpha^{(n)}(t) = e^{-1/t} p_n(1/t). \qquad (1.21)$$

Granted this, we have

$$\lim_{t \to 0^+} \alpha^{(n)}(t) = \lim_{s \to +\infty} \frac{p_n(s)}{e^s} = 0;$$

therefore to get the assertion it suffices to prove (1.21). We argue by induction on n. For $n = 0$, it is enough to take $p_0 \equiv 1$. Assume that (1.21) is verified for a given $n \geq 0$; then

$$\alpha^{(n+1)}(t) = \frac{d}{dt}\left[e^{-1/t} p_n(1/t)\right] = e^{-1/t}\left[\frac{1}{t^2} p_n(1/t) - \frac{1}{t^2} p_n'(1/t)\right],$$

so we may choose $p_{n+1}(s) = s^2\big(p_n(s) - p_n'(s)\big)$. $\qquad \square$

Corollary 1.5.2. *Given two real numbers $a < b$, we may always find a function $\beta\colon \mathbb{R} \to [0,1]$ of class C^∞ such that $\beta(t) = 1$ if and only if $t \leq a$, and $\beta(t) = 0$ if and only if $t \geq b$.*

Proof. It suffices to take

$$\beta(t) = \frac{\alpha(b-t)}{\alpha(b-t) + \alpha(t-a)},$$

where $\alpha\colon \mathbb{R} \to \mathbb{R}$ is the function given by the previous lemma; see Fig. 1.9.(b).
□

Corollary 1.5.3. *For all $p_0 \in \mathbb{R}^n$ and $r > 0$, there exists a function $f\colon \mathbb{R}^n \to [0,1]$ of class C^∞ such that*

$$f^{-1}(1) = \overline{B(p_0, r/2)} \qquad and \qquad f^{-1}(0) = \mathbb{R}^n \setminus B(p_0, r),$$

where $B(p,r) \subset \mathbb{R}^n$ is the open ball with center p and radius r in $\mathbb{R}^n$.

Proof. Let $\beta\colon \mathbb{R} \to [0,1]$ be the function defined in the previous corollary, with $a = r^2/4$ and $b = r^2$. Then, $f(p) = \beta(\|p - p_0\|^2)$ is as required.
□

Lemma 1.5.4. *Let $V \subseteq \mathbb{R}^n$ be an arbitrary open subset. Then we may find a sequence of points having rational coordinates $\{p_k\}_{k \in \mathbb{N}} \subseteq \mathbb{Q}^n \cap V$ and a sequence of rational numbers $\{r_k\}_{k \in \mathbb{N}} \subseteq \mathbb{Q}^+$ such that $V = \bigcup_{k \in \mathbb{N}} B(p_k, r_k)$.*

Proof. Take $p \in V$. Since V is open, there exists $\varepsilon > 0$ such that $B(p, \varepsilon) \subset V$. Choose then $q \in \mathbb{Q}^n \cap V$ and $r \in \mathbb{Q}^+$ such that $\|p - q\| < r < \varepsilon/2$. Clearly, $p \in B(q, r)$; moreover, if $x \in B(q, r)$ we have

$$\|p - x\| \leq \|p - q\| + \|q - x\| < 2r < \varepsilon,$$

and so $B(q, r) \subseteq B(p, \varepsilon) \subset V$. Hence, every point of V belongs to a ball with rational center and radius, completely contained in V; since there are at most countably many such balls, the assertion follows.
□

We can now prove Whitney's theorem:

Theorem 1.1.7 (Whitney). *Let $\Omega \subseteq \mathbb{R}^n$ be an open set. Then a subset $C \subseteq \Omega$ is closed in Ω if and only if there exists a function $f\colon \Omega \to \mathbb{R}$ of class C^∞ such that $C = f^{-1}(0)$.*

Proof. If $C = f^{-1}(0)$, we already know that C is closed in Ω. Conversely, assume that C is closed in Ω; then $V = \Omega \setminus C$ is open in Ω, and consequently in $\mathbb{R}^n$. Lemma 1.5.4 says that we can find a sequence $\{p_k\} \subseteq \mathbb{Q}^k$ and a sequence $\{r_k\} \subseteq \mathbb{Q}^+$ such that $\Omega \setminus C = \bigcup_{k \in \mathbb{N}} B(p_k, r_k)$. For each $k \in \mathbb{N}$ let then $f_k\colon \Omega \to [0,1]$ be the restriction to Ω of the function obtained by applying Corollary 1.5.3 to p_k and r_k.

By construction, f_k and all of its derivatives are zero everywhere except in $B(p_k, r_k)$. Since $\overline{B(p_k, r_k)}$ is a compact set, for all $k \in \mathbb{N}$ we find $c_k > 1$ such that the absolute value of any derivative of f_k of order at most k is bounded above by c_k over Ω. Set then

$$f = \sum_{k=0}^{\infty} \frac{f_k}{2^k c_k}.$$

This series is bounded above by $\sum_k 2^{-k}$, so it converges uniformly on Ω. Given $m \geq 0$, by construction each m-th derivative of the k-th term of the series is bounded from above by 2^{-k} as soon as $k \geq m$; so the series of the m-derivatives converge uniformly too, and hence $f \in C^{\infty}(\Omega)$.

We have still to show that $C = f^{-1}(0)$. If $p \in C$, then $p \notin B(p_k, r_k)$ for all $k \in \mathbb{N}$, so $f_k(p) = 0$ for all $k \in \mathbb{N}$, and $f(p) = 0$. Conversely, if $p \in \Omega \setminus C$, we can find $k_0 \in \mathbb{N}$ such that $p \in B(p_{k_0}, r_{k_0}) \subset \Omega \setminus C$; hence $f_{k_0}(p) > 0$ and $f(p) \geq f_{k_0}(p)/2^{k_0} c_{k_0} > 0$. $\qquad \square$

1.6 Classification of 1-submanifolds

As promised at the end of Section 1.1, we want to discuss now another possible approach to the problem of defining what a curve is. As we shall see, even if in the case of curves this approach will turn out to be too restrictive, for surfaces it will be the correct way to follow (as you shall learn in Section 3.1).

The idea consists in concentrating on the support. The support of a curve has to be a subset of $\mathbb{R}^n$ that looks (at least locally) like an interval of the real line. What we have seen studying curves suggests that a way to give concrete form to the concept of "looking like" consists in using homeomorphisms with the image that are regular curves of class at least C^1 too. So we introduce:

Definition 1.6.1. A 1-*submanifold* of class C^k in $\mathbb{R}^n$ (with $k \in \mathbb{N}^* \cup \{\infty\}$ and $n \geq 2$) is a connected subset $C \subset \mathbb{R}^n$ such that for all $p \in C$ there exist a neighborhood $U \subset \mathbb{R}^n$ of p, an open interval $I \subseteq \mathbb{R}$, and a map $\sigma \colon I \to \mathbb{R}^n$ (called *local parametrization*) of class C^k, such that:

(i) $\sigma(I) = C \cap U$;
(ii) σ is a homeomorphism with its image;
(iii) $\sigma'(t) \neq O$ for all $t \in I$.

If $\sigma(I) = C$, we shall say that σ is a *global parametrization*. A *periodic parametrization* is a map $\sigma \colon \mathbb{R} \to \mathbb{R}^n$ of class C^k which is periodic of period $\ell > 0$, with $\sigma(\mathbb{R}) = C$, and such that for all $t_0 \in \mathbb{R}$ the restriction $\sigma|_{(t_0, t_0 + \ell)}$ is a local parametrization of C having image $C \setminus \{\sigma(t_0)\}$.

Example 1.6.2. The graph $\Gamma_f \subset \mathbb{R}^2$ of a function $f \colon I \to \mathbb{R}$ of class C^k is a 1-submanifold of class C^k. Indeed, (check it!) a global parametrization of Γ_f is the usual map $\sigma \colon I \to \mathbb{R}^2$ given by $\sigma(t) = \bigl(t, f(t)\bigr)$.

Example 1.6.3. If $f: \Omega \to \mathbb{R}$, where $\Omega \subseteq \mathbb{R}^2$ is an open subset, is a function of class C^k with gradient nowhere zero in the points of $C = \{x \in \Omega \mid f(x) = 0\}$, then by Proposition 1.1.18 we may conclude that C is locally a graph, and so a 1-submanifold, even if it is not necessarily provided with a global parametrization. For instance, a circle in the plane is a 1-submanifold, but cannot have a global parametrization (why?); however, it has a periodic parametrization (see Example 1.1.12).

Example 1.6.4. A closed line segment S in $\mathbb{R}^n$ is *not* a 1-submanifold. Indeed, no neighborhood in S of the endpoints is homeomorphic to an open interval of a line (why?), so the endpoints cannot belong to the image of a local parametrization. Analogously, a "figure 8" in a plane is not a 1-submanifold. In this case, the central point of the 8 is the one lacking a neighborhood homeomorphic to an interval (why?).

Example 1.6.5. The graph Γ_f of the absolute value function $f(t) = |t|$ is not a 1-submanifold. Indeed, suppose on the contrary that it is; then there exist an open interval $I \subseteq \mathbb{R}$ — which, without loss of generality, may be assumed to contain 0 (why?) — and a homeomorphism with its image $\sigma: I \to \mathbb{R}^2$ of class at least C^1, satisfying $\sigma(0) = (0,0)$, $\sigma(I) \subseteq \Gamma_f$, and $\sigma'(t) \neq O$ for all $t \in I$. Now, if we write $\sigma = (\sigma_1, \sigma_2)$, saying that the image of σ is included in Γ_f amounts to saying that $\sigma_2(t) = |\sigma_1(t)|$ for all $t \in I$. In particular, the function $t \mapsto |\sigma_1(t)|$ has to be differentiable in 0. Its difference quotient is

$$\frac{\sigma_2(t)}{t} = \frac{|\sigma_1(t)|}{t} = \frac{|\sigma_1(t)|}{\sigma_1(t)} \frac{\sigma_1(t)}{t}. \tag{1.22}$$

The rightmost quotient tends to $\sigma_1'(0)$ when $t \to 0$. If we had $\sigma_1'(0) \neq 0$, in a neighborhood of 0 the function $t \mapsto |\sigma_1(t)|/\sigma_1(t)$ would be continuous and with values in $\{+1, -1\}$, so it would be constant; but this cannot happen, since the image of σ, being a neighborhood of the origin in Γ_f, contains both points with positive abscissa and with negative abscissa. So, $\sigma_1'(0) = 0$; but then (1.22) also implies $\sigma_2'(0) = 0$, against the assumption $\sigma'(0) \neq O$.

Remark 1.6.6. A 1-submanifold C has no interior points. If, by contradiction, $p \in C$ were an interior point, C would contain a ball B with center p; in particular, the set $U \cap C \setminus \{p\}$ would be connected for any connected neighborhood $U \subseteq B$ of p. But if U is as in the definition of a 1-submanifold, then $U \cap C \setminus \{p\}$ is homeomorphic to an open interval without a point, which is a disconnected set, contradiction.

Remark 1.6.7. Conditions (i) and (ii) in the definition of a 1-submanifold say that the set C is, from a topological viewpoint, locally like an interval. Condition (iii), on the other hand, serves three goals: provides the 1-submanifold with a tangent vector, by excluding sharp corners like those in the graph of the absolute value function; ensures an interval-like structure from a differential viewpoint too (as we shall understand better when we shall tackle the analogous problem for surfaces in Chapter 3); and prevents other kinds of singularities, such as the cusp in the image of the map $\sigma(t) = (t^2, t^3)$; see Exercise 1.4 and Fig. 1.2.

The main result of this section states that, to study curves, the definition of a 1-submanifold we have given here is unnecessarily complicated. Indeed, any 1-submanifold is an open Jordan arc or a Jordan curve. More precisely, the following holds:

Theorem 1.6.8. *Every non-compact 1-submanifold has a global parametrization, and every compact 1-submanifold has a periodic parametrization. More precisely, if $C \subset \mathbb{R}^n$ is a 1-submanifold of class C^k, then there exists a map $\hat{\sigma} \colon \mathbb{R} \to \mathbb{R}^n$ of class C^k such that $\hat{\sigma}'(t) \neq O$ for all $t \in \mathbb{R}$ and:*

(a) *if C is not compact, then $\hat{\sigma}$ is a global parametrization of C, and C is homeomorphic to $\mathbb{R}$;*

(b) *if C is compact, then $\hat{\sigma}$ is a periodic parametrization of C, and C is homeomorphic to the circle S^1.*

Proof. The proof consists of several steps.

(1) By Theorem 1.2.11 applied to local parametrizations, every point of C belongs to the image of a local parametrization by arc length; we want to see what happens when two local parametrizations by arc length have intersecting images.

Let $\sigma = (\sigma_1, \ldots, \sigma_n) \colon I_\sigma \to C$ and $\tau = (\tau_1, \ldots, \tau_n) \colon I_\tau \to C$ be two local parametrizations by arc length such that $\sigma(I_\sigma) \cap \tau(I_\tau) \neq \varnothing$; set $J_\sigma = \sigma^{-1}\big(\sigma(I_\sigma) \cap \tau(I_\tau)\big) \subseteq I_\sigma$, $J_\tau = \tau^{-1}\big(\sigma(I_\sigma) \cap \tau(I_\tau)\big) \subseteq I_\tau$, and $h = \tau^{-1} \circ \sigma \colon J_\sigma \to J_\tau$. Clearly, the function h is a homeomorphism between open subsets of $\mathbb{R}$; moreover, it is (at least) of class C^1. Indeed, fix $t_0 \in J_\sigma$. Then, from $\tau \circ h = \sigma$ we get

$$\frac{\sigma(t) - \sigma(t_0)}{t - t_0} = \frac{\tau\big(h(t)\big) - \tau\big(h(t_0)\big)}{h(t) - h(t_0)} \frac{h(t) - h(t_0)}{t - t_0}$$

for all $t \in J_\sigma$. When t approaches t_0, the first ratio tends to $\sigma'(t_0)$, and the second one to $\tau'\big(h(t_0)\big)$. Since τ is a local parametrization, there exists an index j for which $\tau_j'\big(h(t_0)\big) \neq 0$; hence the limit

$$\lim_{t \to t_0} \frac{h(t) - h(t_0)}{t - t_0} = \frac{\sigma_j'(t_0)}{\tau_j'\big(h(t_0)\big)}$$

exists, so h is differentiable. Moreover, the same argument with the same j holds for all t in a neighborhood of t_0, and so we find

$$h' = \frac{\sigma_j'}{\tau_j' \circ h}$$

in a neighborhood of t_0; consequently, h' is continuous.

Now, from $\tau \circ h = \sigma$ we also deduce that $(\tau' \circ h)h' = \sigma'$, and thus

$$|h'| \equiv 1 \,,$$

since σ and τ are parametrized by arc length. So the graph Γ of h consists of line segments having slope ± 1, in the same number as the components of J_σ (and consequently of J_τ). So, in each of these components, we have $h(t) = \pm t + a$, that is, $\sigma(t) = \tau(\pm t + a)$, for a suitable $a \in \mathbb{R}$ (which may depend on the component of J_σ we are considering).

We are not done yet. The graph Γ of h is included in the rectangle $I_\sigma \times I_\tau$; we want to prove now that the endpoints of the line segments of Γ are necessarily on the perimeter of this rectangle. First of all, note that $(s_0, s) \in \Gamma$ if and only if $s = h(s_0)$, so

$$(s_0, s) \in \Gamma \quad \Longrightarrow \quad \tau(s) = \sigma(s_0) \,. \tag{1.23}$$

Assume now, by contradiction, that $(t_0, t) \in I_\sigma \times I_\tau$ is an endpoint of a line segment of Γ lying in the interior of the rectangle (in particular, $\sigma(t_0)$ and $\tau(t)$ exist). From the fact that (t_0, t) is an endpoint, we deduce that $t_0 \in \partial J_\sigma$; but, on the other hand, by continuity, (1.23) implies that $\tau(t) = \sigma(t_0) \in \sigma(I_\sigma) \cap \tau(I_\tau)$, so $t_0 \in J_\sigma$, obtaining a contradiction.

Now, Γ is the graph of an injective function; so, each side of the rectangle can contain at most one endpoint of Γ (why?). But this implies that Γ — and hence J_σ — has at most 2 components; and if it has two, both have the same slope.

Summing up, we have shown that if $\sigma(I_\sigma) \cap \tau(I_\tau) \neq \varnothing$ then there are only three possibilities (see Fig. 1.10):

(i) $\tau(I_\tau) \subseteq \sigma(I_\sigma)$ or $\sigma(I_\sigma) \subseteq \tau(I_\tau)$, in which case Γ consists of a single line segment having slope ± 1 and joining two opposite sides of the rectangle $I_\sigma \times I_\tau$;

(ii) $\sigma(I_\sigma) \cap \tau(I_\tau)$ consists of a single component, different from both $\sigma(I_\sigma)$ and $\tau(I_\tau)$, and Γ consists of a single segment having slope ± 1, joining two adjacent sides of the rectangle $I_\sigma \times I_\tau$;

(iii) $\sigma(I_\sigma) \cap \tau(I_\tau)$ consists of two components, and Γ consists of two segments with equal slope ± 1, joining two adjacent sides of the rectangle $I_\sigma \times I_\tau$.

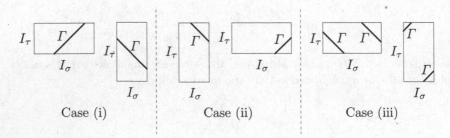

Fig. 1.10.

Finally, note that if Γ has slope -1, setting $\sigma_1(t) = \sigma(-t)$, we get a local parametrization by arc length σ_1 of C, having the same image as σ, but such that the graph of $h_1 = \sigma_1^{-1} \circ \sigma_0$ has slope $+1$ (why?).

(2) Assume now that there exist two local parametrizations by arc length $\sigma : I_\sigma \to C$ and $\tau : I_\tau \to C$ such that $\sigma(I_\sigma) \cap \tau(I_\tau)$ actually has two components; we want to prove that, if this happens, we are necessarily in case (b) of the theorem. Thanks to the previous step, we may assume that the parameter change $h = \tau^{-1} \circ \sigma$ has slope 1 in both components J_σ^1 and J_σ^2 of J_σ. So there exist $a, b \in \mathbb{R}$ such that

$$h(t) = \begin{cases} t + a & \text{if } t \in J_\sigma^1 , \\ t + b & \text{if } t \in J_\sigma^2 ; \end{cases} \qquad (1.24)$$

see Fig. 1.11.

Note now that $I_\sigma \cap (I_\tau - b) = J_\sigma^2$, where $I_\tau - b = \{ t - b \mid t \in I_\tau \}$. So, define $\tilde{\sigma} : I_\sigma \cup (I_\tau - b) \to C$ by setting

$$\tilde{\sigma}(t) = \begin{cases} \sigma(t) & \text{if } t \in I_\sigma , \\ \tau(t + b) & \text{if } t \in I_\tau - b . \end{cases}$$

Since $\sigma(t) = \tau(t + b)$ on $I_\sigma \cap (I_\tau - b)$, the function $\tilde{\sigma}$ is well defined and of class C^k. Moreover, if $t \in J_\sigma^1$, we have

$$\tilde{\sigma}(t) = \sigma(t) = \tau(t + a) = \tilde{\sigma}(t + \ell) ,$$

where $\ell = a - b$, so we may extend $\tilde{\sigma}$ to a mapping $\hat{\sigma} : \mathbb{R} \to C$ of class C^k, periodic with period ℓ. Now, $\hat{\sigma}(\mathbb{R}) = \sigma(I_\sigma) \cup \tau(I_\tau)$ is open in C. But for all $t_0 \in \mathbb{R}$ we have $\hat{\sigma}(\mathbb{R}) = \hat{\sigma}([t_0, t_0 + \ell])$; as a consequence, $\hat{\sigma}(\mathbb{R})$ is compact and therefore closed in C. Since C is connected, we get $\hat{\sigma}(\mathbb{R}) = C$, that is, $\hat{\sigma}$ is surjective and C is compact. Moreover, since σ and τ are local parametrizations, we may immediately deduce (why?) that $\hat{\sigma}$ restricted to $(t_0, t_0 + \ell)$ is a local parametrization, whatever $t_0 \in \mathbb{R}$ is; so $\hat{\sigma}$ is a periodic parametrization.

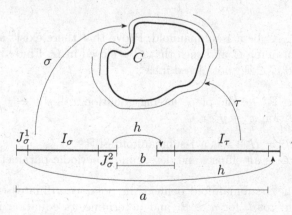

Fig. 1.11. A compact 1-submanifold

We have still to show that C is homeomorphic to S^1. We describe S^1 as the quotient space obtained by identifying the endpoints of the interval $[0, \ell]$. As $\hat\sigma(0) = \hat\sigma(\ell)$, the map $\hat\sigma$ induces a surjective continuous map $f \colon S^1 \to C$. Since $\hat\sigma|_{[0,\ell)}$ is injective, f turns out to be bijective; since S^1 is compact and C is Hausdorff, f is a homeomorphism, as required.

(3) Assume now that the intersection of the images of two local parametrizations by arc length never has two components; we want to show that we are in case (a) of the theorem. Let $\sigma \colon I_\sigma \to C$ be a *maximal* local parametrization, that is, such that it cannot be extended to a local parametrization defined in an open interval strictly larger than I_σ. Assume, by contradiction, that $\sigma(I_\sigma)$ is not the whole C, and take a point p in the boundary of $\sigma(I_\sigma)$ in C. Since C is a 1-submanifold, there exists a local parametrization by arc length $\tau \colon I_\tau \to C$ whose image includes p. In particular, $\tau(I_\tau) \cap \sigma(I_\sigma) \neq \varnothing$, since p is adherent to $\sigma(I_\sigma)$; and $\tau(I_\tau) \not\subset \sigma(I_\sigma)$ since $p \in \tau(I_\tau) \setminus \sigma(I_\sigma)$; and $\sigma(I_\sigma) \not\subset \tau(I_\tau)$ by the maximality of σ (why?).

So we are now in case (ii) of step (1); we may suppose further that the map $h = \tau^{-1} \circ \sigma \colon J_\sigma \to J_\tau$ is of the form $h(t) = t + b$ for some $b \in \mathbb{R}$. Define $\tilde\sigma \colon I_\sigma \cup (I_\tau - b) \to C$ by setting again

$$\tilde\sigma(t) = \begin{cases} \sigma(t) & \text{if } t \in I_\sigma, \\ \tau(t+b) & \text{if } t \in I_\tau - b. \end{cases}$$

But now it is easy to verify (exercise) that $\tilde\sigma$ is a local parametrization by arc length defined over an open interval strictly larger that I_σ, which is impossible.

The contradiction comes from having assumed that σ is not surjective. Hence we must have $\sigma(I_\sigma) = C$, so C has a global parametrization, and is homeomorphic to the interval I_σ. Since every open interval is homeomorphic to $\mathbb{R}$ (see Exercise 1.5), we are done. $\square$

Exercises

1.100. Let $C \subset \mathbb{R}^n$ be a 1-submanifold. Prove that there exists an open subset $\Omega \subset \mathbb{R}^n$ containing C and such that C is closed in Ω. Find an example of a 1-submanifold $C \subset \mathbb{R}^2$ not closed in $\mathbb{R}^2$.

1.101. Let $H \subset \mathbb{R}^3$ be the plane having equation $x + y + z = 1$ and S the cylinder of equation $x^2 + y^2 = 1$.

(i) Prove that $C = H \cap S$ is a 1-submanifold of $\mathbb{R}^3$.
(ii) Prove that C is an ellipse, and determine a periodic parametrization.

1.102. Draw the 1-submanifold defined, in polar coordinates (r, θ), by the equation $r = a/\cos\theta$, for $a \in \mathbb{R}$, and determine its equation in Cartesian coordinates.

1.7 Frenet-Serret formulas in higher dimensions

The goal of this section is to obtain Frenet-Serret formulas, and to prove a theorem analogous to Theorem 1.3.37, for curves in $\mathbb{R}^n$. We have seen that to study curves in $\mathbb{R}^2$ the regularity condition was sufficient, whereas for curves in $\mathbb{R}^3$ it was necessary introduce biregularity. To study curves in $\mathbb{R}^n$ with $n \geq 3$ we need an even stronger condition.

Regularity amounts to assuming that σ' is nowhere zero, while biregularity is equivalent to asking that σ' and σ'' are always linearly independent. For $\mathbb{R}^n$ we shall need $(n-1)$-regularity:

Definition 1.7.1. A curve $\sigma: I \to \mathbb{R}^n$ of class (at least) C^k is *k-regular*, with $1 \leq k \leq n-1$, if

$$\sigma', \sigma'', \ldots, \frac{\mathrm{d}^k \sigma}{\mathrm{d}t^k}$$

are always linearly independent. In particular, 1-regular is equivalent to regular, and 2-regular is equivalent to biregular.

Remark 1.7.2. If $\sigma: I \to \mathbb{R}^n$ is parametrized by arc length, and $2 \leq k \leq n-1$, it is easy to verify by induction (see Exercise 1.103) that σ is k-regular if and only if

$$\mathbf{t}, \dot{\mathbf{t}}, \ldots, \frac{\mathrm{d}^{k-1}\mathbf{t}}{\mathrm{d}s^{k-1}}$$

are always linearly independent.

Now, let $\sigma: I \to \mathbb{R}^n$ be a regular curve parametrized by arc length; for reasons that will be clear shortly, we denote by $\mathbf{t}_1$ its tangent vector $\dot{\sigma}$, and by $\kappa_1 = \|\dot{\mathbf{t}}_1\|$ its curvature. If σ is 2-regular, we have seen that the normal versor $\mathbf{n}$, that we shall denote by $\mathbf{t}_2$, is always a vector orthogonal to $\mathbf{t}_1$; furthermore,

$$\dot{\mathbf{t}}_1 = \kappa_1 \mathbf{t}_2 .$$

In particular, $\mathrm{Span}(\dot{\mathbf{t}}_1, \mathbf{t}_1) = \mathrm{Span}(\mathbf{t}_2, \mathbf{t}_1)$. Differentiating $\langle \mathbf{t}_2, \mathbf{t}_2 \rangle \equiv 1$ and $\langle \mathbf{t}_2, \mathbf{t}_1 \rangle \equiv 0$ we find that $\dot{\mathbf{t}}_2$ is always orthogonal to $\mathbf{t}_2$, and that $\langle \dot{\mathbf{t}}_2, \mathbf{t}_1 \rangle \equiv -\kappa_1$. Moreover,

$$\frac{\mathrm{d}^2 \mathbf{t}_1}{\mathrm{d}s^2} = \kappa_1 \dot{\mathbf{t}}_2 + \dot{\kappa}_1 \mathbf{t}_2 ; \tag{1.25}$$

so $\dot{\mathbf{t}}_2$ is in the subspace spanned by $\mathbf{t}_1$ e $\mathbf{t}_2$ if and only if $\mathrm{d}^2\mathbf{t}_1/\mathrm{d}s^2$ is in it, that is, if and only if σ is not 3-regular. This means that if σ is 3-regular (and $n \geq 4$), the vector $\dot{\mathbf{t}}_2 - \langle \dot{\mathbf{t}}_2, \mathbf{t}_1 \rangle \mathbf{t}_1 - \langle \dot{\mathbf{t}}_2, \mathbf{t}_2 \rangle \mathbf{t}_2 = \dot{\mathbf{t}}_2 + \kappa_1 \mathbf{t}_1$ is never zero, so we can find $\mathbf{t}_3: I \to \mathbb{R}^n$ and $\kappa_2: I \to \mathbb{R}^+$ such that $\mathbf{t}_3$ is always a versor orthogonal to $\mathbf{t}_1$ and $\mathbf{t}_2$, and

$$\dot{\mathbf{t}}_2 = -\kappa_1 \mathbf{t}_1 + \kappa_2 \mathbf{t}_3$$

holds with κ_2 nowhere zero. In particular,

$$\mathrm{Span}(\mathbf{t}_1, \mathbf{t}_2, \mathbf{t}_3) = \mathrm{Span}(\mathbf{t}_1, \mathbf{t}_2, \dot{\mathbf{t}}_2) = \mathrm{Span}\left(\mathbf{t}_1, \dot{\mathbf{t}}_1, \frac{\mathrm{d}^2 \mathbf{t}_1}{\mathrm{d}s^2} \right)$$

and

$$\frac{d^3 t_1}{ds^3} - \kappa_1 \kappa_2 \dot{t}_3 \in \mathrm{Span}(t_1, t_2, t_3) \; ;$$

so $\dot{t}_3$ is in the subspace spanned by t_1, t_2 and t_3 if and only if σ is not 4-regular.

For the next step, differentiate $\langle t_3, t_3 \rangle \equiv 1$ and $\langle t_3, t_1 \rangle \equiv \langle t_3, t_2 \rangle \equiv 0$ to see that the vector $\dot{t}_3$ is always orthogonal to t_3 and to t_1, and that $\langle \dot{t}_3, t_2 \rangle \equiv -\kappa_2$. By the above, if σ is 4-regular (and $n \geq 5$), we can find $\kappa_3 \colon I \to \mathbb{R}^+$ and $t_4 \colon I \to \mathbb{R}^n$ such that t_4 is always a versor orthogonal to t_1, t_2 and t_3, and

$$\dot{t}_3 = -\kappa_2 t_2 + \kappa_3 t_4$$

holds with κ_3 nowhere zero. In particular,

$$\mathrm{Span}(t_1, t_2, t_3, t_4) = \mathrm{Span}(t_1, t_2, t_3, \dot{t}_3) = \mathrm{Span}\left(t_1, \dot{t}_1, \frac{d^2 t_1}{ds^2}, \frac{d^3 t_1}{ds^3}\right)$$

and

$$\frac{d^4 t_1}{ds^4} - \kappa_1 \kappa_2 \kappa_3 \dot{t}_4 \in \mathrm{Span}(t_1, t_2, t_3, t_4) \; ;$$

so $\dot{t}_4$ is in the subspace spanned by t_1, t_2, t_3 and t_4 if and only if σ is not 5-regular.

In an analogous way, if σ is $(n-2)$-regular, we may construct $n-1$ maps $t_1, \ldots, t_{n-1} \colon I \to \mathbb{R}^n$ and $n-2$ positive functions $\kappa_1, \ldots, \kappa_{n-2} \colon I \to \mathbb{R}^+$ such that $t_1, \ldots, t_{n-1}$ are always pairwise orthogonal vectors so that

$$\mathrm{Span}(t_1, \ldots, t_j) = \mathrm{Span}\left(t_1, \dot{t}_1, \ldots, \frac{d^{j-1} t_1}{ds^{j-1}}\right)$$

for $j = 1, \ldots, n-1$, and

$$\dot{t}_j = -\kappa_{j-1} t_{j-1} + \kappa_j t_{j+1}$$

for $j = 1, \ldots, n-2$ (where $\kappa_0 \equiv 0$). Now there exists a unique map $t_n \colon I \to \mathbb{R}^n$ such that $\{t_1, \ldots, t_n\}$ is always an orthonormal basis of $\mathbb{R}^n$ with the same orientation as the canonical basis. If σ is $(n-1)$-regular, then $\dot{t}_{n-1}$ does not belong (why?) to the subspace spanned by $t_1, \ldots, t_{n-1}$; therefore we can find a not necessarily positive function $\kappa_{n-1} \colon I \to \mathbb{R}$ such that

$$\dot{t}_{n-1} = -\kappa_{n-2} t_{n-2} + \kappa_{n-1} t_n \qquad \text{and} \qquad \dot{t}_n = -\kappa_{n-1} t_{n-1} \; .$$

We summarize our work so far with the following:

Definition 1.7.3. Let $\sigma \colon I \to \mathbb{R}^n$ be a $(n-1)$-regular curve of class C^k (with $k \geq n$), parametrized by arc length. The maps $t_1, \ldots, t_n \colon I \to \mathbb{R}^n$ we have just defined constitute the *Frenet frame* of σ. For $j = 1, \ldots, n-1$ the map t_j is of class C^{k-j}, while t_n is of class $C^{k-(n-1)}$. Moreover, for $j = 1, \ldots, n-1$

the function $\kappa_j\colon I \to \mathbb{R}$ is the *j-th curvature* of σ, and is of class $C^{k-(j+1)}$. The curvatures are all positive valued functions, except the last one, which may assume arbitrary real values. The formulas

$$\forall j = 1,\ldots,n \qquad \dot{\mathbf{t}}_j = -\kappa_{j-1}\mathbf{t}_{j-1} + \kappa_j\mathbf{t}_{j+1}$$

(where we have set $\kappa_0 \equiv \kappa_n \equiv 0$) are the *Frenet-Serret formulas* for σ.

Frenet-Serret formulas may be summarised in a single matrix equation, introducing the vector $\mathbf{T}$ and the antisymmetric matrix K given by

$$\mathbf{T} = \begin{vmatrix} \mathbf{t}_1 \\ \vdots \\ \mathbf{t}_n \end{vmatrix} \quad \text{and} \quad \mathsf{K} = \begin{vmatrix} 0 & \kappa_1 & & & \\ -\kappa_1 & 0 & & & \\ & & \ddots & & \\ & & & \ddots & \kappa_{n-1} \\ & & & -\kappa_{n-1} & 0 \end{vmatrix}.$$

With this notation, Frenet-Serret formulas can be expressed by the linear system of ordinary differential equations

$$\frac{\mathrm{d}}{\mathrm{d}s}\mathbf{T} = \mathsf{K}\mathbf{T}. \qquad (1.26)$$

We have now all we need to prove the *fundamental theorem of the local theory of curves in $\mathbb{R}^n$*, mimicking what we did in Theorem 1.3.37:

Theorem 1.7.4. *Given functions $\kappa_1,\ldots,\kappa_{n-2}\colon I \to \mathbb{R}^+$ and $\kappa_{n-1}\colon I \to \mathbb{R}$, where κ_j is of class $C^{k-(j+1)}$, the functions $\kappa_1,\ldots,\kappa_{n-2}$ are always positive, and $k \geq n+1$, there exists a unique (up to rigid motions) $(n-1)$-regular curve $\sigma\colon I \to \mathbb{R}^n$ of class C^k parametrized by arc length, having $\kappa_1,\ldots,\kappa_{n-1}$ as its curvatures.*

Proof. As in the proof of Theorem 1.3.37, the crucial point is that the Frenet-Serret formulas (1.26) are a linear system of ordinary differential equations in $3n$ unknowns, allowing us to apply Theorem 1.3.36.

Let us begin by studying the existence. Fix a point $s_0 \in I$ and an orthonormal basis $\{\mathbf{t}_1^0,\ldots,\mathbf{t}_n^0\}$ of $\mathbb{R}^n$ having the same orientation as the canonical basis. By Theorem 1.3.36, there exists a unique n-tuple of maps $\mathbf{t}_1,\ldots,\mathbf{t}_n\colon I \to \mathbb{R}^n$ verifying (1.26) and such that $\mathbf{t}_j(s_0) = \mathbf{t}_j^0$ for $j = 1,\ldots,n$. Moreover, it is immediate to check that $\mathbf{t}_j$ is of class C^{k-j} for $j = 1,\ldots,n-1$, while $\mathbf{t}_n$ is of class $C^{k-(n-1)}$.

We want to prove that this n-tuple $\{\mathbf{t}_1,\ldots,\mathbf{t}_n\}$ is the Frenet frame of a curve. We begin by showing that it is an orthonormal basis in every point, once we have forced it to be so in s_0. For the sake of simplicity, we introduce the following convention: given two vectors $\mathbf{S},\,\mathbf{T} \in (\mathbb{R}^n)^n$, we denote by $\langle\mathbf{S},\mathbf{T}\rangle \in M_{n,n}(\mathbb{R})$ the matrix having as its element in position (h,k) the scalar product $\langle \mathbf{s}_h, \mathbf{t}_k \rangle$ between the h-th component of $\mathbf{S}$ and the k-th component of $\mathbf{T}$. It is easy to verify that

$$\langle \mathsf{K}\mathbf{T}, \mathbf{T} \rangle = \mathsf{K}\langle \mathbf{T}, \mathbf{T} \rangle \quad \text{and} \quad \langle \mathbf{T}, \mathsf{K}\mathbf{T} \rangle = \langle \mathbf{T}, \mathbf{T} \rangle \mathsf{K}^T = -\langle \mathbf{T}, \mathbf{T} \rangle \mathsf{K} \; ;$$

in the last step we used the fact that K is antisymmetric. Hence, if we take as $\mathbf{T}$ the vector having as its components the maps $\mathbf{t}_1, \ldots \mathbf{t}_n$ that solve (1.26), the matrix of functions $\langle \mathbf{T}, \mathbf{T} \rangle$ satisfies the linear system of ordinary differential equations

$$\frac{\mathrm{d}}{\mathrm{d}s}\langle \mathbf{T}, \mathbf{T} \rangle = \left\langle \frac{\mathrm{d}\mathbf{T}}{\mathrm{d}s}, \mathbf{T} \right\rangle + \left\langle \mathbf{T}, \frac{\mathrm{d}\mathbf{T}}{\mathrm{d}s} \right\rangle = \langle \mathsf{K}\mathbf{T}, \mathbf{T} \rangle + \langle \mathbf{T}, \mathsf{K}\mathbf{T} \rangle$$
$$= \mathsf{K}\langle \mathbf{T}, \mathbf{T} \rangle - \langle \mathbf{T}, \mathbf{T} \rangle \mathsf{K} \; ,$$

with initial conditions $\langle \mathbf{T}, \mathbf{T} \rangle(s_0) = I_n$, where I_n is the identity matrix. But the matrix of functions equal to I_n everywhere satisfies the same linear system of ordinary differential equations; then, by Theorem 1.3.36, we have $\langle \mathbf{T}, \mathbf{T} \rangle \equiv I_n$, which means exactly that $\{\mathbf{t}_1, \ldots, \mathbf{t}_n\}$ is an orthonormal basis in each point. In particular, the function $\det(\mathbf{t}_1, \ldots, \mathbf{t}_n)$ is nowhere zero; being positive in s_0, it is positive everywhere, and the basis $\{\mathbf{t}_1, \ldots, \mathbf{t}_n\}$ has always the same orientation as the canonical basis.

Finally, define the curve $\sigma \colon I \to \mathbb{R}^n$, by setting

$$\sigma(s) = \int_{s_0}^{s} \mathbf{t}_1(t) \, \mathrm{d}t \; .$$

The curve σ is of class C^k; its derivative is $\mathbf{t}_1(s)$, so it is regular, is parametrized by arc length and has tangent versor $\mathbf{t}_1$. By formulas (1.26) we know then that σ is $(n-1)$-regular with curvatures $\kappa_1, \ldots, \kappa_{n-1}$, and existence is achieved.

The uniqueness can be proved as in Theorem 1.3.37. Let $\sigma_1 \colon I \to \mathbb{R}^n$ be another $(n-1)$-regular curve of class C^k parametrized by arc length, having curvatures $\kappa_1, \ldots, \kappa_{n-1}$. Fix $s_0 \in I$; up to a rigid motion, we may assume that $\sigma(s_0) = \sigma_1(s_0)$ and that σ and σ_1 have the same Frenet frame in s_0. By the uniqueness of the solution of (1.26), it follows that σ and σ_1 have the same Frenet frame in every point of I; in particular, $\dot{\sigma} \equiv \dot{\sigma}_1$. But then

$$\sigma(s) = \sigma(s_0) + \int_{s_0}^{s} \dot{\sigma}(t) \, \mathrm{d}t = \sigma_1(s_0) + \int_{s_0}^{s} \dot{\sigma}_1(t) \, \mathrm{d}t = \sigma_1(s) \; ,$$

and $\sigma_1 \equiv \sigma$. $\qquad\square$

Exercises

1.103. Prove Remark 1.7.2.

1.104. Is the curve $\sigma: \mathbb{R} \to \mathbb{R}^4$ given by $\sigma(t) = (t, t^2, t^3, t^4)$ 3-regular?

1.105. Show that a curve $\sigma: [a, b] \to \mathbb{R}^n$ whose support is contained in an affine subspace of dimension k of $\mathbb{R}^n$ has k-th curvature identically zero.

1.106. If $\sigma: \mathbb{R} \to \mathbb{R}^4$ is given by $\sigma(t) = (3t - t^3, 3, 3t^2, 3t + t^3)$, determine the curvatures of σ.

1.107. Determine, for $a \in \mathbb{R}$, the curvatures of the curve $\sigma: \mathbb{R} \to \mathbb{R}^4$, parametrized by $\sigma(t) = \big(a(1 + \cos t), a \sin t, 2a \sin \frac{t}{2}, a\big)$.

1.108. Let $\sigma: (0, +\infty) \to \mathbb{R}^4$ be the curve given by $\sigma(t) = (t, 2t, t^4, t)$.

(i) Prove that σ is a 3-regular curve, determine its Frenet frame, and compute all its curvatures.
(ii) Note that the orthogonal projection of σ on the plane $x_4 = 0$ is a curve contained in an affine space S of dimension 2 (see also Exercise 1.31). Determine explicitly the equations of S.

1.109. Let $\sigma: \mathbb{R} \to \mathbb{R}^n$ be a curve such that $\sigma'' \equiv 0$. What can we say of σ?

2

Global theory of plane curves

In the previous chapter we concentrated our attention on *local* properties of curves, that is, on properties that can be studied looking at the behavior of a curve in the neighborhood of a point. In this chapter, on the contrary, we want to present some results in the global theory of plane curves, that is, results that involve (mainly but not exclusively topological) properties of the support of the curve as a whole.

The first section of the chapter is a short introduction to the theory of the degree for maps with values in the circle S^1. As you will see, it is possible to assign an integer number, the *degree*, to each closed curve having its support in a circle, counting the number of full turns made by the curve. Using the degree, we shall be able to assign to any plane curve two more integer numbers: the *winding number* and the *rotation index*. The winding number is one of the two key ingredients needed to prove, in Section 2.3, the first main result of this chapter, the Jordan curve theorem. The rotation index is the key ingredient for the proof, in Section 2.4, of the second main result of this chapter, Hopf's theorem of turning tangents. Section 2.2 is devoted to the construction of the tubular neighborhood of a simple curve, the second key ingredient for the proof of the Jordan curve theorem. Finally, in the supplementary material of this chapter, you will find further results in global geometry of curves, such as the characterization of curves with oriented curvature of constant sign, the four-vertex theorem, the isoperimetric inequality, and Schönflies' theorem.

2.1 The degree of curves in S^1

Denote by $S^1 \subset \mathbb{R}^2$ the unitary circle in the plane, that is, the set of points $(x, y) \in \mathbb{R}^2$ at distance 1 from the origin:

$$S^1 = \{(x, y) \in \mathbb{R}^2 \mid x^2 + y^2 = 1\}.$$

Abate M., Tovena F.: Curves and Surfaces.
DOI 10.1007/978-88-470-1941-6_2, © Springer-Verlag Italia 2012

By identifying $\mathbb{R}^2$ with the complex plane $\mathbb{C}$ in the usual way, that is, by associating $z = x + iy \in \mathbb{C}$ with the pair consisting of its real and imaginary parts $(\operatorname{Re} z, \operatorname{Im} z) = (x, y) \in \mathbb{R}^2$, we may describe S^1 as the set of complex numbers with modulus 1:

$$S^1 = \{z \in \mathbb{C} \mid |z| = 1\}.$$

In what follows it will be useful to keep in mind both these descriptions.

One of the fundamental tools for the study of the geometry of S^1 is the periodic parametrization $\hat{\pi}\colon \mathbb{R} \to S^1$ given by

$$\hat{\pi}(x) = (\cos x, \sin x),$$

or, if we consider S^1 in the complex plane, by $\hat{\pi}(x) = \exp(ix)$. Note that saying that $\hat{\pi}(x) = v \in S^1$ is equivalent to saying that $x \in \mathbb{R}$ is a possible *determination of the angle* from the x-axis to vector v. In particular,

$$\hat{\pi}(x_1) = \hat{\pi}(x_2) \iff x_1 - x_2 \in 2\pi\mathbb{Z}, \tag{2.1}$$

that is, two determinations of the angle of the same point in S^1 always differ by an integer multiple of 2π.

We shall use the periodic parametrization $\hat{\pi}$ to transfer problems from S^1 to $\mathbb{R}$, lifting maps:

Definition 2.1.1. If $\phi\colon X \to S^1$ is a continuous map from a topological space X with values in S^1, a *lift* of ϕ is a continuous map $\tilde{\phi}\colon X \to \mathbb{R}$ such that $\hat{\pi} \circ \tilde{\phi} = \phi$, that is such that the diagram

is commutative.

In terms of angles, having a lift $\tilde{\phi}$ of a continuous map $\phi\colon X \to S^1$ means we are able to associate in a continuous way to every $x \in X$ a determination of the angle between the x-axis and $\phi(x)$; for this reason we shall sometimes call *continuous determination of the angle* a lift of ϕ.

Not all maps with values in S^1 admit a lift (see Exercise 2.1). However, there is a particularly important case in which lifts always exist: curves. To prove this, we need an interesting General Topology result:

Theorem 2.1.2. *Let* $\mathfrak{U} = \{U_\alpha\}_{\alpha \in A}$ *be an open cover of a compact metric space* (X, d). *Then there exists a number* $\delta > 0$ *such that for all* $x \in X$ *there is* $\alpha \in A$ *such that* $B_d(x, \delta) \subset U_\alpha$, *where* $B_d(x, \delta)$ *is the open ball with center* x *and radius* δ *with respect to distance* d.

Proof. Let $\{U_1, \ldots, U_n\}$ be a fixed finite subcover of $\mathfrak{U}$. For all $\alpha = 1, \ldots, n$ define the continuous function $f_\alpha \colon X \to \mathbb{R}$ by setting

$$f_\alpha(x) = d(x, X \setminus U_\alpha),$$

and set $f = \max\{f_1, \ldots, f_n\}$. The function f is continuous; moreover, for each $x \in X$ we can find $1 \le \alpha \le n$ such that $x \in U_\alpha$, and so $f(x) \ge f_\alpha(x) > 0$. Hence, $f > 0$ everywhere; let $\delta > 0$ be the minimum of f in X. Then for all $x \in X$ we can find $1 \le \alpha \le n$ such that $f_\alpha(x) \ge \delta$, and so the open ball with center x and radius δ is completely contained in U_α, as required. □

Definition 2.1.3. Let $\mathfrak{U} = \{U_\alpha\}_{\alpha \in A}$ be an open cover of a compact metric space X. The largest $\delta > 0$ such that for all $x \in X$ we can find $\alpha \in A$ with $B(x, \delta) \subset U_\alpha$ is the *Lebesgue number* of the cover $\mathfrak{U}$.

We are now able to prove the existence of lifts of curves:

Proposition 2.1.4. *Let $\phi \colon [a, b] \to S^1$ be a continuous curve, and let $x_0 \in \mathbb{R}$ be such that $\hat{\pi}(x_0) = \phi(a)$. Then there exists a unique lift $\tilde{\phi} \colon [a, b] \to \mathbb{R}$ of ϕ such that $\tilde{\phi}(a) = x_0$.*

Proof. We need a preliminary remark. Consider an arbitrary $p \in S^1$, and let $x \in \mathbb{R}$ be such that $\hat{\pi}(x) = p$. Then

$$\hat{\pi}^{-1}(S^1 \setminus \{p\}) = \bigcup_{k \in \mathbb{Z}} (x + 2(k-1)\pi, x + 2k\pi),$$

and $\hat{\pi}|_{(x+2(k-1)\pi, x+2k\pi)} \colon (x + 2(k-1)\pi, x + 2k\pi) \to S^1 \setminus \{p\}$ is a homeomorphism for all $k \in \mathbb{Z}$; this is the crucial property of $\hat{\pi}$ that makes it possible to lift curves.

Set now $p_0 = \phi(a)$ as well as $U = S^1 \setminus \{p_0\}$ and $V = S^1 \setminus \{-p_0\}$. Clearly, $S^1 = U \cup V$, so $\mathfrak{U} = \{\phi^{-1}(U), \phi^{-1}(V)\}$ is an open cover of $[a, b]$; by Theorem 2.1.2 we know that if $\mathcal{P} = (t_0, \ldots, t_n)$ is a partition of $[a, b]$ with $\|\mathcal{P}\|$ less than the Lebesgue number of $\mathfrak{U}$, then $\phi([t_{j-1}, t_j]) \subset U$ or $\phi([t_{j-1}, t_j]) \subset V$ for all $j = 1, \ldots, n$.

Since $\phi(a) = p_0 \notin U$, necessarily $\phi([t_0, t_1]) \subset V$. Now,

$$\hat{\pi}^{-1}(V) = \bigcup_{k \in \mathbb{Z}} (x_0 + (2k-1)\pi, x_0 + (2k+1)\pi);$$

so, the image of a lift $\tilde{\phi}|_{[a, t_1]}$ of $\phi|_{[a, t_1]}$ with $\tilde{\phi}(a) = x_0$, being connected, is necessarily contained in the interval $(x_0 - \pi, x_0 + \pi)$. But $\hat{\pi}$ restricted to $(x_0 - \pi, x_0 + \pi)$ is a homeomorphism between $(x_0 - \pi, x_0 + \pi)$ and V; so a lift $\tilde{\phi}|_{[a, t_1]}$ exists and is necessarily given by

$$\tilde{\phi}|_{[a, t_1]} = \left(\hat{\pi}|_{(x_0 - \pi, x_0 + \pi)}\right)^{-1} \circ \phi|_{[a, t_1]}.$$

Assume now we have proved that there is a unique lift $\tilde{\phi}|_{[a,t_j]}$ of $\phi|_{[a,t_j]}$ with $\tilde{\phi}(a) = x_0$: we are going to show that there is a unique way to extend it to a lift $\tilde{\phi}|_{[a,t_{j+1}]}$ of $\phi|_{[a,t_{j+1}]}$. It suffices to proceed as before: since $\phi([t_j, t_{j+1}]) \subset U$ or $\phi([t_j, t_{j+1}]) \subset V$, the image of a lift $\tilde{\phi}|_{[t_j,t_{j+1}]}$ of $\phi|_{[t_j,t_{j+1}]}$ that extends continuously $\tilde{\phi}|_{[a,t_j]}$ must be contained in an interval I of $\mathbb{R}$ on which $\hat{\pi}$ is a homeomorphism with U or V. It then follows that this lift exists, is unique and is given by $\tilde{\phi}|_{[t_j,t_{j+1}]} = (\hat{\pi}|_I)^{-1} \circ \phi|_{[t_j,t_{j+1}]}$; and so we have extended the lift to the interval $[a, t_{j+1}]$. By proceeding in this way till we reach (in finitely many steps) the endpoint b, we prove the assertion. $\square$

Once we have found one lift of a continuous map, it is easy to find all other lifts:

Proposition 2.1.5. *Assume that $\tilde{\phi}_1$, $\tilde{\phi}_2 \colon X \to \mathbb{R}$ are two lifts of a continuous maps $\phi \colon X \to S^1$, where X is a connected topological space. Then there exists $k \in \mathbb{Z}$ such that $\tilde{\phi}_2 - \tilde{\phi}_1 \equiv 2k\pi$.*

Proof. Since $\tilde{\phi}_1$ and $\tilde{\phi}_2$ are two lifts of ϕ, we have

$$\hat{\pi} \circ \tilde{\phi}_1 \equiv \phi \equiv \hat{\pi} \circ \tilde{\phi}_2 \,.$$

By (2.1), it follows that the continuous map $\tilde{\phi}_2 - \tilde{\phi}_1$ is defined on a connected topological space and has values in $2\pi\mathbb{Z}$, which is a totally disconnected topological space; so it is constant, as required. $\square$

In particular, if $\phi \colon [a, b] \to S^1$ is a continuous curve, then the number

$$\tilde{\phi}(b) - \tilde{\phi}(a)$$

is the same for any lift $\tilde{\phi} \colon [a, b] \to \mathbb{R}$ di ϕ. Moreover, if ϕ is a *closed* curve, that is, such that $\phi(b) = \phi(a)$, then (2.1) implies that $\tilde{\phi}(b) - \tilde{\phi}(a)$ is necessarily an integer multiple of 2π, since

$$\hat{\pi}(\tilde{\phi}(b)) = \phi(b) = \phi(a) = \hat{\pi}(\tilde{\phi}(a)) \,.$$

So we may introduce the following:

Definition 2.1.6. Let $\phi \colon [a, b] \to S^1$ be a closed continuous curve. The *degree* of ϕ is the integer number

$$\deg \phi = \frac{1}{2\pi}(\tilde{\phi}(b) - \tilde{\phi}(a)) \in \mathbb{Z} \,,$$

where $\tilde{\phi} \colon [a, b] \to \mathbb{R}$ is any lift of ϕ.

In other words, the degree measures the number of full turns the curve ϕ makes before closing, where a full turn is counted with a positive sign when ϕ travels counterclockwise, and with a negative sign otherwise.

Example 2.1.7. A constant curve has degree zero, since any lift is constant.

Example 2.1.8. Given $k \in \mathbb{Z}$, let $\phi_k \colon [0,1] \to S^1$ be given by

$$\phi_k(t) = \big(\cos(2k\pi t), \sin(2k\pi t)\big) \, .$$

Clearly, a lift of ϕ_k is given by $\tilde{\phi}_k(t) = 2k\pi t$, so

$$\deg \phi_k = \frac{1}{2\pi}\big(\tilde{\varphi}_k(1) - \tilde{\varphi}_k(0)\big) = k \, .$$

Later (Corollary 2.1.18) we shall see formulas for computing the degree of closed piecewise C^1 curves with values in S^1; but we want to give first an important necessary and sufficient condition for two closed curves in S^1 to have the same degree. To get there, we need the fundamental notion of homotopy between maps.

Definition 2.1.9. A *homotopy* between two continuous maps between topological spaces $\phi_0, \phi_1 \colon Y \to X$ is a continuous map $\Phi \colon [0,1] \times Y \to X$ such that $\Phi(0,\cdot) \equiv \phi_0$ and $\Phi(1,\cdot) \equiv \phi_1$. If a homotopy between ϕ_0 and ϕ_1 exists, ϕ_0 and ϕ_1 are said to be *homotopic*. If $Y = [a,b]$ is a real interval and the curves ϕ_0 and ϕ_1 are closed, that is, $\phi_0(a) = \phi_0(b)$ and $\phi_1(a) = \phi_1(b)$, then we shall always ask Φ to be a homotopy *of closed curves*, that is, with $\Phi(t,a) = \Phi(t,b)$ for all $t \in [0,1]$.

In other words, two maps ϕ_0 and ϕ_1 are homotopic if it is possible to go continuously from one to the other, that is, if ϕ_0 can be deformed in a continuous way to obtain ϕ_1. Note that being homotopic is an equivalence relation (see Exercise 2.2).

Example 2.1.10. If X if a path-connected space, two constant maps with values in X are always homotopic. Indeed, let $\phi_0 \colon Y \to X$ and $\phi_1 \colon Y \to X$ be given by $\phi_j(y) = x_j$ for all $y \in Y$, where x_0 and x_1 are arbitrary points of X. Let $\sigma \colon [0,1] \to X$ be a continuous curve with $\sigma(0) = x_0$ and $\sigma(1) = x_1$; then a homotopy $\Phi \colon [0,1] \times Y \to X$ between ϕ_0 and ϕ_1 is given by

$$\forall t \in [0,1] \, \forall y \in Y \qquad \Phi(t,y) = \sigma(t) \, .$$

Example 2.1.11. Every non-surjective closed continuous curve $\phi_0 \colon [a,b] \to S^1$ is homotopic (as a closed curve) to a constant curve. Indeed, let $p \in S^1$ be a point not in the image of ϕ_0, and consider $x^* \in [0, 2\pi)$ such that $\hat{\pi}(x^*) = p$. Now, choose a point $x_0 \in (x^* - 2\pi, x^*)$ such that $\hat{\pi}(x_0) = \phi_0(a)$, and let $\tilde{\phi}_0 \colon [a,b] \to \mathbb{R}$ be the unique lift of ϕ_0 with $\tilde{\phi}_0(a) = x_0$. Since the image of ϕ_0 does not contain p, the image of $\tilde{\phi}_0$ cannot include x^* or $x^* - 2\pi$; so, it is completely contained in the interval $(x^* - 2\pi, x^*)$ and, in particular, $\phi_0(a) = \phi_0(b)$ implies $\tilde{\phi}_0(b) = x_0$. Let now $\Phi \colon [0,1] \times [a,b] \to S^1$ be defined by

$$\Phi(s,t) = \hat{\pi}\big(\tilde{\phi}_0(t) + s(x_0 - \tilde{\phi}_0(t))\big) \, .$$

It is straightforward to verify that $\Phi(0,\cdot) \equiv \phi_0$, $\Phi(1,\cdot) \equiv \phi_0(a)$, and $\Phi(\cdot,a) \equiv \Phi(\cdot,b) \equiv \phi_0(a)$; hence Φ is a homotopy of closed curves between ϕ_0 and the constant curve $\phi_1 \equiv \phi_0(a)$.

Example 2.1.12. There are surjective closed continuous curves that are homotopic to a constant. For instance, let $\phi_0\colon [0, 4\pi] \to S^1$ be given by

$$\phi_0(t) = \begin{cases} (\cos t, \sin t) & \text{for } t \in [0, \pi]\,, \\ (\cos(2\pi - t), \sin(2\pi - t)) & \text{for } t \in [\pi, 3\pi]\,, \\ (\cos(t - 4\pi), \sin(t - 4\pi)) & \text{for } t \in [3\pi, 4\pi]\,. \end{cases}$$

Then $\Phi\colon [0, 1] \times [0, 4\pi] \to S^1$, given by

$$\Phi(s,t) = \begin{cases} \big(\cos[(1 - s)t], \sin[(1 - s)t]\big) & \text{for } t \in [0, \pi]\,, \\ \big(\cos[(1 - s)(2\pi - t)], \sin[(1 - s)(2\pi - t)]\big) & \text{for } t \in [\pi, 3\pi]\,, \\ \big(\cos[(1 - s)(t - 4\pi)], \sin[(1 - s)(t - 4\pi)]\big) & \text{for } t \in [3\pi, 4\pi]\,, \end{cases}$$

is a homotopy of closed curves between ϕ_0 and the constant curve $\phi_1 \equiv (1, 0)$.

One of the main reasons why the degree is important is that *two closed curves in S^1 are homotopic if and only if they have the same degree.* To prove this, we need the following:

Proposition 2.1.13. *Let $\Phi\colon [0, 1] \times [a, b] \to S^1$ be a homotopy between continuous curves in S^1, and $x_0 \in \mathbb{R}$ such that $\hat{\pi}(x_0) = \Phi(0, a)$. Then there exists a unique lift $\tilde{\Phi}\colon [0, 1] \times [a, b] \to \mathbb{R}$ of Φ such that $\tilde{\Phi}(0, a) = x_0$.*

Proof. The argument is analogous to the one used in the proof of Proposition 2.1.4. Set $p_0 = \Phi(0, a)$, $U = S^1 \setminus \{p_0\}$, and $V = S^1 \setminus \{-p_0\}$. By Theorem 2.1.2, we know that it is possible to partition $[0, 1] \times [a, b]$ in finitely many small squares sent by Φ entirely inside U or V. Define now the lift $\tilde{\Phi}$ one square at a time, starting from the bottom left corner and going from left to right and from the bottom up, using the procedure described in the proof of Proposition 2.1.4, taking always as starting point the bottom left corner of each square. Now, again by Proposition 2.1.4, we know that, having fixed the value in the bottom left corner, the lift is uniquely determined along the bottom side and along the left side of each square; since each square iintersects the previous ones along these two sides only, this ensures (why?) that the lift $\tilde{\Phi}$ thus obtained is globally continuous, as well as unique. $\qquad\square$

Corollary 2.1.14. *Let T be a topological space homeomorphic to a rectangle $[0, 1] \times [a, b]$, and $\Psi\colon T \to S^1$ a continuous map. Then, for all $t_0 \in \partial T$ and all $x_0 \in \mathbb{R}$ such that $\hat{\pi}(x_0) = \Psi(t_0)$ there exists a unique lift $\tilde{\Psi}\colon T \to \mathbb{R}$ of Ψ such that $\tilde{\Psi}(t_0) = x_0$.*

Proof. Indeed, since T is homeomorphic to $[0, 1] \times [a, b]$, we may find (why?) a homeomorphism $h\colon T \to [0, 1] \times [a, b]$ with $h(t_0) = (0, a)$; then it suffices to define $\tilde{\Psi} = \tilde{\Phi} \circ h$, where $\tilde{\Phi}\colon [0, 1] \times [a, b] \to \mathbb{R}$ is the unique lift of $\Phi = \Psi \circ h^{-1}$ with $\tilde{\Phi}(0, a) = x_0$, and the uniqueness of $\tilde{\Psi}$ follows from the uniqueness of $\tilde{\Phi}$. $\qquad\square$

And then, as promised:

Theorem 2.1.15. *Two closed curves ϕ_0, $\phi_1\colon [a,b] \to S^1$ are homotopic (as closed curves) if and only if they have the same degree. In particular, a closed curve is homotopic to a constant if and only if it has degree 0.*

Proof. Let $\Phi\colon [0,1] \times [a,b] \to S^1$ be a homotopy of closed curves between ϕ_0 and ϕ_1, and set $\phi_s(t) = \Phi(s,t)$; in particular, all ϕ_s's are closed curves. Lift Φ to a $\tilde{\Phi}\colon [0,1] \times [a,b] \to \mathbb{R}$. Since the curves ϕ_s are closed, we have $\tilde{\Phi}(s,b) - \tilde{\Phi}(s,a) \in 2\pi\mathbb{Z}$ for all $s \in [0,1]$. In particular, $s \mapsto \tilde{\Phi}(s,b) - \tilde{\Phi}(s,a)$ is a continuous function with values in a totally disconnected space; so it is necessarily constant, and

$$2\pi \deg \phi_0 = \tilde{\Phi}(0,b) - \tilde{\Phi}(0,a) = \tilde{\Phi}(1,b) - \tilde{\Phi}(1,a) = 2\pi \deg \phi_1 \ .$$

For the converse, assume that ϕ_0 and ϕ_1 have the same degree $k \in \mathbb{Z}$. This means that if we take an arbitrary lift $\tilde{\phi}_0$ of ϕ_0 and an arbitrary lift $\tilde{\phi}_1$ of ϕ_1, we have

$$\tilde{\phi}_1(b) - \tilde{\phi}_1(a) = 2k\pi = \tilde{\phi}_0(b) - \tilde{\phi}_0(a) \ ,$$

and so

$$\tilde{\phi}_1(b) - \tilde{\phi}_0(b) = \tilde{\phi}_1(a) - \tilde{\phi}_0(a) \ . \tag{2.2}$$

Define now $\Phi\colon [0,1] \times [a,b] \to S^1$ by setting

$$\Phi(s,t) = \hat{\pi}\big(\tilde{\phi}_0(t) + s\big(\tilde{\phi}_1(t) - \tilde{\phi}_0(t)\big)\big) \ .$$

It is immediate to verify that, thanks to (2.2), Φ is a homotopy of closed curves between ϕ_0 and ϕ_1, as desired. Finally, the last claim follows from Example 2.1.7. $\qquad\square$

We conclude this section by showing how it is possible to give an integral formula for computing the degree of differentiable curves, a formula we shall find very useful later on.

Proposition 2.1.16. *Let $\phi = (\phi_1,\phi_2)\colon [a,b] \to S^1$ be a C^1 curve, and choose $x_0 \in \mathbb{R}$ in such a way that $\phi(a) = (\cos x_0, \sin x_0)$. Then the function $\tilde{\phi}\colon [a,b] \to \mathbb{R}$ given by*

$$\tilde{\phi}(t) = x_0 + \int_a^t (\phi_1\phi_2' - \phi_1'\phi_2)\, ds$$

is the lift of ϕ such that $\tilde{\phi}(a) = x_0$.

Proof. We must show that $\cos\tilde{\phi} \equiv \phi_1$ and $\sin\tilde{\phi} \equiv \phi_2$, that is, that

$$0 \equiv (\phi_1 - \cos\tilde{\phi})^2 + (\phi_2 - \sin\tilde{\phi})^2 = 2 - 2(\phi_1\cos\tilde{\phi} + \phi_2\sin\tilde{\phi}) \ ;$$

so it suffices to verify that $\phi_1 \cos \tilde{\phi} + \phi_2 \sin \tilde{\phi} \equiv 1$. This equality is true when $t = a$; so it is enough to check that the derivative of $\phi_1 \cos \tilde{\phi} + \phi_2 \sin \tilde{\phi}$ is zero everywhere.

Now, differentiating $\phi_1^2 + \phi_2^2 \equiv 1$ we get

$$\phi_1 \phi_1' + \phi_2 \phi_2' \equiv 0 ; \tag{2.3}$$

then

$$
\begin{aligned}
(\phi_1 \cos \tilde{\phi} + \phi_2 \sin \tilde{\phi})' &= \phi_1' \cos \tilde{\phi} - \tilde{\phi}' \phi_1 \sin \tilde{\phi} + \phi_2' \sin \tilde{\phi} + \tilde{\phi}' \phi_2 \cos \tilde{\phi} \\
&= (\phi_1' + \phi_1 \phi_2 \phi_2' - \phi_1' \phi_2^2) \cos \tilde{\phi} + (\phi_2' + \phi_2 \phi_1 \phi_1' - \phi_2' \phi_1^2) \sin \tilde{\phi} \\
&= \phi_1'(1 - \phi_1^2 - \phi_2^2) \cos \tilde{\phi} + \phi_2'(1 - \phi_2^2 - \phi_1^2) \sin \tilde{\phi} \\
&\equiv 0 ,
\end{aligned}
$$

as required. $\square$

Definition 2.1.17. A continuous curve $\sigma \colon [a, b] \to \mathbb{R}^n$ is said to be *piecewise* C^k if there exists a partition $a = t_0 < t_1 < \cdots < t_r = b$ di $[a, b]$ such that $\sigma|_{[t_{j-1}, t_j]}$ is of class C^k for $j = 1, \ldots, r$. Moreover, we say that σ is *regular* if its restriction to each interval $[t_{j-1}, t_j]$ is regular, and that it is *parametrized by arc length* if the restriction to each interval $[t_{j-1}, t_j]$ is parametrized by arc length.

Corollary 2.1.18. *Let* $\phi = (\phi_1, \phi_2) \colon [a, b] \to S^1$ *be a piecewise* C^1 *closed continuous curve. Then*

$$\deg \phi = \frac{1}{2\pi} \int_a^b (\phi_1 \phi_2' - \phi_1' \phi_2) \, dt .$$

Proof. This follows from the previous proposition applied to each interval over which ϕ is of class C^1, and from the definition of degree. $\square$

If we identify $\mathbb{R}^2$ and $\mathbb{C}$, the previous formula has an even more compact expression:

Corollary 2.1.19. *Let* $\phi \colon [a, b] \to S^1 \subset \mathbb{C}$ *be a piecewise* C^1 *closed continuous curve. Then*

$$\deg \phi = \frac{1}{2\pi i} \int_a^b \frac{\phi'}{\phi} \, dt .$$

Proof. Since ϕ has values in S^1, we have $1/\phi = \bar{\phi}$, where $\bar{\phi}$ is the complex conjugate of ϕ. By writing $\phi = \phi_1 + i\phi_2$, we get

$$\phi' \bar{\phi} = (\phi_1 \phi_1' + \phi_2 \phi_2') + i(\phi_1 \phi_2' - \phi_1' \phi_2) = i(\phi_1 \phi_2' - \phi_1' \phi_2) ,$$

by (2.3), and the assertion follows from the previous corollary. $\square$

Example 2.1.20. Let us use this formula to compute again the degree of the curves $\phi_k \colon [0,1] \to S^1$ defined in Example 2.1.8. Using complex numbers, we may write $\phi_k(t) = \exp(2k\pi \mathrm{i} t)$; so $\phi_k'(t) = 2k\pi \mathrm{i} \exp(2k\pi \mathrm{i} t)$, and

$$\deg \phi_k = \frac{1}{2\pi \mathrm{i}} \int_0^1 \frac{2k\pi \mathrm{i} \exp(2k\pi \mathrm{i} t)}{\exp(2k\pi \mathrm{i} t)} \, \mathrm{d}t = k \ .$$

Example 2.1.21. We want to verify that the curve $\phi \colon [0, 4\pi] \to S^1$ of Example 2.1.12 has indeed degree zero. Using complex numbers, we may write

$$\phi_0(t) = \begin{cases} \exp(\mathrm{i} t) & \text{per } t \in [0, \pi] \, , \\ \exp\big(\mathrm{i}(2\pi - t)\big) & \text{per } t \in [\pi, 3\pi] \, , \\ \exp\big(\mathrm{i}(t - 4\pi)\big) & \text{per } t \in [3\pi, 4\pi] \, ; \end{cases}$$

hence,

$$\deg \phi = \frac{1}{2\pi \mathrm{i}} \left(\int_0^\pi \frac{\phi'}{\phi} \, \mathrm{d}t + \int_\pi^{3\pi} \frac{\phi'}{\phi} \, \mathrm{d}t + \int_{3\pi}^{4\pi} \frac{\phi'}{\phi} \, \mathrm{d}t \right) = \frac{1}{2\pi \mathrm{i}} (\pi \mathrm{i} - 2\pi \mathrm{i} + \pi \mathrm{i}) = 0 \ .$$

2.2 Tubular neighborhoods

Now that preliminaries are over we may introduce the class of curves we shall deal with most in this chapter (see also Definition 1.1.28).

Definition 2.2.1. A curve $\sigma \colon [a, b] \to \mathbb{R}^n$ is *simple* if it is injective on $[a, b)$ and on $(a, b]$. A *Jordan curve* is a closed simple continuous plane curve.

Fig. 2.1 shows examples of non-simple, simple and Jordan curves.

Remark 2.2.2. A non-closed simple continuous curve $\sigma \colon [a, b] \to \mathbb{R}^n$ is a homeomorphism with its image, so it is a closed Jordan arc (see Definition 1.1.27). Indeed, it is globally injective, and every bijective continuous map from a compact space to a Hausdorff space is a homeomorphism. Arguing in an analogous way, it is easy to prove (see Exercise 2.12) that the support of a Jordan curve is homeomorphic to the circle S^1.

One of the main properties of Jordan curves is (not surprisingly) the *Jordan curve theorem*, which states that the support of a closed simple plane curve divides the plane in exactly two components (Theorem 2.3.6). This is one of those results that seem obvious at first sight, but is actually very deep and hard to prove (in fact, it is not even completely obvious that the curve in Fig. 2.1.(d) divides the plane in just two parts...).

In this and in next section we shall give a proof of Jordan curve theorem which holds for regular curves of class (at least) C^2, using just tools from differential geometry and topology (a proof for continuous curves is given in Section 2.8 of the supplementary material to this chapter). As you will see,

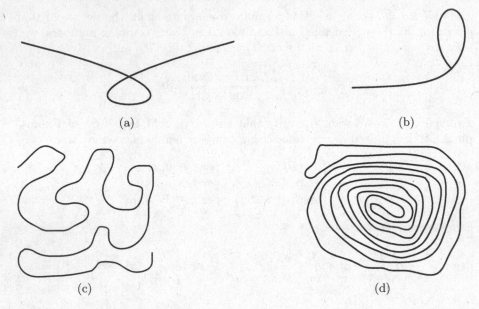

Fig. 2.1. (a), (b) non-simple curves; (c) simple curve; (d) Jordan curve

we shall need two ingredients for the proof: a tubular neighborhood of the support of a curve (to prove that the complement of the support of a Jordan curve has *at most* two components), and the winding number (to prove that the complement of the support of a Jordan curve as *at least* two components). The goal of this section is to introduce the first ingredient.

If $\sigma\colon I \to \mathbb{R}^2$ is a regular plane curve, and $\varepsilon > 0$, for all $t \in I$ we may take a small line segment of length 2ε with center in $\sigma(t)$ and orthogonal to $\sigma'(t)$. The union of all these small segments together always is a neighborhood of the support of the curve (with the possible exception of its endpoints); however, it is well possible that segments for distinct t's intersect. The main result of this section says that, if σ is a simple curve of class C^2, then we can find $\varepsilon > 0$ such that these small segments are pairwise disjoint, thus forming a particularly useful neighborhood of the support of the curve, called *tubular neighborhood* (see Fig. 2.2).

Definition 2.2.3. Let $\sigma\colon [a,b] \to \mathbb{R}^2$ be a simple regular plane curve, with support $C = \sigma([a,b])$. Given $\varepsilon > 0$ and $p = \sigma(t) \in C$, we denote by $I_\sigma(p,\varepsilon)$ the line segment $\sigma(t) + (-\varepsilon, \varepsilon)\tilde{\mathbf{n}}(t)$ having length 2ε with center in p and orthogonal to $\sigma'(t)$, where $\tilde{\mathbf{n}}(t)$ is the oriented normal versor of σ at $\sigma(t)$. Denote further by $N_\sigma(\varepsilon)$ the union of the segments $I_\sigma(p,\varepsilon)$, with $p \in C$. The set $N_\sigma(\varepsilon)$ is a *tubular neighborhood* of σ if ε is such that $I_\sigma(p_1,\varepsilon) \cap I_\sigma(p_2,\varepsilon) = \varnothing$ for all $p_1 \neq p_2 \in C$.

Tubular neighborhoods always exist. To prove it, we need a classical Differential Calculus theorem (see [5, p. 140]).

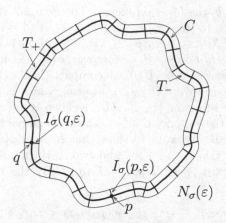

Fig. 2.2. Tubular neighborhood

Theorem 2.2.4 (Inverse function theorem). *Let* $F: \Omega \to \mathbb{R}^n$ *be a map of class* C^k, *with* $k \in \mathbb{N}^* \cup \{\infty\}$, *where* Ω *is an open subset of* $\mathbb{R}^n$. *Let* $p_0 \in \Omega$ *be such that* $\det \operatorname{Jac} F(p_0) \neq 0$, *where* $\operatorname{Jac} F$ *is the Jacobian matrix of* F. *Then there exist a neighborhood* $U \subset \Omega$ *of* p_0 *and a neighborhood* $V \subset \mathbb{R}^n$ *of* $F(p_0)$ *such that* $F|_U: U \to V$ *is a diffeomorphism of class* C^k.

Then:

Theorem 2.2.5 (Existence of tubular neighborhoods). *Let* $C \subset \mathbb{R}^2$ *be the support of a simple regular plane curve* $\sigma: [a, b] \to \mathbb{R}^2$ *of class* C^2. *Then there exists* $\varepsilon_0 > 0$ *such that* $N_\sigma(\varepsilon)$ *is a tubular neighborhood of* σ *for all* $0 < \varepsilon \le \varepsilon_0$. *In particular,* $N_\sigma(\varepsilon)$ *is an open neighborhood of* C *(endpoints excluded if the curve is not closed).*

Proof. First of all, we recall that a curve σ is of class C^2 in $[a, b]$ if it can be extended to a C^2 map defined in an open neighborhood I of $[a, b]$. In particular, if σ is closed we may extend it to a periodic C^2 map defined on all $\mathbb{R}$.

We begin by proving the local existence of the tubular neighborhood. Define a map $F: I \times \mathbb{R} \to \mathbb{R}^2$ by setting

$$F(t, x) = \sigma(t) + x\tilde{\mathbf{n}}(t) , \tag{2.4}$$

in such a way that $N_\sigma(\varepsilon) = F([a, b] \times (-\varepsilon, \varepsilon))$. Being a plane curve, the regularity of the normal versor $\tilde{\mathbf{n}} = (\tilde{n}_1, \tilde{n}_2)$ is equivalent to the regularity of the tangent versor $\mathbf{t}$, which is of class C^1; so the map F is of class C^1. Now, the Jacobian determinant of F in $(t, 0)$ is

$$\det \begin{vmatrix} \sigma_1'(t) & \sigma_2'(t) \\ \tilde{n}_1(t) & \tilde{n}_2(t) \end{vmatrix} \neq 0 ;$$

hence for all $t_0 \in [a, b]$ the inverse function Theorem 2.2.4 yields δ_{t_0}, $\varepsilon_{t_0} > 0$ such that F restricted to $(t_0 - \delta_{t_0}, t_0 + \delta_{t_0}) \times (-\varepsilon_{t_0}, \varepsilon_{t_0})$ is invertible; and this means exactly that $I_\sigma(p_1, \varepsilon_{t_0}) \cap I_\sigma(p_2, \varepsilon_{t_0}) = \varnothing$ for all $p_1 = \sigma(t_1) \neq \sigma(t_2) = p_2$ with $t_1, t_2 \in (t_0 - \delta_{t_0}, t_0 + \delta_{t_0}) = U_{t_0}$. Moreover, $\sigma(U_{t_0})$ is an open subset of C because its complement $\sigma([a, b] \setminus U_{t_0})$ is a compact (hence closed) subset of C.

So we have an open cover $\{U_t\}_{t \in [a,b]}$ of $[a, b]$, which is a compact set; extract from it a finite subcover $\{U_{t_1}, \dots, U_{t_r}\}$. Then $\mathfrak{U} = \{\sigma(U_{t_1}), \dots, \sigma(U_{t_r})\}$ is an open cover of the support C of σ, which is compact; let $\delta > 0$ be the Lebesgue number of $\mathfrak{U}$. We want to show that $\varepsilon_0 = \min\{\varepsilon_{t_1}, \dots, \varepsilon_{t_k}, \delta/2\}$ is as required. Indeed, take $0 < \varepsilon \leq \varepsilon_0$ and two distinct points $p, q \in C$, and assume that there exist $p_0 \in I_\sigma(p, \varepsilon) \cap I_\sigma(q, \varepsilon)$. By the triangle inequality, then, we have

$$\|p - q\| \leq \|p - p_0\| + \|p_0 - q\| < 2\varepsilon \leq \delta ,$$

so p and q have to be in a same $\sigma(U_{t_j})$. But, since F is injective on $U_{t_j} \times (-\varepsilon, \varepsilon)$, the condition $I_\sigma(p, \varepsilon) \cap I_\sigma(q, \varepsilon) \neq \varnothing$ implies $p = q$, which is a contradiction, and we are done.

Now, F is globally injective on $[a, b) \times (-\varepsilon, \varepsilon)$ and on $(a, b] \times (-\varepsilon, \varepsilon)$, and $F(I_0 \times (-\varepsilon, \varepsilon))$ is an open neighborhood of C, where $I_0 = U_{t_1} \cup \cdots \cup U_{t_r}$ contains $[a, b]$. From this immediately follows (why?) that $N_\sigma(\varepsilon)$ is an open neighborhood of the support of σ, with the exception of the endpoints if σ is not closed. $\qquad\square$

Remark 2.2.6. If $N_\sigma(\varepsilon)$ is a tubular neighborhood of a simple regular plane curve $\sigma : [a, b] \to \mathbb{R}^2$ of class C^2 and $q_0 \in N_\sigma(\varepsilon)$, then the point $p_0 = \sigma(t_0)$ of the support C of σ that is closest to q_0 is the unique point $p \in C$ for which $q_0 \in I_\sigma(p, \varepsilon)$. Indeed, if the function $t \mapsto \|q_0 - \sigma(t)\|^2$ has a minimum in t_0, then by differentiating we find $\langle q_0 - \sigma(t_0), \sigma'(t_0) \rangle = 0$, and so $q_0 \in I_\sigma(p_0, \varepsilon)$.

The assumption of C^2-regularity in the theorem is crucial:

Example 2.2.7. Fix $2 < \alpha < 3$, and let $\sigma_\alpha : \mathbb{R} \to \mathbb{R}^2$ be the curve given by $\sigma_\alpha(t) = \big(t, f_\alpha(t)\big)$, where $f_\alpha : \mathbb{R} \to \mathbb{R}$ is the function

$$f_\alpha(t) = \begin{cases} t^\alpha \sin \frac{1}{t} & \text{if } t > 0 , \\ 0 & \text{if } t \leq 0 . \end{cases}$$

Since

$$\frac{d}{dt}\left(t^\alpha \sin \frac{1}{t} \right) = \alpha t^{\alpha-1} \sin \frac{1}{t} - t^{\alpha-2} \cos \frac{1}{t} ,$$

the function f_α and the curve σ_α are of class C^1, but not C^2; we want to show that σ_α (restricted to any closed interval containing the origin) has no tubular neighborhood. First of all, it is easy to see that the oriented normal versor of σ_α is given by

$$\tilde{\mathbf{n}}(t) = \frac{\left(t^{\alpha-2} \left(\cos \frac{1}{t} - \alpha t \sin \frac{1}{t} \right), 1 \right)}{\sqrt{1 + t^{2(\alpha-2)} \left(\cos \frac{1}{t} - \alpha t \sin \frac{1}{t} \right)^2}}$$

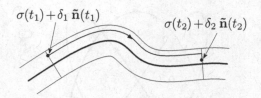

$$\sigma(t_1) + \delta_1\, \tilde{\mathbf{n}}(t_1) \qquad\qquad \sigma(t_2) + \delta_2\, \tilde{\mathbf{n}}(t_2)$$

Fig. 2.3.

for $t \geq 0$, and by $\tilde{\mathbf{n}}(t) = (0,1)$ for $t \leq 0$. If the curve σ_α had a tubular neighborhood, there would exist an $\varepsilon > 0$ such that, for all sufficiently small $t > 0$, the segment parallel to $\tilde{\mathbf{n}}(t)$ and connecting $\sigma_\alpha(t)$ to the y-axis would have length at least ε. But the length of this segment is

$$\ell(t) = t^{3-\alpha}\, \frac{\sqrt{1 + t^{2(\alpha-2)}\left(\cos\frac{1}{t} - \alpha t \sin\frac{1}{t}\right)^2}}{\left|\cos\frac{1}{t} - \alpha t \sin\frac{1}{t}\right|}\,,$$

and for all $\varepsilon > 0$ we may find a value of t arbitrarily close to zero such that $\ell(t) < \varepsilon$; yielding a contradiction.

We may now prove the first part of Jordan curve theorem:

Proposition 2.2.8. *Let $\sigma\colon [a,b] \to \mathbb{R}^2$ be a simple closed regular plane curve of class C^2, and denote by $C = \sigma([a,b])$ its support. Then $\mathbb{R}^2 \setminus C$ has at most two components, and C is the boundary of both.*

Proof. Choose $\varepsilon > 0$ so that $N_\sigma(\varepsilon)$ is a tubular neighborhood of σ. Denote by T_+ (respectively, T_-) the set of points of $N_\sigma(\varepsilon)$ of the form $\sigma(t) + \delta\tilde{\mathbf{n}}(t)$ with $\delta > 0$ (respectively, $\delta < 0$), where $\tilde{\mathbf{n}}$ is as usual the oriented normal versor of σ. It is clear that $N_\sigma(\varepsilon) \setminus C = T_+ \cup T_-$. Moreover, both T_+ and T_- are connected. Indeed, given $\sigma(t_1) + \delta_1\tilde{\mathbf{n}}(t_1)$ and $\sigma(t_2) + \delta_2\tilde{\mathbf{n}}(t_2) \in T_+$, the path (see Fig 2.3) that starting from $\sigma(t_1) + \delta_1\tilde{\mathbf{n}}(t_1)$ moves parallel to σ up to $\sigma(t_2) + \delta_1\tilde{\mathbf{n}}(t_2)$ and then parallel to $\tilde{\mathbf{n}}(t_2)$ up to $\sigma(t_2) + \delta_2\tilde{\mathbf{n}}(t_2)$ is completely within T_+. In an analogous way it can be shown that T_- is (pathwise) connected.

Let now K be a component of $\mathbb{R}^2 \setminus C$; clearly, $\varnothing \neq \partial K \subseteq C$. On the other hand, if $p \in C$ then there is a neighborhood of p contained in $C \cup T_+ \cup T_-$. Hence, either T_+ or T_- (or both) intersect K; being connected, either $K \supset T_+$ or $K \supset T_-$, and in particular $\partial K \supseteq C$. Since two different components are necessarily disjoint, it follows that the complement of the support of σ has at most two components, and the boundary of both is C. $\qquad\square$

2.3 The Jordan curve theorem

In this section we shall complete the proof of the Jordan curve theorem for regular curves, by showing that the complement of the support of a simple

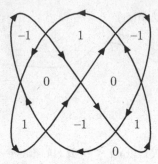

Fig. 2.4. Winding numbers

closed regular plane curve of class C^2 has at least two components. To get there we need a new ingredient, which we shall construct by using the degree introduced in Section 2.1.

Given a continuous closed plane curve, there are (at least) two ways to associate with it a curve with values in S^1, and consequently a degree. In this section we are interested in the first way, while in next section we shall use the second one.

Definition 2.3.1. Let $\sigma\colon [a, b] \to \mathbb{R}^2$ be a continuous closed plane curve. Given a point $p \notin \sigma([a, b])$ we may define $\phi_p\colon [a, b] \to S^1$ by setting

$$\phi_p(t) = \frac{\sigma(t) - p}{\|\sigma(t) - p\|}.$$

The *winding number* $\iota_p(\sigma)$ of σ with respect to p is, by definition, the degree of ϕ_p; it measures the number of times σ goes around the point p.

Fig. 2.4 shows the winding number of a curve with respect to several points, computed as we shall see in Example 2.3.5.

The main properties of the winding number are given in the following:

Lemma 2.3.2. *Let $\sigma\colon [a, b] \to \mathbb{R}^2$ be a continuous closed plane curve, and let K be a component of the open set $U = \mathbb{R}^2 \setminus \sigma([a, b])$. Then:*

(i) $\iota_{p_0}(\sigma) = \iota_{p_1}(\sigma)$ *for every pair of points p_0, $p_1 \in K$;*
(ii) *U has exactly one unbounded component K_0, and $\iota_p(\sigma) = 0$ for all points $p \in K_0$.*

Proof. (i) Let $\eta\colon [0, 1] \to K$ be a curve with $\eta(0) = p_0$ and $\eta(1) = p_1$, and define $\Phi\colon [0, 1] \times [a, b] \to S^1$ by setting

$$\Phi(s, t) = \frac{\sigma(t) - \eta(s)}{\|\sigma(t) - \eta(s)\|}.$$

Since the image of η is disjoint from the support of σ, the map Φ is a homotopy of closed curves between ϕ_{p_0} and ϕ_{p_1}, and so $\iota_{p_0}(\sigma) = \iota_{p_1}(\sigma)$, by Theorem 2.1.15.

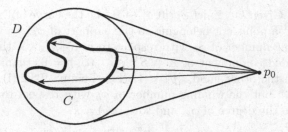

Fig. 2.5.

(ii) Since $[a, b]$ is compact, the support of σ is contained in a closed disk D with center in the origin and sufficiently large radius $R > 0$. Since $\mathbb{R}^2 \setminus D \subset U$ is connected, it is contained in a single component of U; so U has only one unbounded component.

Let now $p_0 \in K_0 \setminus D$; then the line segments joining p_0 to points in the support of σ are all contained in the sector with center p_0 and sides the half-lines issuing from p_0 tangent to D (see Fig. 2.5). This implies that the image of ϕ_{p_0} is contained in a proper subset of S^1, so (Example 2.1.11) ϕ_{p_0} is homotopic to a constant curve. Since (Theorem 2.1.15) the degree of a constant curve is zero, we get $\iota_{p_0}(\sigma) = 0$. □

Identifying as usual $\mathbb{R}^2$ with the complex plane $\mathbb{C}$, Corollary 2.1.19 provides us with an integral formula for computing the winding number of differentiable curves:

Lemma 2.3.3. *Let* $\sigma: [a, b] \to \mathbb{C}$ *be a piecewise* C^1 *closed continuous plane curve, and* $p \notin C = \sigma([a, b])$. *Then the winding number of* σ *with respect to* p *is given by*

$$\iota_p(\sigma) = \frac{1}{2\pi i} \int_a^b \frac{\sigma'}{\sigma - p}\, dt\,.$$

Proof. It is straightforward to check that

$$\frac{\phi_p'}{\phi_p} = i\, \mathrm{Im}\, \frac{\sigma'}{\sigma - p}\,;$$

then, by Corollary 2.1.19, to get the assertion it is sufficient to prove that the integral of the real part of $\sigma'/(\sigma - p)$ is zero. But indeed,

$$\frac{d}{dt} \log \|\sigma(t) - p\| = \mathrm{Re}\, \frac{\sigma'(t)}{\sigma(t) - p}\,,$$

and so

$$\int_a^b \mathrm{Re}\, \frac{\sigma'}{\sigma - p}\, dt = \log \|\sigma(b) - p\| - \log \|\sigma(a) - p\| = 0$$

because σ is closed. □

Example 2.3.4. Given $k \in \mathbb{Z}$, let $\phi_k \colon [0,1] \to \mathbb{R}^2$ be the curve of Example 2.1.8, and $p \in \mathbb{R}^2 \setminus S^1$ a point not belonging to the support of ϕ_k; we want to compute the winding number of ϕ_k with respect to p. If $\|p\| > 1$, then p belongs to the unbounded component of $\mathbb{R}^2 \setminus S^1$, so $\iota_p(\phi_k) = 0$, thanks to Lemma 2.3.2.(ii). If, on the other hand, $\|p\| < 1$, then Lemma 2.3.2.(i) tells us that $\iota_p(\phi_k) = \iota_O(\phi_k)$; but the winding number of ϕ_k with respect to the origin is equal (why?) to the degree of ϕ_k, and so $\iota_p(\phi_k) = k$.

Example 2.3.5. The curve $\sigma \colon [0, 2\pi] \to \mathbb{R}^2$ of Fig. 2.4 is given by

$$\sigma(t) = \big(\sin(2t), \sin(3t)\big) .$$

The support C of σ divides $\mathbb{R}^2$ in 9 components $K_0, \ldots, K_8$, where K_0 is the unbounded component. Take $p_0 \in \mathbb{R}^2 \setminus C$; we want to compute $\iota_{p_0}(\sigma)$. If $p_0 \in K_0$, we already know that $\iota_{p_0}(\sigma) = 0$; assume then that $p_0 = x_0 + iy_0$ belongs to one of the other components. By Lemma 2.3.3 and its proof,

$$
\begin{aligned}
\iota_{p_0}(\sigma) &= \frac{1}{2\pi i} \int_0^{2\pi} \frac{\sigma'}{\sigma - p} \, dt \\
&= \frac{1}{2\pi} \int_0^{2\pi} \frac{3\big(\sin(2t) - x_0\big)\cos(3t) - 2\big(\sin(3t) - y_0\big)\cos(2t)}{\big(\sin(2t) - x_0\big)^2 + \big(\sin(3t) - y_0\big)^2} \, dt .
\end{aligned}
$$

It is possible (with a good deal of patience) to evaluate this integral by means of elementary functions, obtaining the values of the winding number shown in Fig. 2.4.

Finally, we have all we need to complete (following ideas due to Pederson; see [19]) the proof of the Jordan curve theorem:

Theorem 2.3.6 (Jordan theorem for regular curves). *Let $\sigma \colon [a, b] \to \mathbb{R}^2$ be a regular Jordan curve of class C^2, and denote by $C = \sigma([a, b])$ its support. Then $\mathbb{R}^2 \setminus C$ has exactly two components, and C is their common boundary.*

Proof. Choose $\varepsilon > 0$ in such a way that $N_\sigma(\varepsilon)$ is a tubular neighborhood of σ, and denote again by T_+ (respectively, T_-) the set of points of $N_\sigma(\varepsilon)$ of the form $\sigma(t) + \delta \tilde{\mathbf{n}}(t)$ with $\delta > 0$ (respectively, $\delta < 0$), where $\tilde{\mathbf{n}}$ is the oriented normal versor of σ.

We already know (Proposition 2.2.8) that the complement of C has at most two components; to show that it has at least two choose $t_0 \in (a, b)$, and for $0 \le \delta < \varepsilon$ set $p_{\pm\delta} = \sigma(t_0) \pm \delta \tilde{\mathbf{n}}(t_0)$. Clearly, $p_\delta \in T_+$ and $p_{-\delta} \in T_-$; so, as $T_\pm$ are connected, the value of $\iota_{p_{\pm\delta}}(\sigma)$ is independent of δ (Lemma 2.3.2). In particular, the integer number

$$\Delta = \iota_{p_\delta}(\sigma) - \iota_{p_{-\delta}}(\sigma)$$

is independent of $\delta > 0$. So, to conclude the proof it suffices to show that $\Delta \neq 0$; indeed, if this is the case, by Lemma 2.3.2 we know that p_δ and $p_{-\delta}$ have to belong to distinct components of $\mathbb{R}^2 \setminus C$.

Now, identify $\mathbb{R}^2$ with $\mathbb{C}$, and assume σ to be parametrized by arc length. Then the oriented normal versor $\tilde{n}$ can be obtained by rotating $\dot\sigma$ by $\pi/2$ radians, an operation which in the complex plane is equivalent to multiplying by i; so we may write $\tilde{n} = i\dot\sigma$. Hence for all $\delta > 0$ we get

$$\frac{\dot\sigma(t)}{\sigma(t) - p_\delta} - \frac{\dot\sigma(t)}{\sigma(t) - p_{-\delta}} = \frac{2\,i\,\delta\,\dot\sigma(t_0)\,\dot\sigma(t)}{\left(\sigma(t) - \sigma(t_0)\right)^2 + \delta^2\dot\sigma(t_0)^2} .$$

Since σ is of class C^1 and $\dot\sigma(t_0) \neq 0$, we may write

$$\sigma(t) - \sigma(t_0) = (t - t_0)\dot\sigma(t_0)[1 + r(t)] ,$$

where $r(t) \to 0$ when $t \to t_0$. So,

$$\frac{\dot\sigma(t)}{\sigma(t) - p_\delta} - \frac{\dot\sigma(t)}{\sigma(t) - p_{-\delta}}$$

$$= \frac{2i\delta}{(t - t_0)^2[1 + r(t)]^2 + \delta^2}\,\frac{\dot\sigma(t)}{\dot\sigma(t_0)}$$

$$= \frac{2i\delta}{(t - t_0)^2 + \delta^2}\,\frac{(t - t_0)^2 + \delta^2}{(t - t_0)^2[1 + r(t)]^2 + \delta^2}\left[1 + \frac{\dot\sigma(t) - \dot\sigma(t_0)}{\dot\sigma(t_0)}\right]$$

$$= \frac{2i\delta}{(t - t_0)^2 + \delta^2} + R(t) ,$$

with

$$R(t) = \frac{2i\delta}{(t - t_0)^2 + \delta^2}$$

$$\times \left[s(t) - r(t)\bigl(2 + r(t)\bigr)\bigl(1 + s(t)\bigr)\frac{(t - t_0)^2}{(t - t_0)^2[1 + r(t)]^2 + \delta^2}\right] ,$$

where $s(t) = \bigl(\dot\sigma(t) - \dot\sigma(t_0)\bigr)/\dot\sigma(t_0) \to 0$ for $t \to t_0$. In particular, for all $\eta > 0$ there exists $\lambda > 0$ (independent of δ) such that

$$|R(t)| < \eta\,\frac{2\delta}{(t - t_0)^2 + \delta^2}$$

as soon as $|t - t_0| < \lambda$.

Choose then $0 < \eta < 1/8$, and take the corresponding $\lambda > 0$. So we may write

$$\Delta = \iota_{p_\delta}(\sigma) - \iota_{p_{-\delta}}(\sigma)$$

$$= \frac{1}{2\pi i}\int_a^{t_0 - \lambda}\left(\frac{\dot\sigma(t)}{\sigma(t) - p_\delta} - \frac{\dot\sigma(t)}{\sigma(t) - p_{-\delta}}\right)dt$$

$$+ \frac{1}{2\pi i}\int_{t_0 + \lambda}^b\left(\frac{\dot\sigma(t)}{\sigma(t) - p_\delta} - \frac{\dot\sigma(t)}{\sigma(t) - p_{-\delta}}\right)dt$$

$$+ \frac{1}{2\pi i}\int_{t_0 - \lambda}^{t_0 + \lambda}\left(\frac{2i\delta}{(t - t_0)^2 + \delta^2} + R(t)\right)dt .$$

By the above remarks, Δ is an integer number independent on δ. Now, let δ approach zero in the right-hand side. The first two integrals tend to zero, as their argument does not have a singularity for $t \neq t_0$. For the third integral, after the change of variables $t - t_0 = \delta s$, we get first of all that

$$\frac{1}{2\pi i} \int_{t_0-\lambda}^{t_0+\lambda} \frac{2i\delta}{(t-t_0)^2 + \delta^2}\, dt = \frac{1}{\pi} \int_{-\lambda/\delta}^{\lambda/\delta} \frac{1}{1+s^2}\, ds \to \frac{1}{\pi} \int_{-\infty}^{\infty} \frac{1}{1+s^2}\, ds = 1$$

for $\delta \to 0$. Moreover,

$$\left| \frac{1}{2\pi i} \int_{t_0-\lambda}^{t_0+\lambda} R(t)\, dt \right| < \frac{\eta}{\pi} \int_{-\lambda/\delta}^{\lambda/\delta} \frac{1}{1+s^2}\, ds < \frac{\eta}{\pi} \int_{-\infty}^{\infty} \frac{1}{1+s^2}\, ds = \eta.$$

Putting all together, we then get that for a sufficiently small δ we may estimate the difference $\Delta - 1$ as follows:

$$
\begin{aligned}
|\Delta - 1| \leq{} & \frac{1}{2\pi} \left| \int_a^{t_0-\lambda} \left(\frac{\dot\sigma(t)}{\sigma(t) - p_\delta} - \frac{\dot\sigma(t)}{\sigma(t) - p_{-\delta}} \right) dt \right| \\
& + \frac{1}{2\pi} \left| \int_{t_0+\lambda}^{b} \left(\frac{\dot\sigma(t)}{\sigma(t) - p_\delta} - \frac{\dot\sigma(t)}{\sigma(t) - p_{-\delta}} \right) dt \right| \\
& + \left| \frac{1}{2\pi} \int_{t_0-\lambda}^{t_0+\lambda} \frac{2\delta}{(t-t_0)^2 + \delta^2}\, dt - 1 \right| \\
& + \frac{1}{2\pi} \left| \int_{t_0-\lambda}^{t_0+\lambda} R(t)\, dt \right| \leq 4\eta < \frac{1}{2}.
\end{aligned}
$$

But Δ is an integer number; so $\Delta = 1$, and we are done. $\square$

Remark 2.3.7. As we have already remarked, the Jordan curve theorem holds for arbitrary continuous Jordan curves, not necessarily differentiable ones (for some special cases, see Exercises 2.8, 2.9, and 2.11). Moreover, the closure of the bounded component of the complement of the support of a (continuous) Jordan curve is homeomorphic to a closed disk (*Schönflies' theorem*). In Section 2.8 of the supplementary material of this chapter we shall prove both the Jordan curve theorem for continuous curves and Schönflies' theorem.

Remark 2.3.8. Jordan theorem describes a specific property of the topology of the plane, not shared by all surfaces: a closed simple regular curve within a surface S that is not a plane might not partition the surface in exactly two parts. It is possible to adapt (see Exercise 5.11 and Section 4.8) the notion of tubular neighborhood in such a way that the proof of Proposition 2.2.8 works, and so the complement of the support of the curve still has at most two components. But on the other hand, two new phenomena may occur. It might be impossible to define consistently the normal versor to the curve, so T_+

and T_- coincide, and this is what happens in non-orientable surfaces such as Möbius strip (see Example 4.3.11). Or, T_+ and T_- might belong to the same component, which is what happens, for instance, in the torus $S = S^1 \times S^1$ (see Example 3.1.19). In both cases, the complement of the support of the curve is connected.

As previously remarked, the complement of a compact set in the plane has exactly one unbounded component. This fact and the proof of Theorem 2.3.6 suggest how to use the winding number to determine the orientation of a regular plane Jordan curve.

Definition 2.3.9. Let $\sigma\colon [a,b] \to \mathbb{R}^2$ be a (piecewise C^2) Jordan curve in the plane. The unique (see Exercise 2.8) bounded component of the complement of the support of σ is the *interior* of σ. Lemma 2.3.2.(ii) and the proof of Theorem 2.3.6 tell us that the winding number of σ with respect to any point of its interior is equal to ± 1. We say that σ is *positively oriented* (respectively, *negatively oriented*) if this winding number is $+1$ (respectively, -1).

Remark 2.3.10. In the proof of Theorem 2.3.6 we saw that $\iota_{p_\delta}(\sigma) - \iota_{p_{-\delta}}(\sigma)$ is always equal to 1; moreover, $\iota_{p_{\pm\delta}}(\sigma) \neq 0$ if and only if $p_{\pm\delta}$ is in the interior of σ, and in that case we have $\iota_{p_{\pm\delta}}(\sigma) = \pm 1$. Now, p_δ is in the interior of σ if and only if $\tilde{n}(t_0)$ points towards the interior of σ, and this happens if and only if σ is described counterclockwise. Hence, σ is positively (negatively) oriented if and only if it is described counterclockwise (clockwise).

Remark 2.3.11. Warning: the sign of the oriented curvature has nothing to do with the orientation of the curve. The simplest way to realize this is to observe that the oriented curvature of a closed simple curve might well be positive somewhere and negative somewhere else, while the curve cannot be partially positively oriented and partially negatively oriented. In Section 2.5 of the supplementary material of this chapter we shall characterize plane curves having oriented curvature with constant sign.

2.4 The turning tangents theorem

There is another very natural way of associating a S^1-valued curve (and consequently a degree) with a closed regular plane curve.

Definition 2.4.1. Let $\sigma\colon [a,b] \to \mathbb{R}^2$ be a closed regular plane curve of class C^1, and let $\mathbf{t}\colon [a,b] \to S^1$ be its tangent versor, given by

$$\mathbf{t}(t) = \frac{\sigma'(t)}{\|\sigma'(t)\|} \ .$$

The *rotation index* $\rho(\sigma)$ of σ is the degree of the map $\mathbf{t}$; it counts the number of full turns made by the tangent versor to σ.

Corollary 2.1.18 provides us with a simple formula to compute the rotation index:

Proposition 2.4.2. *Let* $\sigma\colon [a,b] \to \mathbb{R}^2$ *be a closed regular plane curve of class* C^1 *with oriented curvature* $\tilde{\kappa}\colon [a,b] \to \mathbb{R}$. *Then*

$$\rho(\sigma) = \frac{1}{2\pi} \int_a^b \tilde{\kappa}\, \|\sigma'\|\, \mathrm{d}t = \frac{1}{2\pi} \int_a^b \frac{\det(\sigma',\sigma'')}{\|\sigma'\|^2}\, \mathrm{d}t \;.$$

Proof. By Corollary 2.1.18,

$$\rho(\sigma) = \frac{1}{2\pi} \int_a^b \det(\mathbf{t}, \mathbf{t}')\, \mathrm{d}t \;.$$

Let $s\colon [a,b] \to [0,\ell]$ be the arc length of σ, measured starting from a; recalling (1.10) and the fact that $\mathrm{d}s/\mathrm{d}t = \|\sigma'\|$ we get

$$\int_a^b \det(\mathbf{t}, \mathbf{t}')\, \mathrm{d}t = \int_a^b \det(\mathbf{t}, \dot{\mathbf{t}})\frac{\mathrm{d}s}{\mathrm{d}t}\, \mathrm{d}t = \int_0^\ell \det(\mathbf{t}, \dot{\mathbf{t}})\, \mathrm{d}s$$

$$= \int_0^\ell \tilde{\kappa}(s)\, \mathrm{d}s = \int_a^b \tilde{\kappa}(t)\|\sigma'(t)\|\, \mathrm{d}t \;,$$

and we are done, thanks to Problem 1.1. $\square$

Example 2.4.3. Let $\phi_k\colon [0,1] \to \mathbb{R}^2$ be the curve in Example 2.1.8, which has oriented curvature $\tilde{\kappa} \equiv 1$. Since $\phi_k'(t) = 2k\pi\big(-\sin(2k\pi t), \cos(2k\pi t)\big)$, we get

$$\rho(\phi_k) = \frac{1}{2\pi} \int_0^1 \tilde{\kappa}(t)\|\phi_k'(t)\|\, \mathrm{d}t = k.$$

Example 2.4.4. Let $\sigma\colon [0,2\pi] \to \mathbb{R}^2$ be the curve in Example 2.3.5. Then

$$\rho(\sigma) = \frac{1}{2\pi} \int_0^{2\pi} \frac{12\sin(2t)\cos(3t) - 18\sin(3t)\cos(2t)}{4\cos^2(2t) + 9\cos^2(3t)}\, \mathrm{d}t = 0 \;.$$

Later on, we shall need the rotation index for piecewise C^1 curves too; to this end, we introduce the following definitions.

Definition 2.4.5. Let $\sigma\colon [a,b] \to \mathbb{R}^2$ be a regular piecewise C^1 plane curve, and choose a partition $a = t_0 < t_1 < \cdots < t_k = b$ of $[a,b]$ such that $\sigma|_{[t_{j-1},t_j]}$ is regular for $j = 1,\ldots,k$. Set

$$\sigma'(t_j^-) = \lim_{t \to t_j^-} \sigma'(t)$$

for $j = 1,\ldots,k$, and

$$\sigma'(t_j^+) = \lim_{t \to t_j^+} \sigma'(t)$$

for $j = 0, \ldots, k - 1$. Moreover, if σ is closed, we also set $\sigma'(t_0^-) = \sigma'(t_k^-)$ and $\sigma'(t_k^+) = \sigma'(t_0^+)$; remark that it may well be that $\sigma'(t_k^-) \neq \sigma'(t_k^+)$. We shall say that t_j is a *cusp* for σ if $\sigma'(t_j^-) = -\sigma'(t_j^+)$. If t_j is not a cusp, the *external angle* $\varepsilon_j \in (-\pi, \pi)$ is the angle between $\sigma'(t_j^-)$ and $\sigma'(t_j^+)$, taken with the positive sign if $\{\sigma'(t_j^-), \sigma'(t_j^+)\}$ is a positive basis of $\mathbb{R}^2$, and negative otherwise. The points where the external angle is not zero are the *vertices* of the curve. Finally, a *curvilinear polygon* of class C^k is a closed simple regular piecewise C^k curve without cusps.

Definition 2.4.6. Let $\sigma \colon [a, b] \to \mathbb{R}^2$ be a regular piecewise C^1 plane curve without cusps, and $a = t_0 < t_1 < \cdots < t_k = b$ a partition of $[a, b]$ such that $\sigma|_{[t_{j-1}, t_j]}$ is regular for $j = 1, \ldots, k$. We define the *angle of rotation* $\theta \colon [a, b] \to \mathbb{R}$ as follows: let $\theta \colon [a, t_1) \to \mathbb{R}$ be the unique lift of $\mathbf{t} \colon [a, t_1) \to S^1$ such that $\theta(a) \in (-\pi, \pi]$. In other words, $\theta|_{[a, t_1)}$ is the continuous determination of the angle between the x-axis and the tangent versor $\mathbf{t}$ with initial value chosen in $(-\pi, \pi]$. Set next

$$\theta(t_1) = \lim_{t \to t_1^-} \theta(t) + \varepsilon_1 \, ,$$

where ε_1 is the external angle of σ in t_1; in particular, $\theta(t_1)$ is a determination of the angle between the x-axis and $\sigma'(t_1^+)$, while $\theta(t_1) - \varepsilon_1$ is a determination of the angle between the x-axis and $\sigma'(t_1^-)$.

Define next $\theta \colon [t_1, t_2) \to \mathbb{R}$ to be the lift of $\mathbf{t} \colon [t_1, t_2) \to S^1$ starting from $\theta(t_1)$, and set again $\theta(t_2) = \lim_{t \to t_2^-} \theta(t) + \varepsilon_2$, where ε_2 is the external angle of σ in t_2. Proceeding in the same way, we define θ on the entire interval $[a, b]$; finally, we set

$$\theta(b) = \lim_{t \to b^-} \theta(t) + \varepsilon_k \, ,$$

where ε_k is the external angle of σ in $b = t_k$ (with $\varepsilon_k = 0$ if σ is not closed). Note that this makes the angle of rotation continuous from the right but not necessarily from the left in the vertices.

Finally, if the curve σ is closed the *rotation index* of σ is the number

$$\rho(\sigma) = \frac{1}{2\pi} \big(\theta(b) - \theta(a) \big) \, .$$

Since $\sigma'(t_k^+) = \sigma'(t_0^+)$, the rotation index is always an integer. Clearly, if we reverse the orientation of the curve the rotation index changes sign.

One of the consequences of our proof of the Jordan curve Theorem 2.3.6 is the fact that the winding number of a Jordan curve with respect to an internal point is always equal to ± 1. The main result of this section is an analogous result for the rotation index, known as *Hopf's turning tangents theorem*:

Theorem 2.4.7 (Hopf's turning tangents theorem, or *Umlaufsatz*). *The rotation index of a curvilinear polygon is always ± 1.*

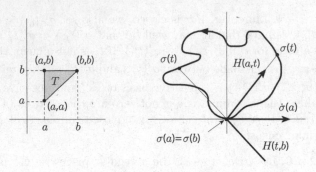

Fig. 2.6.

Proof. Let $\sigma = (\sigma_1, \sigma_2) \colon [a, b] \to \mathbb{R}^2$ be a curvilinear polygon, which we may assume to be parametrized by arc length. We begin by assuming that σ has no vertices; in particular, $\mathbf{t} = \dot{\sigma}$ is continuous and $\dot{\sigma}(a) = \dot{\sigma}(b)$. Since σ is closed, we may extend it by periodicity to a periodic C^1 plane curve, which we shall still denote by σ, defined on the whole $\mathbb{R}$ and with period $b - a$.

If $[\tilde{a}, \tilde{b}] \subset \mathbb{R}$ is an arbitrary interval of length $b - a$, clearly we have $\rho(\sigma|_{[\tilde{a}, \tilde{b}]}) = \rho(\sigma|_{[a,b]})$; hence, up to translations of the parameter we may assume that $\sigma_2(t)$ has a minimum in $t = a$. Moreover, up to translations in the plane, we also assume that $\sigma(a) = O$. Thus the support of σ is contained in the upper half-plane, and $\dot{\sigma}_2(a) = 0$, and so (up to inverting the orientation of the curve, if necessary) we have $\dot{\sigma}(a) = \dot{\sigma}(b) = \mathbf{e}_1$, where $\mathbf{e}_1$ is the first vector of the canonical basis of $\mathbb{R}^2$; see Fig. 2.6.

Let $\theta \colon [a, b] \to \mathbb{R}$ be the angle of rotation of σ, with initial value $\theta(a) = 0$. We want to define a *secant angle* $\eta \colon T \to \mathbb{R}$ on the triangle

$$T = \{(t_1, t_2) \in \mathbb{R}^2 \mid a \le t_1 \le t_2 \le b\}\,,$$

so that it will be a continuous determination of the angle between the x-axis and the vector from $\sigma(t_1)$ to $\sigma(t_2)$. To do so, we define a map $H \colon T \to S^1$ by setting

$$H(t_1, t_2) = \begin{cases} \dfrac{\sigma(t_2) - \sigma(t_1)}{\|\sigma(t_2) - \sigma(t_1)\|} & \text{if } t_1 < t_2 \text{ and } (t_1, t_2) \ne (a, b)\,; \\[2mm] \dot{\sigma}(t_1) & \text{if } t_1 = t_2\,; \\[1mm] -\dot{\sigma}(b) & \text{if } (t_1, t_2) = (a, b)\,. \end{cases}$$

The map H is continuous along the segment $t_1 = t_2$, since

$$\lim_{(t_1, t_2) \to (t, t)} H(t_1, t_2) = \lim_{(t_1, t_2) \to (t, t)} \frac{\sigma(t_2) - \sigma(t_1)}{t_2 - t_1} \bigg/ \left\| \frac{\sigma(t_2) - \sigma(t_1)}{t_2 - t_1} \right\|$$

$$= \frac{\dot{\sigma}(t)}{\|\dot{\sigma}(t)\|} = H(t, t)\,.$$

The continuity of H in (a, b) is proved in an analogous way:

$$\lim_{(t_1,t_2)\to(a,b)} H(t_1,t_2) = \lim_{(t_1,t_2)\to(a,b)} \frac{\sigma(t_2) - \sigma(t_1 + b - a)}{\|\sigma(t_2) - \sigma(t_1 + b - a)\|}$$

$$= \lim_{(s,t_2)\to(b,b)} \frac{\sigma(t_2) - \sigma(s)}{\|\sigma(t_2) - \sigma(s)\|}$$

$$= \lim_{(s,t_2)\to(b,b)} -\frac{\sigma(t_2) - \sigma(s)}{t_2 - s} \Bigg/ \left\|\frac{\sigma(t_2) - \sigma(s)}{t_2 - s}\right\|$$

$$= -\frac{\dot{\sigma}(b)}{\|\dot{\sigma}(b)\|} = H(a, b) \, ,$$

where the minus sign appears because $s = t_1 + b - a > t_2$. Since the triangle T is homeomorphic to the rectangle $[0, 1] \times [a, b]$, by using Corollary 2.1.14 we may lift H to a unique continuous function $\eta \colon T \to \mathbb{R}$ such that $\eta(a, a) = 0$; the function η is our secant angle.

Now, both θ and $t \mapsto \eta(t, t)$ are lifts of $\dot{\sigma}$; since $\theta(a) = 0 = \eta(a, a)$, by the uniqueness of the lift we get that $\theta(t) = \eta(t, t)$ for all t, so

$$\rho(\sigma) = \frac{1}{2\pi}\big(\theta(b) - \theta(a)\big) = \frac{1}{2\pi}\,\eta(b, b) \, .$$

We want to find the value of $\eta(b, b)$ by going along the other two sides of the triangle T. By construction (see Fig. 2.6) the vector $\sigma(t) - \sigma(a)$ always points towards the upper half-plane: so $\eta(a, t) \in [0, \pi]$ for all $t \in [a, b]$. In particular, $H(a, b) = -\dot{\sigma}(b) = -\mathbf{e}_1$ yields $\eta(a, b) = \pi$. Analogously, the vector $\sigma(b) - \sigma(t)$ always points towards the lower half-plane; as $\eta(a, b) = \pi$, we have $\eta(t, b) \in [\pi, 2\pi]$ for all $t \in [a, b]$. In particular, $H(b, b) = \dot{\sigma}(b) = \mathbf{e}_1$ yields $\eta(b, b) = 2\pi$, and the assertion is proved in the case of a C^1 curvilinear polygon.

Suppose now that σ has vertices; to prove the theorem, we just need a curvilinear polygon without vertices having the same angle of rotation as σ. To do so, we shall modify σ near each vertex in such a way to make it regular without changing the angle of rotation.

Let then $\sigma(t_i)$ be a vertex having external angle ε_i, and choose a positive number $\alpha \in \big(0, (\pi - |\varepsilon_i|)/2\big)$; by using the periodicity of σ, up to a change in the domain we may also assume that $t_i \neq a, b$. By our definition of the angle of rotation, we have

$$\lim_{t \to t_i^-} \theta(t) = \theta(t_i) - \varepsilon_i \qquad \text{and} \qquad \lim_{t \to t_i^+} \theta(t) = \theta(t_i) \, .$$

So we may find a $\delta > 0$ less than both $t_i - t_{i-1}$ and $t_{i+1} - t_i$ such that $\big|\theta(t) - \big(\theta(t_i) - \varepsilon_i\big)\big| < \alpha$ for $t \in (t_i - \delta, t_i)$, and $|\theta(t) - \theta(t_i)| < \alpha$ for $t \in (t_i, t_1 + \delta)$. In particular,

$$|\theta(t) - \theta(s)| \leq 2\alpha + |\varepsilon_i| < \pi \tag{2.5}$$

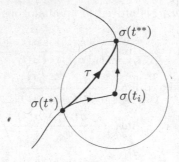

Fig. 2.7.

for all s, $t \in (t_i - \delta, t_i + \delta)$. So the angle of rotation of σ has a variation less than π in this interval.

The set $C = \sigma\big([a,b] \setminus (t_i - \delta, t_i + \delta)\big)$ is a compact set not containing $\sigma(t_i)$; so we may find $r > 0$ such that $C \cap \overline{B\big(\sigma(t_i), r\big)} = \varnothing$. Let t^*, $t^{**} \in (t_i - \delta, t_i + \delta)$ be the first and the last value of t, respectively, for which $\sigma(t) \in \partial B\big(\sigma(t_i), r\big)$; then $\dot{\sigma}(t^*)$ points towards the interior of $\partial B\big(\sigma(t_i), r\big)$, while $\dot{\sigma}(t^{**})$ points towards the exterior of $\partial B\big(\sigma(t_i), r\big)$; see Fig 2.7.

Replace the segment of σ from t^* to t^{**} with (Exercise 2.17) a regular curve $\tau \colon [t^*, t^{**}] \to \mathbb{R}^2$ having support contained in $B\big(\sigma(t_i), r\big)$, tangent to σ at $\sigma(t^*)$ and $\sigma(t^{**})$, and with an angle of rotation ψ satisfying $\psi(t^*) = \theta(t^*)$ and

$$|\psi(s) - \psi(t)| < \pi$$

for all s, $t \in [t^*, t^{**}]$. Then the inequality (2.5) implies (why?) that

$$\theta(t^{**}) - \theta(t^*) = \psi(t^{**}) - \psi(t^*) \, ,$$

and so the curvilinear polygon obtained by inserting τ in the place of $\sigma|_{[t^*, t^{**}]}$ has exactly the same rotation index as σ. By repeating the operation in all vertices of σ, we get a curvilinear polygon of class C^1 with the same rotation index as σ, and we are done.   □

Definition 2.4.8. We shall say that a curvilinear polygon is *positively oriented* if its rotation index is $+1$.

Remark 2.4.9. A regular Jordan curve σ of class C^2 is positively oriented in the sense of Definition 2.3.9 if and only if it is so according to the present definition. Indeed, Remark 2.3.10 tells us that σ is positively oriented according to Definition 2.3.9 if and only if its normal versor points towards its interior. In the situation described at the beginning of the proof of the turning tangents theorem, the interior of σ necessarily lies in the upper half-plane; so σ is positively oriented according to Definition 2.3.9 if and only if the normal versor to σ at $\sigma(a)$ is $\mathbf{e}_2$. But this holds if and only if $\dot{\sigma}(a) = \mathbf{e}_1$ without having to change the orientation, and the proof later shows that this happens if and only if the rotation index of σ is $+1$. So the two definitions are consistent.

Guided problems

Problem 2.1. *Consider the plane curve* $\sigma: [0, 2\pi] \to \mathbb{R}^2$ *parametrized by*

$$\sigma(t) = \big((2\cos t - 1)\cos t, (2\cos t - 1)\sin t\big).$$

(i) *Show that the support of* σ *is defined, in polar coordinates* (r, θ), *by the equation* $r = 2\cos\theta - 1$.

(ii) *Draw the support of* σ, *pointing out the orientation defined by* σ.

(iii) *Compute the winding number of* σ *with respect to the point* $p = (1/2, 0)$.

(iv) *If* $h: [0, 2\pi] \to [0, 2\pi]$ *is given by*

$$h(t) = \begin{cases} t & \text{for } t \in [0, \pi/3] , \\ -t + 2\pi & \text{for } t \in [\pi/3, 5\pi/3] , \\ t & \text{for } t \in [5\pi/3, 2\pi] , \end{cases}$$

determine the parametrization of the curve $\sigma_1 = \sigma \circ h: [0, 2\pi] \to \mathbb{R}^2$, *and compute its winding number with respect to the point* $p = (1/2, 0)$. *Note that* σ_1 *is not equivalent to* σ, *since* h *is not continuous.*

Solution. (i) Easily obtained from $x = r\cos\theta$ and $y = r\sin\theta$.

(ii) See Fig. 2.8.(a).

(iii) Fig. 2.8.(a) suggests that $i_p(\sigma) = 2$, and indeed a calculation shows that

$$i_p(\sigma) = \frac{1}{2\pi} \int_0^{2\pi} \frac{2(\cos t - 1)^2 + \frac{1}{2}\cos t}{2\cos^2 t - 3\cos t + \frac{5}{4}} \, dt = 2 .$$

(iv) The parametrization of σ_1 is given by

$$\sigma_1(t) = \big((2\cos h(t) - 1)\cos h(t), (2\cos h(t) - 1)\sin h(t)\big) ;$$

see Fig. 2.8.(b). Looking at the figure, one suspects that the winding number should be $i_p(\sigma_1) = 0$, and indeed

$$i_p(\sigma_1) = \frac{1}{2\pi} \left[\int_0^{\pi/3} - \int_{\pi/3}^{5\pi/3} + \int_{5\pi/3}^{2\pi} \right] \frac{2(\cos t - 1)^2 + \frac{1}{2}\cos t}{2\cos^2 t - 3\cos t + \frac{5}{4}} \, dt = 0 . \qquad \square$$

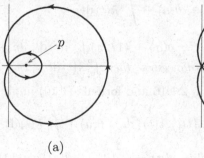
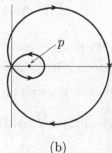

(a)   (b)

Fig. 2.8.

Problem 2.2. *Let σ_0, $\sigma_1\colon [0,1] \to \mathbb{R}^2$ be two closed continuous curves such that $\sigma_0(0) = \sigma_1(0) = q$, and choose $p_0 \in \mathbb{R}^2$ not in the union of the supports of σ_0 and σ_1. Prove that if $\|\sigma_0(t) - \sigma_1(t)\| \le \|\sigma_0(t) - p_0\|$ for all $t \in [0,1]$, then $i_{p_0}(\sigma_0) = i_{p_0}(\sigma_1)$.*

Solution. Consider the map $\Phi\colon [0,1] \times [0,1] \to \mathbb{R}^2$ defined by

$$\Phi(\lambda, t) = (1 - \lambda)\sigma_0(t) + \lambda\sigma_1(t).$$

We want to show that Φ is a homotopy of closed curves in $\mathbb{R}^2 \setminus \{p_0\}$.

Note first that Φ is a continuous map, since so are its components. Clearly, $\Phi(0, \cdot) \equiv \sigma_0$ and $\Phi(1, \cdot) \equiv \sigma_1$; moreover, $\Phi(\lambda, 0) = \Phi(\lambda, 1) = q$ for all $\lambda \in [0,1]$. We have now to show that p_0 does not belong to the image of Φ. If, by contradiction, we had $p_0 = \Phi(\lambda_0, t_0)$ for a suitable $(\lambda_0, t_0) \in [0,1] \times [0,1]$, we would have

$$\sigma_0(t_0) - p_0 = \lambda_0\big(\sigma_0(t_0) - \sigma_1(t_0)\big).$$

Since, by assumption, p_0 is not in the support of σ_1, we deduce that $\lambda_0 \ne 1$, and so

$$\|\sigma_0(t_0) - p_0\| = \lambda_0 \|\sigma_0(t_0) - \sigma_1(t_0)\| < \|\sigma_0(t_0) - \sigma_1(t_0)\| \,,$$

against the assumption on σ_0 and σ_1.

Define now ϕ_0, $\phi_1\colon [0,1] \to S^1$ and $\hat{\Phi}\colon [0,1] \times [0,1] \to S^1$ by setting

$$\phi_j(t) = \frac{\sigma_j(t) - p_0}{\|\sigma_j(t) - p_0\|} \quad \text{and} \quad \hat{\Phi}(\lambda, t) = \frac{\Phi(\lambda, t) - p_0}{\|\Phi(\lambda, t) - p_0\|} \,.$$

Then $\hat{\Phi}$ is a homotopy between ϕ_0 and ϕ_1, and so

$$i_{p_0}(\sigma_0) = \deg \phi_0 = \deg \phi_1 = i_{p_0}(\sigma_1) \,,$$

by Theorem 2.1.15. □

Problem 2.3. *Let $\sigma\colon [a,b] \to \mathbb{R}^2$ be a regular plane curve of class at least C^2, parametrized by arc length, and let $\theta\colon [a,b] \to \mathbb{R}$ be its angle of rotation.*

(i) *Prove that $\dot{\theta} = \tilde{\kappa}$ and*

$$\theta(s) = \theta(a) + \int_0^s \tilde{\kappa}(t)\,\mathrm{d}t \,.$$

(ii) *Show that the rotation index is $\rho(\sigma) = (1/2\pi)\int_a^b \tilde{\kappa}(t)\,\mathrm{d}t$. In particular, if σ is a positively oriented Jordan curve then $\int_a^b \tilde{\kappa}(t)\,\mathrm{d}t = 2\pi$.*

Solution. (i) Indeed, Proposition 2.1.16 and formula (1.10) imply

$$\theta(s) = \theta(a) + \int_a^s \det\big(\mathbf{t}(t), \dot{\mathbf{t}}(t)\big)\,\mathrm{d}t = \theta(a) + \int_a^s \tilde{\kappa}(t)\,\mathrm{d}t \,.$$

(ii) It follows from the equality $\rho(\sigma) = \frac{1}{2\pi}\big(\theta(b) - \theta(a)\big)$, and thus the last assertion is a consequence of Theorem 2.4.7 and Remark 2.4.9. □

Problem 2.4. *Let* $\sigma\colon [0,1] \to \mathbb{R}^2$ *be a regular closed plane curve of class at least* C^2, *and suppose that there exists* $c > 0$ *such that* $|\tilde{\kappa}(t)| \leq c$ *for all* $t \in [0,1]$. *Prove that*

$$L(\sigma) \geq \frac{2\pi|\rho(\sigma)|}{c},$$

where $L(\sigma)$ *is the length of* σ. *Deduce, in particular, that if* σ *is simple then* $L(\sigma) \geq 2\pi/c$.

Solution. Proposition 2.4.2 implies

$$2\pi|\rho(\sigma)| = \left|\int_0^1 \tilde{\kappa}\,\|\sigma'\|\,\mathrm{d}t\right| \leq \int_0^1 |\tilde{\kappa}|\,\|\sigma'\|\,\mathrm{d}t \leq c\int_0^1 \|\sigma'\|\,\mathrm{d}t = c\,L(\sigma)\,.$$

The last claim follows from the fact that if σ is simple then $\rho(\sigma) = \pm 1$ (Theorem 2.4.7).   □

Problem 2.5. *Let* $\sigma\colon [a,b] \to \mathbb{R}^2$ *be a regular closed plane curve with oriented curvature* $\tilde{\kappa}$ *positive everywhere.*

(i) *Show that if* σ *is not simple then* $\rho(\sigma) \geq 2$.
(ii) *Find a regular non-simple closed plane curve having an oriented curvature that changes sign and rotation index equal to zero.*

Solution. (i) Since σ is not simple, there is at least one point $p \in \mathbb{R}^2$ that is image of two distinct values of the parameter. Up to a parameter change for σ and a rigid motion in $\mathbb{R}^2$, we may assume that σ is parametrized by arc length, that $p = \sigma(a) = (0,0)$ and that $\mathbf{t}(0) = (1,0)$. Let $s_0 \in (a,b)$ be the least value of the parameter such that $\sigma(s_0) = p$. Since $\tilde{\kappa}(0) > 0$, for $s > a$ close enough to a the point $\sigma(s) = \big(\sigma_1(s), \sigma_2(s)\big)$ lies in the upper half-plane. So, the second component σ_2 of σ is a continuous function in the compact set $[a, s_0]$ that is not identically zero, vanishes at the endpoints, and in at least one point in the interior of the interval has a strictly positive value. Hence σ_2 reaches a strictly positive local maximum in a point $s_1 \in (a, s_0)$. So, $\dot\sigma_2(s_1) = 0$, and the tangent line to σ at s_1 is horizontal. If $\theta\colon [a,b] \to \mathbb{R}$ is the angle of rotation of σ, we deduce then that $\theta(s_1) = k\,\pi$, for some $k \in \mathbb{Z}$.

Now, in Problem 2.3 we saw that $\dot\theta = \tilde{\kappa}$; so θ is strictly increasing. In particular, $0 = \theta(0) < \theta(s_1) < \theta(s_0) < \theta(b)$. Our goal is equivalent to proving that $\theta(b) > 2\pi$. If $\theta(s_0) \geq 2\pi$, we are done. If, on the contrary, $\theta(s_0) < 2\pi$, we have $\theta(s_1) = \pi$; but arguing as above for the interval $[s_0, b]$ we find a $s_2 \in (s_0, b)$ such that $\theta(s_2) = \theta(s_0) + \pi$, so $\theta(b) > \theta(s_2) > \theta(s_1) + \pi = 2\pi$, as required.

(ii) Let $\tau\colon [0, 2\pi] \to \mathbb{R}^2$ be given by $\tau(t) = \big(\sin t, \sin(2t)\big)$. It is straightforward to see that $\tau(0) = \tau(\pi) = (0,0)$, so τ is not simple. Moreover, $\|\tau'(t+\pi)\| = \|\tau'(t)\|$, while $\det\big(\tau'(t+\pi), \tau''(t+\pi)\big) = -\det\big(\tau'(t), \tau''(t)\big)$. Hence Problem 1.1 yields $\tilde{\kappa}(t+\pi) = -\tilde{\kappa}(t)$, so $\tilde{\kappa}$ changes sign. Finally, using Proposition 2.4.2 we conclude that $\rho(\tau) = 0$, as required.   □

Exercises

HOMOTOPIES

2.1. Prove that there is no lift of the identity map id: $S^1 \to S^1$.

2.2. Given two topological spaces X and Y, prove that "being homotopic" is an equivalence relation on the set of continuous maps from Y to X.

2.3. Consider the curve $\sigma_1: [0, 2\pi] \to \mathbb{R}^2$ defined by

$$\sigma_1(t) = \begin{cases} (\cos(2t), \sin(2t)) & t \in [0, \pi], \\ (1, 0) & t \in [\pi, 2\pi]. \end{cases}$$

Prove that σ_1 is homotopic to the usual parametrization $\sigma_0: [0, 2\pi] \to \mathbb{R}^2$ of the circle given by $\sigma_0(t) = (\cos t, \sin t)$ by explicitly constructing a homotopy.

2.4. Prove that two continuous closed curves σ_0 and $\sigma_1: [a, b] \to D$, where D is an open disk in the plane, are always homotopic (as closed curves) in D.

TUBULAR NEIGHBORHOOD

2.5. Find a tubular neighborhood for the ellipse $\sigma: \mathbb{R} \to \mathbb{R}^2$ parametrized by $\sigma(t) = (\cos t, 2 \sin t)$.

2.6. Find a tubular neighborhood for the arc of parabola $\sigma: [-3, 3] \to \mathbb{R}^2$ parametrized by $\sigma(t) = (t, 2t^2)$.

WINDING NUMBER AND THE JORDAN CURVE THEOREM

2.7. Show that the curve $\sigma: [-2, 2] \to \mathbb{R}^2$ given by $\sigma(t) = (t^3 - 4, t^2 - 4)$ is a Jordan curve.

2.8. Prove that the complement of the support of a regular piecewise C^2 Jordan curve has exactly two components, one of them bounded and the other unbounded.

2.9. Prove the *Jordan arc theorem*: if $C \subset \mathbb{R}^2$ is the support of a non-closed regular piecewise C^2 plane simple curve $\sigma: [a, b] \to \mathbb{R}^2$, then $\mathbb{R}^2 \setminus C$ is connected.

2.10. Is the Jordan arc theorem, stated in Exercise 2.9, still true if σ is defined in the interval $[a, b)$?

2.11. Let $\sigma: \mathbb{R} \to \mathbb{R}^2$ be a piecewise C^2 Jordan arc having support C closed in $\mathbb{R}^2$. Prove that $\mathbb{R}^2 \setminus C$ has exactly two components. Find a regular piecewise C^2 plane simple curve with support C closed in $\mathbb{R}^2$ such that $\mathbb{R}^2 \setminus C$ has more than two components.

2.12. Prove that the support of a continuous Jordan curve is homeomorphic to S^1.

2.13. Let σ_0, $\sigma_1 \colon [a, b] \to D \setminus \{0\}$ be two continuous closed curves, where $D \subset \mathbb{R}^2$ is an open disk with center in the origin. Prove that σ_0 and σ_1 have the same winding number with respect to the origin if and only if there exists a homotopy $\Phi \colon [0, 1] \times [a, b] \to D \setminus \{O\}$ between σ_0 and σ_1.

2.14. Let $\sigma = (\sigma_1, \sigma_2) \colon [a, b] \to \mathbb{R}^2$ be a piecewise C^1 closed curve with support contained in a circle with center $p = (x_1^o, x_2^o) \in \mathbb{R}^2$ and radius $r > 0$. Prove that

$$\int_a^b (\sigma_1 - x_1^o)\sigma_2' \, dt = -\int_a^b \sigma_1'(\sigma_2 - x_2^o) \, dt = \pi r^2 \iota_{p_0}(\sigma) \,.$$

2.15. For all $\lambda \in \mathbb{R}$ let $\sigma_\lambda \colon \mathbb{R} \to \mathbb{R}^3$ be the curve given by

$$\sigma_\lambda(t) = (2\cos(t) + \lambda t, \sin(t) + \lambda t^2, \lambda t^3) \,.$$

(i)   Compute the curvature of $\sigma_\lambda(t)$ at the point $t = \pi$ as a function of the parameter λ.

(ii) For which values of the parameter λ is the curve $\sigma_\lambda(t)$ a plane curve?

(iii) Compute the winding number of σ_0 with respect to the points $(3, 0, 0)$, and $(0, 0, 0)$.

ROTATION INDEX AND THE TURNING TANGENTS THEOREM

2.16. Let $\sigma \colon [a, b] \to \mathbb{R}^2$ be a regular closed plane curve, and let $\tilde{n} \colon [a, b] \to S^1$ be its oriented normal versor. Compute the degree of $\tilde{n}$ in terms of the rotation index of σ.

2.17. Take a number $r > 0$ and two distinct points p_1, $p_2 \in \partial B(O, r) \subset \mathbb{R}^2$. Choose then two vectors v_1, $v_2 \in S^1$ such that $v_1 \neq -v_2$, $\langle v_1, p_1 \rangle \leq 0$, and $\langle v_2, p_2 \rangle \geq 0$. Prove that there exists a regular curve $\tau \colon [a, b] \to \mathbb{R}^2$ parametrized with respect to the arc length, satisfying the following conditions:

(a) $\tau(a) = p_1$, $\dot{\tau}(a) = v_1$, $\tau(b) = p_2$, and $\dot{\tau}(b) = v_2$;

(b) $\tau([a, b]) \subset \overline{B(O, r)}$;

(c) if $\psi \colon [a, b] \to \mathbb{R}$ is the angle of rotation of τ, then $|\psi(s) - \psi(t)| < \pi$ for all $s, t \in [a, b]$.

(*Hint:* in most cases a hyperbola works.)

Supplementary material

2.5 Convex curves

It is easy to construct examples of plane curves whose oriented curvature changes sign (see, for instance, Problem 2.5); in this section we want to characterize the closed curves for which this does *not* happen. The interesting fact is that the characterization involves global properties of the support of the curve.

Definition 2.5.1. A regular plane curve $\sigma\colon [a, b] \to \mathbb{R}^2$ is said to be *convex* if for all $t_0 \in [a, b]$ the support of σ is contained in one of the two closed half-planes bounded by the affine tangent line to σ at $\sigma(t_0)$, that is, if the function

$$\psi_{t_0}(t) = \langle \sigma(t) - \sigma(t_0), \tilde{\mathbf{n}}(t_0) \rangle$$

has a constant sign on $[a, b]$ for all $t_0 \in [a, b]$.

Definition 2.5.2. We shall say that a regular plane curve $\sigma\colon [a, b] \to \mathbb{R}^2$ parametrized by arc length is *repetitive* if it is not simple and for all $a < s_0 < s_1 < b$ such that $\sigma(s_0) = \sigma(s_1)$ there exists $\varepsilon > 0$ such that $\sigma(s_0 + s) = \sigma(s_1 + s)$ for all $s \in (-\varepsilon, \varepsilon)$ or $\sigma(s_0 + s) = \sigma(s_1 - s)$ for all $s \in (-\varepsilon, \varepsilon)$.

Repetitive curve are in a sense periodic in their domain; see Exercise 2.20. We may now prove the announced characterization:

Theorem 2.5.3. *Let* $\sigma\colon [a, b] \to \mathbb{R}^2$ *be a* C^2 *regular closed curve parametrized by arc length having oriented curvature* $\tilde{\kappa}\colon [a, b] \to \mathbb{R}$. *Then* σ *is simple and* $\tilde{\kappa}$ *does not change its sign if and only if* σ *is convex and non-repetitive.*

Proof. Since σ is parametrized by arc length, by (1.10) and Proposition 2.1.16 (see also Problem 2.3) we know that

$$\theta(s) = x_0 + \int_a^s \tilde{\kappa}(t)\,\mathrm{d}t$$

is a lift of $\mathbf{t}$, where $x_0 = \theta(a) \in \mathbb{R}$ is a determination of the angle between the x-axis and $\mathbf{t}(a)$.

Assume now that σ is simple and (up to inverting the orientation) with oriented curvature $\tilde{\kappa} \geq 0$; in particular, θ, having a non-negative derivative, is a not-decreasing function.

If, by contradiction, σ were not convex, we could find s_0, s_1, and $s_2 \in [a, b]$ distinct (and with at least two of them different from a and b) such that

$$\psi_{s_0}(s_1) = \langle \sigma(s_1) - \sigma(s_0), \tilde{\mathbf{n}}(s_0) \rangle < 0 < \langle \sigma(s_2) - \sigma(s_0), \tilde{\mathbf{n}}(s_0) \rangle = \psi_{s_0}(s_2) \;;$$

we may also assume that s_1 is the minimum point and s_2 the maximum point of ψ_{s_0}. In particular, $\psi'_{s_0}(s_1) = \psi'_{s_0}(s_2) = 0$; as $\psi'_{s_0}(s) = \langle \mathbf{t}(s), \tilde{\mathbf{n}}(s_0) \rangle$, we deduce that $\mathbf{t}(s_0)$, $\mathbf{t}(s_1)$ and $\mathbf{t}(s_2)$ are all parallel, so two of them coincide. Assume $\mathbf{t}(s_1) = \mathbf{t}(s_2)$, with $s_1 < s_2$; the proof in the other cases is analogous.

From $\mathbf{t}(s_1) = \mathbf{t}(s_2)$ we deduce that $\theta(s_2) = \theta(s_1)+2k\pi$ for some $k \in \mathbb{Z}$. But θ is not-decreasing and, since σ is simple, $\theta(b) - \theta(a) = 2\pi$ (Theorem 2.4.7). So, necessarily, $\theta(s_2) = \theta(s_1)$, and hence θ and $\mathbf{t}$ are constant on the interval $[s_1, s_2]$. But this would imply that $\sigma|_{[s_1,s_2]}$ is a line segment; in particular, we would have $\sigma(s_2) = \sigma(s_1) + (s_2 - s_1)\mathbf{t}(s_1)$. But, since $\mathbf{t}(s_1) = \pm\mathbf{t}(s_0) \perp \tilde{\mathbf{n}}(s_0)$, this would imply $\psi_{s_0}(s_2) = \psi_{s_0}(s_1)$, contradiction.

To prove the converse, assume that σ is convex and non-repetitive. The function $(s,t) \mapsto \psi_s(t)$ is continuous, and for fixed s has a constant sign. If the sign were positive for some values of s and negative for other values of s, by continuity there should exist (why?) a s_0 for which $\psi_{s_0} \equiv 0$; but this would imply $\sigma(s) = \sigma(s_0) + (s - s_0)\mathbf{t}(s_0)$ for all $s \in [a,b]$, and σ would be convex indeed, but not closed.

So the sign of the function $\psi_s(t)$ is constant; up to inverting the orientation of σ we may assume $\psi_s(t) \geq 0$ everywhere. In particular, every $s_0 \in [a,b]$ is a absolute minimum point for ψ_{s_0}, and so

$$\tilde{\kappa}(s_0) = \psi''_{s_0}(s_0) \geq 0,$$

as required.

We have still to prove that σ is simple. Suppose on the contrary that it is not so, and let s_0, $s_1 \in [a,b)$ be distinct points such that $\sigma(s_0) = \sigma(s_1)$. Since

$$\psi_{s_0}(s_1 + s) = \langle \sigma(s_1 + s) - \sigma(s_0), \tilde{\mathbf{n}}(s_0)\rangle = \langle \mathbf{t}(s_1), \tilde{\mathbf{n}}(s_0)\rangle s + o(s) \,,$$

the convexity of σ implies $\mathbf{t}(s_1) = \pm\mathbf{t}(s_0)$. For $\varepsilon > 0$ small enough, define then $\sigma_1 \colon (-\varepsilon, \varepsilon) \to \mathbb{R}^2$ by setting $\sigma_1(t) = \sigma(s_1 \pm s)$, where the sign is chosen in such a way that $\dot{\sigma}_1(0) = \mathbf{t}(s_0)$. Moreover, since σ is not repetitive, there exist $\delta > 0$ and a sequence $s_\nu \to 0$ such that every $\sigma_1(s_\nu)$ is not of the form $\sigma(s_0 + s)$ for any $s \in (-\delta, \delta)$.

Now, arguing as in the first part of the proof of the existence of tubular neighborhoods, we find that for $\varepsilon > 0$ small enough there exist two functions h, $\alpha \colon (-\varepsilon, \varepsilon) \to \mathbb{R}$ of class C^1 such that it is possible to write

$$\sigma_1(s) = \sigma\big(h(s)\big) + \alpha(s)\tilde{\mathbf{n}}\big(h(s)\big) \,,$$

with $\alpha(0) = 0$, $h(0) = s_0$ and $h(s) \in (s_0 - \delta, s_0 + \delta)$ for all $s \in (-\varepsilon, \varepsilon)$.

Note that

$$\alpha(s) = \langle \sigma_1(s) - \sigma\big(h(s)\big), \tilde{\mathbf{n}}\big(h(s)\big)\rangle = \psi_{h(s)}(s_1 \pm s) \geq 0 \,.$$

In particular, $\dot{\alpha}(0) = 0$; moreover, by construction we have $\alpha(s_\nu) > 0$ for all $\nu \in \mathbb{N}$.

Now,

$$\dot{\sigma}_1(s) = \dot{h}(s)\big[1 - \alpha(s)\tilde{\kappa}\big(h(s)\big)\big]\mathbf{t}\big(h(s)\big) + \dot{\alpha}(s)\tilde{\mathbf{n}}\big(h(s)\big) \,.$$

In particular, $\dot{\sigma}_1(0) = \dot{h}(0)\mathbf{t}(s_0)$, and so $\dot{h}(0) = 1$. Moreover, the oriented normal versor of σ in $s_1 \pm s$ is obtained by rotating $\dot{\sigma}_1(s)$ counterclockwise by $\pi/2$, so

$$\tilde{\mathbf{n}}(s_1 \pm s) = \dot{h}(s)\big[1 - \alpha(s)\tilde{\kappa}\big(h(s)\big)\big]\tilde{\mathbf{n}}\big(h(s)\big) - \dot{\alpha}(s)\mathbf{t}\big(h(s)\big) \ .$$

But then

$$0 \le \psi_{s_1 \pm s_\nu}\big(h(s_\nu)\big) = \big\langle \sigma\big(h(s_\nu)\big) - \sigma_1(s_\nu), \tilde{\mathbf{n}}(s_1 \pm s_\nu)\big\rangle$$
$$= -\alpha(s_\nu)\dot{h}(s_\nu)\big[1 - \alpha(s_\nu)\tilde{\kappa}\big(h(s_\nu)\big)\big]$$

which is negative for ν large enough, yielding a contradiction. $\square$

The next few examples show that it is not possible to relax the hypotheses of this theorem.

Example 2.5.4. The logarithmic spiral $\sigma \colon \mathbb{R} \to \mathbb{R}^2$ defined in Problem 1.4 and given by $\sigma(t) = (e^{-t}\cos t, e^{-t}\sin t)$ is simple and has oriented curvature $\tilde{\kappa}(t) = e^t/\sqrt{2}$ positive everywhere, but it is neither closed nor (globally) convex.

Example 2.5.5. The regular closed curve $\sigma \colon [0, 2\pi] \to \mathbb{R}^2$ introduced in Problem 2.1 (see Fig. 2.8) and given by $\sigma(t) = \big((2\cos t - 1)\cos t, (2\cos t - 1)\sin t\big)$ has oriented curvature $\tilde{\kappa}(t) = (9 - 6\cos t)/(5 - 4\cos t)^{3/2}$ positive everywhere but is neither convex nor simple: $\sigma(\pi/3) = \sigma(5\pi/3) = (0,0)$.

Example 2.5.6. The regular closed curve $\sigma \colon [0, 4\pi] \to \mathbb{R}^2$ parametrized by arc length given by $\sigma(s) = (\cos s, \sin s)$ is convex and repetitive with positive oriented curvature.

Exercises

2.18. Show that the curve $\sigma \colon [-2, 2] \to \mathbb{R}^2$ given by $\sigma(t) = (t^3 - 4, t^2 - 4)$ is convex (see also Exercise 2.7).

2.19. Show that the graph of a function $f \colon [a, b] \to \mathbb{R}$ of class C^2 is convex if and only if f is convex.

2.20. Let $\sigma \colon [a, b] \to \mathbb{R}^2$ be a repetitive regular plane curve parametrized by arc length. Prove that one of the following holds:

(a) there exists $l \in (0, b - a)$ such that $\sigma(a + s) = \sigma(a + l + s)$ for all $s \in [0, b - a - l]$;
(b) there exists $l \in (2a, a + b)$ such that $\sigma(a + s) = \sigma(l - a - s)$ for all $s \in [0, l - 2a]$;
(c) there exists $l \in (a + b, 2b)$ such that $\sigma(a + s) = \sigma(l - a - s)$ for all $s \in [l - a - b, b - a]$.

2.21. The *cissoid of Diocles* is the support of the curve $\sigma\colon \mathbb{R} \to \mathbb{R}^2$ given by

$$\sigma(t) = \left(\frac{2ct^2}{1+t^2}, \frac{2ct^3}{1+t^2} \right),$$

with $c \in \mathbb{R}^*$.

(i)   Show that the curve σ is simple.
(ii) Show that σ is not regular at O.
(iii) Show that for all $0 < a < b$ the restriction of σ to $[a, b]$ is convex.
(iv) Show that, for $t \to +\infty$, the tangent line to σ at $\sigma(t)$ tends to the line having equation $x = 2c$, the *asymptote* of the cissoid.

2.6 The four-vertex theorem

The oriented curvature $\tilde{\kappa}$ of a closed plane curve is a continuous function defined on a compact set, and so it always admits a maximum and a minimum; in particular, $\tilde{\kappa}'$ is zero in at least two points. An unexpected result from the global theory of plane curves is the fact that the derivative of the oriented curvature of a Jordan curve is in fact zero in at least four points, rather than just two.

Definition 2.6.1. A *vertex* of a regular plane curve is a zero of the derivative of its oriented curvature.

Remark 2.6.2. In this section we shall only deal with everywhere regular curves, so no confusion between this notion of a vertex and that of vertex of a piecewise C^1 curve will be possible.

The main goal of this section is to prove the following:

Theorem 2.6.3 (Four-vertex theorem). *A C^2 regular Jordan curve has at least four vertices.*

We shall give a proof based on ideas by Robert Ossermann (see [18]). Let us begin by introducing the notion of circle circumscribing a set.

Lemma 2.6.4. *Let $K \subset \mathbb{R}^2$ be a compact set containing more than one point. Then there exists a unique closed disk with minimum radius that contains K.*

Proof. Let $f\colon \mathbb{R}^2 \to \mathbb{R}$ be the continuous function given by

$$f(p) = \max_{x \in K} \|x - p\|.$$

Since K contains more than one point, we have $f(p) > 0$ for all $p \in \mathbb{R}^2$. Moreover, the closed disk with center p and radius R contains K if and only if $R \geq f(p)$. Lastly, it is clear that $f(p) \to +\infty$ as $\|p\| \to +\infty$; so f has an absolute minimum point $p_0 \in \mathbb{R}^2$, and $R_0 = f(p_0) > 0$ is the minimum radius of a closed disk containing K.

Finally, if K were contained in two distinct closed disks D_1 and D_2 both with radius R_0, we would have $K \subseteq D_1 \cap D_2$. But the intersection of two distinct disks with the same radius always lies in a disk with a smaller radius, against our choice of R_0, so the disk with minimum radius containing K is unique.   □

Definition 2.6.5. Let $K \subset \mathbb{R}^2$ be a compact set containing at least two points. The boundary of the disk with minimum radius containing K is the *circle circumscribed* about K.

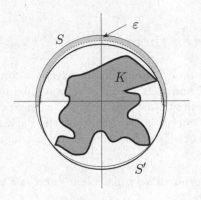

Fig. 2.9.

The circle circumscribed about a compact set intersects it in at least two points:

Lemma 2.6.6. *Let $K \subset \mathbb{R}^2$ be a compact set containing more than one point, and S the circle circumscribed about K. Then K intersects every closed semicircle of S. In particular, $S \cap K$ contains at least two points, and if it contains exactly two points, they are antipodal.*

Proof. We may assume that S has center in the origin and radius $R > 0$; it is sufficient (why?) to prove that K intersects the upper semicircle S_+. Since S_+ and K are compact, if they were disjoint they would be at a strictly positive distance ε. So K would be contained in the set consisting of the disk having S as boundary with a neighborhood of radius ε of the upper semicircle removed; but then it is easy to verify (see Fig. 2.9) that this set is contained within a circle S' with center on the y-axis and radius strictly smaller than R, yielding a contradiction.   □

The idea is that any intersection point between the support of a regular Jordan curve and the circle circumscribed yields at least two vertices. To prove this we need two more lemmas.

Lemma 2.6.7. *Let* $\sigma\colon [a,b] \to \mathbb{R}^2$ *be a regular plane curve of class* C^2 *parametrized by arc length with oriented curvature* $\tilde{\kappa}\colon [a,b] \to \mathbb{R}$ *and support* $C = \sigma([a,b])$, *and let* S *be a positively oriented circle with radius* $R > 0$. *Assume that there exists* $s_0 \in [a,b]$ *such that* $\dot{\sigma}(s_0)$ *coincides with the tangent versor to* S *at* $p_0 = \sigma(s_0) \in S$. *Then:*

(i) *if there exists a neighborhood* U *of* p_0 *such that* $U \cap C$ *lies within the closed disk having* S *as its boundary then* $\tilde{\kappa}(s_0) \geq 1/R$;

(ii) *if there exists a neighborhood* U *of* p_0 *such that* $U \cap C$ *lies in the complement of the open disk having* S *as its boundary then* $\tilde{\kappa}(s_0) \leq 1/R$.

Proof. Assuming that $\dot{\sigma}(s_0)$ coincides with the tangent versor to S at p_0 amounts to saying that $R\,\tilde{\mathbf{n}}(s_0) = -(p_0 - x_0)$, where $x_0 \in \mathbb{R}^2$ is the center of S and $\tilde{\mathbf{n}}$ is the oriented normal versor of σ. Define then the function $h\colon [a,b] \to \mathbb{R}$ by setting $h(s) = \|\sigma(s) - x_0\|^2$. Clearly, we have $h(s_0) = R^2$; moreover, by differentiating we find $h'(s_0) = 2\langle \mathbf{t}(s_0), p_0 - x_0 \rangle = 0$, and $h''(s_0) = 2\big(1 - \tilde{\kappa}(s_0)R\big)$. Since s_0 is, in case (i), a local maximum point for h and, in case (ii), a local minimum point for h, we get the assertion. $\qquad\square$

Lemma 2.6.8. *Let* $\sigma\colon [a,b] \to \mathbb{R}^2$ *be a positively oriented regular Jordan curve of class* C^2 *parametrized by arc length with support* $C = \sigma([a,b])$ *and oriented curvature* $\tilde{\kappa}\colon [a,b] \to \mathbb{R}$. *Let* S *be the circle circumscribed about* C, *having radius* $R > 0$. *Take two distinct points* $p_1 = \sigma(s_1)$, $p_2 = \sigma(s_2) \in C \cap S$, *with* $s_1 < s_2$. *Then either the support of* $\sigma|_{[s_1, s_2]}$ *is contained in* S *or there exists a point* $s_0 \in (s_1, s_2)$ *such that* $\tilde{\kappa}(s_0) < 1/R$.

Proof. We begin with a remark. Let S be positively oriented too; moreover, up to a translation of the plane, we may assume that S has center in the origin. If $h\colon [a,b] \to \mathbb{R}$ is the function $h(s) = \|\sigma(s)\|^2$ then the maxima of h are exactly the points of intersection between C and S. By differentiating h we find that in these points σ has to be tangent to S; moreover, since the interior of σ is contained in the interior of S and both curves are positively oriented, the oriented normal versors (and thus the tangent versors) of σ and of S at these points coincide.

Up to a rotation, we may assume that p_1 and p_2 belong to the same vertical line ℓ, and that p_1 is under p_2. Since S and σ have the same tangent versor in p_1, the support C_1 of $\sigma|_{(s_1, s_2)}$ intersects the right half-plane determined by ℓ. If C_1 is contained in S, we are done; otherwise, there exists a point $q_1 \in C_1$ belonging to the open disk having S as boundary, and the circle through p_1, q_1 and p_2 has radius $R' > R$.

Translate leftwards this circle to get a circle S' that intersects C_1 in a point $q_0 = \sigma(s_0)$, but such that any further left translation would give us a circle disjoint from C_1. In particular, by using an analogue of function h, we immediately see that σ is tangent to S' at q_0.

Moreover, σ and S' (positively oriented) have the same tangent versor at q_0. Indeed, consider the simple curve obtained by joining C_1 to the two half-lines having as endpoints p_1 e p_2. By Exercise 2.11 this curve divides the plane in two components, which we shall call the right and left components. Since σ is positively oriented, the interior of σ is completely contained in the left component. By our construction of S', the interior of S' is contained in the left component as well (since it lies on the left of the half-lines starting from p_1 and p_2). But this implies that in q_0 both the oriented normal versor of S' and the oriented normal versor of σ (which, as we already know, are parallel) point toward the left component, and so coincide.

We can then apply Lemma 2.6.7, obtaining $\tilde{\kappa}(s_0) \leq 1/R' < 1/R$, as required. $\square$

We may now prove Theorem 2.6.3:

Proof (of Theorem 2.6.3). Let $\sigma\colon [a,b] \to \mathbb{R}^2$ be a positively oriented regular Jordan curve of class C^2 parametrized by arc length with support C, and let S be the circle circumscribed about C, positively oriented; up to a translation of the plane, we may assume that S has center in the origin and radius $R > 0$.

Arguing as in the first part of the proof of Lemma 2.6.8, we see that the tangent versors of σ and of S in the points of intersection coincide. Lemma 2.6.7 implies then that the oriented curvature of σ is greater or equal than $1/R$ in every point of intersection.

Let now p_0 and $p_1 \in C \cap S$ be two distinct points of intersection between C and S; by Lemma 2.6.6 we know that they exist. Up to modifying the domain of σ, we may assume $p_0 = \sigma(a)$ and $p_1 = \sigma(s_1)$ for some $s_1 \in (a,b)$. If the support of $\sigma|_{[a,s_1]}$ lies in S, the oriented curvature $\tilde{\kappa}$ of σ is constant on $[a, s_1]$, and so we have infinitely many vertices. Otherwise, Lemma 2.6.8 tells us that the minimum of $\tilde{\kappa}$ in $[a, s_1]$ occurs in a point $s_0 \in (a, s_1)$, and thus we have found a vertex of σ. Now, the same lemma applied to p_1 and $p_0 = \sigma(b)$ instead, provides us with another minimum point $s_0' \in (s_1, b)$, and a second vertex. But between two minimum points a maximum point must occur, different from them since $\tilde{\kappa}(s_1) \geq 1/R$; so we find a third vertex in (s_0, s_0'), and, for the same reason, a fourth vertex in $[a, s_0) \cup (s_0', b]$. $\square$

In Theorem 2.6.3 the hypotheses for the curve to be simple and closed are crucial, as the following examples show.

Example 2.6.9. The curve $\sigma\colon [0, 2\pi] \to \mathbb{R}^2$ of Example 2.5.5 is not simple and has just two vertices, since

$$\tilde{\kappa}'(t) = \frac{12(\cos t - 2)\sin t}{(5 - 4\cos t)^{5/2}}.$$

Example 2.6.10. The parabola $\sigma\colon \mathbb{R} \to \mathbb{R}^2$ given by $\sigma(t) = (t, t^2)$ is simple and has oriented curvature $\tilde{\kappa}(t) = 2/(1 + 4t^2)^{3/2}$, so it has just one vertex.

Finally, there exist curves with exactly four vertices:

Example 2.6.11. Let $\sigma\colon [0, 2\pi] \to \mathbb{R}^2$ be the ellipse having semiaxes $a, b > 0$ parametrized as in Example 1.1.30. Keeping in mind Example 1.3.11, it is straightforward to see that the oriented curvature of σ is

$$\tilde{\kappa}(t) = \frac{ab}{(a^2 \sin^2 t + b^2 \cos^2 t)^{3/2}} \, ,$$

so

$$\tilde{\kappa}'(t) = \frac{3ab(b^2 - a^2) \sin(2t)}{2(a^2 \sin^2 t + b^2 \cos^2 t)^{5/2}} \, ,$$

and σ has exactly 4 vertices as soon as $a \neq b$ (that is, as soon as σ is not a circle).

Exercises

2.22. Let $\sigma\colon \mathbb{R} \to \mathbb{R}^2$ be the curve $\sigma(t) = (3\cos^2 t - t, 4 + 2\sin^2 t, 2 + 2t)$.

(i) Compute the curvature and the torsion of σ.
(ii) Find at least one line tangent to σ at infinitely many points.

2.23. Let $\sigma\colon \mathbb{R} \to \mathbb{R}^2$ be the ellipse given by $\sigma(t) = (4 + 2\cos t, -5 + 3\sin t)$.

(i) Find the circle circumscribed about the ellipse.
(ii) Find the points where the ellipse and the circle touch and show that the two curves have there the same affine tangent line.

2.7 Isoperimetric inequality

In this section we want to prove a result in the global theory of curves with a flavour somewhat different from the previous ones. The question we intend to answer is the following: which regular simple closed plane curve having a fixed length $L > 0$ is the boundary of a region of the largest area? As we shall see, answering this question we shall find a more general result relating the area and the perimeter of any domain bounded by a piecewise C^2 regular simple closed curve.

First of all, we need a formula to compute the area of the interior (see Exercise 2.8) of a piecewise C^2 regular Jordan curve. To find it, we borrow from Differential Calculus the classical *divergence or Gauss-Green theorem* (see [5, pp. 359–360]):

Theorem 2.7.1 (Gauss-Green). *Let $\sigma = (\sigma_1, \sigma_2)\colon [a, b] \to \mathbb{R}^2$ be a positively oriented piecewise C^2 regular Jordan curve, and denote by $D \subset \mathbb{R}^2$ the* *interior of σ. Then, for any pair of functions f_1, $f_2 \in C^1(\bar{D})$ defined and of class C^1 in a neighborhood of D we have*

$$\int_a^b \left[f_1(\sigma(t))\sigma_1'(t) + f_2(\sigma(t))\sigma_2'(t) \right] dt = \int_D \left(\frac{\partial f_2}{\partial x_1} - \frac{\partial f_1}{\partial x_2} \right) dx_1\, dx_2 \,. \quad (2.6)$$

Then:

Lemma 2.7.2. *Let $\sigma = (\sigma_1, \sigma_2) \colon [a, b] \to \mathbb{R}^2$ be a positively oriented piecewise C^2 regular Jordan curve, and denote by $D \subset \mathbb{R}^2$ the interior of σ. Then*

$$\mathrm{Area}(D) = \int_a^b \sigma_1\sigma_2'\, dt = -\int_a^b \sigma_1'\sigma_2\, dt \,.$$

Proof. Let f_1, $f_2 \colon \mathbb{R}^2 \to \mathbb{R}$ be given by $f_1(x) = -x_2$ and $f_2(x) = x_1$. Then (2.6) yields

$$\mathrm{Area}(D) = \int_D dx_1\, dx_2 = \frac{1}{2} \int_D \left(\frac{\partial f_2}{\partial x_1} - \frac{\partial f_1}{\partial x_2} \right) dx_1\, dx_2$$

$$= \frac{1}{2} \int_a^b (\sigma_1\sigma_2' - \sigma_1'\sigma_2)\, dt \,.$$

Now,

$$\int_a^b (\sigma_1\sigma_2' + \sigma_1'\sigma_2)\, dt = \int_a^b (\sigma_1\sigma_2)'\, dt = \sigma_1(b)\sigma_2(b) - \sigma_1(a)\sigma_2(a) = 0\,;$$

so

$$\int_a^b \sigma_1\sigma_2'\, dt = -\int_a^b \sigma_1'\sigma_2\, dt \,,$$

and we are done.   □

We are now able to prove, following an idea by Peter Lax (see [9]) the main result of this section, the *isoperimetric inequality*.

Theorem 2.7.3. *Let $\sigma \colon [a, b] \to \mathbb{R}^2$ be a piecewise C^2 regular Jordan curve having length $L > 0$, and denote by $D \subset \mathbb{R}^2$ its interior. Then*

$$4\pi \mathrm{Area}(D) \le L^2 \,. \quad (2.7)$$

Moreover, the equality holds if and only if the support of σ is a circle.

Proof. Given $r > 0$, the length of the curve $\sigma^r = r\sigma$ (which is the curve obtained by applying to σ a homothety with ratio r) is rL, while (why?) the area of the interior of σ^r is $r^2\mathrm{Area}(D)$.

So, up to replacing σ by $\sigma^{2\pi/L}$, we may assume $L = 2\pi$, and it is enough to prove that

$$\mathrm{Area}(D) \le \pi \,,$$

with the equality holding if and only if the support of σ is a circle.

Clearly, we may assume σ to be parametrized by arc length, with $a = 0$ and $b = 2\pi$, and the starting point is chosen so that $\sigma(0)$ and $\sigma(\pi)$ are not vertices of σ (in other words, σ is of class C^2 in a neighborhood of 0 and of π). Moreover, up to a rigid motion of the plane, we may also assume that $\sigma_1(0) = \sigma_1(\pi) = 0$.

By Lemma 2.7.2,

$$\text{Area}(D) = \int_0^{2\pi} \sigma_1 \dot{\sigma}_2 \, ds \; ;$$

so it will be enough to prove that both the integral from 0 to π and the integral from π to 2π of $\sigma_1 \dot{\sigma}_2$ are at most $\pi/2$, with the equality holding if and only if the support of σ is a circle.

First of all, we have

$$\int_0^\pi \sigma_1 \dot{\sigma}_2 \, ds \leq \frac{1}{2} \int_0^\pi (\sigma_1^2 + \dot{\sigma}_2^2) \, ds = \frac{1}{2} \int_0^\pi (1 + \sigma_1^2 - \dot{\sigma}_1^2) \, ds \, , \qquad (2.8)$$

where the last equality holds because σ is parametrized by arc length.

Now, since $\sigma_1(0) = \sigma_1(\pi) = 0$, there exists a piecewise C^1 function $u \colon [0, L/2] \to \mathbb{R}$ such that

$$\sigma_1(s) = u(s) \sin s \, .$$

In particular, $\dot{\sigma}_1 = \dot{u} \sin s + u \cos s$, so (2.8) gives

$$\int_0^\pi \sigma_1 \dot{\sigma}_2 \, ds \leq \frac{1}{2} \int_0^\pi \left(1 - \dot{u}^2 \sin^2 s + u^2(\sin^2 s - \cos^2 s) - 2u\dot{u} \sin s \cos s\right) ds \, .$$

But

$$\int_0^\pi \left(u^2(\sin^2 s - \cos^2 s) - 2u\dot{u} \sin s \cos s\right) ds = -\int_0^\pi \frac{d}{ds}\left[u^2 \sin s \cos s\right] ds = 0 \; ;$$

so

$$\int_0^\pi \sigma_1 \dot{\sigma}_2 \, ds \leq \frac{1}{2} \int_0^\pi (1 - \dot{u}^2 \sin^2 s) \, ds \leq \frac{\pi}{2} \, , \qquad (2.9)$$

as required. An analogous argument applies to the integral from π to 2π, and so (2.7) is proved.

If the support of σ s a circle, equality in (2.7) is well known. Conversely, if in (2.7) the equality holds, it has to hold in (2.8) and (2.9) as well. Equality in the latter implies $\dot{u} \equiv 0$, so $\sigma_1(s) = c \sin s$ for a suitable $c \in \mathbb{R}$. But equality in (2.8) implies $\sigma_1 \equiv \dot{\sigma}_2$, so $\sigma_2(s) = -c \cos s + d$, for a suitable $d \in \mathbb{R}$, and the support of σ is a circle. $\qquad\square$

Remark 2.7.4. Theorem 2.7.3 actually holds for all rectifiable continuous Jordan curves, but with a rather different proof; see [2] and [17].

We conclude this section with a straightforward corollary of the isoperimetric inequality, which, in particular, answers the question we asked at the beginning of this section:

Corollary 2.7.5. *Among the piecewise C^2 regular Jordan curves having a fixed length the circle is the one enclosing the largest area. Conversely, a disk, among the domains with fixed area bounded by a piecewise C^2 regular Jordan curve, has the shortest perimeter.*

Proof. If the length is fixed to be equal to L, then by Theorem 2.7.3 the area may be at most $L^2/4\pi$, and this value is only attained by the circle. Conversely, if the area is fixed to be equal to A, then by Theorem 2.7.3 the circumference is at least equal to $\sqrt{4\pi A}$, and this value is only attained by a disk. $\qquad\square$

Exercises

2.24. Let $\sigma\colon [0,b] \to \mathbb{R}^2$ be a positively oriented convex regular plane curve of class C^∞ parametrized by arc length. Having fixed a real number $\lambda > 0$, the map $\sigma_\lambda\colon [0,b] \to \mathbb{R}^2$ defined by

$$\sigma_\lambda(s) = \sigma(s) - \lambda\tilde{\mathbf{n}}(s)$$

is called *curve parallel to* σ .

(i)   Show that σ_λ is a regular curve.
(ii) Show that the curvature κ_λ of σ_λ is given by $\kappa_\lambda = \kappa/(1 + \lambda\kappa)$, where $\kappa = \kappa_0$ is the curvature of σ.
(iii) Draw the supports of σ_λ and of σ in the case of the unit circle $\sigma\colon [0,2\pi] \to \mathbb{R}^2$ parametrized by $\sigma(t) = (\cos t, \sin t)$.
(iv) Show that if σ is as in (iii) then the lengths of σ and σ_λ satisfy $L(\sigma_\lambda) = L(\sigma) + 2\lambda\pi$, while the areas of the interiors D of σ and D_λ of σ_λ satisfy $\mathrm{Area}(D_\lambda) = \mathrm{Area}(D) + 2\lambda\pi + \pi\lambda^2$.
(v)   Are the formulas for the length and the area found in (iv) still true for general σ?
(vi) Show that there exists $\lambda < 0$ such that σ_λ is still regular.

2.8 Schönflies' theorem

In this section we give an elementary proof due to Thomassen (see [23]) of the Schönflies theorem for Jordan curves mentioned in Remark 2.3.7. Along the way we shall also give a proof of the Jordan curve theorem for continuous curves.

Remark 2.8.1. In this section, with a slight abuse of language, we shall call Jordan arcs and curves what we have been calling supports of Jordan arcs and curves.

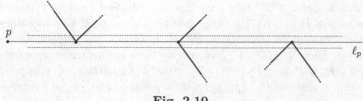

Fig. 2.10.

Definition 2.8.2. A *simple polygonal arc* in the plane is a Jordan arc consisting of finitely many line segments. Analogously, a *simple polygonal closed curve* is a plane Jordan curve consisting of finitely many line segments.

We begin by proving the Jordan curve theorem for simple polygonal closed curves.

Lemma 2.8.3. *If $C \subset \mathbb{R}^2$ is a simple polygonal closed curve, then $\mathbb{R}^2 \setminus C$ consists of exactly two components having C as their common boundary.*

Proof. We begin by showing that $\mathbb{R}^2 \setminus C$ has at most two components. Assume, by contradiction, that p_1, p_2, $p_3 \in \mathbb{R}^2 \setminus C$ belong to distinct components of $\mathbb{R}^2 \setminus C$, and choose an open disk $D \subset \mathbb{R}^2$ such that $D \cap C$ is a line segment (so that $D \setminus C$ has just two components). Since each component of $\mathbb{R}^2 \setminus C$ has C as its boundary, for $j = 1, 2, 3$ we may find a curve starting from p_j, arriving as close to C as we want, and then going parallel to C till it meets D. But $D \setminus C$ has just two components; so at least two of the p_j's can be connected by a curve, against the hypothesis that they belong to distinct components.

We have now to show that $\mathbb{R}^2 \setminus C$ is disconnected. Up to a rotation, we may assume that no horizontal line contains more than one vertex of C. Then define a function $i \colon \mathbb{R}^2 \setminus C \to \{0, 1\}$ as follows: given $p \in \mathbb{R}^2 \setminus C$, denote by ℓ_p the horizontal right half-line issuing from p. If ℓ_p does not contain vertices of C, then $i(p)$ is equal to the number (taken modulo 2) of intersections of ℓ_p with C. If, instead, ℓ_p contains a vertex of C, then $i(p)$ equals the number (taken modulo 2) of intersections with C of a right half-line slightly above (or slightly below) ℓ_p. Fig. 2.10 shows the cases that may happen, and it is clear that $i(p)$ is well defined.

The function i is clearly continuous, so it is constant on each component of $\mathbb{R}^2 \setminus C$. It takes the value 0 for points sufficiently above C; to prove that $\mathbb{R}^2 \setminus C$ is disconnected it suffices to show that it takes the value 1 too. To this end, take a point $p \in \mathbb{R}^2 \setminus C$ such that the line ℓ_p intersects C in points that are not vertices, and let $q_0 \in \ell_p \cap C$ be the rightmost intersection point. Then $i(q) = 1$ for all points of ℓ_p slightly on the left of q_0, and we are done. $\quad\square$

Definition 2.8.4. The *exterior* $\text{ext}(C)$ of a Jordan curve $C \subset \mathbb{R}^2$ is the unique unbounded connected component of $\mathbb{R}^2 \setminus C$, while the *interior* $\text{int}(C)$ is the union of the bounded components of $\mathbb{R}^2 \setminus C$.

Lemma 2.8.5. *Let $C \subseteq \mathbb{R}^2$ be a simple polygonal closed curve. Choose a simple polygonal arc $P \subset \overline{\mathrm{int}(C)}$ joining two points p_1, $p_2 \in C$ and intersecting C only in the endpoints. Denote by P_1, $P_2 \subset C$ the two polygonal arcs in C from p_1 to p_2. Then $\mathbb{R}^2 \setminus (C \cup P)$ has exactly three connected components, with boundary C, $P_1 \cup P$ and $P_2 \cup P$, respectively. In particular, every curve in $\overline{\mathrm{int}(C)}$ joining a point of $P_1 \setminus \{p_1, p_2\}$ with a point of $P_2 \setminus \{p_1, p_2\}$ intersects P.*

Proof. One of the components is clearly $\mathrm{ext}(C)$; so we have to prove that $\mathrm{int}(C) \setminus P$ has exactly two components.

The fact that there are at most two of them is shown as in the proof of previous lemma. The same proof also shows that there are at least two: indeed, choose a line segment $\ell \subset \mathrm{int}(C)$ that intersects P in a single point in $P \cap \mathrm{int}(C)$. Then the endpoints of ℓ belong to different components of $\mathbb{R}^2 \setminus (P \cup P_1)$, and so to different components of $\mathbb{R}^2 \setminus (P \cup C)$. $\square$

Before going on, we need some definitions from graph theory.

Definition 2.8.6. A *graph* G is a pair $(V(G), E(G))$ consisting of a finite set $V(G)$ of points, called *vertices* of the graph, and a set $E(G)$ of (unordered) pairs of points of $V(G)$, called *edges* of the graph. If $e = \{v, w\} \in E(G)$ is an edge of the graph G, we shall say that *e joins* the vertices v and w, or that it is *incident* to v and w. A *subgraph* H of a graph G is given by subsets $V(H) \subseteq V(G)$ and $E(H) \subseteq E(G)$ such that all edges in $E(H)$ join vertices in $V(H)$. A *graph isomorphism* is a bijection between the sets of vertices that induces a bijection between the sets of edges. A *path* L in a graph G is a finite ordered sequence $v_1, \ldots, v_k \in V(G)$ of vertices, with $v_1, \ldots, v_{k-1}$ distinct, and such that $\{v_1, v_2\}, \ldots, \{v_{k-1}, v_k\}$ are edges of G (and we shall say that L *joins* v_1 and v_k). If $v_k = v_1$, we say that L is a *cycle*. If $A \subseteq V(G) \cup E(G)$, we shall denote by $G - A$ the graph obtained by removing all vertices in A and all edges either in A or incident to vertices in A.

Remark 2.8.7. In this definition each pair of vertices of a graph may be joined by at most one edge; our graphs do not admit multiple edges.

Definition 2.8.8. A *realization* of a graph G is a topological space X together with a finite subset of points $V_X(G)$ in a 1-1 correspondence with the vertices of G, and a finite set $E_X(G)$ of Jordan arcs in X, in a 1-1 correspondence with the edges of G, satisfying the following properties:

(a) X is given by the union of the (supports of the) Jordan arcs in $E_X(G)$;
(b) if $\ell \in E_X(G)$ corresponds to $\{v, w\} \in E(G)$, then ℓ is a Jordan arc joining the point $p_v \in V_X(G)$ corresponding to v with the point $p_w \in V_X(G)$ corresponding to w, and we shall write $\ell = p_v p_w$;
(c) two distinct elements of $E_X(G)$ intersect at most in the endpoints.

We shall say that a graph G is *planar* if it can be realized as a subset of the plane $\mathbb{R}^2$. In this case, with a slight abuse of language, we shall often identify a planar graph with its plane realization.

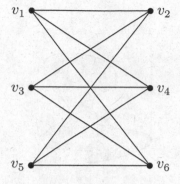

Fig. 2.11. $K_{3,3}$

Example 2.8.9. A main role in the following will be played by the graph $K_{3,3}$: it has six vertices, subdivided in two subsets of three, and all the edges join vertices of the first subset with those of the second one; see Fig. 2.11. However, note that the drawing in Fig. 2.11 is *not* a realization of $K_{3,3}$ in the plane, since the edges intersect in points that are not vertices. As we shall see, one of the key steps for the proof of Schönflies' theorem is exactly the fact that $K_{3,3}$ is not a planar graph.

Every planar graph can be realized by using polygonal arcs:

Lemma 2.8.10. *Every planar graph G has a realization $X_0 \subset \mathbb{R}^2$ such that the elements of $E_X(G)$ are polygonal arcs.*

Proof. Let $X \subset \mathbb{R}^2$ be a realization of G, and for all $p \in V_X(G)$ choose a disk $D_p \subset \mathbb{R}^2$ with center in p, which only intersects X in the edges containing p. Since $V_X(G)$ is finite, we may assume that $D_q \cap D_p = \varnothing$ if $p \neq q$.

For all $pq \in E_X(G)$, let $C_{pq} \subset pq$ be a Jordan arc joining ∂D_p with ∂D_q and only intersecting $\partial D_p \cup \partial D_q$ in the endpoints. Then, first of all, replace $pq \setminus C_{pq}$ with two segments, one joining p with $C_{pq} \cap \partial D_p$, and the other joining q with $C_{pq} \cap \partial D_q$. Since the distance between two distinct arcs of the form C_{pq} is strictly positive we may replace them with disjoint polygonal arcs contained in $\mathbb{R}^2 \setminus \bigcup_{p \in V_X(G)} D_p$, and we have the realization X_0 we were looking for.   □

We are now able to prove the first key result of this section:

Proposition 2.8.11. $K_{3,3}$ *is not planar.*

Proof. The edges of $K_{3,3}$ can be obtained by taking those in the cycle $v_1 v_2 v_3 v_4 v_5 v_6 v_1$, plus $\{v_1, v_4\}$, $\{v_2, v_5\}$ and $\{v_3, v_6\}$; see Fig. 2.11.

Assume that there is a plane realization $X \subset \mathbb{R}^2$ of $K_{3,3}$; by the previous lemma, we may assume that the edges of X are polygonal arcs. Then the edges in the cycle would form a simple closed polygonal closed curve C, and there is no way of inserting the other three edges without violating Lemma 2.8.5.   □

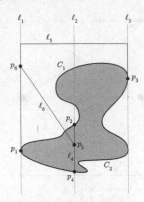

Fig. 2.12.

It is not hard to use this proposition to show that Jordan curves disconnect the plane:

Corollary 2.8.12. *If $C \subset \mathbb{R}^2$ is a Jordan curve, then $\mathbb{R}^2 \setminus C$ is disconnected.*

Proof. Denote by ℓ_1 (respectively, ℓ_3; the rationale behind this numbering will be revealed at the end of the proof) a vertical straight line intersecting C and such that C is contained in the closed right (respectively, left) half-plane having ℓ_1 (respectively, ℓ_3) as its boundary. Let $p_j \in C \cap \ell_j$ be the point having maximum y-coordinate, and denote by C_1 and C_2 the two arcs on C between p_1 and p_3. Let now ℓ_2 be a vertical line between ℓ_1 and ℓ_3. Since $C_1 \cap \ell_2$ and $C_2 \cap \ell_2$ are compact and disjoint, we can find within ℓ_2 a line segment ℓ_4 joining C_1 with C_2 and intersecting C only in its endpoints, p_2 and p_4. Let then ℓ_5 be a simple polygonal arc joining p_1 to p_3 going upwards along ℓ_1 up to a point of $\text{ext}(C)$ above C, going to ℓ_3 with a horizontal line segment, and then going along ℓ_3 down to p_3. Clearly ℓ_5 is contained (except for its endpoints) in $\text{ext}(C)$.

If ℓ_4 (except its endpoints) were contained in $\text{ext}(C)$ as well, we might find a simple arc ℓ_6 joining an interior point p_5 of ℓ_4 with an interior point p_6 of ℓ_5 in $\text{ext}(C)$. But then (see Fig. 2.12) $C \cup \ell_4 \cup \ell_5 \cup \ell_6$ would be a realization of $K_{3,3}$ in the plane (with vertices $\{p_1, p_2, p_3, p_4, p_5, p_6\}$; the numbering has been chosen to be coherent with Fig. 2.11) which is impossible. Hence the interior points of ℓ_4 have to be in $\text{int}(C)$, and thus $\mathbb{R}^2 \setminus C$ is disconnected.   □

We shall need some more results in graph theory.

Definition 2.8.13. A graph G is said to be *connected* if each pair of vertices in G can be joined by a path in G. We shall say that G is *2-connected* if $G - \{v\}$ is connected for all $v \in V(G)$.

Remark 2.8.14. Clearly, a graph is connected if and only if each of its realizations is path-connected.

Lemma 2.8.15. *Let H be a 2-connected subgraph (with at least two vertices) of a 2-connected graph G. Then it is possible to construct a finite sequence $H = G_0, G_1, \ldots, G_k = G$ of subgraphs of G such that G_j is obtained from G_{j-1} by adding a path that joins two distinct vertices of G_{j-1} while the remaining vertices of the path do not belong to G_{j-1}.*

Proof. We argue by induction on the number of edges in $E(G) \setminus E(H)$. If this number is zero then $G = H$ and there is nothing to prove. Assume then that $G \neq H$, and that the lemma has been proved for all pair (G', H') such that the cardinality of $E(G') \setminus E(H')$ is strictly less than the cardinality of $E(G) \setminus E(H)$.

Let H' be a maximal 2-connected proper subgraph of G containing H. If $H' \neq H$, we may apply the inductive hypothesis to (H', H) and to (G, H'); so we may assume that $H' = H$. Since G is connected, there exists an edge $v_1 v_2 \in E(G) \setminus E(H)$ such that $v_1 \in V(H)$. But $G - \{v_1\}$ is connected too; so there is a path $v_2 \ldots v_k$ in $G - \{v_1\}$ such that v_k belongs to $V(H)$ while the v_j's for $2 \leq j < k$ do not (and we are allowing $k = 2$, that is $v_2 \in H$). Since the subgraph G_1 obtained by adding the path $v_1 v_2 \ldots v_k$ to H is still 2-connected, by maximality we have $G_1 = G$, and we are done.   □

Lemma 2.8.16. *Let G be a 2-connected planar graph, and denote by $e(G)$ and $v(G)$ the number of edges and of vertices of G. If $X \subset \mathbb{R}^2$ is a realization of G whose edges are polygonal arcs, then $\mathbb{R}^2 \setminus X$ has exactly $e(G) - v(G) + 2$ connected components, each of which is bounded by a simple polygonal closed curve in X.*

Proof. Let C be a simple polygonal closed curve in X. By Lemma 2.8.3, the claim is true if $X = C$. Otherwise, we can get X from C by adding paths as in the previous lemma. Each path increases the difference between the number of edges and the number of vertices by 1, and subdivides (by Lemma 2.8.5) an existing component in two components in such a way that the boundary of each is still a simple polygonal closed curve in X. The assertion then follows by induction.   □

Definition 2.8.17. If $X \subset \mathbb{R}^2$ is the realization of a planar graph, the connected components of $\mathbb{R}^2 \setminus X$ are the *faces* of X. In particular, the unbounded face is the *outer face* and, if X is 2-connected, the boundary of the outer face is the *outer cycle*.

Remark 2.8.18. If X_1 and X_2 are planar graphs whose edges are realized as polygonal arcs, then it is easy to see that $X_1 \cup X_2$ is (the plane realization of) a third planar graph, having as vertices the intersections of the edges of X_1 and X_2 (in addition to the vertices of X_1 and X_2), and this is the graph we shall refer to when we shall consider unions of planar graphs. Moreover, if X_1 and X_2 are 2-connected and have at least 2 points in common, then $X_1 \cup X_2$ is 2-connected as well.

Lemma 2.8.19. *Let $X_1, \ldots, X_k \subset \mathbb{R}^2$ be 2-connected planar graphs with edges realized as polygonal arcs. Assume that, for $j = 2, \ldots, k-1$, each X_j intersects in at least two points both X_{j-1} and X_{j+1}, and does not meet the other X_i's. Assume furthermore that $X_1 \cap X_k = \varnothing$. Then the intersection of the outer faces of $X_1 \cup X_2$, $X_2 \cup X_3, \ldots, X_{k-1} \cup X_k$ is contained in the outer face of $X_1 \cup \cdots \cup X_k$.*

Proof. Let p be a point in a bounded face of $X = X_1 \cup \cdots \cup X_k$. Since X is 2-connected, by Lemma 2.8.16 there exists a simple polygonal closed curve C in X such that $p \in \text{int}(C)$. Choose C in such a way that it lies in a union $X_i \cup X_{i+1} \cup \cdots \cup X_j$ with $j - i$ minimum. We claim that $j - i \leq 1$. Assume by contradiction that $j - i \geq 2$; we may assume that among all the simple polygonal closed curves in $X_i \cup X_{i+1} \cup \cdots \cup X_j$ having p in their interior, the number of edges in C and not in X_{j-1} is the least possible. Since C intersects both X_j and X_{j-2} (which are disjoint), C must contain at least two disjoint maximal paths in X_{j-1}. Let L be one of those, and let L' be the shortest path in X_{j-1} from L to $C \setminus V(L)$. The endpoints of L' split C in two polygonal arcs C_1 and C_2, each of which contains segments not in X_{j-1}. But one of the two polygonal closed curves $L' \cup C_1$ and $L' \cup C_2$ contains p in its interior, and has a lower number of edges in X_{j-1} than C has, a contradiction.

Hence, C has to lie in a union of the form $X_j \cup X_{j+1}$, and so p is in a bounded face of a $X_j \cup X_{j+1}$. $\square$

We are now able to prove the *Jordan arc theorem*:

Theorem 2.8.20 (Jordan arc theorem). *If $L \subset \mathbb{R}^2$ is a Jordan arc, then $\mathbb{R}^2 \setminus L$ is connected. More precisely, each pair of points $p, q \in \mathbb{R}^2 \setminus L$ can be joined by a simple polygonal arc in $\mathbb{R}^2 \setminus L$.*

Proof. Let $p, q \in \mathbb{R}^2 \setminus L$, and let $0 < 3\delta < \min\{d(p, L), d(q, L)\}$, where d is the Euclidean distance. As L is the support of a uniformly continuous curve, we may divide L into a finite number of arcs $L_1, \ldots, L_k$ each with diameter less than δ; and, for $j = 1, \ldots, k-1$, denote by p_j and p_{j+1} the endpoints of L_j (in the obvious order).

Let δ' be the least distance between L_i and L_j for $|i - j| \geq 2$; clearly, $0 < \delta' \leq \delta$. Divide each L_i into sub-arcs $L_{i1}, \ldots, L_{ik_i}$ with a diameter less than $\delta'/4$, and denote by $p_{i,j}$ and $p_{i,j+1}$ the endpoints of L_{ij}.

Let X_i be the planar graph consisting of the union of the boundaries of the squares with center $p_{i,j}$ and sides of length $\delta'/2$. Then the graphs $X_1, \ldots, X_k$ satisfy the hypotheses of Lemma 2.8.19. Moreover, $X_i \cup X_{i+1}$ is in the interior of the disk with center p_i and radius $5\delta'/2$, while both p and q are out of the disk with center p_i and radius 3δ; so, by Lemma 2.8.19, we know that p and q belong to the outer face of $X_1 \cup \cdots \cup X_k$, and thus they can be connected with a simple polygonal arc not intersecting L. $\square$

To deduce from this result the Jordan curve theorem we need a last definition and a last lemma.

Definition 2.8.21. Let $C \subset \mathbb{R}^2$ be a closed subset, and Ω a connected component of $\mathbb{R}^2 \setminus C$. We shall say that a point $p \in C$ is *accessible* from Ω if there exists a simple polygonal arc from a point $q \in \Omega$ to p contained in Ω (except for the endpoints).

Lemma 2.8.22. *Let $C \subset \mathbb{R}^2$ be a Jordan curve, and Ω a component of $\mathbb{R}^2 \setminus C$. Then the set of points of C accessible from Ω is dense in C.*

Proof. Choose a point $q \in \Omega$, and let $C_1 \subset C$ be an open subarc of C. Theorem 2.8.20 implies that $\mathbb{R}^2 \setminus (C \setminus C_1)$ is connected; so there exists a simple polygonal arc in $\mathbb{R}^2 \setminus (C \setminus C_1)$ from q to a point lying in a component of $\mathbb{R}^2 \setminus C$ distinct from Ω. Then this arc has to intersect C_1 in a point accessible from Ω. So every open subarc of C contains accessible points, and we are done. $\square$

So we have reached the proof of:

Theorem 2.8.23 (Jordan curve theorem). *Let $C \subset \mathbb{R}^2$ be a Jordan curve. Then $\mathbb{R}^2 \setminus C$ has exactly two components, and C is their common boundary.*

Proof. Assume, by contradiction, that $\mathbb{R}^2 \setminus C$ has at least three components Ω_1, Ω_2 and Ω_3, and choose $p_j \in \Omega_j$, for $j = 1, 2, 3$. Let then C_1, C_2 and C_3 be three pairwise disjoint arcs of C. By the previous lemma, for $i, j = 1, 2, 3$ we can find a simple polygonal arc L_{ij} from p_i to C_j. Moreover, we may also assume that $L_{ij} \cap L_{ij'} = \{p_i\}$ if $j \neq j'$. Indeed, if following L_{i2} starting from C_2 we intersect L_{i1} in a point $p_i' \neq p_i$, we may modify L_{i2} in such a way that its final section runs close to the section of L_{i1} from p_i' to p_i and that the new L_{i2} intersects L_{i1} only in p_i. We modify analogously L_{i3} if necessary.

Clearly, $L_{ij} \cap L_{i'j} = \varnothing$ if $i \neq i'$. But then the planar graph obtained by adding suitable subarcs of the C_j's to the L_{ij}'s is a planar realization of $K_{3,3}$, contradicting Proposition 2.8.11. $\square$

As usual, the bounded component of $\mathbb{R}^2 \setminus C$ will be called *interior* of C.

To prove Schönflies' theorem we need a generalization of Lemmas 2.8.5 and 2.8.16:

Lemma 2.8.24. *Let $C \subset \mathbb{R}^2$ be a Jordan curve, and $P \subset \overline{\mathrm{int}(C)}$ a simple polygonal arc joining two points p_1, $p_2 \in C$ and intersecting C only in the endpoints. Denote by C_1, $C_2 \subset C$ the two arcs in C from p_1 to p_2. Then $\mathbb{R}^2 \setminus (C \cup P)$ has exactly three components, having as boundary C, $C_1 \cup P$ and $C_2 \cup P$, respectively. In particular, every curve in $\overline{\mathrm{int}(C)}$ that joins a point of $C_1 \setminus \{p_1, p_2\}$ with a point of $C_2 \setminus \{p_1, p_2\}$ intersects P.*

Proof. As in the proof of Lemma 2.8.5, the only non-trivial part is to prove that $\mathrm{int}(C) \setminus P$ has at least two components.

Let $\ell \subset \mathrm{int}(C)$ be a line segment intersecting P in a single point p, not a vertex of P. If the endpoints of ℓ were in the same component Ω of $\mathbb{R}^2 \setminus (C \cup P)$, then we could find in Ω a simple polygonal arc L such that $L \cup \ell$ is a simple polygonal closed curve. But in this case the endpoints of the segment of P containing p should be in distinct components of $\mathbb{R}^2 \setminus (L \cup \ell)$. On the other hand, they are joined by a simple curve (contained in $P \cup C$) not intersecting $L \cup \ell$, a contradiction. □

Corollary 2.8.25. *Let G be a 2-connected planar graph, and denote by $e(G)$ and $v(G)$ the number of edges and of vertices of G. If $X \subset \mathbb{R}^2$ is a realization of G whose edges are polygonal arcs, except possibly for the outer circle which is a Jordan curve, then X has exactly $l(G) - v(G) + 2$ faces, each of which has as its boundary a cycle in X.*

Proof. It works exactly as the proof of Lemma 2.8.16, using Lemma 2.8.24 instead of Lemma 2.8.5. · □

Last definitions:

Definition 2.8.26. Two 2-connected planar graphs X, $X' \subset \mathbb{R}^2$ are said to be $\mathbb{R}^2$-*isomorphic* if there exists a graph isomorphism g between X and X' mapping cycles that are boundaries of faces to cycles with the same property and the outer cycle to the outer cycle. The isomorphism g is called a $\mathbb{R}^2$-*isomorphism*.

Remark 2.8.27. Clearly, a $\mathbb{R}^2$-isomorphism between planar graphs X and X' can be extended to an homeomorphism between X and X' as topological spaces, but unless otherwise stated we shall consider it just as a vertex-to-vertex map between graphs.

Definition 2.8.28. A *subdivision* of a graph G is a graph G' obtained by replacing some (or all) the edges of G by paths with the same endpoints.

At last, we are ready for *Schönflies' theorem*:

Theorem 2.8.29 (Schönflies). *Let $C \subset \mathbb{R}^2$ be a Jordan curve. Then $\overline{\mathrm{int}(C)}$ is homeomorphic to a closed disk.*

Proof. By assumption, we have a homeomorphism $f : C \to S^1$; we want to extend f to a homeomorphism between $\overline{\mathrm{int}(C)}$ and $\overline{\mathrm{int}(S^1)} = \overline{D}$.

Let $B \subset \mathrm{int}(C)$ be a countable dense subset, and $A \subset C$ a countable dense set of points accessible from $\mathrm{int}(C)$, which exists by Lemma 2.8.22. Choose a sequence $\{p_n\} \subset A \cup B$ such that every point of $A \cup B$ appears infinitely many times; we may also assume that $p_0 \in A$.

Let $X_0 \subset \overline{\mathrm{int}(C)}$ be 2-connected planar graph consisting of C and a simple polygonal arc joining p_0 with another point of C as in Lemma 2.8.24. Clearly, we may find a 2-connected planar graph $X_0' \subset \overline{D}$, consisting of S^1 and a simple polygonal arc, and a $\mathbb{R}^2$-isomorphism $g_0 : X_0 \to X_0'$ that coincides with f on the vertices of X_0 in C.

Our first goal is to construct two sequences of 2-connected planar graphs $X_0, X_1, \ldots \subset \overline{\mathrm{int}(C)}$ and $X_0', X_1', \ldots \subset \overline{D}$ satisfying the following properties:

(i) X_n (respectively, X'_n) contains as a subgraph a subdivision of X_{n-1} (respectively, X'_{n-1});

(ii) there exists a $\mathbb{R}^2$-isomorphism $g_n\colon X_n \to X'_n$ that coincides with g_{n-1} on the vertices of X_{n-1}, and with f on the vertices of X_n contained in C;

(iii) X_n (respectively, X'_n) consists of the union of C (respectively, S^1) with polygonal arcs contained in $\mathrm{int}(C)$ (respectively, in $\overline{D}$);

(iv) $p_n \in X_n$;

(v) $X'_n \setminus S^1$ is connected.

We proceed by induction; so assume we have already defined $X_1, \ldots, X_{n-1}$ and $X'_1, \ldots, X'_{n-1}$ in such a way that (i)–(v) hold.

If $p_n \in A$, let P be a polygonal simple arc from p_n to a point $q_n \in X_{n-1} \setminus C$ (chosen in such a way that $p_n q_n$ is not already an edge of X_{n-1}) such that $X_{n-1} \cap P \subseteq \{p_n, q_n\}$, and set $X_n = X_{n-1} \cup P$. Let $S \subset X_{n-1}$ be the cycle (Corollary 2.8.25) bounding the face of X_{n-1} containing P. Then add to X'_{n-1} a simple polygonal arc P' contained in the face bounded by $g_{n-1}(S)$ and joining $f(p_n)$ with $g_{n-1}(q_n)$ if q_n was a vertex of X_{n-1}, or else with another point of $g_{n-1}(\ell)$, where ℓ is the edge of X_{n-1} containing q_n. Set $X'_n = X'_{n-1} \cup P'$, and define the $\mathbb{R}^2$-isomorphism $g_n\colon X_n \to X'_n$ in the obvious way.

If, on the other hand, $p_n \in B$, we have to work slightly more. Consider the largest square with center p_n and whose sides are vertical and horizontal, contained in $\mathrm{int}(C)$. Within this square (which we shall not take as is, since its sides may contain infinitely many points of C) we draw a new square with center p_n, and vertical and horizontal sides at a distance less than $1/n$ from the corresponding sides of the larger square. Divide this second square with horizontal and vertical lines in regions having a diameter less than $1/n$, and in such a way that both a horizontal line segment and a vertical one go through p_n. Denote by Y_n the union of X_{n-1} with all these horizontal and vertical segments, and possibly an additional simple polygonal arc in $\mathrm{int}(C)$ in such a way that Y_n is 2-connected and $Y_n \setminus C$ is connected. By Lemma 2.8.15, we know that Y_n is obtained starting from X_{n-1} and adding paths contained in faces. Add the corresponding paths to X'_{n-1} so as to obtain a graph Y'_n that is $\mathbb{R}^2$-isomorphic to Y_n. Then add horizontal and vertical segments in $\overline{D}$ to Y'_n in such a way that the (bounded) faces of the resulting graph have all a diameter less than $1/2n$. If necessary, move slightly these segments to have them intersect S^1 only in $f(A)$ and Y'_{n-1} in finitely many points, keeping the diameter of the bounded faces less than $1/n$; in this way, we get a graph we shall call X'_n. Since X'_n is 2-connected, we may obtain it by starting with Y'_n and adding paths (in this case, line segments) contained in faces. Do the same on Y_n, in such a way to obtain a graph X_n that is $\mathbb{R}^2$-isomorphic to X'_n, and (i)–(v) are satisfied.

So we have extended f to a bijective map defined on $C \cup V(X_0) \cup V(X_1) \cup \cdots$ with values in $S^1 \cup V(X'_0) \cup V(X'_1) \cup \cdots$. These sets are dense in $\overline{\mathrm{int}(C)}$ and $\overline{D}$, respectively; we want to show that f admits a bijective continuous extension from $\overline{\mathrm{int}(C)}$ to $\overline{D}$, and this will be the required homeomorphism.

Pick $p \in \mathrm{int}(C)$ where f is not yet defined, and choose a sequence $\{q_k\} \subset V(X_0) \cup V(X_1) \cup \cdots$ converging to p. Let $\delta = d(p, C)$, and $p_n \in B$ such that $\|p_n - p\| < \delta/3$. If n is large enough, p is contained in the square with center p_n used to construct Y_n. In particular, p lies in a face of X_n bounded by a cycle S such that both S and $f(S)$ are contained in a disk with radius less than $1/n$. Since f preserves the faces of X_n, it maps the interior of S to the interior of $f(S)$. But $\{q_k\}$ is in the interior of S for k large enough; so, $\{f(q_k)\}$ is in the interior of $f(S)$ for k large enough. It follows that the sequence $\{f(q_k)\}$ is Cauchy, so it converges. In an analogous way it can be shown that the limit does not depend on the sequence we have chosen, so we have extended f to the whole interior of C. Moreover, this construction clearly shows that f is injective and continuous in the interior of C; it is bijective too, since $V(X_0') \cup V(X_1') \cup \cdots$ is dense in D. Moreover, f^{-1} is analogously continuous on D. To conclude the proof, it is enough to show that f is continuous on C too, since a bijective continuous map between a compact set and a Hausdorff space is a homeomorphism.

Choose a sequence $\{q_k\} \subset \mathrm{int}(C)$ converging to a point $q \in C$; we have to show that $\{f(q_k)\}$ converges to $f(q)$. Suppose it does not. Since $\overline{D}$ is compact, up to taking a subsequence, we may assume that $\{f(q_k)\}$ converges to a point $q' \neq f(q)$. Since f^{-1} is continuous in D, it follows that $q' \in S^1$. Since A is dense in C, its image $f(A)$ is dense in S^1; so, in both arcs S_1 and S_2 from q' to $f(q)$ in S^1 we may find a point $f(p_j) \in S_j \cap f(A)$. By construction, there exists a n such that X_n contains a path P from p_1 to p_2 that intersects C only in the endpoints. By Lemma 2.8.24, P divides $\mathrm{int}(C)$ in two connected components. The map f maps these components to two components of $D \setminus f(P)$. One of these contains $f(q_k)$ for k large enough, while the other one has $f(q)$ on its boundary (but not on the section of boundary common to both components). But since the q_k's converge to q, we may find a connected set (an infinite union of polygonal curves) in $\mathrm{int}(C)$ containing $\{q_k\}$ and having q as unique accumulation point in C; so any accumulation point of $\{f(q_k)\}$ has to lie in the intersection of S^1 and the boundary of the component containing $f(q)$, a contradiction. □

Remark 2.8.30. We have actually proved that it is possible to extend any homeomorphism f between C and S^1 to a homeomorphism between the closures of the interiors. With slightly more effort, it can be proved that it is also possible extend it to a homeomorphism of the whole $\mathbb{R}^2$ to itself; see [23].

3

Local theory of surfaces

The rest of this book is devoted to the study of surfaces in space. As we did for
the curves, we shall begin by trying to understand how best define a surface;
but, unlike what happened for curves, for surfaces it will turn out to be more
useful to work with subsets of $\mathbb{R}^3$ that locally look like an open subset of the
plane, instead of working with maps from an open subset of $\mathbb{R}^2$ to $\mathbb{R}^3$ having
an injective differential.

When we say that a surface locally looks like an open subset of the plane,
we are not (only) talking about its topological structure, but (above all) about
its differential structure. In other words, it must be possible to differentiate
functions on a surface exactly as we do on open subsets of the plane: computing a partial derivative is a purely local operation, so it is has to be possible to
perform similar operation in every object that locally looks (from a differential
viewpoint) like an open subset of the plane.

To carry out this program, after presenting in Section 3.1 the official definition of what a surface is, in Section 3.2 we shall define precisely the family
of functions that are smooth on a surface, that is, the functions we shall be
able to differentiate; in Section 3.4 we shall show how to differentiate them,
and we shall define the notion of differential of a smooth map between surfaces. Furthermore, in Sections 3.3 and 3.4, we shall introduce the tangent
vectors to a surface and we shall explain why they are an embodiment of partial derivatives. Finally, in the supplementary material we shall prove (Section
3.5) Sard's theorem, an important result about critical points of smooth functions, and we shall see (Section 3.6) how to extend smooth functions from a
surface to the whole of $\mathbb{R}^3$.

3.1 How to define a surface

As we did for curves, we begin by discussing the question of the correct definition of what a surface is. Our experience from the one-dimensional case

Abate M., Tovena F.: Curves and Surfaces.
DOI 10.1007/978-88-470-1941-6_3, © Springer-Verlag Italia 2012

suggests two possible approaches: we might define surfaces as subsets of the space with some properties, or we can define them as maps from an open subset of the plane to the space, satisfying suitable regularity properties.

Working with curves we preferred this second approach, since the existence of parameterizations by arc length allowed us to directly relate the geometric properties of the support of the curve with the differential properties of the curve itself.

As we shall see, in the case of surfaces the situation is significantly more complex. The approach that emphasizes maps will be useful to study local questions; but from a global viewpoint it will be more effective to privilege the other approach.

But let us not disclose too much too soon. Let us instead start by introducing the obvious generalization of the notion of a regular curve:

Definition 3.1.1. An *immersed* (or *parametrized*) *surface* in space is a map $\varphi: U \to \mathbb{R}^3$ of class C^∞, where $U \subseteq \mathbb{R}^2$ is an open set, such that the differential $d\varphi_x: \mathbb{R}^2 \to \mathbb{R}^3$ is injective (that is, has rank 2) in every point $x \in U$. The image $\varphi(U)$ of φ is the *support* of the immersed surface.

Remark 3.1.2. For reasons that will become clear in Section 3.4 (see Remark 3.4.20), when studying surfaces we shall only use C^∞ maps, and we shall not discuss lower regularity issues.

Remark 3.1.3. The differential $d\varphi_x$ of $\varphi = (\varphi_1, \varphi_2, \varphi_3)$ in $x \in U$ is represented by the Jacobian matrix

$$\text{Jac}\,\varphi\,(x) = \begin{vmatrix} \frac{\partial \varphi_1}{\partial x_1}(x) & \frac{\partial \varphi_1}{\partial x_2}(x) \\ \frac{\partial \varphi_2}{\partial x_1}(x) & \frac{\partial \varphi_2}{\partial x_2}(x) \\ \frac{\partial \varphi_3}{\partial x_1}(x) & \frac{\partial \varphi_3}{\partial x_2}(x) \end{vmatrix} \in M_{3,2}(\mathbb{R})\,.$$

As for curves, in this definition the emphasis is on the map rather than on its image. Moreover, we are not asking for the immersed surfaces to be a homeomorphism with their images or to be injective (see Example 3.1.6); both these properties are nevertheless locally true. To prove this, we need a lemma, somewhat technical but extremely useful.

Lemma 3.1.4. *Let $\varphi: U \to \mathbb{R}^3$ be an immersed surface, where $U \subseteq \mathbb{R}^2$ is open. Then for all $x_0 \in U$ there exist an open set $\Omega \subseteq \mathbb{R}^3$ of $(x_0, 0) \in U \times \mathbb{R}$, an open neighborhood $W \subseteq \mathbb{R}^3$ of $\varphi(x_0)$, and a diffeomorphism $G: \Omega \to W$ such that $G(x, 0) = \varphi(x)$ for all $(x, 0) \in \Omega \cap (U \times \{0\})$.*

Proof. By definition of immersed surface, the differential in x_0 of the map $\varphi = (\varphi_1, \varphi_2, \varphi_3)$ has rank 2; so the Jacobian matrix of φ computed in x_0 has a 2×2 minor with nonzero determinant. Up to reordering the coordinates, we may assume that this minor is obtained by discarding the third row, that is, we can assume that

$$\det\left(\frac{\partial \varphi_i}{\partial x_j}(x_0)\right)_{i,j=1,2} \neq 0\,.$$

Let then $G: U \times \mathbb{R} \to \mathbb{R}^3$ be given by

$$G(x_1, x_2, t) = \varphi(x_1, x_2) + (0, 0, t) \; ;$$

note that if to find the minor with nonzero determinant, we had discarded the j-th row, then G would be defined by adding $t\mathbf{e}_j$ to φ, where $\mathbf{e}_j$ is the j-th vector of the canonical basis of $\mathbb{R}^3$.

Clearly, $G(x, 0) = \varphi(x)$ for all $x \in U$, and

$$\det \text{Jac}\, G\,(x_0, O) = \det \left(\frac{\partial \varphi_i}{\partial x_j}(x_0) \right)_{i,j=1,2} \neq 0 \; ;$$

So the inverse function theorem (Theorem 2.2.4) gives us a neighborhood $\Omega \subseteq U \times \mathbb{R}$ of (x_0, O) and a neighborhood $W \subseteq \mathbb{R}^3$ of $\varphi(x_0)$ such that $G|_\Omega$ is a diffeomorphism between Ω and W, as required. $\qquad\square$

In particular, we have:

Corollary 3.1.5. *Let $\varphi: U \to \mathbb{R}^3$ be an immersed surface. Then every $x_0 \in U$ has a neighborhood $U_1 \subseteq U$ such that $\varphi|_{U_1}: U_1 \to \mathbb{R}^3$ is a homeomorphism with its image.*

Proof. Let $G: \Omega \to W$ be the diffeomorphism provided by the previous lemma, $\pi: \mathbb{R}^3 \to \mathbb{R}^2$ the projection on the first two coordinates, and set $U_1 = \pi(\Omega \cap (U \times \{0\}))$. Then $\varphi(x) = G(x, 0)$ for all $x \in U_1$, and so $\varphi|_{U_1}$ is a homeomorphism with its image, as required. $\qquad\square$

However, it is important to remember that, in general, immersed surfaces are *not* homeomorphisms with their images:

Example 3.1.6. For $U = (-1, +\infty) \times \mathbb{R}$, let $\varphi: U \to \mathbb{R}^3$ be given by

$$\varphi(x, y) = \left(\frac{3x}{1 + x^3}, \frac{3x^2}{1 + x^3}, y \right) \; ;$$

see Fig. 3.1. It is easy to verify that φ is an injective immersed surface, but is not a homeomorphism with its image, as $\varphi\big((-1, 1) \times (-1, 1)\big)$ is not open in $\varphi(U)$.

A careful consideration of the material seen in the previous chapters will show that the approach based on defining a curve as an equivalence class of maps was effective because of the existence of a canonical representative element defined from such a fundamental geometric concept as length. The drawback (or the advantage, depending on which half of the glass you prefer) of the theory of surfaces with respect to the theory of curves is that for surfaces this cannot be done, and cannot be done because of intrinsic, unavoidable reasons.

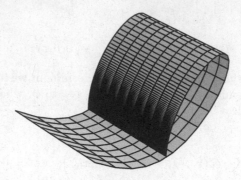

Fig. 3.1.

Of course, we may well define two immersed surfaces $\varphi\colon U \to \mathbb{R}^3$ and $\psi\colon V \to \mathbb{R}^3$ to be *equivalent* if there exists a diffeomorphism $h\colon U \to V$ such that $\varphi = \psi \circ h$. However, the problem we face with such a definition is that the procedure we followed in the case of curves to choose in each equivalence class an (essentially) unique representative element does not work anymore.

In the case of curves, we have chosen a canonical representative element, the parametrization by arc length, by using the geometric notion of length. Two equivalent parametrizations by arc length have to differ by a diffeomorphism h that preserves lengths; and this implies (see the proof of Proposition 1.2.11) that $|h'| \equiv 1$, so h is an affine isometry and the parametrization by arc length is unique up to parameter translations (and orientation changes).

In the case of surfaces, it is natural to try using area instead of length. Two equivalent "parametrizations by area" should differ by a diffeomorphism h of open sets in the plane that preserves areas. But Calculus experts teach us that a diffeomorphism h preserves areas if and only if $|\det \mathrm{Jac}(h)| \equiv 1$, which is a far weaker condition than $|h'| \equiv 1$. For instance, *all* diffeomorphisms of the form $h(x,y) = \big(x + f(y), y\big)$, where f is *any* smooth function of one variable, preserve areas; so using this method there is no hope of identifying an essentially unique representative element.

But the obstruction is even more fundamental than this. Arc length parametrization works because it is a (local) isometry between an interval and the curve; on the other hand, we shall see towards the end of Chapter 4 (with Gauss' theorema egregium Theorem 4.6.11) that, except for very particular cases, *isometries between an open set in the plane and a surface do not exist*. A notion equivalent to parametrization by arc length to study the metric structure of surface cannot possibly exist. Moreover, even the topological structure of surface is far more complex than that of open subsets of the plane (see Remark 3.1.21); to try and study it by using a single map would be hopeless.

Historically, the most successful — for its effectiveness both in dealing with local questions and in studying global problems — definition of a surface tries, in a sense, to take the best from both worlds. It emphasizes the support, that is, the subset of $\mathbb{R}^3$ considered as such; but the idea that a surface has to be a set locally built like an open subset of the plane is made concrete and formal by using immersed surfaces (which work well locally, as we have seen).

Enough chatting: it is now time to give the official definition of surface in space.

Definition 3.1.7. A connected subset $S \subset \mathbb{R}^3$ is a *(regular* or *embedded) surface* in space if for all $p \in S$ there exists a map $\varphi: U \to \mathbb{R}^3$ of class C^∞, where $U \subseteq \mathbb{R}^2$ is an open subset, such that:

(a) $\varphi(U) \subseteq S$ is an open neighborhood of p in S (or, equivalently, there exists an open neighborhood $W \subseteq \mathbb{R}^3$ of p in $\mathbb{R}^3$ such that $\varphi(U) = W \cap S$);
(b) φ is a homeomorphism with its image;
(c) the differential $\mathrm{d}\varphi_x: \mathbb{R}^2 \to \mathbb{R}^3$ is injective (that is, it has maximum rank, i.e., 2) for all $x \in U$.

Any map φ satisfying (a)–(c) is a *local* (or *regular*) *parametrization* in p; if $O \in U$ and $\varphi(O) = p$ we say that the local parametrization is *centered* in p. The inverse map $\varphi^{-1}: \varphi(U) \to U$ is called *local chart* in p; the neighborhood $\varphi(U)$ of p in S is called a *coordinate neighborhood*, the coordinates $(x_1(p), x_2(p)) = \varphi^{-1}(p)$ are called *local coordinates* of p; and, for $j = 1, 2$, the curve $t \mapsto \varphi(x_o + t\mathbf{e}_j)$ is the j-th *coordinate curve* (or *line*) through $\varphi(x_o)$.

Definition 3.1.8. An *atlas* for a regular surface $S \subset \mathbb{R}^3$ is a family $\mathcal{A} = \{\varphi_\alpha\}$ of local parametrizations $\varphi_\alpha: U_\alpha \to S$ such that $S = \bigcup_\alpha \varphi_\alpha(U_\alpha)$.

Remark 3.1.9. Clearly, a local parametrization $\varphi: U \to \mathbb{R}^3$ of a surface S carries the topology of the open subset U of the plane to the topology of the open set $\varphi(U)$ of S, since φ is a homeomorphism between U and $\varphi(U)$. But to work with surfaces it is important to keep in mind that φ carries another fundamental thing from U to S: a coordinate system. As shown in Fig. 3.2, the local parametrization φ assigns to each point $p \in \varphi(U)$ a pair of real numbers $(x, y) = \varphi^{-1}(p) \in U$, which will play the role of coordinates of p in S, in analogy to the role played by the usual Cartesian coordinates for points in the plane. In a sense, choosing a local parametrization of a surface amounts to constructing a geographical map of a part of the surface; and this is the reason (historically too) of the use of geographical terminology in this context. *Warning:* different local parametrizations provide different local coordinates (charts)! In the next section we shall describe the connection between coordinates induced by different parametrizations (Theorem 3.2.3).

Remark 3.1.10. If $\varphi: U \to S$ is a local parametrization of a surface $S \subset \mathbb{R}^3$, and $\chi: U_1 \to U$ is a diffeomorphism, where U_1 is another open subset of $\mathbb{R}^2$, then $\tilde{\varphi} = \varphi \circ \chi$ is another local parametrization of S (why?). In particular, if $p = \varphi(x_0) \in S$ and χ is the translation $\chi(x) = x + x_0$ then $\tilde{\varphi} = \varphi \circ \chi$ is a local parametrization of S centered at p.

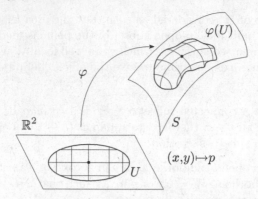

Fig. 3.2. A local parametrization

Remark 3.1.11. If $\varphi\colon U \to S$ is a local parametrization of a surface $S \subset \mathbb{R}^3$, and $V \subset U$ is an open subset of $\mathbb{R}^2$ then $\varphi|_V$ also is a local parametrization of S (why?). In particular, we may find local parametrizations with arbitrarily small domain.

As we shall see, the philosophy beneath the theory of surfaces is to use local parametrizations to transfer notions, properties, and proofs from open subsets of the plane to open sets of the surfaces, and vice versa. But let us see for now some examples of surface.

Example 3.1.12. The plane $S \subset \mathbb{R}^3$ through $p_0 \in \mathbb{R}^3$ and parallel to the linearly independent vectors $\mathbf{v}_1$, $\mathbf{v}_2 \in \mathbb{R}^3$ is a regular surface, with an atlas consisting of a single local parametrization $\varphi\colon \mathbb{R}^2 \to \mathbb{R}^3$ given by $\varphi(x) = p_0 + x_1\mathbf{v}_1 + x_2\mathbf{v}_2$.

Example 3.1.13. Let $U \subseteq \mathbb{R}^2$ be an open set, and $f \in C^\infty(U)$ an arbitrary function. Then the *graph* $\Gamma_f = \{(x, f(x)) \in \mathbb{R}^3 \mid x \in U\}$ of f is a regular surface, with an atlas consisting of a single local parametrization $\varphi\colon U \to \mathbb{R}^3$ given by $\varphi(x) = (x, f(x))$. Indeed, condition (a) of the definition of a surface is clearly satisfied. The restriction to Γ_f of the projection on the first two coordinates is the (continuous) inverse of φ, so condition (b) is satisfied as well. Finally,

$$\operatorname{Jac}\varphi(x) = \begin{vmatrix} 1 & 0 \\ 0 & 1 \\ \frac{\partial f}{\partial x_1}(x) & \frac{\partial f}{\partial x_2}(x) \end{vmatrix}$$

has rank 2 everywhere, and we are done.

Example 3.1.14. The support S of an immersed surface φ that is a homeomorphism with its image is a regular surface with atlas $\mathcal{A} = \{\varphi\}$. In this case we shall say that φ is a *global parametrization* of S.

Example 3.1.15. We want to show that the *sphere*

$$S^2 = \{p \in \mathbb{R}^3 \mid \|p\| = 1\}$$

with center in the origin and radius 1 is a regular surface by finding an atlas. Let $U = \{(x, y) \in \mathbb{R}^2 \mid x^2 + y^2 < 1\}$ be the open unit disc in the plane, and define $\varphi_1, \ldots, \varphi_6 \colon U \to \mathbb{R}^3$ by setting

$$\varphi_1(x, y) = \left(x, y, \sqrt{1 - x^2 - y^2}\right), \quad \varphi_2(x, y) = \left(x, y, -\sqrt{1 - x^2 - y^2}\right),$$

$$\varphi_3(x, y) = \left(x, \sqrt{1 - x^2 - y^2}, y\right), \quad \varphi_4(x, y) = \left(x, -\sqrt{1 - x^2 - y^2}, y\right),$$

$$\varphi_5(x, y) = \left(\sqrt{1 - x^2 - y^2}, x, y\right), \quad \varphi_6(x, y) = \left(-\sqrt{1 - x^2 - y^2}, x, y\right).$$

Arguing as in Example 3.1.13, it is easy to see that all the maps φ_j are local parametrizations of S^2; moreover, $S^2 = \varphi_1(U) \cup \cdots \cup \varphi_6(U)$, and so $\{\varphi_1, \ldots, \varphi_6\}$ is an atlas for S^2. Note that if we omit even one of these local parametrizations we do not cover the whole sphere.

Example 3.1.16. We now describe another atlas for the sphere. Set

$$U = \{(\theta, \phi) \in \mathbb{R}^2 \mid 0 < \theta < \pi, 0 < \phi < 2\pi\},$$

and let $\varphi_1 \colon U \to \mathbb{R}^3$ be given by

$$\varphi_1(\theta, \phi) = (\sin\theta\cos\phi, \sin\theta\sin\phi, \cos\theta);$$

we want to prove that φ_1 is a local parametrization of the sphere. The parameter θ is usually called *colatitude* (the *latitude* is $\pi/2 - \theta$), while ϕ is the *longitude*. The local coordinates (θ, ϕ) are called *spherical coordinates*; see Fig. 3.3.

First of all,

$$\varphi_1(U) = S^2 \setminus \{(x, y, z) \in \mathbb{R}^3 \mid y = 0, x \geq 0\}$$

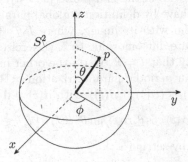

Fig. 3.3. Spherical coordinates

is an open subset of S^2, so condition (a) is satisfied. Next,

$$\mathrm{Jac}\,\varphi_1(\theta,\phi) = \begin{vmatrix} \cos\theta\cos\phi & -\sin\theta\sin\phi \\ \cos\theta\sin\phi & \sin\theta\cos\phi \\ -\sin\theta & 0 \end{vmatrix} ,$$

and it is straightforward to verify that this matrix has rank 2 everywhere (since $\sin\theta \neq 0$ when $(\theta,\phi) \in U$), so condition (c) is satisfied. Moreover, if we take an arbitrary $(x,y,z) = \varphi(\theta,\phi) \in \varphi_1(U)$, we find $\theta = \arccos z \in (0,\pi)$; being $\sin\theta \neq 0$, we recover $(\cos\phi,\sin\phi) \in S^1$ and consequently $\phi \in (0,2\pi)$ in terms of x, y and z, so φ_1 is globally injective. To conclude, we should prove that φ_1 is a homeomorphism with its image (i.e., that φ_1^{-1} is continuous); but we shall see shortly (Proposition 3.1.31) that this is a consequence of the fact that we already know that S^2 is a surface, so we leave this as an exercise (but see also Example 3.1.18). Finally, let $\varphi_2 \colon U \to \mathbb{R}^3$ be given by

$$\varphi_2(\theta,\phi) = (-\sin\theta\cos\phi, \cos\theta, -\sin\theta\sin\phi) .$$

Arguing as above, we see that φ_2 is also a local parametrization, with $\varphi_2(U) = S^2 \setminus \{(x,y,z) \in \mathbb{R}^3 \mid z = 0, x \leq 0\}$, so $\{\varphi_1,\varphi_2\}$ is an atlas for S^2.

Exercise 3.4 describes a third possible atlas for the sphere.

Example 3.1.17. Let $S \subset \mathbb{R}^3$ be a surface, and $S_1 \subseteq S$ an open subset of S. Then S_1 is a surface as well. Indeed, choose $p \in S_1$ and let $\varphi \colon U \to \mathbb{R}^3$ be a local parametrization of S at p. Then $U_1 = \varphi^{-1}(S_1)$ is open in $\mathbb{R}^2$ and $\varphi_1 = \varphi|_{U_1} \colon U_1 \to \mathbb{R}^3$ is a local parametrization of S_1 at p.

If $\chi \colon \Omega \to \mathbb{R}^3$ is a diffeomorphism with its image defined on an open neighborhood Ω of S, then $\chi(S)$ is a surface. Indeed, if φ is a local parametrization of S at $p \in S$, the map $\chi \circ \varphi$ is a local parametrization of $\chi(S)$ at $\chi(p)$.

Example 3.1.18 (Surfaces of revolution). Let $H \subset \mathbb{R}^3$ be a plane, $C \subset H$ the support of an open Jordan arc or of a Jordan curve of class C^∞, and $\ell \subset H$ a straight line disjoint from C. We want to prove that the set $S \subset \mathbb{R}^3$ obtained by rotating C around ℓ is a regular surface, called *surface of revolution* having C as its *generatrix* and ℓ as its *axis*.

Without loss of generality, we may assume that H is the plane xz, that ℓ is the z-axis, and that C lies in the half-plane $\{x > 0\}$. If C is the support of an open Jordan arc, we have by definition a global parametrization $\sigma \colon I \to \mathbb{R}^3$ that is a homeomorphism with its image, where $I \subseteq \mathbb{R}$ is an open interval. Since all open intervals are diffeomorphic to $\mathbb{R}$ (Exercise 1.5), we may assume without loss of generality that $I = \mathbb{R}$. If, on the other hand, C is the support of a Jordan curve, take a periodic parametrization $\sigma \colon \mathbb{R} \to \mathbb{R}^3$ of C. In both cases, we may write $\sigma(t) = \big(\alpha(t), 0, \beta(t)\big)$ with $\alpha(t) > 0$ for all $t \in \mathbb{R}$, so

$$S = \big\{ \big(\alpha(t)\cos\theta, \alpha(t)\sin\theta, \beta(t)\big) \mid t, \theta \in \mathbb{R} \big\} .$$

Define now $\varphi \colon \mathbb{R}^2 \to \mathbb{R}^3$ by setting

$$\varphi(t,\theta) = \big(\alpha(t)\cos\theta, \alpha(t)\sin\theta, \beta(t)\big) ,$$

so that $S = \varphi(\mathbb{R}^2)$. If we fix $t_0 \in \mathbb{R}$, the curve $\theta \mapsto \varphi(t_0, \theta)$ is a *parallel* of S; it is the circle with radius $\alpha(t_0)$ obtained by rotating the point $\sigma(t_0)$ around ℓ. If we fix $\theta_0 \in \mathbb{R}$, the curve $t \mapsto \varphi(t, \theta_0)$ is a *meridian* of S; it is obtained rotating C by an angle θ_0 around ℓ.

Now we have

$$\operatorname{Jac}\varphi(t, \theta) = \begin{vmatrix} \alpha'(t)\cos\theta & -\alpha(t)\sin\theta \\ \alpha'(t)\sin\theta & \alpha(t)\cos\theta \\ \beta'(t) & 0 \end{vmatrix}.$$

So $\operatorname{Jac}\varphi(t, \theta)$ has rank less than 2 if and only if

$$\begin{cases} \alpha'(t)\alpha(t) = 0\,, \\ \alpha(t)\beta'(t)\sin\theta = 0\,, \\ \alpha(t)\beta'(t)\cos\theta = 0\,, \end{cases}$$

and this never happens since α is always positive and σ is regular. In particular, φ is an immersed surface having S as its support.

This, however, is not enough to prove that S is a regular surface. To conclude, we have to consider two cases.

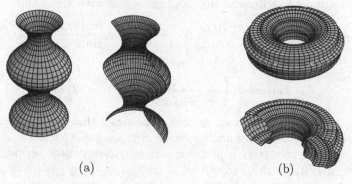

(a)   (b)

Fig. 3.4. Surfaces of revolution (whole and sections)

(a) *C is not compact, and σ is a global parametrization*: see Fig. 3.4.(a). In this case, we set $\varphi_1 = \varphi|_{\mathbb{R}\times(0,2\pi)}$ and $\varphi_2 = \varphi|_{\mathbb{R}\times(-\pi,\pi)}$; since the union of the supports of φ_1 and φ_2 is S, if we prove that φ_1 and φ_2 are local parametrizations we are done. Since

$$\varphi_1\big(\mathbb{R} \times (0, 2\pi)\big) = S \setminus \{(x, y, z) \in \mathbb{R}^3 \mid y = 0, x \geq 0\}$$

is open in S and φ_1 is the restriction of an immersed surface, to show that φ_1 is a local parametrization it suffices to prove that it is a homeomorphism with its image. From $\varphi_1(t, \theta) = (x, y, z)$ we find $\beta(t) = z$ and

$\alpha(t) = \sqrt{x^2 + y^2}$. As σ is injective, from this we can find a unique $t \in I$, and hence a unique $\theta \in (0, 2\pi)$, such that $x = \alpha(t)\cos\theta$ and $y = \alpha(t)\sin\theta$; thus, φ_1 is invertible. Furthermore, since σ is a homeomorphism with its image, the coordinate t depends continuously on z and $\sqrt{x^2 + y^2}$; if we prove that θ also depends continuously on (x, y, z) we have proved that φ_1^{-1} is continuous. Now, if $(x, y, z) \in S$ is such that $y > 0$ we have

$$0 < \frac{y}{x + \sqrt{x^2 + y^2}} = \frac{y/\alpha(t)}{1 + x/\alpha(t)} = \frac{\sin\theta}{1 + \cos\theta} = \frac{\sin(\theta/2)}{\cos(\theta/2)} = \tan\frac{\theta}{2} ,$$

so

$$\theta = 2\arctan\left(\frac{y}{x + \sqrt{x^2 + y^2}}\right) \in (0, \pi)$$

depends continuously on (x, y, z). Analogously, is $(x, y, z) \in S$ is such that $y < 0$ we find

$$\theta = 2\pi + 2\arctan\left(\frac{y}{x + \sqrt{x^2 + y^2}}\right) \in (\pi, 2\pi) ,$$

and in this case too we are done. Finally, in order to verify that φ_1^{-1} is continuous in a neighborhood of a point $(x_0, 0, z_0) \in \varphi_1\big(\mathbb{R} \times (0, 2\pi)\big)$ note that $x_0 < 0$ necessarily, and that if $(x, y, z) \in S$ with $x < 0$ then

$$\frac{y}{\sqrt{x^2 + y^2} - x} = \frac{y/\alpha(t)}{1 - x/\alpha(t)} = \frac{\sin\theta}{1 - \cos\theta} = \frac{\cos(\theta/2)}{\sin(\theta/2)} = \cotan\frac{\theta}{2} ,$$

so

$$\theta = 2\,\text{arccotan}\left(\frac{y}{-x + \sqrt{x^2 + y^2}}\right) \in (\pi/2, 3\pi/2) ,$$

and in this case we are done as well. The proof that φ_2 is a local parametrization is completely analogous, so S is a regular surface.

(b) C is compact, and σ is a periodic parametrization with period $2r > 0$; see Fig. 3.4.(b). In this case set $\varphi_1 = \varphi|_{(0,2r) \times (0,2\pi)}$, $\varphi_2 = \varphi|_{(0,2r) \times (-\pi,\pi)}$, $\varphi_3 = \varphi|_{(-r,r) \times (0,2\pi)}$, and $\varphi_4 = \varphi|_{(-r,r) \times (-\pi,\pi)}$; then, arguing as in the previous case, we immediately see that $\{\varphi_1, \varphi_2, \varphi_3, \varphi_4\}$ is an atlas for S.

Another way to prove that surfaces of revolution are regular surfaces is outlined in Exercise 3.23.

Example 3.1.19. A *torus* is a surface obtained by rotating a circle around an axis (contained in the plane of the circle) not intersecting it. For instance, if C is the circle with center $(x_0, 0, z_0)$ and radius $0 < r_0 < |x_0|$ in the xz-plane, then the torus obtained by rotating C around the z-axis is the support of the immersed surface $\varphi \colon \mathbb{R}^2 \to \mathbb{R}^3$ given by

$$\varphi(t, \theta) = \big((r\cos t + x_0)\cos\theta, (r\cos t + x_0)\sin\theta, r\sin t + z_0\big) ;$$

see Fig. 3.5.(a).

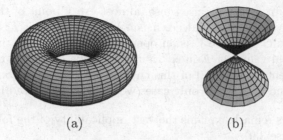

(a)   (b)

Fig. 3.5. (a) a torus; (b) a two-sheeted cone

Example 3.1.20. Let us see now an example of a subset of $\mathbb{R}^3$ that is not a regular surface. The *two-sheeted (infinite) cone* is the set

$$S = \{(x, y, z) \in \mathbb{R}^3 \mid x^2 + y^2 = z^2\} \,;$$

see Fig. 3.5.(b). The set S cannot be a regular surface: indeed, if the origin $O \in S$ had in S a neighborhood homeomorphic to an open subset of the plane, then $S \setminus \{O\}$ should be connected (why?), but this is not the case. We shall see shortly (Example 3.1.30) that the one-sheeted infinite cone $S \cap \{z \geq 0\}$ is not a regular surface too, whereas either connected component of $S \setminus \{O\}$ is (Exercise 3.9).

Remark 3.1.21. We can now show that there are non-compact surfaces that cannot be the support of a single immersed surface which is also a homeomorphism with its image. In other words, there exist non-compact regular surfaces not homeomorphic to an open subset of the plane. Let $S \subset \mathbb{R}^3$ be the non-compact surface obtained by removing a point from a torus (Examples 3.1.17 and 3.1.19). Then S contains Jordan curves (the meridians of the torus) that do not disconnect it, so it cannot be homeomorphic to an open subset of the plane without contradicting the Jordan curve theorem (Theorem 2.3.6).

We give now a general procedure for building regular surfaces. Let us begin with a definition:

Definition 3.1.22. Let $V \subseteq \mathbb{R}^n$ be an open set, and $F \colon V \to \mathbb{R}^m$ a C^∞ map. We shall say that $p \in V$ is a *critical point* of F if $dF_p \colon \mathbb{R}^n \to \mathbb{R}^m$ is not surjective. We shall denote the set of critical points of F by $\mathrm{Crit}(F)$. If $p \in V$ is a critical point $F(p) \in \mathbb{R}^m$ will be called a *critical value*. A point $y \in F(V) \subseteq \mathbb{R}^m$ that is not a critical value is a *regular value*.

Remark 3.1.23. If $f \colon V \to \mathbb{R}$ is a C^∞ function defined on an open subset $V \subset \mathbb{R}^n$ and $p \in V$ then $df_p \colon \mathbb{R}^n \to \mathbb{R}$ is not surjective if and only if it is everywhere zero. In other words, $p \in V$ is a critical point of f if and only if the gradient of f is zero in p.

Remark 3.1.24. In a very precise sense, almost every point of the image of a C^∞ map is a regular value. Indeed, it can be shown that if $F\colon V \to \mathbb{R}^m$ is a function of class C^∞, where V is an open subset of $\mathbb{R}^n$, then the measure of the set of critical values of F in $\mathbb{R}^m$ is zero (*Sard's theorem*). In Section 3.5 of the supplementary material of this chapter we shall prove Sard's theorem for $1 \le n \le 3$ and $m = 1$, the only cases we shall be interested in here.

The previous remark explains the vast applicability of the following result (see also Exercise 3.9):

Proposition 3.1.25. *Let $V \subseteq \mathbb{R}^3$ be an open set, and $f \in C^\infty(V)$. If $a \in \mathbb{R}$ is a regular value of f then every connected component of the level set $f^{-1}(a) = \{p \in V \mid f(p) = a\}$ is a regular surface.*

Proof. Let $p_0 = (x_0, y_0, z_0) \in f^{-1}(a)$. Since a is a regular value for f, the gradient of f is not zero in p_0; so, up to permuting the coordinates, we may assume that $\partial f/\partial z(p_0) \ne 0$. Let now $F\colon V \to \mathbb{R}^3$ be given by $F(x, y, z) = \bigl(x, y, f(x, y, z)\bigr)$. Clearly,

$$\det \operatorname{Jac} F(p_0) = \frac{\partial f}{\partial z}(p_0) \ne 0 .$$

Thus we may apply the inverse function theorem (Theorem 2.2.4) to get neighborhoods $\tilde{V} \subseteq V$ of p_0 and $W \subseteq \mathbb{R}^3$ of $F(p_0)$ such that $F|_{\tilde{V}}\colon \tilde{V} \to W$ is a diffeomorphism. Setting $G = (g_1, g_2, g_3) = F^{-1}$ we have

$$(u, v, w) = F \circ G(u, v, w) = \bigl(g_1(u, v, w), g_2(u, v, w), f\bigl(G(u, v, w)\bigr)\bigr) .$$

So $g_1(u, v, w) \equiv u$, $g_2(u, v, w) \equiv v$, and

$$\forall (u, v, w) \in W \qquad f\bigl(G(u, v, w)\bigr) \equiv w . \tag{3.1}$$

Clearly, the set $U = \{(u, v) \in \mathbb{R}^2 \mid (u, v, a) \in W\}$ is an open subset of $\mathbb{R}^2$, and we may define $\varphi\colon U \to \mathbb{R}^3$ with

$$\varphi(u, v) = G(u, v, a) = \bigl(u, v, g_3(u, v, a)\bigr) .$$

By (3.1), we know (why?) that $\varphi(U) = f^{-1}(a) \cap \tilde{V}$, and it is straightforward to verify that φ is a local parametrization of $f^{-1}(a)$ at p_0. $\qquad\square$

Definition 3.1.26. Let $V \subseteq \mathbb{R}^3$ be an open set and $f \in C^\infty(V)$. Every component of $f^{-1}(a)$, where $a \in \mathbb{R}$ is a regular value for f, is a *level surface* of f.

In the supplementary material of the next chapter we shall prove (Corollary 4.8.7) that, conversely, every closed surface in $\mathbb{R}^3$ is a level surface; see also Exercise 3.21.

Example 3.1.27. The *ellipsoid* having equation

$$\frac{x^2}{a^2} + \frac{y^2}{b^2} + \frac{z^2}{c^2} = 1$$

with a, b, $c > 0$ is a regular surface. Indeed, it is of the form $f^{-1}(1)$, where $f \colon \mathbb{R}^3 \to \mathbb{R}$ is given by

$$f(x, y, z) = \frac{x^2}{a^2} + \frac{y^2}{b^2} + \frac{z^2}{c^2} \, .$$

Since $\nabla f = \left(2x/a^2, 2y/b^2, 2z/c^2\right)$, the only critical point of f is the origin, the only critical value of f is 0, and so $f^{-1}(1)$ is a level surface.

Example 3.1.28. More in general, a *quadric* is the subset of $\mathbb{R}^3$ of the points that are solutions of an equation of the form $p(x, y, z) = 0$, where p is a polynomial of degree 2. Not all quadrics are regular surfaces (see Example 3.1.20 and Problem 3.4), but the components of those that are provide a good repertory of examples of surfaces. Besides the ellipsoid, we have the *two-sheeted* (or *elliptic*) *hyperboloid* of equation $(x/a)^2 + (y/b)^2 \ (z/c)^2 + 1 = 0$, the *one sheeted* (or *hyperbolic*) *hyperboloid* of equation $(x/a)^2 + (y/b)^2 - (z/c)^2 - 1 = 0$, the *elliptic paraboloid* having equation $(x/a)^2 + (y/b)^2 - z = 0$, the *hyperbolic paraboloid* having equation $(x/a)^2 - (y/b)^2 - z = 0$, and cylinders having a conic section as generatrix (see Problem 3.3). Fig. 3.6 shows some quadrics.

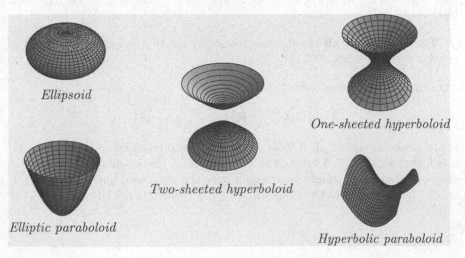

Ellipsoid

One-sheeted hyperboloid

Two-sheeted hyperboloid

Elliptic paraboloid

Hyperbolic paraboloid

Fig. 3.6. Quadrics

We end this section with two general results.

Proposition 3.1.29. *Every regular surface is locally a graph. In other words, if $S \subset \mathbb{R}^3$ is a regular surface and $p \in S$ then there exists a local parametrization $\varphi : U \to S$ in p of one of the following forms:*

$$\varphi(x,y) = \begin{cases} \big(x,y,f(x,y)\big), & or \\ \big(x,f(x,y),y\big), & or \\ \big(f(x,y),x,y\big), \end{cases}$$

for a suitable $f \in C^\infty(U)$. In particular, there is always an open neighborhood $\Omega \subseteq \mathbb{R}^3$ of S such that S is closed in Ω.

Proof. Let $\psi = (\psi_1, \psi_2, \psi_3) : U_1 \to \mathbb{R}^3$ be a local parametrization centered at p. Up to permuting coordinates, we may assume that

$$\det \left(\frac{\partial \psi_h}{\partial x_k}(O) \right)_{h,k=1,2} \neq 0 \, ;$$

so, setting $F = (\psi_1, \psi_2)$ we may find a neighborhood $V \subseteq U_1$ of O and a neighborhood $U \subseteq \mathbb{R}^2$ of $F(O)$ such that $F|_V : V \to U$ is a diffeomorphism. Let $F^{-1} : U \to V$ be the inverse map, and set $f = \psi_3 \circ F^{-1} : U \to \mathbb{R}$. Since we have $F \circ F^{-1} = \mathrm{id}_U$, we get

$$\psi \circ F^{-1}(u,v) = \big(u,v,f(u,v)\big) \, ,$$

so $\varphi = \psi \circ F^{-1} : U \to \mathbb{R}^3$ is a local parametrization of S at p of the required form. Finally, for all $p \in S$ let $W_p \subset \mathbb{R}^3$ be an open neighborhood of p such that $W_p \cap S$ is a graph. Then $W_p \cap S$ is closed in W_p, and so S is closed (why?) in $\Omega = \bigcup_{p \in S} W_p$. □

The converse of this result holds too: every set that is locally a graph is a regular surface (Exercise 3.11).

Example 3.1.30. The one-sheeted infinite cone

$$S = \{(x,y,z) \in \mathbb{R}^3 \mid z = \sqrt{x^2 + y^2}\}$$

is not a regular surface. If it were, it should be the graph of a C^∞ function in a neighborhood of $(0,0,0)$. As the projections on the xz-plane and yz-plane are not injective, it should be a graph over the xy-plane; but in this case it should be the graph of the function $\sqrt{x^2 + y^2}$, which is not of class C^∞.

And at last, here is the result promised in Example 3.1.16:

Proposition 3.1.31. *Let $S \subset \mathbb{R}^3$ be a regular surface, $U \subseteq \mathbb{R}^2$ an open subset, and $\varphi\colon U \to \mathbb{R}^3$ an immersed surface with support contained in S. Then:*

(i) *$\varphi(U)$ is open in S;*

(ii) *if φ is injective then for all $p \in \varphi(U)$ there exist a neighborhood $W \subset \mathbb{R}^3$ of p in $\mathbb{R}^3$ with $W \cap S \subseteq \varphi(U)$, and a map $\Phi\colon W \to \mathbb{R}^2$ of class C^∞ such that $\Phi(W) \subseteq U$ and $\Phi|_{W \cap S} \equiv \varphi^{-1}|_{W \cap S}$. In particular, $\varphi^{-1}\colon \varphi(U) \to U$ is continuous, so φ is a local parametrization of S.*

Proof. Let $p = \varphi(x_0, y_0) \in \varphi(U)$. As S is a surface, we can find a neighborhood W_0 of p in $\mathbb{R}^3$ such that $W_0 \cap S$ is a graph; to fix ideas, say that $W_0 \cap S$ is the graph over the xy-plane of a function f. If $\pi\colon \mathbb{R}^3 \to \mathbb{R}^2$ is the projection on the xy-plane, set $U_0 = \varphi^{-1}(W_0) \subseteq U$ and $h = \pi \circ \varphi\colon U_0 \to \mathbb{R}^2$. For $(x, y) \in U_0$ we have $\varphi_3(x, y) = f(\varphi_1(x, y), \varphi_2(x, y))$, and so the third row of the Jacobian matrix of φ in (x, y) is a linear combination of the first two. Since the differential of φ is supposed to have rank 2 everywhere, it follows that the first two rows of the Jacobian matrix of φ have to be linearly independent, and so $\operatorname{Jac} h(x, y)$ is invertible. The inverse function theorem (Theorem 2.2.4) then yields a neighborhood $U_1 \subseteq U_0$ of (x_0, y_0) and a neighborhood $V_1 \subseteq \mathbb{R}^2$ of $h(x_0, y_0) = \pi(p)$ such that $h|_{U_1}\colon U_1 \to V_1$ is a diffeomorphism. In particular, $\varphi(U_1) = \varphi \circ h|_{U_1}^{-1}(V_1) = (\pi|_{S \cap W_0})^{-1}(V_1)$ is open in S, so $\varphi(U)$ is a neighborhood of p in S. Since p is arbitrary, it follows that $\varphi(U)$ is open in S, and (i) is proved.

Suppose now that φ is injective, so $\varphi^{-1}\colon \varphi(U) \to U$ is defined. As $\varphi(U)$ is open in S, up to restricting W_0 we may assume that $W_0 \cap S \subseteq \varphi(U)$. Set $W = W_0 \cap \pi^{-1}(V_1)$ and $\Phi = h|_{U_1}^{-1} \circ \pi$; to prove (ii) it remains to show that $\Phi|_{W \cap S} \equiv \varphi^{-1}|_{W \cap S}$.

Let $q \in W \cap S$. As $q \in W_0 \cap \pi^{-1}(V_1)$, we can find a point $(u, v) \in V_1$ such that $q = (u, v, f(u, v))$. On the other hand, being $q \in \varphi(U)$ there is a unique point $(x, y) \in U$ such that $q = \varphi(x, y)$. But then $(u, v) = \pi(q) = h(x, y)$; so $(x, y) = h|_{U_1}^{-1}(u, v) \in U_1$ and $\varphi^{-1}(q) = (x, y) = h|_{U_1}^{-1} \circ \pi(q) = \Phi(q)$, as required. $\qquad\square$

In other words, if we already know that S is a surface, to verify whether a map $\varphi\colon U \to \mathbb{R}^3$ from an open subset U of $\mathbb{R}^2$ to S is a local parametrization it suffices to check that φ is injective and that $d\varphi_x$ has rank 2 for all $x \in U$.

Remark 3.1.32. The previous proposition and Lemma 3.1.4 might suggest that a claim along the following lines might be true: "Let $\varphi\colon U \to \mathbb{R}^3$ be an injective immersed surface with support $S = \varphi(U)$. Then for all $p \in \varphi(U)$ we can find a neighborhood $W \subset \mathbb{R}^3$ of p in $\mathbb{R}^3$ and a map $\Phi\colon W \to \mathbb{R}^2$ of class C^∞ such that $\Phi(W) \subseteq U$ and $\Phi|_{W \cap S} \equiv \varphi^{-1}|_{W \cap S}$. In particular, $\varphi^{-1}\colon \varphi(U) \to U$ is continuous, and S is a regular surface." We have even a "proof" of this claim: "Since, by assumption, φ is an immersed surface, we may apply Lemma 3.1.4. Let $p = \varphi(x_0) \in \varphi(U)$, and $G\colon \Omega \to W$ the diffeomorphism provided by

Lemma 3.1.4; up to restricting Ω, we may also assume that $\Omega = U_1 \times (-\delta, \delta)$, where $\delta > 0$ and $U_1 \subseteq U$ is a suitable neighborhood of x_0. Then $\Phi = \pi \circ G^{-1}$, where $\pi: \mathbb{R}^3 \to \mathbb{R}^2$ is the projection on the first two coordinates, is as required. Indeed, for all $q \in W \cap \varphi(U)$ the point $G^{-1}(q) = (y, t) \in \Omega$ is the only one satisfying $G(y, t) = q$. But $G(\varphi^{-1}(q), 0) = \varphi(\varphi^{-1}(q)) = q$, so $G^{-1}(q) = (\varphi^{-1}(q), 0)$, and we are done." However, *the claim is false and this proof is wrong*.

The (subtle) error in the proof is that if $q \in W \cap \varphi(U)$ then $\varphi^{-1}(q)$ *does not necessarily belong to* U_1, and so $(\varphi^{-1}(q), 0)$ does not belong to the domain of G; hence we cannot say that $G(\varphi^{-1}(q), 0) = \varphi(\varphi^{-1}(q)) = q$ or deduce that $G^{-1}(q) = (\varphi^{-1}(q), 0)$. Of course, the claim might be true even if this particular proof is wrong. But the claim is false indeed, and in Example 3.1.33 you'll find a counterexample.

Summing up, we may deduce the continuity of the inverse of a globally injective immersed surface φ only *if we already know that the image of φ lies within a regular surface*; otherwise, it might be false.

Example 3.1.33. Let $\varphi: (-1, +\infty) \times \mathbb{R} \to \mathbb{R}^3$ be the immersed surface of Example 3.1.6. We have already remarked that φ is an injective immersed surface that is not a homeomorphism with its image, and it is immediate to notice that its support S is not a regular surface, since in a neighborhood of the point $(0, 0, 0) \in S$ none of the three projections on the coordinate planes is injective, and so S cannot be locally a graph.

3.2 Smooth functions

Local parametrizations are the tool that allows us to give concrete form to the idea that a surface locally resembles an open subset of the plane; let us see how to use them to determine when a function defined on a surface is smooth.

Definition 3.2.1. Let $S \subset \mathbb{R}^3$ be a surface, and $p \in S$. A function $f: S \to \mathbb{R}$ is of *class C^∞* (or *smooth*) at p if there exists a local parametrization $\varphi: U \to S$ at p such that $f \circ \varphi: U \to \mathbb{R}$ is of class C^∞ in a neighborhood of $\varphi^{-1}(p)$. We shall say that f is of *class C^∞* (or *smooth*) if it is so at every point. The space of C^∞ functions on S will be denoted by $C^\infty(S)$.

Remark 3.2.2. A smooth function $f: S \to \mathbb{R}$ is automatically continuous. Indeed, let $I \subseteq \mathbb{R}$ be an open interval, and $p \in f^{-1}(I)$. By assumption, there is a local parametrization $\varphi: U \to S$ at p such that $f \circ \varphi$ is of class C^∞ (and thus continuous) in a neighborhood of $\varphi^{-1}(p)$. Then $(f \circ \varphi)^{-1}(I) = \varphi^{-1}(f^{-1}(I))$ is a neighborhood of $\varphi^{-1}(p)$. But φ is a homeomorphism with its image; so $f^{-1}(I)$ has to be a neighborhood of $\varphi(\varphi^{-1}(p)) = p$. Since p was arbitrary, it follows that $f^{-1}(I)$ is open in S, and so f is continuous.

A possible problem with this definition is that it might depend on the particular local parametrization we have chosen: a priori, there might be another local parametrization ψ at p such that $f \circ \psi$ is *not* smooth in $\psi^{-1}(p)$. Luckily, the following theorem implies that this cannot happen.

Theorem 3.2.3. *Let S be a surface, and let $\varphi: U \to S$, $\psi: V \to S$ be two local parametrizations with $\Omega = \varphi(U) \cap \psi(V) \neq \varnothing$. Then the map $h = \varphi^{-1} \circ \psi|_{\psi^{-1}(\Omega)}: \psi^{-1}(\Omega) \to \varphi^{-1}(\Omega)$ is a diffeomorphism.*

Proof. The map h is a homeomorphism, as it is a composition of homeomorphisms; we have to show that it and its inverse are of class C^∞.

Let $x_0 \in \psi^{-1}(\Omega)$, $y_0 = h(x_0) \in \varphi^{-1}(\Omega)$, and $p = \psi(x_0) = \varphi(y_0) \in \Omega$. Proposition 3.1.31 provides us with a neighborhood W of $p \in \mathbb{R}^3$ and a map $\Phi: W \to \mathbb{R}^2$ of class C^∞ such that $\Phi|_{W \cap S} \equiv \varphi^{-1}$. Now, by the continuity of ψ, there is a neighborhood $V_1 \subset \psi^{-1}(\Omega)$ of x_0 such that $\psi(V_1) \subset W$. Then $h|_{V_1} = \Phi \circ \psi|_{V_1}$, and so h is of class C^∞ in x_0. Since x_0 is an arbitrary element, h is of class C^∞ everywhere. In an analogous way it can be proved that h^{-1} is of class C^∞, and so h is a diffeomorphism. $\square$

Corollary 3.2.4. *Let $S \subset \mathbb{R}^3$ be a surface, $f: S \to \mathbb{R}$ a function, and $p \in S$. If there is a local parametrization $\varphi: U \to S$ at p such that $f \circ \varphi$ is of class C^∞ in a neighborhood of $\varphi^{-1}(p)$, then $f \circ \psi$ is of class C^∞ in a neighborhood of $\psi^{-1}(p)$ for all local parametrization $\psi: V \to S$ of S at p.*

Proof. We may write

$$f \circ \psi = (f \circ \varphi) \circ (\varphi^{-1} \circ \psi) \,,$$

and thus the previous theorem implies that $f \circ \psi$ is of class C^∞ in a neighborhood of $\psi^{-1}(p)$ if and only if $f \circ \varphi$ is of class C^∞ in a neighborhood of $\varphi^{-1}(p)$. $\square$

So the being smooth on a surface is a property of the function, and does not depend on local parametrizations; to test whether a function is smooth we may use an arbitrary local parametrization.

Using the same approach, we may define the notion of smooth map between two surfaces:

Definition 3.2.5. If S_1, $S_2 \subset \mathbb{R}^3$ are two surfaces, we shall say that a map $F: S_1 \to S_2$ is of *class C^∞* (or *smooth*) at $p \in S_1$ if there exist a local parametrization $\varphi_1: U_1 \to S_1$ in p and a local parametrization $\varphi_2: U_2 \to S_2$ in $F(p)$ such that $\varphi_2^{-1} \circ F \circ \varphi_1$ is of class C^∞ (where defined). We shall say that F is *of class C^∞* (or *smooth*) if it so at every point. If F is of class C^∞ and invertible with inverse of class C^∞ we shall say that F is a *diffeomorphism*, and that S_1 and S_2 are *diffeomorphic*.

Remark 3.2.6. The notion of smooth map defined on an open subset of $\mathbb{R}^n$ with values in a surface, or from a surface with values in $\mathbb{R}^n$, can be introduced in an analogous way (see Exercise 3.20).

It is easy to prove that the definition of smooth map does not depend on the local parametrizations used (Exercise 3.41), that smooth maps are continuous (Exercise 3.40), and that a composition of smooth maps is smooth:

Proposition 3.2.7. *If* $F: S_1 \to S_2$ *and* $G: S_2 \to S_3$ *are smooth maps between surfaces, then the composition* $G \circ F: S_1 \to S_3$ *is smooth as well.*

Proof. Fix $p \in S_1$ and choose an arbitrary local parametrization $\varphi_1: U_1 \to S_1$ of S_1 at p, a local parametrization $\varphi_2: U_2 \to S_2$ of S_2 at $F(p)$, and a local parametrization $\varphi_3: U_3 \to S_3$ of S_3 at $G(F(p))$. Then

$$\varphi_3^{-1} \circ (G \circ F) \circ \varphi_1 = (\varphi_3^{-1} \circ G \circ \varphi_2) \circ (\varphi_2^{-1} \circ F \circ \varphi_1)$$

is of class C^∞ where defined, and we are done. $\qquad\square$

Example 3.2.8. A local parametrization $\varphi: U \to \varphi(U) \subset S$ is a diffeomorphism between U and $\varphi(U)$. Indeed, first of all, it is invertible by definition. Next, to test the differentiability of φ and φ^{-1} we can use the identity map id as local parametrization of U, and φ itself as local parametrization of S. So it suffices to verify that $\varphi^{-1} \circ \varphi \circ \mathrm{id}$ and $\mathrm{id} \circ \varphi^{-1} \circ \varphi$ are of class C^∞, which is straightforward.

Example 3.2.9. If $U \subset \mathbb{R}^n$ is open and $F: U \to \mathbb{R}^3$ is a C^∞ map whose image is contained in a surface S then F is of class C^∞ as an S-valued map as well. Indeed, let ψ be a local parametrization at a point $p \in F(U)$; Proposition 3.1.31 tells us that there exists a function Ψ of class C^∞ defined in a neighborhood of p such that $\psi^{-1} \circ F = \Psi \circ F$, and the latter composition is of class C^∞.

Example 3.2.10. If $S \subset \mathbb{R}^3$ is a surface, then the inclusion $\iota: S \hookrightarrow \mathbb{R}^3$ is of class C^∞. Indeed, saying that ι is of class C^∞ is exactly equivalent (why?) to saying that local parametrizations are of class C^∞ when considered as maps with values in $\mathbb{R}^3$.

Example 3.2.11. If $\Omega \subseteq \mathbb{R}^3$ is an open subset of $\mathbb{R}^3$ that contains the surface S, and $\tilde{f} \in C^\infty(\Omega)$, then the restriction $f = \tilde{f}|_S$ is of class C^∞ on S. Indeed, $f \circ \varphi = \tilde{f} \circ \varphi$ is of class C^∞ for every local parametrization φ.

Actually, in Section 3.6 of the supplementary material of this chapter we shall prove (Theorem 3.6.7) that the previous example provides all C^∞ functions on a surface S. However, for our purposes, it is sufficient a local version of this result:

Proposition 3.2.12. *Let* $S \subset \mathbb{R}^3$ *be a surface, and* $p \in S$*. Then a function* $f: S \to \mathbb{R}$ *is of class* C^∞ *at* p *if and only if there exist an open neighborhood* $W \subseteq \mathbb{R}^3$ *of* p *in* $\mathbb{R}^3$ *and a function* $\tilde{f} \in C^\infty(W)$ *such that* $\tilde{f}|_{W \cap S} \equiv f|_{W \cap S}$.

Proof. One implication is given by Example 3.2.11. For the converse, suppose that f is of class C^∞ at p, and let $\varphi: U \to S$ be a local parametrization centered at p. Proposition 3.1.31.(ii) provides us with a neighborhood W of p in $\mathbb{R}^3$ and a map $\Phi: W \to \mathbb{R}^2$ of class C^∞ such that $\Phi(W) \subseteq U$ and $\Phi_{W \cap S} \equiv \varphi^{-1}|_{W \cap S}$. Then the function $\tilde{f} = (f \circ \varphi) \circ \Phi \in C^\infty(W)$ is as required. $\qquad\square$

3.3 Tangent plane

We have seen that tangent vectors play a major role in the study of curves. In this section we intend to define the notion of a tangent vector to a surface at a point. The geometrically simplest way is as follows:

Definition 3.3.1. Let $S \subseteq \mathbb{R}^3$ be a set, and $p \in S$. A *tangent vector* to S at p is a vector of the form $\sigma'(0)$, where $\sigma: (-\varepsilon, \varepsilon) \to \mathbb{R}^3$ is a curve of class C^∞ whose support lies in S and such that $\sigma(0) = p$. The set of all possible tangent vectors to S at p is the *tangent cone* T_pS to S at p.

Remark 3.3.2. A *cone* (with the origin as vertex) in a vector space V is a subset $C \subseteq V$ such that $av \in C$ for all $a \in \mathbb{R}$ and $v \in C$. It is not difficult to verify that the tangent cone to a set is in fact a cone in this sense. Indeed, first of all, the zero vector is the tangent vector to a constant curve, so $O \in T_pS$ for all $p \in S$. Next, if $a \in \mathbb{R}^*$ and $O \neq v \in T_pS$, if we choose a curve $\sigma: (-\varepsilon, \varepsilon) \to S$ with $\sigma(0) = p$ and $\sigma'(0) = v$, then the curve $\sigma_a: (-\varepsilon/|a|, \varepsilon/|a|) \to S$ given by $\sigma_a(t) = \sigma(at)$ is such that $\sigma_a(0) = p$ and $\sigma_a'(0) = av$; so $av \in T_pS$ as required by the definition of cone.

Example 3.3.3. If $S \subset \mathbb{R}^3$ is the union of two straight lines through the origin, it is straightforward to verify (check it) that $T_OS = S$.

The advantage of this definition of tangent vector is the clear geometric meaning. If S is a surface, however, our geometric intuition tells us that T_pS should be a plane, not just a cone. Unfortunately, this is not so evident from the definition: the sum of two curves in S is not necessarily a curve in S, and so the "obvious" way of proving that the sum of two tangent vectors is a tangent vector does not work. On the other hand, the previous examples shows that if S is not a surface the tangent cone has no reason to be a plane; so, in order to get such a result, we have to fully exploit the definition of a surface, that is, we must involve local parametrizations.

Let us begin by seeing what happens in the simplest case, that of open sets in the plane:

Example 3.3.4. Let $U \subseteq \mathbb{R}^2$ be an open set, and $p \in U$. Every curve contained in U is plane, and so the tangent vectors to U at p lie necessarily in $\mathbb{R}^2$. Conversely, if $v \in \mathbb{R}^2$ then the curve $\sigma: (-\varepsilon, \varepsilon) \to V$ given by $\sigma(t) = p + tv$ has its support within U for ε small enough, and has v as its tangent vector. So we have proved that $T_pU = \mathbb{R}^2$.

Applying the usual strategy of using local parametrizations to carry notions from open subsets of the plane to surfaces, we get the following:

Proposition 3.3.5. *Let $S \subset \mathbb{R}^3$ be a surface, $p \in S$, and $\varphi: U \to S$ a local parametrization at p with $\varphi(x_o) = p$. Then $\mathrm{d}\varphi_{x_o}$ is an isomorphism between $\mathbb{R}^2$ and T_pS. In particular, $T_pS = \mathrm{d}\varphi_{x_o}(\mathbb{R}^2)$ is always a vector space of dimension 2, and $\mathrm{d}\varphi_{x_o}(\mathbb{R}^2)$ does not depend on φ but only on S and p.*

Proof. Given $v \in \mathbb{R}^2$, we may find $\varepsilon > 0$ such that $x_o + tv \in U$ for all $t \in (-\varepsilon, \varepsilon)$; so the curve $\sigma_v: (-\varepsilon, \varepsilon) \to S$ given by $\sigma_v(t) = \varphi(x_o + tv)$ is well defined. Since $\sigma_v(0) = p$ and $\sigma'_v(0) = d\varphi_{x_o}(v)$, it follows that $d\varphi_{x_o}(\mathbb{R}^2) \subseteq T_pS$.

Vice versa, let $\sigma: (-\varepsilon, \varepsilon) \to S$ be a curve such that $\sigma(0) = p$; up to taking a smaller ε, we may assume that the support of σ is contained in $\varphi(U)$. Proposition 3.1.31.(ii) ensures that the composition $\sigma_o = \varphi^{-1} \circ \sigma$ is a C^∞ curve in U such that $\sigma_o(0) = x_o$; set $v = \sigma'_o(0) \in \mathbb{R}^2$. Then

$$d\varphi_{x_o}(v) = \frac{d(\varphi \circ \sigma_o)}{dt}(0) = \sigma'(0) \, ,$$

and so $T_pS \subseteq d\varphi_{x_o}(\mathbb{R}^2)$. Hence, $d\varphi_{x_o}: \mathbb{R}^2 \to T_pS$ is surjective; since it is injective too, it is an isomorphism between $\mathbb{R}^2$ and T_pS. $\square$

Definition 3.3.6. Let $S \subset \mathbb{R}^3$ be a surface, and $p \in S$. The vector space $T_pS \subset \mathbb{R}^3$ is the *tangent plane* to S at p.

Remark 3.3.7. Warning: according to our definition, the tangent plane is a vector subspace of $\mathbb{R}^3$, and so it passes through the origin, no matter where the point $p \in S$ is. When we draw the tangent plane as a plane resting on the surface, we are not actually drawing T_pS, but rather the plane $p + T_pS$ parallel to it, which is the *affine tangent plane* through p.

Remark 3.3.8. It can be shown (see Exercise 3.37) that the affine tangent plane $p + T_pS$ is the plane best approximating the surface S in p, in the sense discussed in Section 1.4.

Remark 3.3.9. It is apparent from the definition that if $S \subset \mathbb{R}^3$ is a surface, $p \in S$ and $U \subseteq S$ is an open subset of S containing p, then $T_pU = T_pS$. In particular, if $S = \mathbb{R}^2$ then $T_pU = T_p\mathbb{R}^2 = \mathbb{R}^2$ for every open set U of the plane and every $p \in U$.

The isomorphism between $\mathbb{R}^2$ and T_pS provided by the local parametrizations allows us to consider special bases of the tangent plane:

Definition 3.3.10. Let $S \subset \mathbb{R}^3$ be a surface, and $p \in S$. If $\varphi: U \to S$ is a local parametrization centered at p, and $\{\mathbf{e}_1, \mathbf{e}_2\}$ is the canonical basis of $\mathbb{R}^2$, then the tangent vectors $\partial/\partial x_1|_p$, $\partial/\partial x_2|_p \in T_pS$ (the reasons behind this notation will be made clear in Remark 3.4.17) are defined by setting

$$\left.\frac{\partial}{\partial x_j}\right|_p = d\varphi_O(\mathbf{e}_j) = \frac{\partial \varphi}{\partial x_j}(O) = \begin{vmatrix} \frac{\partial \varphi_1}{\partial x_j}(O) \\ \frac{\partial \varphi_2}{\partial x_j}(O) \\ \frac{\partial \varphi_3}{\partial x_j}(O) \end{vmatrix} \, .$$

We shall often write $\partial_j|_p$ (or even, when no confusion may arise, simply ∂_j) rather than $\partial/\partial x_j|_p$. Clearly, $\{\partial_1|_p, \partial_2|_p\}$ is a basis of T_pS, the basis *induced* by the local parametrization φ. Note that $\partial_1|_p$ and $\partial_2|_p$ are just the two columns of the Jacobian matrix of φ computed in $O = \varphi^{-1}(p)$. Finally, a curve in S tangent to $\partial_j|_p$ is the j-th coordinate curve $\sigma: (-\varepsilon, \varepsilon) \to S$ given by $\sigma(t) = \varphi(t\mathbf{e}_j)$ for ε small enough.

We have seen that a possible way to define surfaces is as level surfaces of a smooth function. The following proposition tells us how to find the tangent plane in this case:

Proposition 3.3.11. *Let $U \subseteq \mathbb{R}^3$ an open set, and $a \in \mathbb{R}$ a regular value of a function $f \in C^\infty(U)$. If S is a connected component of $f^{-1}(a)$ and $p \in S$, the tangent plane T_pS is the subspace of $\mathbb{R}^3$ orthogonal to $\nabla f(p)$.*

Proof. Take $v = (v_1, v_2, v_3) \in T_pS$ and let $\sigma: (-\varepsilon, \varepsilon) \to S$ be a curve with $\sigma(0) = p$ and $\sigma'(0) = v$. Differentiating $f \circ \sigma \equiv a$ and evaluating in 0 we find

$$\frac{\partial f}{\partial x_1}(p)v_1 + \frac{\partial f}{\partial x_2}(p)v_2 + \frac{\partial f}{\partial x_3}(p)v_3 = 0 \ ,$$

and so v is orthogonal to $\nabla f(p)$. Hence T_pS is contained in the subspace orthogonal to $\nabla f(p)$; but both spaces have dimension 2, and so they coincide. $\square$

Let us now see some examples of tangent planes.

Example 3.3.12. Let $H \subset \mathbb{R}^3$ be a plane through a point $p_0 \in \mathbb{R}^3$, and $H_0 = H - p_0 \subset \mathbb{R}^3$ the plane through the origin and parallel to H. Since the tangent vectors to curves with support in H must belong to H_0 (see the proof of Proposition 1.3.25), we obtain $T_{p_0}H = H_0$.

Example 3.3.13. Let $p_0 = (x_0, y_0, z_0) \in S^2$ be a point of the unit sphere S^2 (see Example 3.1.15). If we set $f(x, y, z) = x^2 + y^2 + z^2$, by Proposition 3.3.11 $T_{p_0}S^2$ is the subspace orthogonal to $\nabla f(p_0) = (2x_0, 2y_0, 2z_0) = 2p_0$. So, *the tangent plane to a sphere at a point is always orthogonal to the radius in that point.* If $z_0 > 0$, by using the local parametrization φ_1 from Example 3.1.15, we find that a basis of $T_{p_0}S^2$ consists of the vectors

$$\left. \frac{\partial}{\partial x} \right|_{p_0} = \frac{\partial \varphi_1}{\partial x}(x_0, y_0) = \begin{vmatrix} 1 \\ 0 \\ \dfrac{-x_0}{\sqrt{1 - x_0^2 - y_0^2}} \end{vmatrix} \ , \quad \left. \frac{\partial}{\partial y} \right|_{p_0} = \frac{\partial \varphi_1}{\partial y}(x_0, y_0) = \begin{vmatrix} 0 \\ 1 \\ \dfrac{-y_0}{\sqrt{1 - x_0^2 - y_0^2}} \end{vmatrix} \ .$$

The basis induced by the local parametrization given by the spherical coordinates (Example 3.1.16), on the other hand, consists of the vectors

$$\left. \frac{\partial}{\partial \theta} \right|_{p_0} = \begin{vmatrix} \cos\theta \cos\phi \\ \cos\theta \sin\phi \\ -\sin\theta \end{vmatrix} \quad \text{and} \quad \left. \frac{\partial}{\partial \phi} \right|_{p_0} = \begin{vmatrix} -\sin\theta \sin\phi \\ \sin\theta \cos\phi \\ 0 \end{vmatrix} \ .$$

Example 3.3.14. Let $\Gamma_f \subset \mathbb{R}^3$ be the graph of a function $f \in C^\infty(U)$. Using Proposition 3.3.5 and the local parametrization from Example 3.1.13 we see that a basis of the tangent plane to Γ_f at the point $p = (x_1, x_2, f(x_1, x_2)) \in \Gamma_f$ consist of the vectors

$$\frac{\partial}{\partial x_1}\bigg|_p = \begin{vmatrix} 1 \\ 0 \\ \frac{\partial f}{\partial x_1}(x_1, x_2) \end{vmatrix}, \qquad \frac{\partial}{\partial x_2}\bigg|_p = \begin{vmatrix} 0 \\ 1 \\ \frac{\partial f}{\partial x_2}(x_1, x_2) \end{vmatrix}.$$

Example 3.3.15. Let $S \subset \mathbb{R}^3$ be the surface of revolution obtained by rotating around the z-axis a Jordan curve (or open arc) C contained in the right half-plane of the xz-plane. Let $\sigma\colon \mathbb{R} \to \mathbb{R}^3$ be a global or periodic parametrization of C of the form $\sigma(t) = (\alpha(t), 0, \beta(t))$, and $\varphi\colon \mathbb{R}^2 \to \mathbb{R}^3$ the immersed surface with support S introduced in Example 3.1.18. By Proposition 3.1.31.(ii), we know that every restriction of φ to an open set on which it is injective is a local parametrization of S; so Proposition 3.3.5 implies that $T_pS = \mathrm{d}\varphi_{(t,\theta)}(\mathbb{R}^2)$ for all $p = \varphi(t, \theta) \in S$. In particular, a basis of the tangent plane at p consists of the vectors

$$\frac{\partial}{\partial t}\bigg|_p = \frac{\partial\varphi}{\partial t}(t, \theta) = \begin{vmatrix} \alpha'(t)\cos\theta \\ \alpha'(t)\sin\theta \\ \beta'(t) \end{vmatrix}, \qquad \frac{\partial}{\partial\theta}\bigg|_p = \frac{\partial\varphi}{\partial\theta}(t, \theta) = \begin{vmatrix} -\alpha(t)\sin\theta \\ \alpha(t)\cos\theta \\ 0 \end{vmatrix}.$$

Example 3.3.16. A second degree polynomial p in three variables can always be written in the form $p(x) = x^T A x + 2b^T x + c$, where $A = (a_{ij}) \in M_{3,3}(\mathbb{R})$ is a symmetric matrix, $b \in \mathbb{R}^3$ (we are writing vectors in $\mathbb{R}^3$ as column vectors), and $c \in \mathbb{R}$. In particular, $\nabla p(x) = 2(Ax + b)$. So, if $S \subset \mathbb{R}^3$ is the component of the quadric having equation $p(x) = 0$ that contains the point $x_0 \notin \mathrm{Crit}(p)$, the tangent plane $T_{x_0}S$ to the surface S (see Exercise 3.9) at x_0 is given by

$$T_{x_0}S = \{v \in \mathbb{R}^3 \mid \langle Ax_0 + b, v \rangle = 0\}.$$

For instance, the tangent plane at the point $x_0 = (1, 0, 1)$ to the one-sheeted hyperboloid having equation $x^2 + y^2 - z^2 - 1 = 0$ is the plane $\{v = (v_1, v_2, v_3) \in \mathbb{R}^3 \mid v_1 = v_3\}$.

3.4 Tangent vectors and derivations

Definition 3.3.6 of tangent plane is not completely satisfactory: it strongly depends on the fact that the surface S is contained in $\mathbb{R}^3$, while it would be nice to have a notion of tangent vector intrinsic to S, independent of its embedding in the Euclidean space. In other words, we would like to have a definition of T_pS not as a subspace of $\mathbb{R}^3$, but as an abstract vector space, depending only on S and p. Moreover, since we are dealing with "differential geometry", sooner or later we shall have to find a way to differentiate on a surface.

Surprisingly enough, we may solve both these problems at the same time. The main idea is contained in the following example.

Example 3.4.1. Let $U \subseteq \mathbb{R}^2$ be an open set, and $p \in U$. Then we can associate with each tangent vector $v \in T_p U = \mathbb{R}^2$ a partial derivative:

$$v = (v_1, v_2) \mapsto \left.\frac{\partial}{\partial v}\right|_p = v_1 \left.\frac{\partial}{\partial x_1}\right|_p + v_2 \left.\frac{\partial}{\partial x_2}\right|_p ,$$

and all partial derivatives are of this kind. So, in a sense, we may identify $T_p U$ with the set of partial derivatives.

Our aim will then be to find a way for identifying, for general surfaces, tangent vectors with the right kind of partial derivative. To do so, we must first of all understand better which objects we want to differentiate. The key observation is that to differentiate a function in a point it suffices to know its behaviour in a neighborhood of the point; if our goal is just to differentiate at p, two functions that coincide in some neighborhood of p are completely equivalent. This remark suggests the following

Definition 3.4.2. Let $S \subset \mathbb{R}^3$ be a surface, and $p \in S$. Denote by $\mathcal{F}$ the set of pairs (U, f), where $U \subseteq S$ is an open neighborhood of p in S and $f \in C^\infty(U)$. We define an equivalence relation $\sim$ on $\mathcal{F}$ as follows: $(U, f) \sim (V, g)$ if there exists an open neighborhood $W \subseteq U \cap V$ of p such that $f|_W \equiv g|_W$. The quotient space $C^\infty(p) = \mathcal{F}/\sim$ is the *space (or stalk) of germs of C^∞ functions* at p, and an element $\mathbf{f} \in C^\infty(p)$ is a *germ* at p. An element (U, f) of the equivalence class $\mathbf{f}$ is a *representative* of $\mathbf{f}$. If it is necessary to remind the surface on which we are working, we shall write $C^\infty_S(p)$ rather than $C^\infty(p)$.

Remark 3.4.3. If $U \subseteq S$ is an open subset of a surface S and $p \in U$ then we clearly have $C^\infty_U(p) = C^\infty_S(p)$.

So, what we really want to differentiate are germs of C^∞ functions. Before seeing how to do this, note that $C^\infty(p)$ has a natural algebraic structure.

Definition 3.4.4. An *algebra* over a field $\mathbb{K}$ is a set A equipped with an addition $+$, a multiplication $\cdot$ and a multiplication by scalars $\lambda\cdot$, such that $(A, +, \cdot)$ is a ring, $(A, +, \lambda\cdot)$ is a vector space, and the associative property $(\lambda f)g = \lambda(fg) = f(\lambda g)$ holds, for all $\lambda \in \mathbb{K}$ and $f, g \in A$.

Lemma 3.4.5. *Let $S \subset \mathbb{R}^3$ be a surface, $p \in S$, and $\mathbf{f}, \mathbf{g} \in C^\infty(p)$ two germs at p. Let also (U_1, f_1), (U_2, f_2) be two representatives of $\mathbf{f}$, and (V_1, g_1), (V_2, g_2) two representatives of $\mathbf{g}$. Then:*

(i) *$(U_1 \cap V_1, f_1 + g_1)$ is equivalent to $(U_2 \cap V_2, f_2 + g_2)$;*
(ii) *$(U_1 \cap V_1, f_1 g_1)$ is equivalent to $(U_2 \cap V_2, f_2 g_2)$;*
(iii) *$(U_1, \lambda f_1)$ is equivalent to $(U_2, \lambda f_2)$ for all $\lambda \in \mathbb{R}$;*
(iv) *$f_1(p) = f_2(p)$.*

Proof. Let us begin with (i). Since $(U_1, f_1) \sim (U_2, f_2)$, there exists an open neighborhood $W_f \subseteq U_1 \cap U_2$ of p such that $f_1|_{W_f} \equiv f_2|_{W_f}$. Analogously, since $(V_1, g_1) \sim (V_2, g_2)$, there exists an open neighborhood $W_g \subseteq V_1 \cap V_2$ di p such that $g_1|_{W_g} \equiv g_2|_{W_g}$. But then $(f_1 + f_2)|_{W_f \cap W_g} \equiv (g_1 + g_2)|_{W_f \cap W_g}$, and so $(U_1 \cap V_1, f_1 + g_1) \sim (U_2 \cap V_2, f_2 + g_2)$ as $W_f \cap W_g \subseteq U_1 \cap V_1 \cap U_2 \cap V_2$.

The proof of (ii) is analogous, and (iii) and (iv) are straightforward.   □

Definition 3.4.6. Let $\mathbf{f}, \mathbf{g} \in C^\infty(p)$ be two germs at a point $p \in S$. We shall denote by $\mathbf{f} + \mathbf{g} \in C^\infty(p)$ the germ represented by $(U \cap V, f + g)$, where (U, f) is an arbitrary representative of $\mathbf{f}$ and (V, g) is an arbitrary representative of $\mathbf{g}$. Analogously, we denote by $\mathbf{fg} \in C^\infty(p)$ the germ represented by $(U \cap V, fg)$, and, given $\lambda \in \mathbb{R}$, by $\lambda\mathbf{f} \in C^\infty(p)$ the germ represented by $(U, \lambda f)$. Lemma 3.4.5 tells us that these objects are well defined, and it is straightforward (why?) to verify that $C^\infty(p)$ with these operations is an algebra. Finally, for all $\mathbf{f} \in C^\infty(p)$ we define its value $\mathbf{f}(p) \in \mathbb{R}$ in p by setting $\mathbf{f}(p) = f(p)$ for an arbitrary representative (U, f) of $\mathbf{f}$; Lemma 3.4.5 again implies that $\mathbf{f}(p)$ is well defined.

The fact that the composition of smooth maps is itself a smooth map allows us to compare stalks in different points of different surfaces. Indeed, let $F: S_1 \to S_2$ be a C^∞ map between surfaces, and let (V_1, g_1) and (V_2, g_2) be two representatives of a germ $\mathbf{g} \in C^\infty(F(p))$. Then it is clear (exercise) that $(F^{-1}(V_1), g_1 \circ F)$ and $(F^{-1}(V_2), g_2 \circ F)$ represent the same germ at p, which, then, depends on $\mathbf{g}$ (and on F) only.

Definition 3.4.7. Let $F: S_1 \to S_2$ be a smooth map between surfaces, and $p \in S_1$. We shall denote by $F_p^*: C^\infty_{S_2}(F(p)) \to C^\infty_{S_1}(p)$ the map associating with a germ $\mathbf{g} \in C^\infty_{S_2}(F(p))$ having (V, g) as a representative the germ $F_p^*(\mathbf{g}) \in C^\infty_{S_1}(p)$ having $(F^{-1}(V), g \circ F)$ as a representative. We shall sometimes write $\mathbf{g} \circ F$ rather than $F_p^*(\mathbf{g})$. It is immediate to see (exercise) that F_p^* is an algebra homomorphism.

Remark 3.4.8. A very common (and very useful) convention in contemporary mathematics consists in denoting by a star written as a superscript (as in F_p^*) a map associated in a canonical way with a given map, but going in the opposite direction: F is a function from S_1 to S_2, whereas F^* is a function from the germs in S_2 to the germs in S_1. The same convention uses a star as a subscript (as in F_*) to denote an associated map going in the same direction as the given one (see for instance Definitions 3.4.12 and 3.4.21 later on).

Lemma 3.4.9. (i)   *We have* $(\mathrm{id}_S)_p^* = \mathrm{id}$ *for all points p of a surface S.*

(ii)   *Let $F: S_1 \to S_2$ and $G: S_2 \to S_3$ be two C^∞ maps between surfaces. Then* $(G \circ F)_p^* = F_p^* \circ G_{F(p)}^*$ *for all $p \in S_1$.*

(iii)   *If $F: S_1 \to S_2$ is a diffeomorphism, then $F_p^*: C^\infty(F(p)) \to C^\infty(p)$ is an algebra isomorphism for all $p \in S_1$. In particular, if $\varphi: U \to S$ is a local parametrization with $\varphi(x_o) = p \in S$, then $\varphi_{x_o}^*: C^\infty_S(p) \to C^\infty_U(x_o)$ is an algebra isomorphism.*

Proof. (i) Obvious.

(ii) Follows immediately (exercise) from the equality $g \circ (G \circ F) = (g \circ G) \circ F$.

(iii) Indeed $(F^{-1})^*_{F(p)}$ is the inverse of F^*_p, by (i) and (ii). $\square$

Now we can define what we mean by a partial derivative on a surface.

Definition 3.4.10. Let $S \subset \mathbb{R}^3$ be a surface, and $p \in S$. A *derivation* at p is a $\mathbb{R}$-linear function $D \colon C^\infty(p) \to \mathbb{R}$ satisfying a Leibniz (or product) rule:

$$D(\mathbf{fg}) = \mathbf{f}(p)D(\mathbf{g}) + \mathbf{g}(p)D(\mathbf{f}) \ .$$

It is immediate to verify (exercise) that the set $\mathcal{D}(C^\infty(p))$ of derivations of $C^\infty(p)$ is a vector subspace of the dual space (as a vector space) of $C^\infty(p)$.

Example 3.4.11. Let $U \subset \mathbb{R}^2$ be an open subset of the plane, and $p \in U$. We have already remarked that $T_pU = \mathbb{R}^2$. On the other hand, the partial derivatives at p are clearly derivations of $C^\infty(p)$; so we may introduce a natural linear map $\alpha \colon T_pU \to \mathcal{D}(C^\infty(p))$ by setting

$$\alpha(v) = \left.\frac{\partial}{\partial v}\right|_p = v_1 \left.\frac{\partial}{\partial x_1}\right|_p + v_2 \left.\frac{\partial}{\partial x_2}\right|_p \ .$$

The key point here is that the map α is actually an isomorphism between T_pU and $\mathcal{D}(C^\infty(p))$. Moreover, we shall show that T_pS and $\mathcal{D}(C^\infty_S(p))$ are canonically isomorphic for every surface S and for every $p \in S$, and this fact will provide us with the desired intrinsic characterization of the tangent plane. To prove all this we need one more definition and a lemma.

Definition 3.4.12. Let $S \subset \mathbb{R}^3$ be a surface, and $p \in S$. Given a local parametrization $\varphi \colon U \to S$ in p with $\varphi(x_o) = p \in S$, we define a map $\varphi_* \colon \mathcal{D}(C^\infty(x_o)) \to \mathcal{D}(C^\infty(p))$ by setting $\varphi_*(D) = D \circ \varphi^*_{x_o}$, that is,

$$\varphi_*(D)(\mathbf{f}) = D(\mathbf{f} \circ \varphi)$$

for all $\mathbf{f} \in C^\infty(p)$ and $D \in \mathcal{D}(C^\infty(x_o))$. It is immediate to verify (check it!) that $\varphi_*(D)$ is a derivation, since $\varphi^*_{x_o}$ is an algebra isomorphism, and so the image of φ_* is actually contained in $\mathcal{D}(C^\infty(p))$. Moreover, it is easy to see (exercise) that φ_* is a vector space isomorphism, with $(\varphi_*)^{-1}(D) = D \circ (\varphi^{-1})^*_p$ as its inverse.

Remark 3.4.13. We shall see later on (Remark 3.4.16) that φ_* can be canonically identified with the differential of the local parametrization.

Lemma 3.4.14. *Let $U \subseteq \mathbb{R}^n$ be an open domain star-shaped with respect to $x^o \in \mathbb{R}^n$. Then for all $f \in C^\infty(U)$ there exist $g_1, \ldots, g_n \in C^\infty(U)$ such that $g_j(x^o) = \frac{\partial f}{\partial x_j}(x^o)$ and*

$$f(x) = f(x^o) + \sum_{j=1}^n (x_j - x_j^o)g_j(x)$$

for all $x \in U$.

Proof. We have

$$f(x) - f(x^o) = \int_0^1 \frac{\partial}{\partial t} f(x^o + t(x - x^o)) \, \mathrm{d}t$$

$$= \sum_{j=1}^n (x_j - x_j^o) \int_0^1 \frac{\partial f}{\partial x_j} (x^o + t(x - x^o)) \, \mathrm{d}t \,,$$

so it suffices to define

$$g_j(x) = \int_0^1 \frac{\partial f}{\partial x_j} (x^o + t(x - x^o)) \, \mathrm{d}t \,.$$

$\square$

We may now prove the characterization of the tangent plane we promised:

Theorem 3.4.15. *Let $S \subset \mathbb{R}^3$ be a surface, and $p \in S$. Then the tangent plane T_pS is canonically isomorphic to the space $\mathcal{D}(C^\infty(p))$ of derivations of $C^\infty(p)$.*

Proof. Let $\varphi \colon U \to S$ be a local parametrization centered at p. Let us begin by writing the following commutative diagram:

$$\begin{array}{ccc} T_O U = \mathbb{R}^2 & \xrightarrow{\alpha} & \mathcal{D}(C^\infty(O)) \\ \mathrm{d}\varphi_O \Big\downarrow & & \Big\downarrow \varphi_* \\ T_p S & \xrightarrow{\beta} & \mathcal{D}(C^\infty(p)) \end{array} \qquad (3.2)$$

where α is the map defined in Example 3.4.11, and $\beta = \varphi_* \circ \alpha \circ (\mathrm{d}\varphi_O)^{-1}$.

We shall proceed in two steps: first of all, we shall show that α is an isomorphism. Since $\mathrm{d}\varphi_O$ and φ_* are isomorphisms, this will imply that β is an isomorphism too. We shall prove next that it is possible to express β in a way independent of φ; so β will be a canonical isomorphism, independent of arbitrary choices, and we shall be done.

Let us prove that α is an isomorphism. As it is obviously linear, it suffices to show that it is injective and surjective. If $v = (v_1, v_2) \in \mathbb{R}^2 = T_O U$, we have

$$v_j = v_j \frac{\partial x_j}{\partial x_j}(O) = \alpha(v)(\mathbf{x}_j)$$

for $j = 1, 2$, where $\mathbf{x}_j$ is the germ at the origin of the coordinate function x_j. Innparticular, if $v_j \neq 0$ we have $\alpha(v)(\mathbf{x}_j) \neq 0$; so $v \neq O$ implies $\alpha(v) \neq O$ and α is injective. To show that it is surjective, take $D \in \mathcal{D}(C^\infty(O))$; we claim that $D = \alpha(v)$, where $v = (D\mathbf{x}_1, D\mathbf{x}_2)$. First of all, note that

$$D\mathbf{1} = D(\mathbf{1} \cdot \mathbf{1}) = 2D\mathbf{1} \,,$$

so $D\mathbf{c} = 0$ for any constant $c \in \mathbb{R}$, where $\mathbf{c}$ is the germ represented by $(\mathbb{R}^2, c)$. Take now an arbitrary $\mathbf{f} \in C^\infty(O)$. By applying Lemma 3.4.14, we find

$$D\mathbf{f} = D\big(\mathbf{f}(O)\big) + D\big(\mathbf{x}_1\mathbf{g}_1 + \mathbf{x}_2\mathbf{g}_2\big) \tag{3.3}$$

$$= \sum_{j=1}^{2}\big[\mathbf{x}_j(O)D\mathbf{g}_j + \mathbf{g}_j(O)D\mathbf{x}_j\big] = \sum_{j=1}^{2} D\mathbf{x}_j \frac{\partial \mathbf{f}}{\partial x_j}(O) = \alpha(v)(\mathbf{f}),$$

where $v = (D\mathbf{x}_1, D\mathbf{x}_2)$ as claimed, and we are done.

So, α and β are isomorphisms; to complete the proof, we only have to show that β does not depend on φ but only on S and p. Let $v \in T_pS$, and choose a curve $\sigma: (-\varepsilon, \varepsilon) \to S$ such that $\sigma(0) = p$ and $\sigma'(0) = v$. We want to show that

$$\beta(v)(\mathbf{f})' = (f \circ \sigma)'(0) \tag{3.4}$$

for all $\mathbf{f} \in C^\infty(p)$ and any representative $(U, f) \in \mathbf{f}$. If we prove this, we are done: indeed, the left-hand side of (3.4) does not depend on σ nor on the chosen representative of $\mathbf{f}$, while the right-hand side does not depend on any local parametrization. So β does not depend on φ or on σ, and thus it is the canonical isomorphism we were looking for.

Let us then prove (3.4). Write $\sigma = \varphi \circ \sigma_o$ as in the proof of Proposition 3.3.5, so that $v = d\varphi_O(v^o) = v_1^o \partial_1|_p + v_2^o \partial_2|_p$ and $v^o = (v_1^o, v_2^o) = \sigma_o'(0) \in \mathbb{R}^2$. Then

$$\beta(v)(\mathbf{f}) = \big(\varphi_* \circ \alpha \circ (d\varphi_O)^{-1}\big)(v)(\mathbf{f}) = (\varphi_* \circ \alpha)(v^o)(\mathbf{f})$$

$$= \alpha(v^o)\big(\varphi_O^*(\mathbf{f})\big) = \alpha(v^o)(\mathbf{f} \circ \varphi)$$

$$= v_1^o \frac{\partial(f \circ \varphi)}{\partial x_1}(O) + v_2^o \frac{\partial(f \circ \varphi)}{\partial x_2}(O) \tag{3.5}$$

$$= (\sigma_o')_1(0)\frac{\partial(f \circ \varphi)}{\partial x_1}(O) + (\sigma_o')_2(0)\frac{\partial(f \circ \varphi)}{\partial x_2}(O)$$

$$= \big((f \circ \varphi) \circ \sigma_o\big)'(0) = (f \circ \sigma)'(0),$$

and we are done. $\qquad\square$

Remark 3.4.16. A consequence of diagram (3.2) is that, as anticipated in Remark 3.4.13, the map φ_* is the exact analogue of the differential of φ when we interpret tangent planes as spaces of derivations.

From now on, we shall always identify T_pS and $\mathcal{D}\big(C^\infty(p)\big)$ without (almost ever) explicitly mentioning the isomorphism β; a tangent vector will be considered both as a vector of $\mathbb{R}^3$ and as a derivation of the space of germs at p without further remarks.

Remark 3.4.17. Let $\varphi: U \to S$ be a local parametrization centered at a point $p \in S$, and take a tangent vector $v = v_1\partial_1|_p + v_2\partial_2|_p \in T_pS$. Then (3.5) tells us that the action of v as a derivation is given by

$$v(\mathbf{f}) = v_1 \frac{\partial(f \circ \varphi)}{\partial x_1}(O) + v_2 \frac{\partial(f \circ \varphi)}{\partial x_2}(O),$$

for all germs $\mathbf{f} \in C^\infty(p)$ and all representatives (V, f) of $\mathbf{f}$. In particular,

$$\left.\frac{\partial}{\partial x_j}\right|_p (\mathbf{f}) = \frac{\partial(f \circ \varphi)}{\partial x_j}(O) \,,$$

a formula which explains the notation introduced in Definition 3.3.10. As a consequence, for any $p \in \mathbb{R}^2$ we shall always identify the vectors $\mathbf{e}_1, \mathbf{e}_2$ of the canonical basis of $\mathbb{R}^2$ with the partial derivatives $\partial/\partial x_1|_p, \partial/\partial x_2|_p \in T_p\mathbb{R}^2$.

Remark 3.4.18. In the previous remark we have described the action of a tangent vector on a germ by expressing the tangent vector in terms of the basis induced by a local parametrization. If, on the other hand, we consider $v = (v_1, v_2, v_3) \in T_pS$ as a vector of $\mathbb{R}^3$, we may describe its action as follows: given $\mathbf{f} \in C^\infty(p)$, choose a representative (V, f) of $\mathbf{f}$ and extend it using Proposition 3.2.12 to a smooth function $\tilde{f}$ defined in a neighborhood W of p in $\mathbb{R}^3$. Finally, let $\sigma: (-\varepsilon, \varepsilon) \to S$ be a curve with $\sigma(0) = p$ and $\sigma'(0) = v$. Then:

$$v(\mathbf{f}) = (f \circ \sigma)'(0) = (\tilde{f} \circ \sigma)'(0) = \sum_{j=1}^{3} v_j \frac{\partial \tilde{f}}{\partial x_j}(p) \,.$$

Warning: while the linear combination in the right-hand side of this formula is well defined and only depends on the tangent vector v and on the germ $\mathbf{f}$, the partial derivatives $\partial \tilde{f}/\partial x_j(p)$ taken on their own depend on the particular extension $\tilde{f}$ and not only on $\mathbf{f}$, and thus they have nothing to do with the surface S.

Remark 3.4.19. If we have two local parametrizations $\varphi: U \to S$ and $\hat{\varphi}: \hat{U} \to S$ centered at $p \in S$, we obtain two bases $\{\partial_1, \partial_2\}$ and $\{\hat{\partial}_1, \hat{\partial}_2\}$ of T_pS, where $\hat{\partial}_j = \partial\hat{\varphi}/\partial\hat{x}_j(O)$, and $(\hat{x}_1, \hat{x}_2)$ are the coordinates in $\hat{U}$; we want to compute the change of basis matrix. If $h = \hat{\varphi}^{-1} \circ \varphi$ is the change of coordinates, we have $\varphi = \hat{\varphi} \circ h$, and so

$$\partial_j = \frac{\partial\varphi}{\partial x_j}(O) = \frac{\partial\hat{\varphi}}{\partial x_1}\big(h(O)\big)\frac{\partial h_1}{\partial x_j}(O) + \frac{\partial\hat{\varphi}}{\partial x_2}\big(h(O)\big)\frac{\partial h_2}{\partial x_j}(O)$$

$$= \frac{\partial\hat{x}_1}{\partial x_j}(O)\hat{\partial}_1 + \frac{\partial\hat{x}_2}{\partial x_j}(O)\hat{\partial}_2 \,,$$

where, to make the formula easier to remember, we have written $\partial\hat{x}_i/\partial x_j$ rather than $\partial h_i/\partial x_j$. So *the change of basis matrix is the Jacobian matrix of the change of coordinates.*

Remark 3.4.20. The identification of tangent vectors and derivations only holds when working with functions and local parametrizations of class C^∞. The reason is Lemma 3.4.14. Indeed, if $f \in C^k(U)$ with $k < \infty$, the same proof provides functions $g_1, \ldots, g_n$ which are in $C^{k-1}(U)$ but a priori might not be in $C^k(U)$, and the computation made in (3.3) might become meaningless. This is an insurmountable obstacle: in fact, the space of derivations of

C^k germs with $1 \le k < \infty$ is of infinite dimension (Exercise 3.35), and thus it cannot be isomorphic to a plane.

The way we have introduced the map φ_*, together with its relation with the usual differential, suggests the following definition of a differential for an arbitrary C^∞ map between surfaces:

Definition 3.4.21. Let $F\colon S_1 \to S_2$ be a C^∞ map between two surfaces. The *differential* of F at $p \in S_1$ is the linear map $\mathrm{d}F_p\colon T_pS_1 \to T_{F(p)}S_2$ defined by

$$\mathrm{d}F_p(D) = D \circ F_p^*$$

for any derivation $D \in T_pS$ of $C^\infty(p)$. We may also write $(F_*)_p$ instead of $\mathrm{d}F_p$.

It is not difficult to see how the differential looks like when applied to vectors seen as tangent vectors to a curve:

Lemma 3.4.22. *Let $F\colon S_1 \to S_2$ be a C^∞ map between surfaces, and $p \in S_1$. If $\sigma\colon (-\varepsilon,\varepsilon) \to S_1$ is a curve with $\sigma(0) = p$ and $\sigma'(0) = v$, then*

$$\mathrm{d}F_p(v) = (F \circ \sigma)'(0) . \tag{3.6}$$

Proof. Set $w = (F \circ \sigma)'(0) \in T_{F(p)}S_2$. Using the notation introduced in the proof of Theorem 3.4.15, we have to show that $\mathrm{d}F_p\big(\beta(v)\big) = \beta(w)$. But for each $\mathbf{f} \in C^\infty\big(F(p)\big)$ we have

$$\mathrm{d}F_p\big(\beta(v)\big)(\mathbf{f}) = \beta(v)\big(F_p^*(\mathbf{f})\big) = \beta(v)(\mathbf{f} \circ F)$$
$$= \big((f \circ F) \circ \sigma\big)'(0) = \big(f \circ (F \circ \sigma)\big)'(0) = \beta(w)(\mathbf{f}) ,$$

where (U, f) is a representative of $\mathbf{f}$, and we have used (3.4). □

As for the tangent plane, we then have two different ways to define the differential, each one with its own strengths and weaknesses. Formula (3.6) underlines the geometric meaning of differential, showing how it acts on tangent vectors to curves; Definition 3.4.21 highlights instead its algebraic properties, such as the fact that the differential is a linear map between tangent planes, and makes it (far) easier to prove its properties. For instance, we obtain a one-line proof of the following proposition:

Proposition 3.4.23. (i) *We have $\mathrm{d}(\mathrm{id}_S)_p = \mathrm{id}$ for every surface S and every $p \in S$.*

(ii) *Let $F\colon S_1 \to S_2$ and $G\colon S_2 \to S_3$ C^∞ be maps between surfaces, and $p \in S_1$. Then $\mathrm{d}(G \circ F)_p = \mathrm{d}G_{F(p)} \circ \mathrm{d}F_p$.*

(iii) *If $F\colon S_1 \to S_2$ is a diffeomorphism then $\mathrm{d}F_p\colon T_pS_1 \to T_{F(p)}S_2$ is invertible and $(\mathrm{d}F_p)^{-1} = \mathrm{d}(F^{-1})_{F(p)}$ for all $p \in S_1$.*

Proof. It is an immediate consequence of Lemma 3.4.9 and of the definition of differential. □

Formula (3.6) also suggests how to define the differential of a C^∞ map defined on a surface but with values in $\mathbb{R}^n$:

Definition 3.4.24. If $F: S \to \mathbb{R}^n$ is a C^∞ map, and $p \in S$, the *differential* $\mathrm{d}F_p: T_pS \to \mathbb{R}^n$ of F at p is defined by setting $\mathrm{d}F_p(v) = (F \circ \sigma)'(0)$ for all $v \in T_pS$, where $\sigma: (-\varepsilon, \varepsilon) \to S$ is an arbitrary curve in S with $\sigma(0) = p$ and $\sigma'(0) = v$. It is not hard (exercise) to verify that $\mathrm{d}F_p(v)$ only depends on v and not on the curve σ, and that $\mathrm{d}F_p$ is a linear map.

Remark 3.4.25. In particular, if $f \in C^\infty(S)$ and $v \in T_pS$ then we have

$$\mathrm{d}f_p(v) = (f \circ \sigma)'(0) = v(\mathbf{f}) \,,$$

where $\mathbf{f}$ is the germ represented by (S, f) at p. This formula shows that the action of the differential of functions on tangent vectors is dual to the action of tangent vectors on functions.

Remark 3.4.26. If $F: S \to \mathbb{R}^n$ is of class C^∞ and $\varphi: U \to S$ is a local parametrization centered at $p \in S$, it is immediate (why?) to see that

$$\mathrm{d}F_p(\partial_j) = \frac{\partial(F \circ \varphi)}{\partial x_j}(O)$$

for $j = 1, 2$, where $\{\partial_1, \partial_2\}$ is the basis of T_pS induced by φ. In particular, if $\tilde{\varphi}$ is another local parametrization of S centered at p and $F = \tilde{\varphi} \circ \varphi^{-1}$, then

$$\mathrm{d}F_p(\partial_j) = \tilde{\partial}_j \tag{3.7}$$

for $j = 1, 2$, where $\{\tilde{\partial}_1, \tilde{\partial}_2\}$ is the basis of T_pS induced by $\tilde{\varphi}$.

Let us see now how to express the differential in local coordinates. Given a smooth map $F: S_1 \to S_2$ between surfaces, choose a local parametrization $\varphi: U \to S_1$ centered at $p \in S_1$, and a local parametrization $\hat{\varphi}: \hat{U} \to S_2$ centered at $F(p) \in S_2$ with $F(\varphi(U)) \subseteq \hat{\varphi}(\hat{U})$. By definition, *the expression of F in local coordinates* is the map $\hat{F} = (\hat{F}_1, \hat{F}_2): U \to \hat{U}$ given by

$$\hat{F} = \hat{\varphi}^{-1} \circ F \circ \varphi \,.$$

We want to find the matrix that represents $\mathrm{d}F_p$ with respect to the bases $\{\partial_1, \partial_2\}$ of T_pS_1 (induced by φ) and $\{\hat{\partial}_1, \hat{\partial}_2\}$ of $T_{F(p)}S_2$ (induced by $\hat{\varphi}$); recall that the columns of this matrix contain the coordinates with respect to the new basis of the images under $\mathrm{d}F_p$ of the vectors of the old basis. We may proceed in any of two ways: either by using curves, or by using derivations.

A curve in S_1, tangent to ∂_j at p, is $\sigma_j(t) = \varphi(t\mathbf{e}_j)$; so

$$\mathrm{d}F_p(\partial_j) = (F \circ \sigma_j)'(0) = \frac{\mathrm{d}}{\mathrm{d}t}\left(\hat{\varphi} \circ \hat{F}(t\mathbf{e}_j)\right)\bigg|_{t=0} = \frac{\partial \hat{F}_1}{\partial x_j}(O)\hat{\partial}_1 + \frac{\partial \hat{F}_2}{\partial x_j}(O)\hat{\partial}_2 \,.$$

Hence, *the matrix that represents* $\mathrm{d}F_p$ *with respect to the bases induced by two local parametrizations is exactly the Jacobian matrix of the expression* $\hat{F}$ *of* F *in local coordinates.* In particular, the differential as we have defined it really is a generalization to surfaces of the usual differential of C^∞ maps between open subsets of the plane.

Let us now get again the same result by using derivations. We want to find $a_{ij} \in \mathbb{R}$ such that $\mathrm{d}F_p(\partial_j) = a_{1j}\hat{\partial}_1 + a_{2j}\hat{\partial}_2$ for $j = 1, 2$. If we set $\hat{\varphi}^{-1} = (\hat{x}_1, \hat{x}_2)$, it is immediate to verify that

$$\hat{\partial}_h(\hat{\mathbf{x}}_k) = \delta_{hk} = \begin{cases} 1 & \text{if } h = k\,, \\ 0 & \text{if } h \neq k\,, \end{cases}$$

where $\hat{\mathbf{x}}_k$ is the germ at p of the function $\hat{x}_k$. Hence,

$$a_{ij} = \mathrm{d}F_p(\partial_j)(\hat{\mathbf{x}}_i) = \partial_j\big(F_p^*(\hat{\mathbf{x}}_i)\big) = \frac{\partial(\hat{x}_i \circ F \circ \varphi)}{\partial x_j}(O) = \frac{\partial \hat{F}_i}{\partial x_j}(O)\,,$$

in accord with what we have already obtained.

Remark 3.4.27. Warning: the matrix representing the differential of a map between surfaces is a 2×2 matrix (and not a 3×3, or 3×2 or 2×3 matrix), because tangent planes have dimension 2.

We conclude this chapter remarking that the fact that the differential of a map between surfaces is represented by the Jacobian matrix of the map expressed in local coordinates allows us to easily transfer to surfaces classical calculus results. For instance, here is the inverse function theorem (for other results of this kind, see Exercises 3.19, 3.31, and 3.22):

Corollary 3.4.28. *Let* $F\colon S_1 \to S_2$ *be a smooth map between surfaces, and* $p \in S_1$ *a point such that* $\mathrm{d}F_p\colon T_pS_1 \to T_{F(p)}S_2$ *is an isomorphism. Then there exist a neighborhood* $V \subseteq S_1$ *of* p *and a neighborhood* $\hat{V} \subseteq S_2$ *of* $F(p)$ *such that* $F|_V\colon V \to \hat{V}$ *is a diffeomorphism.*

Proof. Let $\varphi\colon U \to S_1$ be a local parametrization at p, and $\hat{\varphi}\colon \hat{U} \to S_2$ a local parametrization at $F(p)$ with $F\big(\varphi(U)\big) \subseteq \hat{\varphi}(\hat{U})$. Then the assertion immediately follows (why?) from the classical inverse function theorem (Theorem 2.2.4) applied to $\hat{\varphi}^{-1} \circ F \circ \varphi$. $\qquad\square$

Guided problems

Definition 3.P.1. The *catenoid* is a surface of revolution having a catenary (see Example 1.2.16) as its generatrix, and axis disjoint from the support of the catenary; see Fig. 3.7.(a).

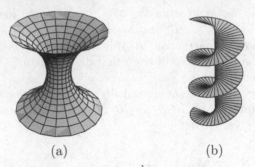

(a)   (b)

Fig. 3.7. (a) a catenoid; (b) a helicoid

Problem 3.1. *Let $\sigma\colon \mathbb{R} \to \mathbb{R}^3$ be the parametrization $\sigma(v) = (a\cosh v, 0, av)$ of a catenary, and let S be the catenoid obtained by rotating this catenary around the z-axis.*

(i) *Determine an immersed surface whose support is the catenoid S.*
(ii) *Determine for each point p of S a basis of the tangent plane T_pS.*

Solution. Proceeding as in Example 3.1.18 we find that an immersed surface $\varphi\colon \mathbb{R}^2 \to \mathbb{R}^3$ having the catenoid as support is

$$\varphi(u, v) = (a\cosh v\cos u, a\cosh v\sin u, av).$$

Since σ is a global parametrization of a regular curve whose support does not meet the z-axis, the catenoid is a regular surface. In particular, for all $(u_0, v_0) \in \mathbb{R}^2$ the restriction of φ to a neighborhood of (u_0, v_0) is a local parametrization of S, and so in the point $p = \varphi(u_0, v_0)$ of the catenoid a basis of the tangent plane is given by

$$\partial_1|_p = \frac{\partial\varphi}{\partial u}(u_0, v_0) = \begin{vmatrix} -a\cosh v_0\sin u_0 \\ a\cosh v_0\cos u_0 \\ 0 \end{vmatrix},$$

$$\partial_2|_p = \frac{\partial\varphi}{\partial v}(u_0, v_0) = \begin{vmatrix} a\sinh v_0\cos u_0 \\ a\sinh v_0\sin u_0 \\ a \end{vmatrix}.$$

$\square$

Definition 3.P.2. Given a circular helix in $\mathbb{R}^3$, the union of the straight lines issuing from a point of the helix and intersecting orthogonally the axis of the helix is the *helicoid* associated with the given helix; see Fig. 3.7.(b).

Problem 3.2. *Given $a \neq 0$, let $\sigma\colon \mathbb{R} \to \mathbb{R}^3$ be the circular helix parametrized by $\sigma(u) = (\cos u, \sin u, au)$.*

(i) *Prove that the helicoid associated with σ is the support of the map $\varphi \colon \mathbb{R}^2 \to \mathbb{R}^3$ given by*

$$\varphi(u, v) = (v \cos u, v \sin u, au) .$$

(ii) *Show that φ is a global parametrization and that the helicoid is a regular surface.*

(iii) *Determine, for every point of the helicoid, a basis of the tangent plane.*

Solution. (i) Indeed the straight line issuing from a point $(x_0, y_0, z_0) \in \mathbb{R}^3$ and intersecting orthogonally the z-axis is parametrized by $v \mapsto (vx_0, vy_0, z_0)$.

(ii) The map φ is clearly of class C^∞. Moreover, its differential is injective in every point; indeed, $\frac{\partial \varphi}{\partial u} = (-v \sin u, v \cos u, a)$ and $\frac{\partial \varphi}{\partial v} = (\cos u, \sin u, 0)$, and so

$$\frac{\partial \varphi}{\partial u} \wedge \frac{\partial \varphi}{\partial v} = (-a \sin u, a \cos u, -v)$$

has absolute value $\sqrt{a^2 + v^2}$ nowhere zero. Finally, φ is injective and is a homeomorphism with its image. Indeed, the continuous inverse can be constructed as follows: if $(x, y, z) = \varphi(u, v)$, then $u = z/a$ and $v = x/\cos(z/a)$, or $v = y/\sin(z/a)$ if $\cos(z/a) = 0$.

(iii) In the point $p = \varphi(u_0, v_0)$ of the helicoid a basis of the tangent plane is given by

$$\partial_1|_p = \frac{\partial \varphi}{\partial u}(u_0, v_0) = \begin{vmatrix} -v_0 \sin u_0 \\ v_0 \cos u_0 \\ a \end{vmatrix} \quad , \quad \partial_2|_p = \frac{\partial \varphi}{\partial v}(u_0, v_0) = \begin{vmatrix} \cos u_0 \\ \sin u_0 \\ 0 \end{vmatrix} .$$

$\square$

Definition 3.P.3. Let $H \subset \mathbb{R}^3$ be a plane, $\ell \subset \mathbb{R}^3$ a straight line not contained in H, and $C \subseteq H$ a subset of H. The *cylinder* with *generatrix* C and *directrix* ℓ is the subset of $\mathbb{R}^3$ consisting of the lines parallel to ℓ issuing from the points of C. If ℓ is orthogonal to H, the cylinder is said to be *right*.

Problem 3.3. *Let $C \subset \mathbb{R}^2$ be the support of a Jordan curve (or open arc) of class C^∞ contained in the xy-plane, and $\ell \subset \mathbb{R}^3$ a straight line transversal to the xy-plane. Denote by $S \subset \mathbb{R}^3$ the cylinder having C as generatrix and ℓ as directrix.*

(i) *Show that S is a regular surface.*

(ii) *Determine an atlas for S when ℓ is the z-axis and C is the circle of equation $x^2 + y^2 = 1$ in the plane $H = \{z = 0\}$.*

(iii) *If S is as in (ii), show that the map $G \colon \mathbb{R}^3 \to \mathbb{R}^2$ defined by*

$$G(x, y, z) = (e^z x, e^z y)$$

induces a diffeomorphism $G|_S \colon S \to \mathbb{R}^2 \setminus \{(0, 0)\}$.

Solution. (i) Let $\mathbf{v}$ be a versor parallel to ℓ, denote by H the xy-plane, and let $\sigma\colon\mathbb{R}\to C$ be a global or periodic parametrization of C (see Example 3.1.18). A point $p = (x, y, z)$ belongs to S if and only if there exist a point $p_0 \in C$ and a real number v such that $p = p_0 + v\mathbf{v}$. Define then $\varphi\colon\mathbb{R}^2 \to \mathbb{R}^3$ by setting

$$\varphi(t, v) = \sigma(t) + v\mathbf{v}\,.$$

Since $\partial\varphi/\partial t(t, v) = \sigma'(t) \in H$ and $\partial\varphi/\partial v(t, v) = \mathbf{v}$, the differential of φ has rank 2 everywhere, and so φ is an immersed surface with support S.

If C is an open Jordan arc, then σ is a homeomorphism with its image, so we obtain a continuous inverse of φ as follows: if $p = \varphi(t, v)$ then

$$v = \frac{\langle p, \mathbf{w}\rangle}{\langle \mathbf{v}, \mathbf{w}\rangle} \quad\text{and}\quad t = \sigma^{-1}(p - v\mathbf{v})\,,$$

where $\mathbf{w}$ is a versor orthogonal to H, and $\langle \mathbf{v}, \mathbf{w}\rangle \neq 0$ because $\mathbf{v}$ is transversal to H. So in this case φ is a global parametrization of the regular surface S.

If C is a Jordan curve, the same argument shows that if (a, b) is an interval where σ is a homeomorphism with its image then φ restricted to $(a, b) \times \mathbb{R}$ is a homeomorphism with its image; so, as seen for surfaces of revolution, it turns out that S is a regular surface with an atlas consisting of two charts, obtained by restricting φ to suitable open subsets of the plane.

(ii) In (i) we already constructed an atlas; let us find another one. This particular cylinder is the level surface $f^{-1}(0)$ of the function $f\colon\mathbb{R}^3 \to \mathbb{R}$ given by $f(x, y, z) = x^2 + y^2 - 1$; note that 0 is a regular value of f because the gradient $\nabla f = (2x, 2y, 0)$ of f is nowhere zero on S. We shall find an atlas for S by following the proof of Proposition 3.1.25, where it is shown that maps of the form $\varphi(u, v) = (u, g(u, v), v)$ with g solving the equation $f(u, g(u, v), v) = 0$ are local parametrizations at points $p_0 \in S$ where $\partial f/\partial y(p_0) \neq 0$, and that maps of the form $\varphi(u, v) = (g(u, v), u, v)$ with g solving the equation $f(g(u, v), u, v) = 0$ are local parametrizations at points $p_0 \in S$ where $\partial f/\partial x(p_0) \neq 0$.

In our case, $\nabla f = (2x, 2y, 0)$. So if $y_0 \neq 0$ we must solve the equation $u^2 + g(u, v)^2 - 1 = 0$; therefore $g(u, v) = \pm\sqrt{1 - u^2}$, and setting $U = \{(u, v) \in \mathbb{R}^2 \mid -1 < u < 1\}$ we get the parametrizations $\varphi_+, \varphi_-\colon U \to \mathbb{R}^3$ at points in $S \cap \{y \neq 0\}$ by setting

$$\varphi_+(u, v) = (u, \sqrt{1 - u^2}, v)\,, \quad \varphi_-(u, v) = (u, -\sqrt{1 - u^2}, v)\,.$$

Analogously, we construct the local parametrizations $\psi_+, \psi_-\colon U \to \mathbb{R}^3$ at points in $S \cap \{x \neq 0\}$ by setting

$$\psi_+(u, v) = (\sqrt{1 - u^2}, u, v)\,, \quad \psi_-(u, v) = (-\sqrt{1 - u^2}, u, v)\,.$$

It is then easy to see that $\{\varphi_+, \varphi_-, \psi_+, \psi_-\}$ is an atlas of S, because every point of S is contained in the image of at least one of them.

(iii) The map $G|_S$ is the restriction to S of the map G which is of class C^∞ on the whole $\mathbb{R}^3$, so it is of class C^∞ on S. So, to prove that $G|_S$ is a diffeomorphism it suffices to find a map $H\colon \mathbb{R}^2 \setminus \{(0,0)\} \to S$ of class C^∞ that is the inverse of $G|_S$. First of all, note that the image of $G|_S$ lies in $\mathbb{R}^2 \setminus \{(0,0)\}$. Moreover, for all $a, b \in \mathbb{R}$ with $a^2 + b^2 = 1$ the restriction of $G|_S$ to the straight line $\{(a,b,v) \in S \mid v \in \mathbb{R}\} \subset S$ is a bijection with the half-line $\{e^v(a,b) \mid v \in \mathbb{R}\} \subset \mathbb{R}^2 \setminus \{(0,0)\}$. So $G|_S$ is a bijection between S and $\mathbb{R}^2 \setminus \{(0,0)\}$, and the inverse H we are looking for is given by

$$H(a,b) = \left(\frac{a}{\sqrt{a^2 + b^2}}, \frac{b}{\sqrt{a^2 + b^2}}, \log \sqrt{a^2 + b^2} \right) .$$

Note that H is of class C^∞, since it is a C^∞ map from $\mathbb{R}^2 \setminus \{(0,0)\}$ to $\mathbb{R}^3$ having S as its image. $\square$

Problem 3.4. *As in Example 3.3.16, write a quadratic polynomial p in three variables in the form $p(x) = x^T A x + 2b^T x + c$, where $A = (a_{ij}) \in M_{3,3}(\mathbb{R})$ is a symmetric matrix, $b \in \mathbb{R}^3$ (we are writing the elements of $\mathbb{R}^3$ as column vectors), and $c \in \mathbb{R}$. Let now S be the quadric in $\mathbb{R}^3$ defined by the equation $p(x) = 0$. Remember that the quadric S is said to be* central *if the linear system $Ax + b = O$ has a solution (called* center *of the quadric), and is a* paraboloid *otherwise (see [1, p. 149]).*

(i) *Prove that (the connected components of) the paraboloids and the central quadrics not containing any of their centers are regular surfaces.*

(ii) *Given the symmetric matrix*

$$B = \begin{pmatrix} A & b \\ b^T & c \end{pmatrix} \in M_{4,4}(\mathbb{R}) ,$$

show that a point $x \in \mathbb{R}^3$ belongs to the quadric if and only if

$$(x^T \quad 1)\, B \begin{pmatrix} x \\ 1 \end{pmatrix} = 0 . \tag{3.8}$$

(iii) *Prove that if $\det B \neq 0$ then the connected components of the quadric S are (either empty or) regular surfaces.*

(iv) *Show that if S is a central quadric containing one of its centers, then its components are regular surfaces if and only if S is a plane if and only if $\operatorname{rg} A = 1$.*

Solution. (i) In Example 3.3.16 we saw that $\nabla p(x) = 2(Ax+b)$, so the critical points of f are exactly the centers of S. So, if S is a paraboloid or it does not contain its centers then 0 is a regular value of p, and the components of $S = p^{-1}(0)$ (if non-empty) are regular surfaces by Proposition 3.1.25.

(ii) The product in the left-hand side of (3.8) is exactly equal to $p(x)$.

(iii) Assume by contradiction that S is a central quadric containing a center x_0. From $p(x_0) = 0$ and $\mathsf{A}x_0 + b = O$ we immediately deduce (because $x_0^T b = b^T x_0$) that $\begin{pmatrix} x_0 \\ 1 \end{pmatrix}$ is a non-zero element of the kernel of B, and hence $\det \mathsf{B} = 0$. The assertion then follows from (i).

(iv) Suppose that $x_0 \in S$ is a center of S. Since the centers of S are exactly the critical points of p, the property of containing one of its own centers is preserved under translations or linear transformations on $\mathbb{R}^3$; hence, up to a translation, we may assume without loss of generality that $x_0 = O$. Now, the origin is a center if and only if $b = O$, and it belongs to S if and only if $c = O$. This means that $O \in S$ is a center of S if and only if $p(x) = x^T \mathsf{A}x$. By Sylvester's law of inertia (see [1, Vol. II, Theorem 13.4.7, p. 98]), we only have the following cases:

(a) if $\det \mathsf{A} \neq 0$, then up to a linear transformation we may assume that $p(x) = x_1^2 + x_2^2 \pm x_3^2$, so either S is a single point or it is a two-sheeted cone, and in both cases it is not a regular surface;

(b) if $\operatorname{rg} \mathsf{A} = 2$, then up to a linear transformation we may assume that $p(x) = x_1^2 \pm x_2^2$, so either S is a straight line or it is the union of two incident planes, and in both cases it is not a regular surface;

(c) if $\operatorname{rg} \mathsf{A} = 1$, then up to a linear transformation we have $p(x) = x_1^2$, and so S is a plane.   □

Exercises

IMMERSED SURFACES AND REGULAR SURFACES

3.1. Show that the map $\varphi \colon \mathbb{R}^2 \to \mathbb{R}^3$ defined by

$$\varphi(u, v) = \left(u - \frac{u^3}{3} + uv^2, v - \frac{v^3}{3} + vu^2, u^2 - v^2 \right).$$

is an injective immersed surface (*Enneper's surface*; see Fig. 3.8). Is it a homeomorphism with its image as well?

3.2. Prove that the map $\varphi \colon \mathbb{R}^2 \to \mathbb{R}^3$ defined by

$$\varphi(u, v) = \left(\frac{u + v}{2}, \frac{u - v}{2}, uv \right)$$

is a global parametrization of the one-sheeted hyperboloid, and describe its coordinate curves, $v \mapsto \varphi(u_0, v)$ with u_0 fixed and $u \mapsto \varphi(u, v_0)$, with v_0 fixed.

3.3. Let $U = \{(u, v) \in \mathbb{R}^2 \mid u > 0\}$. Show that the map $\varphi \colon U \to \mathbb{R}^3$ given by $\varphi(u, v) = (u + v \cos u, u^2 + v \sin u, u^3)$ is an immersed surface.

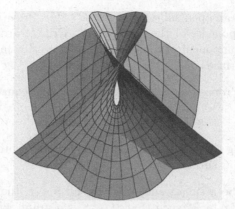

Fig. 3.8. Enneper's surface

3.4. Let $S^2 \subset \mathbb{R}^3$ be the sphere of equation $x^2 + y^2 + z^2 = 1$, denote by N the point of coordinates $(0, 0, 1)$, and let H be the plane of equation $z = 0$, which we shall identify with $\mathbb{R}^2$ by the projection $(u, v, 0) \mapsto (u, v)$. The *stereographic projection* $\pi_N : S^2 \setminus \{N\} \to \mathbb{R}^2$ from the point N onto the plane H maps $p = (x, y, z) \in S^2 \setminus \{N\}$ to the intersection point $\pi_N(p)$ between H and the line joining N and p.

(i) Show that the map π_N is bijective and continuous, with continuous inverse $\pi_N^{-1} : \mathbb{R}^2 \to S \setminus \{N\}$ given by

$$\pi_N^{-1}(u, v) = \left(\frac{2u}{u^2 + v^2 + 1}, \frac{2v}{u^2 + v^2 + 1}, \frac{u^2 + v^2 - 1}{u^2 + v^2 + 1} \right).$$

(ii) Show that π_N^{-1} is a local parametrization of S^2.
(iii) Determine, in an analogous way, the stereographic projection π_S of S^2 from the point $S = (0, 0, -1)$ onto the plane H.
(iv) Show that $\{\pi_N^{-1}, \pi_S^{-1}\}$ is an atlas for S^2 consisting of two charts.

3.5. Let $S \subset \mathbb{R}^3$ be a connected subset of $\mathbb{R}^3$ such that there exists a family $\{S_\alpha\}$ of surfaces with $S = \bigcup_\alpha S_\alpha$ and such that every S_α is open in S. Prove that S is a surface.

3.6. Find an atlas for the ellipsoid of equation $(x/a)^2 + (y/b)^2 + (z/c)^2 = 1$ (see also Example 3.1.27).

3.7. Consider a Jordan curve C of class C^∞ contained in a plane $H \subset \mathbb{R}^3$, take a straight line $\ell \subset H$ not containing C, and suppose that C is symmetric with respect to ℓ (that is, $\rho(C) = C$, where $\rho : H \to H$ is the reflection with respect to ℓ). Prove that the set obtained by rotating C around ℓ is a regular surface. In particular, this shows again that the sphere is a regular surface.

3.8. Prove that the set of critical points of a map $F : U \to \mathbb{R}^m$ of class C^∞, where $U \subset \mathbb{R}^n$ is open, is a closed subset of U.

3.9. Let $V \subseteq \mathbb{R}^3$ be an open subset and $f \in C^\infty(V)$. Prove that for all $a \in \mathbb{R}$ the connected components of the set $f^{-1}(a) \setminus \mathrm{Crit}(f)$ are regular surfaces. Deduce that each component of complement of the vertex in a double-sheeted cone is a regular surface.

3.10. Prove using Proposition 3.1.25 that the torus of equation

$$z^2 = r^2 - (\sqrt{x^2 + y^2} - a)^2,$$

obtained by rotating the circle with radius $r < a$ and center $(a, 0, 0)$ around the z-axis is a regular surface.

3.11. Let $S \subset \mathbb{R}^3$ be a subset such that for all $p \in S$ there exists an open neighborhood W of p in $\mathbb{R}^3$ such that $W \cap S$ is a graph with respect to one of the three coordinate planes. Prove that S is a regular surface.

3.12. Show that if $\sigma: I \to U \subset \mathbb{R}^2$ is the parametrization of a regular C^∞ curve whose support is contained in an open set $U \subset \mathbb{R}^2$, and if $\varphi: U \to S$ is a local parametrization of a surface S then the composition $\varphi \circ \sigma$ parametrizes a C^∞ curve in S.

3.13. Prove that the set $S = \{(x, y, z) \in \mathbb{R}^3 \mid x^2 + y^2 - z^3 = 1\}$ is a regular surface, and find an atlas for it.

3.14. Let $\varphi: \mathbb{R} \times (0, \pi) \to S^2$ be the immersed surface given by

$$\varphi(u, v) = (\cos u \sin v, \sin u \sin v, \cos v),$$

and let $\sigma: (0, 1) \to S^2$ be the curve defined by $\sigma(t) = \varphi(\log t, 2 \arctan t)$. Show that the tangent vector to σ at $\sigma(t)$ forms a constant angle of $\pi/4$ with the tangent vector to the meridian passing through $\sigma(t)$, where the meridians are characterized by the condition $u = \mathrm{const}$.

3.15. Show that the surface $S_1 \subset \mathbb{R}^3$ of equation $x^2 + y^2 z^2 = 1$ is not compact, while the surface $S_2 \subset \mathbb{R}^3$ of equation $x^2 + y^4 + z^6 = 1$ is compact.

3.16. Consider a map $f: U \to \mathbb{R}^3$, defined on an open subset $U \subset \mathbb{R}^2$ and of class C^∞, and let $\varphi: \mathbb{R}^2 \to R^3$ be given by $\varphi(u, v) = (u, v, f(u, v))$. Prove that φ is a diffeomorphism between U and $S = \varphi(U)$.

3.17. Show that, for any real numbers $a, b, c > 0$, the following maps are local parametrizations of quadrics in $\mathbb{R}^3$, with equations analogous to those given in Example 3.1.28:

$$\begin{array}{ll}
\varphi_1(u, v) = (a \sin u \cos v, b \sin u \sin v, c \cos u) & \text{ellipsoid,} \\
\varphi_2(u, v) = (a \sinh u \cos v, b \sinh u \sin v, c \cosh u) & \text{two-sheeted hyperboloid,} \\
\varphi_3(u, v) = (a \sinh u \sinh v, b \sinh u \cosh v, c \sinh u) & \text{one-sheeted hyperboloid,} \\
\varphi_4(u, v) = (a u \cos v, b u \sin v, u^2) & \text{elliptic paraboloid,} \\
\varphi_5(u, v) = (a u \cosh v, b u \sinh v, u^2) & \text{hyperbolic paraboloid.}
\end{array}$$

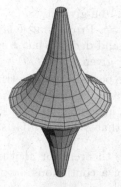

Fig. 3.9. The pseudosphere

Is it possible to choose a, b, c in such a way that the surface is a surface of revolution with respect to one of the coordinate axes? Consider each case separately.

3.18. Let $S \subset \mathbb{R}^3$ be the set (called *pseudosphere*) obtained by rotating around the z-axis the support of the tractrix $\sigma \colon (0, \pi) \to \mathbb{R}^3$ given by

$$\sigma(t) = \left(\sin t, 0, \cos t + \log \tan(t/2) \right) ;$$

see Fig. 3.9. Denote by $H \subset \mathbb{R}^3$ the plane $\{z = 0\}$. Prove that S is not a regular surface, whereas each connected component of $S \setminus H$ is; see Example 4.5.23 and Problem 4.8.

SMOOTH FUNCTIONS

3.19. Let $S \subset \mathbb{R}^3$ be a surface. Prove that if $p \in S$ is a local minimum or a local maximum of a function $f \in C^\infty(S)$ then $\mathrm{d}f_p \equiv 0$.

3.20. Define the notions of a C^∞ map from an open subset of $\mathbb{R}^n$ to a surface, and of a C^∞ map from a surface to an Euclidean space $\mathbb{R}^m$.

3.21. Let $S \subset \mathbb{R}^3$ be a surface, and $p \in S$. Prove that there exists an open set $W \subseteq \mathbb{R}^3$ of p in $\mathbb{R}^3$, a function $f \in C^\infty(W)$ and a regular value $a \in \mathbb{R}$ of f such that $S \cap W = f^{-1}(a)$.

3.22. Given a surface $S \subset \mathbb{R}^3$, take a function $f \in C^\infty(S)$ and a regular value $a \in \mathbb{R}$ of f, in the sense that $\mathrm{d}f_p \not\equiv O$ for all $p \in f^{-1}(a)$. Prove that $f^{-1}(a)$ is locally the support of a simple curve of class C^∞.

3.23. Let $C \subset \mathbb{R}^2$ be the support of a Jordan curve (or open arc) of class C^∞ contained in the half-plane $\{x > 0\}$. Identify $\mathbb{R}^2$ with the xz-plane in $\mathbb{R}^3$, and let S be the set obtained by rotating C around the z-axis, which we shall denote by ℓ.

(i)   Let $\Phi \colon \mathbb{R}^+ \times \mathbb{R} \times S^1 \to \mathbb{R}^3$ be given by $\Phi\big(x, z, (s,t)\big) = (xs, xt, z)$ for all $x > 0$, $z \in \mathbb{R}$ and $(s,t) \in S^1$. Prove that Φ is a homeomorphism between $\mathbb{R}^+ \times \mathbb{R} \times S^1$ and $\mathbb{R}^3 \setminus \ell$, and deduce that S is homeomorphic to $C \times S^1$.

(ii) Let $\Psi \colon \mathbb{R}^2 \to C \times S^1$ be given by $\Psi(t, \theta) = \big(\sigma(t), (\cos\theta, \sin\theta)\big)$, where $\sigma \colon \mathbb{R} \to \mathbb{R}^2$ is a global or periodic parametrization of C, and let $I \subseteq \mathbb{R}$ be an open interval where σ is injective. Prove that $\Psi|_{I \times (\theta_0, \theta_0 + 2\pi)}$ is a homeomorphism with its image for all $\theta_0 \in \mathbb{R}$.

(iii) Use (i) and (ii) to prove that S is a regular surface.

3.24. Using the Jordan curve theorem for continuous curves prove that the complement of the support of a continuous Jordan curve in S^2 has exactly two components.

TANGENT PLANE

3.25. Let $S \subset \mathbb{R}^3$ be a surface, $p \in S$ and $\{v_1, v_2\}$ a basis of $T_p S$. Prove that there is a local parametrization $\varphi \colon U \to S$ centered at p such that $\partial_1|_p = v_1$ and $\partial_2|_p = v_2$.

3.26. Given an open set $W \subseteq \mathbb{R}^3$ and a function $f \in C^\infty(W)$, take $a \in \mathbb{R}$ and let S be a connected component of $f^{-1}(a) \setminus \mathrm{Crit}(f)$. Prove that for all $p \in S$ the tangent plane $T_p S$ coincides with the subspace of $\mathbb{R}^3$ orthogonal to $\nabla f(p)$.

3.27. Show that the tangent plane at a point $p = (x_0, y_0, z_0)$ of a level surface $f(x, y, z) = 0$ corresponding to the regular value 0 of a C^∞ function $f \colon \mathbb{R}^3 \to \mathbb{R}$ is given by the equation

$$\frac{\partial f}{\partial x}(p)\, x + \frac{\partial f}{\partial y}(p)\, y + \frac{\partial f}{\partial z}(p)\, z = 0 \,,$$

while the equation of the affine tangent plane, parallel to the tangent plane and passing through p, is given by

$$\frac{\partial f}{\partial x}(p)\, (x - x_0) + \frac{\partial f}{\partial y}(p)\, (y - y_0) + \frac{\partial f}{\partial z}(p)\, (z - z_0) = 0 \,.$$

3.28. Determine the tangent plane at every point of the hyperbolic paraboloid with global parametrization $\varphi \colon \mathbb{R}^2 \to \mathbb{R}^3$ given by $\varphi(u, v) = (u, v, u^2 - v^2)$.

3.29. Let $S \subset \mathbb{R}^3$ be a surface, and $p \in S$. Prove that

$$\mathfrak{m} = \{\mathbf{f} \in C^\infty(p) \mid \mathbf{f}(p) = 0\}$$

is the unique maximal ideal of $C^\infty(p)$, and that $T_p S$ is canonically isomorphic to the dual (as vector space) of $\mathfrak{m}/\mathfrak{m}^2$.

3.30. Let $\varphi: \mathbb{R}^2 \to \mathbb{R}^3$ be given by $\varphi(u,v) = (u, v^3, u - v)$, and let $\sigma: \mathbb{R} \to \mathbb{R}^3$ be the curve parametrized by $\sigma(t) = (3t, t^6, 3t - t^2)$.

(i)   Prove that $S = \varphi(\mathbb{R}^2)$ is a regular surface.
(ii) Show that σ is regular and has support contained in S.
(iii) Determine the curve $\sigma_o: \mathbb{R} \to \mathbb{R}^2$ such that $\sigma = \varphi \circ \sigma_o$.
(iv) Write the tangent versor to σ at $O = \sigma(0)$ as a combination of the basis ∂_1 and ∂_2 of the tangent plane $T_O S$ to S at O induced by φ.

3.31. Let $S \subset \mathbb{R}^3$ be a surface. Prove that a C^∞ function $F: S \to \mathbb{R}^m$ satisfies $dF_p \equiv O$ for all $p \in S$ if and only if F is constant.

3.32. Two surfaces $S_1, S_2 \subset \mathbb{R}^3$ are *transversal* if $S_1 \cap S_2 \neq \varnothing$ and $T_p S_1 \neq T_p S_2$ for all $p \in S_1 \cap S_2$. Prove that if S_1 and S_2 are transversal, then each component of $S_1 \cap S_2$ is locally the support of a simple regular C^∞ curve.

3.33. Let $H \subset \mathbb{R}^3$ be a plane, $\ell \subset \mathbb{R}^3$ a straight line not contained in H, and $C \subseteq H$ a subset of H. Consider the cylinder S with generatrix C and directrix ℓ. Show that the tangent plane to S is constant at the points of S belonging to a line parallel to the directrix ℓ.

3.34. Let $\varphi: \mathbb{R}^3 \to \mathbb{R}^3$ be the global parametrization of the regular surface $S = \varphi(\mathbb{R}^2)$ given by $\varphi(u,v) = (u - v, u^2 + v, u - v^3)$. Determine the Cartesian equation of the tangent plane to S at $p = (0, 2, 0) = \varphi(1, 1)$.

3.35. Prove that the space of derivations of germs of C^k functions (with $1 \leq k < \infty$) has infinite dimension.

3.36. Prove that the space of derivations of germs of continuous functions consists of just the zero derivation.

Definition 3.E.1. Let S_1 and S_2 be regular surfaces in $\mathbb{R}^3$ having in common a point p. We say that S_1 and S_2 have *contact of order at least* 1 at p if there exist parametrizations φ_1 of S_1 and φ_2 of S_2, centered at p, with $\partial \varphi_1 / \partial u(O) = \partial \varphi_2 / \partial u(O)$ and $\partial \varphi_1 / \partial v(O) = \partial \varphi_2 / \partial v(O)$. Moreover, the surfaces are said to have *contact of order at least* 2 at p if there is a pair of parametrizations centered at p for which all the second order partial derivatives coincide too.

3.37. Show that two surfaces have contact of order at least 1 at p if and only if they have same tangent plane at p. In particular, the tangent plane at p is the only plane having contact of order at least 1 with a regular surface.

3.38. Show that if the intersection between a regular surface S and a plane H consists of a single point p_0, then H is the tangent plane to S at p_0.

3.39. Let $\Omega \subseteq \mathbb{R}^3$ be an open subset, $f \in C^\infty(\Omega)$ such that 0 is a regular value of f, and S a connected component of $f^{-1}(0)$. Given $p_0 \in S$, let $\sigma: (-\varepsilon, \varepsilon) \to \mathbb{R}^3$ be a curve with $\sigma(0) = p_0$. Prove that $\sigma'(0) \in T_{p_0} S$ if and only if f and σ have contact of order at least 1 in the sense of Definition 1.E.2.

SMOOTH MAPS BETWEEN SURFACES

3.40. Prove that a smooth map between surfaces is necessarily continuous.

3.41. Let $F: S_1 \to S_2$ be a map between surfaces, and $p \in S_1$. Prove that if there exist a local parametrization $\varphi_1: U_1 \to S_1$ at p and a local parametrization $\varphi_2: U_2 \to S_2$ at $F(p)$ such that $\varphi_2^{-1} \circ F \circ \varphi_1$ is of class C^∞ in a neighborhood of $\varphi_1^{-1}(p)$, then $\psi_2^{-1} \circ F \circ \psi_1$ is of class C^∞ in a neighborhood of $\psi_1^{-1}(p)$ for any local parametrization $\psi_1: V_1 \to S_1$ of S at p and any local parametrization $\psi_2: V_2 \to S_2$ of S at $F(p)$.

3.42. Show that the relation "S_1 is diffeomorphic to S_2" is an equivalence relation on the set of regular surfaces in $\mathbb{R}^3$.

3.43. Let $F: S^2 \to \mathbb{R}^3$ be defined by

$$F(p) = (x^2 - y^2, xy, yz)$$

for all $p = (x, y, z) \in S^2$. Set $N = (0, 0, 1)$ and $E = (1, 0, 0)$.

(i)   Prove that dF_p is injective on $T_p S^2$ for all $p \in S^2 \setminus \{\pm N, \pm E\}$.
(ii) Prove that $S_1 = F(S^2 \setminus \{y = 0\})$ is a regular surface, and find a basis of $T_q S_1$ for all $q \in S_1$.
(iii) Given $p = (0, 1, 0)$ and $q = F(p)$, choose a local parametrization of S^2 at p, a local parametrization of S_1 at q, and write the matrix representing the linear map $dF_p: T_p S^2 \to T_q S_1$ with respect to the bases of $T_p S^2$ and $T_q S_1$ determined by the local coordinates you have chosen.

3.44. Show that the antipodal map $F: S^2 \to S^2$ defined by $F(p) = -p$ is a diffeomorphism.

3.45. Determine an explicit diffeomorphism between the portion of a cylinder defined by $\{(x, y, z) \in \mathbb{R}^3 \mid x^2 + y^2 = 1,\ -1 < z < 1\}$ and $S^2 \setminus \{N, S\}$, where $N = (0, 0, 1)$ and $S = (0, 0, -1)$.

3.46. Determine a diffeomorphism between the unit sphere $S^2 \subset \mathbb{R}^3$ and the ellipsoid of equation $4x^2 + 9y^2 + 25z^2 = 1$.

3.47. Let C_1 and C_2 be supports of two regular curves contained in a surface S that are tangent at a point p_0, that is, having the same tangent line at a common point p_0. Show that if $F: S \to S$ is a diffeomorphism then $F(C_1)$ and $F(C_2)$ are the supports of regular curves tangent at $F(p_0)$.

3.48. Let $f: S_1 \to S_2$ be a smooth map between connected regular surfaces. Show that f is constant if and only if $df \equiv 0$.

3.49. Prove that every surface of revolution having as its generatrix an open Jordan arc is diffeomorphic to a circular cylinder.

3.50. Show that a rotation of an angle θ of $\mathbb{R}^3$ around the z-axis induces a diffeomorphism on a regular surface of revolution obtained by rotating a curve around the z-axis.

3.51. Let $S \subset \mathbb{R}^n$ be a regular surface and $p_0 \notin S$. Prove that the function $d \colon S \to \mathbb{R}$ defined by $d(p) = \|p - p_0\|$, i.e., the distance from p_0, is of class C^∞.

3.52. Construct an explicit diffeomorphism F between the one-sheeted hyperboloid of equation $(x/a)^2 + (y/b)^2 - (z/c)^2 = 1$ and the right circular cylinder of equation $x^2 + y^2 = 1$, determine its differential $\mathrm{d}F_p$ at every point, and describe the inverse of F in local coordinates.

3.53. Construct a diffeomorphism between the right circular cylinder of equation $x^2 + y^2 = 4$ and the plane $\mathbb{R}^2$ with the origin removed.

Supplementary material

3.5 Sard's theorem

In this section we shall describe a proof found by P. Holm (see [6]) of Sard's theorem (mentioned in Remark 3.1.24), but only in the cases we are interested in, that is, for functions of at most three variables; you may read in [6] the proof for the general case. We begin by explaining what it means for a subset of a line to have measure zero.

Definition 3.5.1. We shall say that a subset $A \subset \mathbb{R}$ has *measure zero* if for all $\varepsilon > 0$ there exists a countable family $\{I_\nu\}$ of intervals whose union covers A and such that $\sum_\nu \operatorname{diam}(I_\nu) < \varepsilon$, where $\operatorname{diam}(I_\nu)$ is the diameter of I_ν.

Clearly, a set of measure zero cannot contain any interval, so its complement is dense. Moreover, it is easy to see (exercise) that the union of a countable family of sets of measure zero is again a set of measure zero.

Theorem 3.5.2 (Sard). *Let $V \subseteq \mathbb{R}^n$ be an open set, with $1 \le n \le 3$, and $f \in C^\infty(V)$. Then the set of critical values of f has measure zero in $\mathbb{R}$.*

Proof. As you will see, this is basically a proof by induction on the dimension. Let us begin with the case $n = 1$; since every open subset of $\mathbb{R}$ is a union of at most countably many open intervals, without loss of generality we may assume that V is an open interval $I \subseteq \mathbb{R}$.

If $x \in \operatorname{Crit}(f)$, then by Taylor's formula we know that for all $y \in I$ there exists ξ between x and y such that

$$f(y) - f(x) = \frac{1}{2} f''(\xi)(y - x)^2 .$$

Fix now $\varepsilon > 0$, and choose a countable family $\{K_\nu\}$ of compact intervals included in I whose union covers $\mathrm{Crit}(f)$; let ℓ_ν be the length of K_ν.

Set $M_\nu = \sup_{x \in K_\nu} |f''(x)|$ and divide K_ν in k_ν small intervals K_ν^j of length ℓ_ν/k_ν, where $k_\nu > 2^\nu M_\nu \ell_\nu^2/\varepsilon$. If x, $y \in \mathrm{Crit}(f) \cap K_\nu^j$, we have

$$|f(y) - f(x)| \le \frac{M_\nu}{2}|y - x|^2 \le \frac{M_\nu \ell_\nu^2}{2k_\nu^2} \,,$$

so $f\big(\mathrm{Crit}(f) \cap K_\nu^j\big)$ has diameter less or equal than $M_\nu \ell_\nu^2/2k_\nu^2$. If we repeat this operation for all small intervals K_ν^j, we find that $f\big(\mathrm{Crit}(f) \cap K_\nu\big)$ is contained in a union of intervals the sum of whose diameters is less than

$$k_\nu \frac{M_\nu \ell_\nu^2}{2k_\nu^2} < \frac{\varepsilon}{2^{\nu+1}} \,.$$

Repeat this procedure for all K_ν to get that $f\big(\mathrm{Crit}(f)\big)$ is contained in a union of intervals having sum of diameters less than

$$\varepsilon \sum_{\nu=0}^{\infty} \frac{1}{2^{\nu+1}} = \varepsilon \,,$$

and we are done.

Let us now consider the case $n = 2$. Set $C_1 = \mathrm{Crit}(f)$ and let $C_2 \subseteq C_1$ be the set of all points of V in which all first and second derivatives of f are zero. Clearly, $\mathrm{Crit}(f) = C_2 \cup (C_1 \setminus C_2)$; hence, it suffices to prove that both $f(C_2)$ and $f(C_1 \setminus C_2)$ have measure zero in $\mathbb{R}$.

To prove that $f(C_2)$ has measure zero, we remark that this time Taylor's formula yields, for each compact convex set $K \subset V$ with $\mathrm{Crit}(f) \cap K \ne \varnothing$, a number $M > 0$ depending only on K and f such that

$$|f(y) - f(x)| \le M\|y - x\|^3$$

for all $x \in \mathrm{Crit}(f) \cap K$ e $y \in K$. Therefore, proceeding as above but using squares instead of intervals, we obtain (exercise) that $f(C_2)$ has measure zero.

Take now $x_0 \in C_1 \setminus C_2$. In particular, there is a second order derivative of f that is non-zero in x_0; so there exists a first order derivative of f, call it g, with non-zero gradient in x_0. Then Proposition 1.1.18 tells us that there exists a neighborhood K of x_0 whose compact closure is disjoint from C_2 and is such that $g^{-1}(0) \cap K$ is a graph. Choose a regular parametrization $\sigma \colon (-\varepsilon, \varepsilon) \to \mathbb{R}^2$ of $g^{-1}(0) \cap K$ with $\sigma(0) = x_0$. Then $\sigma^{-1}\big((C_1 \setminus C_2) \cap K\big) \subseteq \mathrm{Crit}(f \circ \sigma)$, so $f\big((C_1 \setminus C_2) \cap K\big)$ is contained in the set of critical values of $f \circ \sigma$, which has measure zero thanks to the case $n = 1$.

Proceeding in this way, we find a countable family $\{K_\nu\}$ of open sets with compact closure contained in V that cover $C_1 \setminus C_2$ and such that $f\big((C_1 \setminus C_2) \cap K_\nu\big)$ has measure zero for all $\nu \in \mathbb{N}$; so $f(C_1 \setminus C_2)$ has measure zero, and we are done.

Let us now deal with $n = 3$. Define C_1 and C_2 as above, introduce the set C_3 of points of V in which all first, second and third derivatives of f are zero, and write $\mathrm{Crit}(f) = C_3 \cup (C_2 \setminus C_3) \cup (C_1 \setminus C_2)$. Arguing as in the previous cases, using small cubes instead of squares, it is straightforward to see that $f(C_3)$ has measure zero. Take now $x_0 \in C_2 \setminus C_3$; then we can find a second derivative of f, call it g, with non-zero gradient in x_0 and such that $C_2 \setminus C_3 \subseteq g^{-1}(0)$. Proposition 3.1.25 (see also Exercise 3.9) gives us a neighborhood K of x_0, with compact closure disjoint from C_3, such that $g^{-1}(0) \cap K$ is a regular surface. Up to restricting K, we may also assume that $g^{-1}(0) \cap K$ is the image of a local parametrization $\varphi \colon U \to \mathbb{R}^3$ centered at x_0. But now we immediately see that $\varphi^{-1}((C_2 \setminus C_3) \cap K) \subseteq \mathrm{Crit}(f \circ \varphi)$, so $f((C_2 \setminus C_3) \cap K)$ is contained in the set of critical values of $f \circ \varphi$, which has measure zero thanks to the case $n = 2$. Arguing as above, we deduce that $f(C_2 \setminus C_3)$ has measure zero. An analogous argument, using first derivatives, also shows that $f(C_1 \setminus C_2)$ has measure zero, and we are done. $\qquad\square$

3.6 Partitions of unity

The goal of this section is to prove that every C^∞ function over a surface can be extended to a C^∞ function defined in an open neighborhood (in $\mathbb{R}^3$) of the surface. We have seen (Proposition 3.2.12) that this is possible locally; so we need a technique for glueing together local extensions in order to get a global extension.

The main tool used for glueing local objects to get a global object is given by the *partitions of unity*. To introduce them, we need some definitions and a lemma.

Definition 3.6.1. A *cover* of a topological space X is a family of subsets of X whose union is the space X itself; the cover is *open* if all subsets in the family are open. We shall say that a cover $\mathfrak{U} = \{U_\alpha\}_{\alpha \in A}$ of a topological space X is *locally finite* if each $p \in X$ has a neighborhood $U \subseteq X$ such that $U \cap U_\alpha \neq \varnothing$ for finitely many indices α only. A cover $\mathfrak{V} = \{V_\beta\}_{\beta \in B}$ is a *refinement* of $\mathfrak{U}$ if for all $\beta \in B$ there exists an $\alpha \in A$ such that $V_\beta \subseteq U_\alpha$.

Lemma 3.6.2. *Let $\Omega \subseteq \mathbb{R}^n$ be an open set, and $\mathfrak{U} = \{U_\alpha\}_{\alpha \in A}$ an open cover of Ω. Then there exists a locally finite open cover $\mathfrak{V} = \{V_\beta\}_{\beta \in B}$ of Ω such that:*

(i) *$\mathfrak{V}$ is a refinement of $\mathfrak{U}$;*
(ii) *for all $\beta \in B$ there exist $p_\beta \in \Omega$ and $r_\beta > 0$ such that $V_\beta = B(p_\beta, r_\beta)$;*
(iii) *setting $W_\beta = B(p_\beta, r_\beta/2)$, then $\mathfrak{W} = \{W_\beta\}_{\beta \in B}$ is also a cover of Ω.*

Proof. The open set Ω is locally compact and has a countable basis; so we may find a countable basis $\{P_j\}$ for the topology consisting of open sets with compact closure. We shall now define inductively an increasing family $\{K_j\}$

of compact sets. Set $K_1 = \overline{P_1}$. Once we have defined K_j, let $r \geq j$ be the least integer such that $K_j \subset \bigcup_{i=1}^r P_i$, and set

$$K_{j+1} = \overline{P_1} \cup \cdots \cup \overline{P_r} \, .$$

Proceeding in this way, we have $K_j \subset \mathring{K}_{j+1}$ (where $\mathring{K}$ denotes the interior of K) and $\Omega = \bigcup_j K_j$.

Now, for all $p \in (\mathring{K}_{j+2} \setminus K_{j-1}) \cap U_\alpha$ choose $r_{\alpha,j,p} > 0$ so that the ball $V_{\alpha,j,p}$ with center p and radius $r_{\alpha,j,p}$ is contained in $(\mathring{K}_{j+2} \setminus K_{j-1}) \cap U_\alpha$, and define $W_{\alpha,j,p} = B(p, r_{\alpha,j,p}/2)$. Now, considering all α's and p's, the open sets $W_{\alpha,j,p}$ form an open cover of $K_{j+1} \setminus \mathring{K}_j$, which is compact; so we may extract a finite subcover $\{W_{j,l}\}$. Taking the union of these covers for all j we get a countable open cover $\{W_\beta\}$ of Ω; if we denote by V_β the ball corresponding to W_β, to conclude we only have to prove that the open cover $\{V_\beta\}$ is locally finite. But indeed, for each $p \in \Omega$ we may find an index j such that $p \in \mathring{K}_j$, and by construction only finitely many V_β intersect $\mathring{K}_j$. $\square$

Definition 3.6.3. Let $f \colon X \to \mathbb{R}$ be a continuous function defined on a topological space X. The *support* of f is the set $\mathrm{supp}(f) = \overline{\{x \in X \mid f(x) \neq 0\}}$.

Definition 3.6.4. A *partition of unity* on an open set $\Omega \subseteq \mathbb{R}^n$ is a family $\{\rho_\alpha\}_{\alpha \in A} \subset C^\infty(\Omega)$ such that:

(a) $\rho_\alpha \geq 0$ on Ω for all $\alpha \in A$;
(b) $\{\mathrm{supp}(\rho_\alpha)\}$ is a locally finite cover of Ω;
(c) $\sum_\alpha \rho_\alpha \equiv 1$.

We shall further say that the partition of unity $\{\rho_\alpha\}$ is *subordinate* to the open cover $\mathfrak{U} = \{U_\alpha\}_{\alpha \in A}$ if we have $\mathrm{supp}(\rho_\alpha) \subset U_\alpha$ for all indices $\alpha \in A$.

Remark 3.6.5. Property (b) of the definition of a partition of unity implies that in a neighborhood of each point of Ω only finitely many elements of the partition of unity are different from zero. In particular, the sum in (c) is well defined, as in each point of Ω only finitely many summands are non zero. Moreover, since Ω has a countable basis, property (b) also implies (why?) that $\mathrm{supp}(\rho_\alpha) \neq \varnothing$ for at most countably many indices α. In particular, if the partition of unity is subordinate to a cover consisting of more than countably many open sets, then $\rho_\alpha \equiv 0$ for all indices except at most countably many. This is not surprising, since in a second-countable topological space every open cover has a countable subcover (*Lindelöf property*); see [7, p. 49].

Theorem 3.6.6. *Let $\Omega \subseteq \mathbb{R}^n$ be an open subset, and $\mathfrak{U} = \{U_\alpha\}_{\alpha \in A}$ an open cover of Ω. Then there exists a partition of unity subordinate to $\mathfrak{U}$.*

Proof. Let $\mathfrak{V} = \{V_\beta\}_{\beta \in B}$ be the refinement of $\mathfrak{U}$ given by Lemma 3.6.2, and, having set $V_\beta = B(p_\beta, r_\beta)$, denote by $f_\beta \in C^\infty(\mathbb{R}^n)$ the function given by

Corollary 1.5.3 applied to $B(p_\beta, r_\beta)$. In particular, $\{\text{supp}(f_\beta)\}$ is a locally finite cover of Ω refining $\mathfrak{U}$, and the sum

$$F = \sum_{\beta \in B} f_\beta$$

defines (why?) a C^∞ function on Ω. So, if we set $\tilde{\rho}_\beta = f_\beta / F$, we get a partition of unity $\{\tilde{\rho}_\beta\}_{\beta \in B}$ such that for all $\beta \in B$ there exists an $\alpha(\beta) \in A$ for which $\text{supp}(\tilde{\rho}_\beta) \subset U_{\alpha(\beta)}$. Define now $\rho_\alpha \in C^\infty(\Omega)$ by

$$\rho_\alpha = \sum_{\substack{\beta \in B \\ \alpha(\beta) = \alpha}} \tilde{\rho}_\beta \ ;$$

it is straightforward to verify (exercise) that $\{\rho_\alpha\}_{\alpha \in A}$ is a partition of unity subordinate to $\mathfrak{U}$, as required. $\qquad\square$

Let us see now how to use partitions of unity to glue local extensions:

Theorem 3.6.7. *Let $S \subset \mathbb{R}^3$ be a surface, and $\Omega \subseteq \mathbb{R}^3$ an open neighborhood of S such that $S \subset \Omega$ is closed in Ω. Then a function $f \colon S \to \mathbb{R}$ is of class C^∞ on S if and only if there exists $\tilde{f} \in C^\infty(\Omega)$ such that $\tilde{f}|_S \equiv f$.*

Proof. One implication is in Example 3.2.11. Conversely, take $f \in C^\infty(S)$. Proposition 3.2.12 tells us that for all $p \in S$ we may find an open neighborhood $W_p \subseteq \Omega$ of p and a function $\tilde{f}_p \in C^\infty(W_p)$ such that $\tilde{f}_p|_{W_p \cap S} \equiv f|_{W_p \cap S}$. Then $\mathfrak{U} = \{W_p\}_{p \in S} \cup \{\Omega \setminus S\}$ is an open cover of Ω; by Theorem 3.6.6 there exists a partition of unity $\{\rho_p\}_{p \in S} \cup \{\rho_{\Omega \setminus S}\}$ subordinate to $\mathfrak{U}$. In particular, if for $p \in S$ we extend $\rho_p \tilde{f}_p$ by putting it equal to zero outside the support of ρ_p, we get (why?) a C^∞ function in the whole Ω. Moreover, $\text{supp}(\rho_{\Omega \setminus S}) \subset \Omega \setminus S$; hence $\rho_{\Omega \setminus S}|_S \equiv 0$ and thus $\sum_{p \in S} \rho_p|_S \equiv 1$. Set then

$$\tilde{f} = \sum_{p \in S} \rho_p \tilde{f}_p. \qquad\qquad (3.9)$$

Since in a neighborhood of any point of Ω only finitely many summands of (3.9) are non zero, it is immediate to see that $\tilde{f} \in C^\infty(\Omega)$. Finally, since all functions $\tilde{f}_\alpha$ are extensions of the same f and $\{\rho_\alpha\}$ is a partition of unity, we clearly get $\tilde{f}|_S \equiv f$, as required. $\qquad\square$

It is easy to see that if the surface S is not closed in $\mathbb{R}^3$, then C^∞ functions on S that do not extend to C^∞ functions on all $\mathbb{R}^3$ may exist:

Example 3.6.8. Let $S = \{(x, y, z) \in \mathbb{R}^3 \mid z = 0, (x, y) \neq (0, 0)\}$ the xy-plane with the origin removed; since it is an open subset of a surface (the xy-plane), S itself is a surface. The function $f \colon S \to \mathbb{R}$ given by $f(x, y, z) = 1/(x^2 + y^2)$ is of class C^∞ on S but there exists (why?) no function $\tilde{f} \in C^\infty(\mathbb{R}^3)$ such that $\tilde{f}|_S \equiv f$. Note that S is not a closed subset of $\mathbb{R}^3$, but is a closed subset of the open set $\Omega = \{(x, y, z) \in \mathbb{R}^3 \mid (x, y) \neq (0, 0)\}$, and a C^∞ extension of f to Ω (given by the same formula) does exist.

Exercises

3.54. Prove that if $\sigma\colon (a,b) \to \mathbb{R}^2$ is an open Jordan arc of class C^2 then there exists a continuous function $\varepsilon\colon (a,b) \to \mathbb{R}^+$ such that

$$I_\sigma\big(\sigma(t_1), \varepsilon(t_1)\big) \cap I_\sigma\big(\sigma(t_2), \varepsilon(t_2)\big) = \varnothing$$

for all $t_1 \neq t_2$, where $I_\sigma\big(\sigma(t), \delta\big)$ is the line segment of length 2δ centered at $\sigma(t)$ and orthogonal to $\sigma'(t)$ introduced in Definition 2.2.3 (see also Theorem 4.8.2).

4

Curvatures

One of the main goals of differential geometry consists in finding an effective and meaningful way of measuring the curvature of non-flat objects (curves and surfaces). For curves we have seen that it is sufficient to measure the changes in tangent versors: in the case of surface things are, understandably, more complicated. The first obvious problem is that a surface can curve differently in different directions; so we need a measure of curvature related to tangent directions, that is, a way of measuring the variation of tangent planes.

To solve this problem we have to introduce several new tools.[1] First of all, we need to know the length of vectors tangent to the surface. As explained in Section 4.1, for this it suffices to restrict to each tangent plane the canonical scalar product in $\mathbb{R}^3$. In this way, we get a positive definite quadratic form on each tangent plane (the *first fundamental form*), which allows us to measure the length of tangent vectors to the surface (and, as we shall see in Section 4.2, the area of regions of the surfaces as well). It is worthwhile to notice right now that the first fundamental form is an *intrinsic* object associated with the surface: we may compute it while remaining within the surface itself, without having to go out to $\mathbb{R}^3$.

A tangent plane, being a plane in $\mathbb{R}^3$, is completely determined as soon as we know an orthogonal versor. So a family of tangent planes can be described by the *Gauss map*, associating each point of the surface with a versor normal to the tangent plane at that point. In Section 4.3 we shall see that the Gauss map always exists locally, and exists globally only on orientable surfaces (that is, surfaces where we can distinguish an interior and an exterior).

In Section 4.4 we shall at last define the curvature of a surface along a tangent direction. We shall do so in two ways: geometrically (as the curvature of the curve obtained by intersecting the surface with an orthogonal plane) and analytically, by using the differential of the Gauss map and an associated quadratic form (the *second fundamental form*). In particular, in Section 4.5 we shall introduce the *Gaussian curvature* of a surface as the determinant of the differential of the Gauss map, and we shall see that the Gaussian cur-

Abate M., Tovena F.: Curves and Surfaces.
DOI 10.1007/978-88-470-1941-6_4, © Springer-Verlag Italia 2012

vature summarize the main curvature properties of a surface. Furthermore, in Section 4.6, we shall prove Gauss' *Theorema Egregium*, showing that, although the definition involves explicitly the ambient space $\mathbb{R}^3$, the Gaussian curvature actually is an *intrinsic* quantity, that is, it can be measured while remaining inside the surface. This, for instance, allows us to determine that the Earth is not flat without resorting to satellite photos, since it is possible to ascertain that the Earth has non zero Gaussian curvature with measurements made at sea level.

Finally, in the supplementary material of this chapter we shall prove that every closed surface in $\mathbb{R}^3$ is orientable (Section 4.7); that a surface S closed in an open subset $\Omega \subseteq \mathbb{R}^3$ is a level surface if and only if S is orientable and $\Omega \setminus S$ is disconnected (Section 4.8); and the *fundamental theorem of the local theory of surfaces*, the less powerful and more complicated analogue of the fundamental theorem of the local theory of curves (Section 4.9).

4.1 The first fundamental form

As mentioned in the introduction to this chapter, we begin our journey among surfaces' curvatures by measuring the length of tangent vectors.

The Euclidean space $\mathbb{R}^3$ is intrinsically provided with the canonical scalar product. If $S \subset \mathbb{R}^3$ is a surface, and $p \in S$, the tangent plane T_pS may be thought of as a vector subspace of $\mathbb{R}^3$, and so we may compute the canonical scalar product of two tangent vectors to S at p.

Definition 4.1.1. Let $S \subset \mathbb{R}^3$ be a surface. For all $p \in S$ we shall denote by $\langle \cdot, \cdot \rangle_p$ the positive definite scalar product on T_pS induced by the canonical scalar product of $\mathbb{R}^3$. The *first fundamental form* $I_p: T_pS \to \mathbb{R}$ is the (positive definite) quadratic form associated with this scalar product:

$$\forall v \in T_pS \qquad I_p(v) = \langle v, v \rangle_p \geq 0 \,.$$

Remark 4.1.2. The knowledge of the first fundamental form I_p is equivalent to the knowledge of the scalar product $\langle \cdot, \cdot \rangle_p$: indeed,

$$\langle v, w \rangle_p = \frac{1}{2}\big[I_p(v+w) - I_p(v) - I_p(w)\big] = \frac{1}{4}\big[I_p(v+w) - I_p(v-w)\big] \,.$$

If we forget that the surface lives in the ambient space $\mathbb{R}^3$, and that the first fundamental form is induced by the constant canonical scalar product of $\mathbb{R}^3$, limiting ourselves to try and understand what can be seen from within the surface, we immediately notice that it is natural to consider $\langle \cdot, \cdot \rangle_p$ as a scalar product defined on the tangent plane T_pS which varies with p (and with the tangent plane).

A way to quantify this variability consists in using local parametrizations and the bases they induce on the tangent planes to deduce the (variable!)

matrix representing this scalar product. Let then $\varphi: U \to S$ be a local parametrization at $p \in S$, and $\{\partial_1, \partial_2\}$ the basis of T_pS induced by φ. If we take two tangent vectors v, $w \in T_pS$ and we write them as linear combination of basis vectors, that is $v = v_1\partial_1 + v_2\partial_2$ and $w = w_1\partial_1 + w_2\partial_2 \in T_pS$, we may express $\langle v, w \rangle_p$ in coordinates:

$$\langle v, w \rangle_p = v_1 w_1 \langle \partial_1, \partial_1 \rangle_p + [v_1 w_2 + v_2 w_1] \langle \partial_1, \partial_2 \rangle_p + v_2 w_2 \langle \partial_2, \partial_2 \rangle_p .$$

Definition 4.1.3. Let $\varphi: U \to S$ be a local parametrization of a surface S. Then the *metric coefficients* of S with respect to φ are the functions E, F, $G: U \to \mathbb{R}$ given by

$$E(x) = \langle \partial_1, \partial_1 \rangle_{\varphi(x)} , \quad F(x) = \langle \partial_1, \partial_2 \rangle_{\varphi(x)} , \quad G(x) = \langle \partial_2, \partial_2 \rangle_{\varphi(x)} ,$$

for all $x \in U$.

Clearly, the metric coefficients are (why?) C^∞ functions on U, and they completely determine the first fundamental form:

$$I_p(v) = E(x)v_1^2 + 2F(x)v_1 v_2 + G(x)v_2^2 = \begin{vmatrix} v_1 & v_2 \end{vmatrix} \begin{vmatrix} E(x) & F(x) \\ F(x) & G(x) \end{vmatrix} \begin{vmatrix} v_1 \\ v_2 \end{vmatrix}$$

for all $p = \varphi(x) \in \varphi(U)$ and $v = v_1\partial_1 + v_2\partial_2 \in T_pS$.

Remark 4.1.4. The notation E, F and G, which we shall systematically use, was introduced by Gauss in the early 19th century. In a more modern notation we may write $E = g_{11}$, $F = g_{12} = g_{21}$ and $G = g_{22}$, so as to get

$$\langle v, w \rangle_p = \sum_{h,k=1}^{2} g_{hk}(p)v_h w_k .$$

Remark 4.1.5. We have introduced E, F and G as functions defined on U. However, it will sometimes be more convenient to consider them as functions defined on $\varphi(U)$, that is, to replace them with $E \circ \varphi^{-1}$, $F \circ \varphi^{-1}$ and $G \circ \varphi^{-1}$, respectively. You might have noticed that we have performed just this substitution in the last formula.

Remark 4.1.6. Warning: the metric coefficients depend strongly on the local parametrization! Example 4.1.10 will show how much they can change, even in a very simple case, when choosing a different local parametrization.

Example 4.1.7. Let $S \subset \mathbb{R}^3$ be the plane passing through $p_0 \in \mathbb{R}^3$ and parallel to the linearly independent vectors $\mathbf{v}_1$, $\mathbf{v}_2 \in \mathbb{R}^3$. In Example 3.1.12 we have seen that a local parametrization of S is the map $\varphi: \mathbb{R}^2 \to \mathbb{R}^3$ given by $\varphi(x_1, x_2) = p_0 + x_1\mathbf{v}_1 + x_2\mathbf{v}_2$. For all $p \in S$ the basis of T_pS induced by φ is $\partial_1 = \mathbf{v}_1$ and $\partial_2 = \mathbf{v}_2$, so the metric coefficients of the plane with respect to φ are given by $E \equiv \|\mathbf{v}_1\|^2$, $F \equiv \langle \mathbf{v}_1, \mathbf{v}_2 \rangle$ and $G \equiv \|\mathbf{v}_2\|^2$. In particular, if $\mathbf{v}_1$ and $\mathbf{v}_2$ are orthonormal versors, we find

$$E \equiv 1 , \quad F \equiv 0 , \quad G \equiv 1 .$$

Example 4.1.8. Let $U \subseteq \mathbb{R}^2$ be an open set, $h \in C^{\infty}(U)$, and $\varphi \colon U \to \mathbb{R}^3$ the local parametrization of the graph Γ_h given by $\varphi(x) = \big(x, h(x)\big)$. Recalling Example 3.3.14 we see that the metric coefficients of Γ_h with respect to φ are given by

$$E = 1 + \left| \frac{\partial h}{\partial x_1} \right|^2 , \quad F = \frac{\partial h}{\partial x_1} \frac{\partial h}{\partial x_2} , \quad G = 1 + \left| \frac{\partial h}{\partial x_2} \right|^2 .$$

Example 4.1.9. Let $S \subset \mathbb{R}^3$ be the right circular cylinder with radius 1 centered on the z-axis. A local parametrization $\varphi \colon (0, 2\pi) \times \mathbb{R} \to \mathbb{R}^3$ is given by $\varphi(x_1, x_2) = (\cos x_1, \sin x_1, x_2)$. The basis induced by this parametrization is $\partial_1 = (-\sin x_1, \cos x_1, 0)$ e $\partial_2 = (0, 0, 1)$, and so

$$E \equiv 1 , \quad F \equiv 0 , \quad G \equiv 1 .$$

Example 4.1.10. Using the local parametrization $\varphi \colon U \to \mathbb{R}^3$ of the unit sphere S^2 given by $\varphi(x, y) = \big(x, y, \sqrt{1 - x^2 - y^2}\big)$ and recalling the local basis computed in Example 3.3.13, we get

$$E = \frac{1 - y^2}{1 - x^2 - y^2} , \quad F = \frac{xy}{1 - x^2 - y^2} , \quad G = \frac{1 - x^2}{1 - x^2 - y^2} .$$

On the other hand, the second local basis in Example 3.3.13 computed using the parametrization $\psi(\theta, \phi) = (\sin \theta \cos \phi, \sin \theta \sin \phi, \cos \theta)$ gives us

$$E \equiv 1 , \quad F \equiv 0 , \quad G = \sin^2 \theta .$$

Example 4.1.11. Let $S \subset \mathbb{R}^3$ be the helicoid with the local parametrization $\varphi \colon \mathbb{R}^2 \to \mathbb{R}^3$ given by $\varphi(x, y) = (y \cos x, y \sin x, ax)$ for some $a \in \mathbb{R}^*$. Then, recalling the local basis computed in Problem 3.2, we find

$$E = y^2 + a^2 , \quad F \equiv 0 , \quad G \equiv 1 .$$

Example 4.1.12. Let $S \subset \mathbb{R}^3$ be the catenoid with the local parametrization $\psi \colon \mathbb{R} \times (0, 2\pi) \to \mathbb{R}^3$ given by $\psi(x, y) = (a \cosh x \cos y, a \cosh x \sin y, ax)$ for some $a \in \mathbb{R}^*$. Then, recalling the local basis computed in Problem 3.1, we find

$$E = a^2 \cosh^2 x , \quad F \equiv 0 , \quad G = a^2 \cosh^2 x .$$

Example 4.1.13. More in general, let $\varphi \colon I \times J \to \mathbb{R}^3$, given by

$$\varphi(t, \theta) = \big(\alpha(t) \cos \theta, \alpha(t) \sin \theta, \beta(t)\big) ,$$

be a local parametrization of a surface of revolution S obtained as described in Example 3.1.18 (where I and J are suitable open intervals). Then, using the local basis computed in Example 3.3.15, we get

$$E = (\alpha')^2 + (\beta')^2 , \quad F \equiv 0 , \quad G = \alpha^2 .$$

For instance, if S is the torus studied in Example 3.1.19 then

$$E \equiv r^2 , \quad F \equiv 0 , \quad G = (r \cos t + x_0)^2 .$$

The first fundamental form allows us to compute the length of curves on the surface. Indeed, if $\sigma: [a, b] \to S$ is a curve whose image is contained in the surface S, we have

$$L(\sigma) = \int_a^b \sqrt{I_{\sigma(t)}\big(\sigma'(t)\big)} \, dt .$$

Conversely, if we can compute the length of curves with support on the surface S, we may retrieve the first fundamental form as follows: given $p \in S$ and $v \in T_pS$ let $\sigma: (-\varepsilon, \varepsilon) \to S$ be a curve with $\sigma(0) = p$ and $\sigma'(0) = v$, and set $\ell(t) = L(\sigma|_{[0,t]})$. Then (check it!):

$$I_p(v) = \left| \frac{d\ell}{dt}(0) \right|^2 .$$

So, in a sense, the first fundamental form is related to the *intrinsic* metric properties of the surface, properties that do not depend on the way the surface is immersed in $\mathbb{R}^3$. Staying within the surface, we may measure the length of a curve, and so we may compute the first fundamental form, without having to pop our head into $\mathbb{R}^3$; moreover, a diffeomorphism that preserves the length of curves also preserves the first fundamental form. For this reason, properties of the surface that only depend on the first fundamental form are called *intrinsic properties*. For instance, we shall see in the next few sections that the value of a particular curvature (the Gaussian curvature) is an intrinsic property which will allow us to determine, without leaving our planet, whether the Earth is flat or not.

The maps between surfaces preserving the first fundamental form deserve a special name:

Definition 4.1.14. Let $H: S_1 \to S_2$ be a C^∞ map between two surfaces. We say that H is an *isometry at* $p \in S_1$ if for all $v \in T_pS_1$ we have

$$I_{H(p)}\big(dH_p(v)\big) = I_p(v) ;$$

clearly (why?) this implies that

$$\big\langle dH_p(v), dH_p(w) \big\rangle_{H(p)} = \langle v, w \rangle_p$$

for all $v, w \in T_pS_1$. We say that H is a *local isometry* at $p \in S_1$ if p has a neighborhood U such that H is an isometry at each point of U; and that H is a *local isometry* if it is so at each point of S_1. Finally, we say that H is an *isometry* if it is both a global diffeomorphism and a local isometry.

Remark 4.1.15. If $H: S_1 \to S_2$ is an isometry at $p \in S_1$, the differential of H at p is invertible, and so H is a diffeomorphism of a neighborhood of p in S_1 with a neighborhood of $H(p)$ in S_2.

Remark 4.1.16. Isometries preserve the lenght of curves, and consequently all intrinsic properties of surfaces.

Example 4.1.17. Denote by $S_1 \subset \mathbb{R}^3$ the plane $\{z = 0\}$, by $S_2 \subset \mathbb{R}^3$ the cylinder of equation $x^2 + y^2 = 1$, and let $H: S_1 \to S_2$ be the map given by $H(x, y, 0) = (\cos x, \sin x, y)$. As seen in Example 3.3.12, the tangent plane to S_1 at any of its points is S_1 itself. Moreover, we have

$$\mathrm{d}H_p(v) = v_1 \frac{\partial H}{\partial x}(p) + v_2 \frac{\partial H}{\partial y}(p) = (-v_1 \sin x, v_1 \cos x, v_2)$$

for all $p = (x, y, 0) \in S_1$ and $v = (v_1, v_2, 0) \in T_p S_1$. Hence,

$$I_{H(p)}\big(\mathrm{d}H_p(v)\big) = \|\mathrm{d}H_p(v)\|^2 = v_1^2 + v_2^2 = \|v\|^2 = I_p(v) \,,$$

and so H is a local isometry. On the other hand, H is not an isometry, because it is not injective.

Definition 4.1.18. We shall say that a surface S_1 is *locally isometric* to a surface S_2 if for all $p \in S_1$ there exists an isometry of a neighborhood of p in S_1 with an open subset of S_2.

Remark 4.1.19. Warning: being locally isometric is *not* an equivalence relation; see Exercise 4.8.

Two surfaces are locally isometric if and only if they have (in suitable local parametrizations) the same metric coefficients:

Proposition 4.1.20. *Let S, $\tilde{S} \subset \mathbb{R}^3$ be two surfaces. Then S is locally isometric to $\tilde{S}$ if and only if for every point $p \in S$ there exist a point $\tilde{p} \in \tilde{S}$, an open subset $U \subseteq \mathbb{R}^2$, a local parametrization $\varphi: U \to S$ of S centered at p, and a local parametrization $\tilde{\varphi}: U \to \tilde{S}$ of $\tilde{S}$ centered at $\tilde{p}$ such that $E \equiv \tilde{E}$, $F \equiv \tilde{F}$ and $G \equiv \tilde{G}$, where E, F, G (respectively $\tilde{E}$, $\tilde{F}$, $\tilde{G}$) are the metric coefficients of S with respect to φ (respectively, of $\tilde{S}$ with respect to $\tilde{\varphi}$).*

Proof. Assume that S is locally isometric to $\tilde{S}$. Then, given $p \in S$, we may find a neighborhood V of p and an isometry $H: V \to H(V) \subseteq \tilde{S}$. Let $\varphi: U \to S$ a local parametrization centered at p and such that $\varphi(U) \subset V$; then $\tilde{\varphi} = H \circ \varphi$ is a local parametrization of $\tilde{S}$ centered at $\tilde{p} = H(p)$ with the required properties (check, please).

Conversely, assume that there exist two local parametrizations φ and $\tilde{\varphi}$ as stated, and set $H = \tilde{\varphi} \circ \varphi^{-1}: \varphi(U) \to \tilde{\varphi}(U)$. Clearly, H is a diffeomorphism with its image; we have to prove that it is an isometry. Take $q \in \varphi(U)$ and $v \in T_q S$, and write $v = v_1 \partial_1 + v_2 \partial_2$. By construction (see Remark 3.4.26) we have $\mathrm{d}H_q(\partial_j) = \tilde{\partial}_j$; so $\mathrm{d}H_q(v) = v_1 \tilde{\partial}_1 + v_2 \tilde{\partial}_2$; hence

$$\begin{aligned}
I_{H(q)}\big(\mathrm{d}H_q(v)\big) &= v_1^2 \tilde{E}\big(\tilde{\varphi}^{-1} \circ H(q)\big) + 2v_1 v_2 \tilde{F}\big(\tilde{\varphi}^{-1} \circ H(q)\big) + v_2^2 \tilde{G}\big(\tilde{\varphi}^{-1} \circ H(q)\big) \\
&= v_1^2 E\big(\varphi^{-1}(q)\big) + 2v_1 v_2 F\big(\varphi^{-1}(q)\big) + v_2^2 G\big(\varphi^{-1}(q)\big) = I_q(v) \,,
\end{aligned}$$

and so H is an isometry, as required. $\square$

Example 4.1.21. As a consequence, a plane and a right circular cylinder are locally isometric, due to the previous proposition and Examples 4.1.7 and 4.1.9 (see also Example 4.1.17). On the other hand, they cannot be globally isometric, since they are not even homeomorphic (a parallel of the cylinder disconnects it into two components neither of which has compact closure, a thing impossible in the plane due to the Jordan curve theorem).

If you are surprised to find out that the plane and the cylinder are locally isometric, wait till you see next example:

Example 4.1.22. Every helicoid is locally isometric to a catenoid. Indeed, let S be a helicoid parametrized as in Example 4.1.11, and let $\tilde{S}$ be the catenoid corresponding to the same value of the parameter $a \in \mathbb{R}^*$, parametrized as in Example 4.1.12. Choose a point $p_0 = \varphi(x_0, y_0) \in S$, and let $\chi \colon \mathbb{R} \times (0, 2\pi) \to \mathbb{R}^2$ be given by $\chi(x, y) = (y - \pi + x_0, a \sinh x)$. Clearly, χ is a diffeomorphism with its image, so $\varphi \circ \chi$ is a local parametrization at p of the helicoid. The metric coefficients with respect to this parametrization are

$$E = a^2 \cosh^2 x, \quad F = 0, \quad G = a^2 \cosh^2 x \, ,$$

so Proposition 4.1.20 ensures that the helicoid is locally isometric to the catenoid. In an analogous way (exercise) it can be proved that the catenoid is locally isometric to the helicoid.

So surfaces having a completely different appearance from outside may well be isometric, and so intrinsically indistinguishable. But if so, how do we tell that two surfaces are *not* locally isometric? Could even the plane and the sphere turn out to be locally isometric? One of the main goals of this chapter is to give a first answer to such questions: we shall construct a function, the Gaussian curvature, defined independently of any local parametrization, measuring intrinsic properties of the surface, so surfaces with significantly different Gaussian curvatures cannot be even locally isometric. In the supplementary material of next chapter we shall complete the discussion by giving a necessary and sufficient criterion not requiring the choice of a particular local parametrization for two surfaces to be locally isometric (Corollary 5.5.6).

By the way, we would like to remind you that one of the problems that prompted the development of differential geometry was the creation of geographical maps. In our language, a geographical map is a diffeomorphism between an open subset of a surface and an open subset of the plane (in other words, the inverse of a local parametrization) preserving some metric properties of the surface. For instance, a geographical map with a 1:1 scale (a full-scale map) is an isometry of an open subset of the surface with an open subset of the plane. Of course, full-scale maps are not terribly practical; usually we prefer smaller-scale maps. This suggests the following:

Definition 4.1.23. A *similitude* with scale factor $r > 0$ between two surfaces is a diffeomorphism $H \colon S_1 \to S_2$ such that

$$I_{H(p)}\big(\mathrm{d}H_p(v)\big) = r^2 I_p(v)$$

for all $p \in S_1$ and $v \in T_p S_1$.

A similitude multiplies the length of curves by a constant factor, the scale factor, and so it is ideal for road maps. Unfortunately, as we shall see (Corollary 4.6.12), similitudes between open subsets of surfaces and open subsets of the plane are very rare. In particular, we shall prove that there exist no similitudes between open subsets of the sphere and open subsets of the plane, so a perfect road map is impossible (the maps we normally use are just approximations). A possible replacement (which is actually used in map-making) is given by *conformal maps*, that is, diffeomorphisms preserving angles; see Exercises 4.62 and 4.63.

While we are at it, let us conclude this section talking about angles:

Definition 4.1.24. Let $S \subset \mathbb{R}^3$ be a surface, and $p \in S$. A *determination of the angle* between two tangent vectors $v_1, v_2 \in T_p S$ is a $\theta \in \mathbb{R}$ such that

$$\cos\theta = \frac{\langle v_1, v_2 \rangle_p}{\sqrt{I_p(v_1) I_p(v_2)}} \ .$$

Moreover, if $\sigma_1, \sigma_2 \colon (-\varepsilon, \varepsilon) \to S$ are curves with $\sigma_1(0) = \sigma_2(0) = p$, we shall call (determination of the) *angle* between σ_1 and σ_2 at p the angle between $\sigma_1'(0)$ and $\sigma_2'(0)$.

In the plane the Cartesian axes meet (usually) at a right angle. Local parametrizations with an analogous property are very useful, and deserve a special name:

Definition 4.1.25. We shall say that a local parametrization φ of a surface S is *orthogonal* if its coordinate curves meet at a right angle, that is, if $\partial_1|_p$ and $\partial_2|_p$ are orthogonal for each p in the image of φ.

Remark 4.1.26. The tangent vectors to coordinate curves are ∂_1 and ∂_2; so the cosine of the angle between two coordinate curves is given by $F/\sqrt{EG}$, and a local parametrization is orthogonal if and only if $F \equiv 0$. In the next chapter we shall see that orthogonal parametrizations always exist (Corollary 5.3.21).

Example 4.1.27. Parallels and meridians are the coordinate curves of the local parametrizations of the surfaces of revolution seen in Example 3.1.18, and so these parametrizations are orthogonal thanks to Example 4.1.13.

4.2 Area

The first fundamental form also allows us to compute the area of bounded regions of a regular surface. For the sake of simplicity, we shall confine our treatment to the case of regions contained in the image of a local parametrization, postponing the general case to Section 6.3.

Let us begin by defining the regions whose area we want to measure.

Definition 4.2.1. Let $\sigma\colon [a, b] \to S$ be a piecewise regular curve parametrized by arc length in a surface $S \subset \mathbb{R}^3$, and let $a = s_0 < s_1 < \cdots < s_k = b$ be a partition of $[a, b]$ such that $\sigma|_{[s_{j-1}, s_j]}$ is regular for $j = 1, \ldots, k$. As for plane curves, we set

$$\dot{\sigma}(s_j^-) = \lim_{s \to s_j^-} \dot{\sigma}(s) \qquad \text{and} \qquad \dot{\sigma}(s_j^+) = \lim_{s \to s_j^+} \dot{\sigma}(s) \, ;$$

$\dot{\sigma}(s_j^-)$ and $\dot{\sigma}(s_j^+)$ are (in general) distinct vectors of $T_{\sigma(s_j)} S$. Of course, $\dot{\sigma}(s_0^-)$ and $\dot{\sigma}(s_k^+)$ are not defined unless the curve is closed, in which case we set $\dot{\sigma}(s_0^-) = \dot{\sigma}(s_k^-)$ and $\dot{\sigma}(s_k^+) = \dot{\sigma}(s_0^+)$. We shall say that $\sigma(s_j)$ is a *vertex* of σ if $\dot{\sigma}(s_j^-) \neq \dot{\sigma}(s_j^+)$, and that it is a *cusp* of σ if $\dot{\sigma}(s_j^-) = -\dot{\sigma}(s_j^+)$. A *curvilinear polygon* in S is a closed simple piecewise regular curve parametrized by arc length without cusps.

Definition 4.2.2. A *regular region* $R \subseteq S$ of a surface S is a connected compact subset of S obtained as the closure of its interior $\mathring{R}$ and whose boundary is parametrized by finitely many curvilinear polygons with disjoint supports. If S is compact, then $R = S$ is a regular region of S with empty boundary.

To define the length of a curve we approximated it with a polygonal closed curve; to define the area of a region we shall proceed in a similar way.

Definition 4.2.3. Let $R \subseteq S$ be a regular region of a surface S. A *partition* of R is a finite family $\mathcal{R} = \{R_1, \ldots, R_n\}$ of regular regions contained in R with $R_i \cap R_j \subseteq \partial R_i \cap \partial R_j$ for all $1 \leq i \neq j \leq n$ and such that $R = R_1 \cup \cdots \cup R_n$. The *diameter* $\|\mathcal{R}\|$ of a partition $\mathcal{R}$ is the maximum of the diameters (in $\mathbb{R}^3$) of the elements of $\mathcal{R}$. Another partition $\hat{\mathcal{R}} = \{\tilde{R}_1, \ldots, \tilde{R}_m\}$ of R is said to be a *refinement* of $\mathcal{R}$ if for all $i = 1, \ldots, m$ there exists a $1 \leq j \leq n$ such that $\tilde{R}_i \subseteq R_j$. Finally, a *pointed partition* of R is given by a partition $\mathcal{R} = \{R_1, \ldots, R_n\}$ of R and a n-tuple $\mathbf{p} = (p_1, \ldots, p_n)$ of points of R such that $p_j \in R_j$ for $j = 1, \ldots, n$.

Definition 4.2.4. Let $R \subseteq S$ be a regular region of a surface S, and $(\mathcal{R}, \mathbf{p})$ a pointed partition of R. For all $R_j \in \mathcal{R}$, denote by $\overline{R_j}$ the orthogonal projection of R_j on the affine tangent plane $p_j + T_{p_j} S$ (see Fig. 4.1), and by $\text{Area}(\overline{R_j})$ its area. The *area* of the pointed partition $(\mathcal{R}, \mathbf{p})$ is defined as

$$\text{Area}(\mathcal{R}, \mathbf{p}) = \sum_{R_j \in \mathcal{R}} \text{Area}(\overline{R_j}) \, .$$

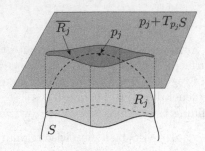

Fig. 4.1.

We say that the region R is *rectifiable* if the limit

$$\text{Area}(R) = \lim_{\|\mathcal{R}\| \to 0} \text{Area}(\mathcal{R}, \mathbf{p})$$

exists and is finite. This limit shall be the *area* of R.

To prove that every regular region contained in the image of a local parametrization is rectifiable we shall need the classical Change of Variables Theorem for multiple integrals (see [5, Theorem 5.8, p. 211]):

Theorem 4.2.5. *Let $h: \tilde{\Omega} \to \Omega$ be a diffeomorphism between open sets of $\mathbb{R}^n$. Then, for each regular region $R \subset \Omega$ and each continuous function $f: R \to \mathbb{R}$ we have*

$$\int_{h^{-1}(R)} (f \circ h) |\det \text{Jac}(h)| \, dx_1 \cdots dx_n = \int_R f \, dx_1 \cdots dx_n \, .$$

Then:

Theorem 4.2.6. *Let $R \subseteq S$ be a regular region contained in the image of a local parametrization $\varphi: U \to S$ of a surface S. Then R is a rectifiable and*

$$\text{Area}(R) = \int_{\varphi^{-1}(R)} \sqrt{EG - F^2} \, dx_1 \, dx_2 \, . \tag{4.1}$$

Proof. Let $R_0 \subseteq R$ be a regular region contained in R, and consider a point $p_0 \in R_0$; our first goal is to describe the orthogonal projection $\overline{R_0}$ of R_0 in $p_0 + T_{p_0} S$. If $p_0 = \varphi(x_0)$, an orthonormal basis of $T_{p_0} S$ is given by the vectors

$$\epsilon_1 = \frac{1}{\sqrt{E(x_0)}} \partial_1(x_0) \, ,$$

$$\epsilon_2 = \sqrt{\frac{E(x_0)}{E(x_0)G(x_0) - F(x_0)^2}} \left(\partial_2(x_0) - \frac{F(x_0)}{E(x_0)} \partial_1(x_0) \right) \, .$$

It follows (exercise) that the orthogonal projection $\pi_{x_0}\colon \mathbb{R}^3 \to p_0 + T_{p_0}S$ is given by the formula

$$\pi_{x_0}(q) = p_0 + \frac{1}{\sqrt{E(x_0)}}\langle q - p_0, \partial_1(x_0)\rangle \epsilon_1$$

$$+ \sqrt{\frac{E(x_0)}{E(x_0)G(x_0) - F(x_0)^2}}\left\langle q - p_0, \partial_2(x_0) - \frac{F(x_0)}{E(x_0)}\partial_1(x_0)\right\rangle \epsilon_2 \,.$$

Denote now by $\psi_{x_0}\colon p_0 + T_{p_0}S \to \mathbb{R}^2$ the map sending each $p \in p_0 + T_{p_0}S$ to the coordinates of $p - p_0$ with respect to the basis $\{\epsilon_1, \epsilon_2\}$; since the latter is an orthonormal basis, ψ_{x_0} preserves areas.

Set now $h_x = \psi_x \circ \pi_x \circ \varphi$, and let $\varPhi\colon U \times U \to \mathbb{R}^2 \times U$ be the map $\varPhi(x, y) = \big(h_x(y), x\big)$. It is immediate to verify that

$$\det \mathrm{Jac}(\varPhi)(x_0, x_0) = \det \mathrm{Jac}(h_{x_0})(x_0) = \sqrt{E(x_0)G(x_0) - F(x_0)^2} > 0 \,; \quad (4.2)$$

so, for all $x_0 \in U$ there exists a neighborhood $V_{x_0} \subseteq U$ of x_0 such that $\varPhi|_{V_{x_0} \times V_{x_0}}$ is a diffeomorphism with its image. Recalling the definition of $\varPhi$, this implies that $h_x|_{V_{x_0}}$ is a diffeomorphism with its image for all $x \in V_{x_0}$. In particular, if $R_0 = \varphi(Q_0) \subset \varphi(U)$ is a regular region with $Q_0 \subset V_{x_0}$ and $x \in Q_0$ then the orthogonal projection $\overline{R_0}$ of R_0 on $\varphi(x) + T_{\varphi(x)}S$ is given by $\pi_x \circ \varphi(Q_0)$ and, since ψ_x preserves areas, Theorem 4.2.5 implies

$$\mathrm{Area}(\overline{R_0}) = \mathrm{Area}\big(h_x(Q_0)\big) - \int_{Q_0} |\det \mathrm{Jac}(h_x)|\, \mathrm{d}y_1 \mathrm{d}y_2 \,. \quad (4.3)$$

Let now $R \subset \varphi(U)$ be an arbitrary regular region, and $Q = \varphi^{-1}(R)$. Given $\varepsilon > 0$, we want to find a $\delta > 0$ such that for each pointed partition $(\mathcal{R}, \mathbf{p})$ of R with diameter less than δ we have

$$\left|\mathrm{Area}(\mathcal{R}, \mathbf{p}) - \int_Q \sqrt{EG - F^2}\, \mathrm{d}y_1\, \mathrm{d}y_2\right| < \varepsilon \,.$$

The family $\mathfrak{V} = \{V_x \mid x \in Q\}$ is an open cover of the compact set Q; let $\delta_0 > 0$ be the Lebesgue number (Theorem 2.1.2) of $\mathfrak{V}$. Let now $\varPsi\colon Q \times Q \to \mathbb{R}$ be given by

$$\varPsi(x, y) = |\det \mathrm{Jac}(h_x)(y)| - \sqrt{E(y)G(y) - F(y)^2} \,.$$

By (4.2) we know that $\varPsi(x, x) \equiv 0$; so the uniform continuity provides us with a $\delta_1 > 0$ such that

$$|y - x| < \delta_1 \quad \Longrightarrow \quad |\varPsi(x, y)| < \varepsilon/\mathrm{Area}(Q) \,.$$

Finally, the uniform continuity of $\varphi^{-1}|_R$ provides us with a $\delta > 0$ such that if $R_0 \subseteq R$ has diameter less than δ then $\varphi^{-1}(R_0)$ has diameter less than $\min\{\delta_0, \delta_1\}$.

Let then $(\mathcal{R}, \mathbf{p})$, with $\mathcal{R} = \{R_1, \ldots, R_n\}$ and $\mathbf{p} = (p_1, \ldots, p_n)$, be a pointed partition of R with diameter less than δ, and set $Q_j = \varphi^{-1}(R_j)$ and $x_j = \varphi^{-1}(p_j)$. Since each Q_j has diameter less than δ_0, we may use formula (4.3) to compute the area of each $\overline{R}_j$. Hence,

$$
\left| \text{Area}(\mathcal{R}, \mathbf{p}) - \int_Q \sqrt{EG - F^2} \, dy_1 \, dy_2 \right|
$$

$$
= \left| \sum_{j=1}^n \int_{Q_j} |\det \text{Jac}(h_{x_j})| \, dy_1 dy_2 - \int_Q \sqrt{EG - F^2} \, dy_1 \, dy_2 \right|
$$

$$
\leq \sum_{j=1}^n \int_{Q_j} |\Psi(x_j, y)| \, dy_1 \, dy_2 < \sum_{j=1}^n \frac{\varepsilon}{\text{Area}(Q)} \text{Area}(Q_j) = \varepsilon \,,
$$

since each Q_j has diameter less than δ_1, and we are done.   □

A consequence of this result is that the value of the integral in the right hand side of (4.1) is independent of the local parametrization whose image contains R. We are going to conclude this section by generalizing this result in a way that will allow us to integrate functions on a surface. We shall need a lemma containing two formulas that will be useful again later on:

Lemma 4.2.7. *Let $\varphi \colon U \to S$ be a local parametrization of a surface S. Then*

$$
\|\partial_1 \wedge \partial_2\| = \sqrt{EG - F^2} \,, \tag{4.4}
$$

where $\wedge$ is the vector product in $\mathbb{R}^3$. Moreover, if $\hat{\varphi} \colon \hat{U} \to S$ is another local parametrization with $V = \hat{\varphi}(\hat{U}) \cap \varphi(U) \neq \varnothing$, and $h = \hat{\varphi}^{-1} \circ \varphi|_{\varphi^{-1}(V)}$, then

$$
\partial_1 \wedge \partial_2|_{\varphi(x)} = \det \text{Jac}(h)(x) \, \hat{\partial}_1 \wedge \hat{\partial}_2|_{\hat{\varphi} \circ h(x)} \tag{4.5}
$$

for all $x \in \varphi^{-1}(V)$, where $\{\hat{\partial}_1, \hat{\partial}_2\}$ is the basis induced by $\hat{\varphi}$.

Proof. Formula (4.4) follows from equality

$$
\|\mathbf{v} \wedge \mathbf{w}\|^2 = \|\mathbf{v}\|^2 \|\mathbf{w}\|^2 - |\langle \mathbf{v}, \mathbf{w} \rangle|^2,
$$

which holds for any pair $\mathbf{v}$, $\mathbf{w}$ of vectors of $\mathbb{R}^3$.

Furthermore, we have seen (Remark 3.4.19) that

$$
\partial_j|_{\varphi(x)} = \frac{\partial \hat{x}_1}{\partial x_j} \hat{\partial}_1|_{\varphi(x)} + \frac{\partial \hat{x}_2}{\partial x_j} \hat{\partial}_2|_{\varphi(x)} \,,
$$

and so (4.5) immediately follows.   □

As a consequence we find:

Proposition 4.2.8. *Let* $R \subseteq S$ *be a regular region of a surface* S, *and* $f: R \to \mathbb{R}$ *a continuous function. Assume that there exists a local parametrization* $\varphi: U \to S$ *of* S *such that* $R \subset \varphi(U)$. *Then the integral*

$$\int_{\varphi^{-1}(R)} (f \circ \varphi) \sqrt{EG - F^2} \, dx_1 \, dx_2$$

does not depend on φ.

Proof. Assume that $\hat{\varphi}: \tilde{U} \to S$ is another local parametrization such that $R \subset \hat{\varphi}(\tilde{U})$, and set $h = \hat{\varphi}^{-1} \circ \varphi$. Then the previous lemma and Theorem 4.2.5 yield

$$\int_{\varphi^{-1}(R)} (f \circ \varphi) \sqrt{EG - F^2} \, dx_1 \, dx_2 = \int_{\varphi^{-1}(R)} (f \circ \varphi) \|\partial_1 \wedge \partial_2\| \, dx_1 \, dx_2$$

$$= \int_{\varphi^{-1}(R)} \left[(f \circ \hat{\varphi}) \|\hat{\partial}_1 \wedge \hat{\partial}_2\| \right] \circ h \, |\det \operatorname{Jac}(h)| \, dx_1 \, dx_2$$

$$= \int_{\hat{\varphi}^{-1}(R)} (f \circ \hat{\varphi}) \sqrt{\hat{E}\hat{G} - \hat{F}^2} \, dx_1 \, dx_2 \; .$$

$\square$

We may then give the following definition of integral on a surface:

Definition 4.2.9. Let $R \subseteq S$ be a regular region of a surface S contained in the image of a local parametrization $\varphi: U \to S$. Then for all continuous functions $f: R \to \mathbb{R}$ the *integral of* f *on* R is the number

$$\int_R f \, d\nu = \int_{\varphi^{-1}(R)} (f \circ \varphi) \sqrt{EG - F^2} \, dx_1 \, dx_2 \; .$$

In Section 6.3 of Chapter 6 we shall extend this definition so as to be able to compute the integral of continuous functions on arbitrary regular regions; here we instead conclude this section by proving an analogue for surfaces of the Change of Variables Theorem for multiple integrals.

Proposition 4.2.10. *Let* $F: \tilde{S} \to S$ *be a diffeomorphism between surfaces, and* $R \subseteq S$ *a regular region contained in the image of a local parametrization* $\varphi: U \to S$ *and such that* $F^{-1}(R)$ *is also contained in the image of a local parametrization* $\tilde{\varphi}: \tilde{U} \to \tilde{S}$. *Then, for all continuous* $f: R \to \mathbb{R}$, *we have*

$$\int_{F^{-1}(R)} (f \circ F) |\det dF| \, d\tilde{\nu} = \int_R f \, d\nu \; .$$

Proof. Set $\Omega = U$ and $\tilde{\Omega} = \tilde{\varphi}^{-1}(F^{-1}(\varphi(U)))$, so as to get $\varphi^{-1}(R) \subset \Omega$ and $\tilde{\varphi}^{-1}(F^{-1}(R)) \subset \tilde{\Omega}$; moreover, $h = \varphi^{-1} \circ F \circ \tilde{\varphi}: \tilde{\Omega} \to \Omega$ is a diffeomorphism. Set

$\hat{\varphi} = F \circ \tilde{\varphi}$. Then $\hat{\varphi}$ is a local parametrization of S, whose local basis $\{\hat{\partial}_1, \hat{\partial}_2\}$ can be obtained from the local basis $\{\tilde{\partial}_1, \tilde{\partial}_2\}$ by the formula $\hat{\partial}_j = dF(\tilde{\partial}_j)$. In particular,

$$\|\tilde{\partial}_1 \wedge \tilde{\partial}_2\| \, |\det dF| \circ \tilde{\varphi} = \|\hat{\partial}_1 \wedge \hat{\partial}_2\| \circ \hat{\varphi} = |\det \operatorname{Jac}(h)| \, \|\partial_1 \wedge \partial_2\| \circ h \, ,$$

by (4.5). Then Theorem 4.2.5 and (4.4) imply

$$\int_{F^{-1}(R)} (f \circ F) \, |\det dF| \, d\tilde{\nu}$$

$$= \int_{\hat{\varphi}^{-1}(R)} (f \circ \hat{\varphi}) |\det dF| \circ \tilde{\varphi} \sqrt{\tilde{E}\tilde{G} - \tilde{F}^2} \, dx_1 dx_2$$

$$= \int_{h^{-1}(\varphi^{-1}(R))} (f \circ \varphi \circ h) |\, \|\partial_1 \wedge \partial_2\| \circ h| \det \operatorname{Jac}(h)| \, dx_1 \, dx_2$$

$$= \int_{\varphi^{-1}(R)} (f \circ \varphi) \sqrt{EG - F^2} \, dx_1 dx_2 = \int_R f \, d\nu \, .$$

$\square$

4.3 Orientability

Orientability is an important notion in the theory of surfaces. To put it simply, a surface is orientable if it has two faces, an internal one and an external one, like the sphere, whereas it is non orientable, like the Möbius band (see Example 4.3.11), if it has only one face, and no well-defined exterior or interior.

There are (at least) two ways to define precisely the notion of orientation: an intrinsic one, and one depending on the embedding of the surface in $\mathbb{R}^3$. To describe the first one, recall that *orienting* a plane means choosing an ordered basis for it (that is, fixing a preferred rotation direction for the angles); two bases determine the same orientation if and only if the change of basis matrix has positive determinant (see [21, p. 167] or [1, p. 57]). So the idea is that a surface is orientable if we may orient in a consistent way all its tangent planes. Locally this is not a problem: just choose a local parametrization and orient each tangent plane of the support by taking as orientation the one given by the (ordered) basis $\{\partial_1, \partial_2\}$ induced by the parametrization. Since the vectors ∂_1 and ∂_2 vary in a C^∞ way, we can sensibly say that all tangent planes of the support of the parametrization are now oriented consistently. Another parametrization induces the same orientation if and only if the change of basis matrix (that is, the Jacobian matrix of the change of coordinates; see Remark 3.4.19) has positive determinant. So the following definition becomes natural:

Definition 4.3.1. Let $S \subset \mathbb{R}^3$ be a surface. We say that two local parametrizations $\varphi_\alpha \colon U_\alpha \to S$ and $\varphi_\beta \colon U_\beta \to S$ *determine the same orientation* (or are *equioriented*) if either $\varphi_\alpha(U_\alpha) \cap \varphi_\beta(U_\beta) = \varnothing$ or $\det \operatorname{Jac}(\varphi_\beta^{-1} \circ \varphi_\alpha) > 0$ where it is defined, that is, on $\varphi_\alpha^{-1}(\varphi_\alpha(U_\alpha) \cap \varphi_\beta(U_\beta))$. If, on the other hand, $\det \operatorname{Jac}(\varphi_\beta^{-1} \circ \varphi_\alpha) < 0$ where it is defined, the two local parametrizations *determine opposite orientations*. The surface S is said to be *orientable* if there exists an atlas $\mathcal{A} = \{\varphi_\alpha\}$ for S consisting of·local parametrizations pairwise equioriented (and we shall say that the atlas itself is *oriented*). If we fix such an atlas $\mathcal{A}$, we say that the surface S is *oriented* by the atlas $\mathcal{A}$.

Remark 4.3.2. Warning: it may happen that a pair of local parametrizations neither determine the same orientation nor opposite orientations. For instance, it may happen that $\varphi_\alpha(U_\alpha) \cap \varphi_\beta(U_\beta)$ has two connected components with $\det \operatorname{Jac}(\varphi_\beta^{-1} \circ \varphi_\alpha)$ positive on one of them and negative on the other one; see Example 4.3.11.

Recalling what we said, we may conclude that a surface S is orientable if and only if we may simultaneously orient all its tangent planes in a consistent way.

Example 4.3.3. A surface admitting an atlas consisting of a single local parametrization is clearly orientable. For instance, all surfaces given as graphs of functions are orientable.

Example 4.3.4. If a surface has an atlas consisting of two local parametrizations whose images have connected intersection, it is orientable. Indeed, the determinant of the Jacobian matrix of the change of coordinates has (why?) constant sign on the intersection, so up to exchanging the coordinates in the domain of one of the parametrizations (an operation that changes the sign of the determinant of the Jacobian matrix of the change of coordinates), we may always assume that both parametrizations determine the same orientation. For instance, the sphere is orientable (see Example 3.1.16).

Remark 4.3.5. Orientability is a *global* property: we cannot verify if a surface is orientable just by checking what happens on single local parametrizations. The image of a single local parametrization is always orientable; the obstruction to orientability (if any) is related to the way local parametrizations are joined.

This definition of orientation is purely intrinsic: it does not depend on the way the surface is immersed in $\mathbb{R}^3$. In particular, if two surfaces are diffeomorphic, the first one is orientable if and only if the other one is (exercise). As already mentioned, the second definition of orientation will be instead extrinsic: it will strongly depend on the fact that a surface is contained in $\mathbb{R}^3$.

When we studied Jordan curves in the plane, we saw that the normal versor allowed us to distinguish the interior of the curve from its exterior. So, it is natural to try and introduce the notions of interior and exterior of a surface by using normal versors:

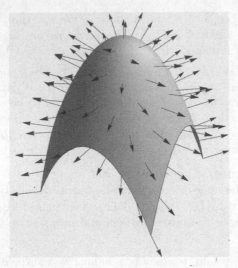

Fig. 4.2. A normal vector field

Definition 4.3.6. A *normal vector field* on a surface $S \subset \mathbb{R}^3$ is a map $N: S \to \mathbb{R}^3$ of class C^∞ such that $N(p)$ is orthogonal to T_pS for all $p \in S$; see Fig. 4.2. If, moreover, $\|N\| \equiv 1$ we shall say that N is *normal versor field* to S.

If N is a normal versor field on a surface S, we may intuitively say that N indicates the external face of the surface, while $-N$ indicates the internal face. But, in contrast to what happens for curves, not every surface as a normal vector field:

Proposition 4.3.7. *A surface $S \subset \mathbb{R}^3$ is orientable if and only if there exists a normal versor field on S.*

Proof. We begin with a general remark. Let $\varphi_\alpha: U_\alpha \to S$ be a local parametrization of a surface S; for all $p \in \varphi_\alpha(U_\alpha)$ set

$$N_\alpha(p) = \frac{\partial_{1,\alpha} \wedge \partial_{2,\alpha}}{\|\partial_{1,\alpha} \wedge \partial_{2,\alpha}\|}(p) \,,$$

where $\partial_{j,\alpha} = \partial\varphi_\alpha/\partial x_j$, as usual. Since $\{\partial_{1,\alpha}, \partial_{2,\alpha}\}$ is a basis of T_pS, the versor $N_\alpha(p)$ is well defined, different from zero, and orthogonal to T_pS; moreover, it clearly is of class C^∞, and so N_α is a normal versor field on $\varphi_\alpha(U_\alpha)$. Notice furthermore that if $\varphi_\beta: U_\beta \to S$ is another local parametrization with $\varphi_\alpha(U_\alpha) \cap \varphi_\beta(U_\beta) \neq \varnothing$ then (4.5) implies

$$N_\alpha = \mathrm{sgn}\big(\det \mathrm{Jac}(\varphi_\beta^{-1} \circ \varphi_\alpha)\big) \, N_\beta \qquad (4.6)$$

on $\varphi_\alpha(U_\alpha) \cap \varphi_\beta(U_\beta)$.

Assume now S to be orientable, and let $\mathcal{A} = \{\varphi_\alpha\}$ be an oriented atlas. If $p \in \varphi_\alpha(U_\alpha) \cap \varphi_\beta(U_\beta)$, with φ_α, $\varphi_\beta \in \mathcal{A}$, equality (4.6) tells us that $N_\alpha(p) = N_\beta(p)$; so the map $p \mapsto N_\alpha(p)$ does not depend on the particular local parametrization we have chosen, and defines a normal versor field on S.

Conversely, let $N \colon S \to \mathbb{R}^3$ be a normal versor field on S, and let $\mathcal{A} = \{\varphi_\alpha\}$ be an arbitrary atlas of S such that the domain U_α of each φ_α is connected. By definition of vector product, $N_\alpha(p)$ is orthogonal to T_pS for all $p \in \varphi_\alpha(U_\alpha)$ and $\varphi_\alpha \in \mathcal{A}$; so $\langle N, N_\alpha \rangle \equiv \pm 1$ on each U_α. Since U_α is connected, up to modifying φ_α by exchanging coordinates in U_α, we may assume that all these scalar products are identically equal to 1. Hence,

$$N_\alpha \equiv N$$

on each U_α, and (4.6) implies that the atlas is oriented. □

Definition 4.3.8. Let $S \subset \mathbb{R}^3$ be a surface oriented by an atlas $\mathcal{A}$. A normal versor field N will be said to *determine the (assigned) orientation* if $N = \partial_1 \wedge \partial_2/\|\partial_1 \wedge \partial_2\|$ for any local parametrization $\varphi \in \mathcal{A}$.

A consequence of the latter proposition is that if S is an oriented surface then there exists always (why?) a *unique* normal versor field determining the assigned orientation.

Example 4.3.9. Every surface of revolution S is orientable. Indeed, we may define a normal versor field $N \colon S \to S^2$ by setting

$$N(p) = \left.\frac{\partial}{\partial t}\right|_p \wedge \left.\frac{\partial}{\partial \theta}\right|_p \Big/ \left\| \left.\frac{\partial}{\partial t}\right|_p \wedge \left.\frac{\partial}{\partial \theta}\right|_p \right\|$$

$$= \frac{1}{\sqrt{(\alpha'(t))^2 + (\beta'(t))^2}} \begin{vmatrix} -\beta'(t)\cos\theta \\ -\beta'(t)\sin\theta \\ \alpha'(t) \end{vmatrix}$$

for all $p = \varphi(t, \theta) \in S$, where $\varphi \colon \mathbb{R}^2 \to S$ is the immersed surface with support S defined in Example 3.1.18, and we used Example 3.3.15.

Definition 4.3.10. Let $S \subset \mathbb{R}^3$ be an oriented surface, and $N \colon S \to S^2$ a normal versor field that determines the assigned orientation. If $p \in S$, we shall say that a basis $\{v_1, v_2\}$ of T_pS is *positive* (respectively, *negative*) if the basis $\{v_1, v_2, N(p)\}$ of $\mathbb{R}^3$ has the same orientation (respectively, the opposite orientation) as the canonical basis of $\mathbb{R}^3$.

In particular, a local parametrization $\varphi \colon U \to S$ determines the orientation assigned on S if and only if (why?) $\{\partial_1|_p, \partial_2|_p\}$ is a positive basis of T_pS for all $p \in \varphi(U)$.

As mentioned above, not every surface is orientable. The most famous example of non orientable surface is the Möbius band.

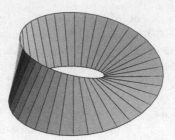

Fig. 4.3. The Möbius band

Example 4.3.11 (The Möbius band). Let C be the circle in the xy-plane with center in the origin and radius 2, and ℓ_0 the line segment in the yz-plane given by $y = 2$ and $|z| < 1$, with center in the point $c = (0, 2, 0)$. Denote by ℓ_θ the line segment obtained by rotating c clock-wise along C by an angle θ and simultaneously rotating ℓ_0 around c by an angle $\theta/2$. The union $S = \bigcup_{\theta \in [0, 2\pi]} \ell_\theta$ is the *Möbius band* (Fig. 4.3); we are going to prove that it is a non orientable surface.

Set $U = \{(u, v) \in \mathbb{R}^2 \mid 0 < u < 2\pi, -1 < v < 1\}$, and define $\varphi, \hat{\varphi} \colon U \to S$ by

$$\varphi(u, v) = \left(\left(2 - v \sin \frac{u}{2}\right) \sin u, \left(2 - v \sin \frac{u}{2}\right) \cos u, v \cos \frac{u}{2}\right),$$

$$\hat{\varphi}(u, v) = \left(\left(2 - v \sin \frac{2u + \pi}{4}\right) \cos u, \left(-2 + v \sin \frac{2u + \pi}{4}\right) \sin u, v \cos \frac{2u + \pi}{4}\right).$$

It is straightforward to verify (exercise) that $\{\varphi, \hat{\varphi}\}$ is an atlas for S, consisting of two local parametrizations whose images have *disconnected* intersection: indeed, $\varphi(U) \cap \hat{\varphi}(U) = \varphi(W_1) \cup \varphi(W_2)$, with

$$W_1 = \{(u, v) \in U \mid \pi/2 < u < 2\pi\} \quad \text{and} \quad W_2 = \{(u, v) \in U \mid 0 < u < \pi/2\}.$$

Now, if $(u, v) \in W_1$ we have $\varphi(u, v) = \hat{\varphi}(u - \pi/2, v)$, while if $(u, v) \in W_2$ we have $\varphi(u, v) = \hat{\varphi}(u + 3\pi/2, -v)$; so

$$\hat{\varphi}^{-1} \circ \varphi(u, v) = \begin{cases} (u - \pi/2, v) & \text{if } (u, v) \in W_1, \\ (u + 3\pi/2, -v) & \text{if } (u, v) \in W_2. \end{cases}$$

In particular,

$$\det \operatorname{Jac}(\hat{\varphi}^{-1} \circ \varphi) \equiv \begin{cases} +1 & \text{on } W_1, \\ -1 & \text{on } W_2. \end{cases}$$

Now, assume by contradiction that S is orientable, and let N be a normal versor field on S. Up to inverting the sign of N, we may assume that N is given by $\partial_u \wedge \partial_v / \|\partial_u \wedge \partial_v\|$ on $\varphi(U)$, where $\partial_u = \partial\varphi/\partial u$ and $\partial_v = \partial\varphi/\partial v$. On the other hand, we have $N = \pm\hat{\partial}_u \wedge \hat{\partial}_v / \|\hat{\partial}_u \wedge \hat{\partial}_v\|$ on $\hat{\varphi}(U)$, where $\hat{\partial}_u = \partial\hat{\varphi}/\partial u$ and $\hat{\partial}_v = \partial\hat{\varphi}/\partial v$, and the sign is constant because U is connected. But (4.6) applied to W_1 tells us that the sign should be $+1$, whereas applied to W_2 yields -1, contradiction.

Let us remark explicitly that the Möbius band is *not* a closed surface in $\mathbb{R}^3$. This is crucial: indeed, in the supplementary material of this chapter we shall prove that every closed surface in $\mathbb{R}^3$ is orientable (Theorem 4.7.15).

Finally, a large family of orientable surfaces is provided by the following

Corollary 4.3.12. *Let $a \in \mathbb{R}$ be a regular value for a function $f : \Omega \to \mathbb{R}$ of class C^∞, where $\Omega \subseteq \mathbb{R}^3$ is an open set. Then every connected component S of $f^{-1}(a)$ is orientable, and a normal versor field is given by $N = \nabla f / \|\nabla f\|$.*

Proof. It immediately follows from Proposition 3.3.11. □

In the supplementary material of this chapter we shall prove a converse of this corollary: if $S \subset \mathbb{R}^3$ is an orientable surface and $\Omega \subseteq \mathbb{R}^3$ an open set containing containing S such that S is closed in Ω with $\Omega \setminus S$ disconnected then there exists a function $f \in C^\infty(\Omega)$ such that S is a level surface for f (Theorem 4.8.6).

4.4 Normal curvature and second fundamental form

As you have undoubtedly already imagined, one of the main questions differential geometry has to answer is how to measure the curvature of a surface. The situation is quite a bit more complicated than for curves, and as a consequence the answer is not only more complex, but it is not even unique: there are several meaningful ways to measure the curvature of a surface, and we shall explore them in detail in the rest of this chapter.

The first natural remark is that the curvature of a surface, whatever it might be, is not constant in all directions. For instance, a circular cylinder is not curved in the direction of the generatrix, whereas it curves along the directions tangent to the parallels. So it is natural to say that the curvature of the cylinder should be zero in the direction of the generatrix, whereas the curvature in the direction of the parallels should be the same as that of the parallels themselves, that is, the inverse of the radius. And what about other directions? Looking at the cylinder, we would guess that its curvature is maximal in the direction of the parallel, minimal in the direction of the generatrix, and takes intermediate values in the other directions. To compute it, we might for instance consider a curve contained in the surface, tangent to the direction we have chosen; at the very least, this is an approach that works for generatrices and parallels. The problem is: which curve? A priori (and a posteriori too, as we shall see), if we choose a random curve the curvature might depend on some property of the curve and not only on the surface S and on the tangent direction v we are interested in. So we need a procedure yielding a curve depending only on S and v and representing appropriately the geometry of the surface along that direction. The next lemma describes how to do this:

Lemma 4.4.1. *Let S be a surface, $p \in S$ and choose a versor $N(p) \in \mathbb{R}^3$ orthogonal to T_pS. Given $v \in T_pS$ of length 1, let H_v be the plane passing through p and parallel to v and $N(p)$. Then the intersection $H_v \cap S$ is, at least in a neighborhood of p, the support of a regular curve.*

Proof. The plane H_v has equation $\langle x - p, v \wedge N(p) \rangle = 0$. So if $\varphi \colon U \to S$ is a local parametrization centered at p, a point $\varphi(y) \in \varphi(U)$ belongs to $H_v \cap S$ if and only if $y \in U$ satisfies the equation $f(y) = 0$, where

$$f(y) = \langle \varphi(y) - p, v \wedge N(p) \rangle .$$

If we prove that $C = \{y \in U \mid f(y) = 0\}$ is the support of a regular curve σ near O, we are done, as $H_v \cap \varphi(U) = \varphi(C)$ is in this case the support of the regular curve $\varphi \circ \sigma$ near p.

Now,

$$\frac{\partial f}{\partial y_i}(O) = \langle \partial_i|_p, v \wedge N(p) \rangle ;$$

so if O were a critical point of f, the vector $v \wedge N(p)$ would be orthogonal to both $\partial_1|_p$ and $\partial_2|_p$, and hence orthogonal to T_pS, that is, parallel to $N(p)$, whereas it is not. So O is not a critical point of f, and by Proposition 1.1.18 we know that C is a graph in a neighborhood of O. □

Definition 4.4.2. Let S be a surface. Given $p \in S$, choose a versor $N(p) \in \mathbb{R}^3$ orthogonal to T_pS. Take $v \in T_pS$ of length one, and let H_v be the plane through p and parallel to v and $N(p)$. The regular curve σ, parametrized by arc length, with $\sigma(0) = p$ whose support is the intersection $H_v \cap S$ in a neighborhood of p is the *normal section* of S at p along v (see Fig. 4.4). Since $\mathrm{Span}\{v, N(p)\} \cap T_pS = \mathbb{R}v$, the tangent versor of the normal section at p has to be $\pm v$; we shall orient the normal section curve so that $\dot{\sigma}(0) = v$. In particular, σ is uniquely defined in a neighborhood of 0 (why?).

The normal section is a curve that only depends on the geometry of the surface S in the direction of the tangent versor v; so we may try and use it to give a geometric definition of the curvature of a surface.

Definition 4.4.3. Let S be a surface, $p \in S$ and let $N(p) \in \mathbb{R}^3$ be a versor orthogonal to T_pS. Given $v \in T_pS$ of length 1, orient the plane H_v by choosing $\{v, N(p)\}$ as positive basis. The *normal curvature* of S at p along v is the oriented curvature at p of the normal section of S at p along v (considerered as a plane curve contained in H_v) .

Remark 4.4.4. Clearly, the normal section curve does not depend on the choice of the particular versor $N(p)$ orthogonal to T_pS. The normal curvature, on the other hand, does: if we substitute $-N(p)$ for $N(p)$, the normal curvature changes sign (why?).

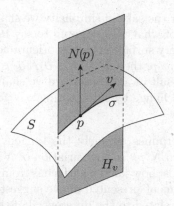

Fig. 4.4. Normal section

It is straightforward to verify (do it!) that the normal curvature of a right circular cylinder with radius $r > 0$ is actually zero along the directions tangent to the generatrices, and equals $\pm 1/r$ along the directions tangent to the parallels (which are normal sections); computing the curvature along other directions is however more complicated. For the cylinder the other normal sections are ellipses, so it can somehow be done; but for an arbitrary surface the problem becomes harder, as normal section curves are defined only implicitly (as the intersection of a plane and a surface), and so computing their oriented curvature might not be easy.

To solve this problem (and, as you will see, we shall solve it and obtain simple explicit formulas to compute normal curvatures), let us introduce a second way to study the curvature of a surface. In a sense, the curvature of a curve is a measure of the variation of its tangent line; the curvature of a surface might then be a measure of the variation of its tangent plane. Now, the tangent line to a curve is determined by the tangent versor, that is, by a vector-valued map, uniquely defined up to sign, so measuring the variation of the tangent line is equivalent to differentiating this map. Instead, at first glance we might think that to determine the tangent plane might be necessary to choose a basis, and this choice is anything but unique. But, since we are talking about surfaces in $\mathbb{R}^3$, the tangent plane actually is also determined by the normal versor, which is unique up to sign; so we may try to measure the variation of the tangent plane by differentiating the normal versor.

Let us now try and make this argument formal and rigorous. As we shall see, we shall actually obtain an effective way of computing the normal curvature; but to get there we shall need a bit of work.

We begin with a crucial definition.

Definition 4.4.5. Let $S \subset \mathbb{R}^3$ be an oriented surface. The *Gauss map* of S is the normal versor field $N \colon S \to S^2$ that identifies the given orientation.

Remark 4.4.6. Even if for the sake of simplicity we shall often work only with oriented surfaces, much of what we are going to say in this chapter holds for every surface. Indeed, every surface is locally orientable: if $\varphi: U \to S$ is a local parametrization at a point p, then $N = \partial_1 \wedge \partial_2 / \|\partial_1 \wedge \partial_2\|$ is a Gauss map of $\varphi(U)$. Therefore every result of a local nature we shall prove by using Gauss maps and that does not change by substituting $-N$ for N actually holds for an arbitrary surface.

The Gauss map determines uniquely the tangent planes to the surface, since $T_p S$ is orthogonal to $N(p)$; so the variation of N measures how the tangent planes change, that is, how far the surface is from being a plane (see also Exercise 4.18). The argument presented above suggests that the curvature of a surface might then be related to the differential of the Gauss map, just like the curvature of a curve was related to the derivative of the tangent versor. To verify the correctness of this guess, let us study some examples.

Example 4.4.7. In a plane parametrized as in Example 3.1.12 we have

$$N \equiv \frac{\mathbf{v}_1 \wedge \mathbf{v}_2}{\|\mathbf{v}_1 \wedge \mathbf{v}_2\|} \, ,$$

so N is constant and $dN \equiv O$.

Example 4.4.8. Let $S = S^2$. By using any of the parametrizations described in Example 4.1.10 we find $N(p) = p$, a result consistent with Example 3.3.13. So the Gauss map of the unit sphere is the identity map and, in particular, $dN_p = \mathrm{id}$ for all $p \in S^2$.

Example 4.4.9. Let $S \subset \mathbb{R}^3$ be a right circular cylinder of equation $x_1^2 + x_2^2 = 1$. Corollary 4.3.12 tells us that a Gauss map of S is given by

$$N(p) = \begin{vmatrix} p_1 \\ p_2 \\ 0 \end{vmatrix}$$

for all $p = (p_1, p_2, p_3) \in S$. In particular,

$$T_p S = N(p)^\perp = \{v \in \mathbb{R}^3 \mid v_1 p_1 + v_2 p_2 = 0\} \, .$$

Moreover, as N is the restriction to S of a linear map of $\mathbb{R}^3$ in itself, we get (why?) $dN_p(v) = (v_1, v_2, 0)$ for all $v = (v_1, v_2, v_3) \in T_p S$. In particular, $dN_p(T_p S) \subseteq T_p S$, and as an endomorphism of $T_p S$ the differential of the Gauss map has an eigenvalue equal to zero and one equal to 1. The eigenvector corresponding to the zero eigenvalue is $(0, 0, 1)$, that is, the direction along which we already know the cylinder has zero normal curvature; the eigenvector corresponding to the eigenvalue 1 is tangent to the parallels of the cylinder, so it is exactly the direction along which the cylinder has normal curvature 1. As we shall see, this is not a coincidence.

Example 4.4.10. Let $\Gamma_h \subset \mathbb{R}^3$ be the graph of a function $h: U \to \mathbb{R}$, where $U \subset \mathbb{R}^2$ is open, and let $\varphi: U \to \Gamma_h$ be the usual parametrization $\varphi(x) = (x, h(x))$ of Γ_h. Example 3.3.14 tells us that a Gauss map $N: \Gamma_h \to S^2$ of Γ_h is given by

$$N \circ \varphi = \frac{\partial_1 \wedge \partial_2}{\|\partial_1 \wedge \partial_2\|} = \frac{1}{\sqrt{1 + \|\nabla h\|^2}} \begin{vmatrix} -\partial h/\partial x_1 \\ -\partial h/\partial x_2 \\ 1 \end{vmatrix}.$$

Let us compute how the differential of N acts on the tangent planes of Γ_h. Choose $p = \varphi(x) \in \Gamma_h$; recalling Remark 3.4.26 we get

$$\mathrm{d}N_p(\partial_j) = \frac{\partial(N \circ \varphi)}{\partial x_j}(x)$$

$$= \frac{1}{(1 + \|\nabla h\|^2)^{3/2}} \left\{ \left[\frac{\partial h}{\partial x_1} \frac{\partial h}{\partial x_2} \frac{\partial^2 h}{\partial x_j \partial x_2} - \left(1 + \left(\frac{\partial h}{\partial x_2} \right)^2 \right) \frac{\partial^2 h}{\partial x_j \partial x_1} \right] \partial_1 \right.$$

$$\left. + \left[\frac{\partial h}{\partial x_1} \frac{\partial h}{\partial x_2} \frac{\partial^2 h}{\partial x_j \partial x_1} - \left(1 + \left(\frac{\partial h}{\partial x_1} \right)^2 \right) \frac{\partial^2 h}{\partial x_j \partial x_2} \right] \partial_2 \right\};$$

in particular, $\mathrm{d}N_p(T_p\Gamma_h) \subseteq T_p\Gamma_h$ for all $p \in \Gamma_h$.

Example 4.4.11. Let S be a helicoid, parametrized as in Example 4.1.11. Then

$$(N \circ \varphi)(x, y) = \frac{1}{\sqrt{a^2 + y^2}} \begin{vmatrix} -a \sin x \\ a \cos x \\ -y \end{vmatrix}.$$

Let now $p = \varphi(x_0, y_0) \in S$, and take $v = v_1 \partial_1 + v_2 \partial_2 \in T_p S$. Arguing as in the previous example we find

$$\mathrm{d}N_p(v) = v_1 \frac{\partial(N \circ \varphi)}{\partial x}(x_0, y_0) + v_2 \frac{\partial(N \circ \varphi)}{\partial y}(x_0, y_0)$$

$$= -\frac{a}{(a^2 + y_0^2)^{3/2}} v_2 \partial_1 - \frac{a}{(a^2 + y_0^2)^{1/2}} v_1 \partial_2.$$

In particular, $\mathrm{d}N_p(T_p S) \subseteq T_p S$ in this case too.

Example 4.4.12. Let $S \subset \mathbb{R}^3$ be a catenoid, parametrized as in Example 4.1.12. Then

$$(N \circ \psi)(x, y) = \frac{1}{\cosh x} \begin{vmatrix} -\cos y \\ -\sin y \\ \sinh x \end{vmatrix}.$$

Let now $p = \psi(x_0, y_0) \in S$, and take $w = w_1 \partial_1 + w_2 \partial_2 \in T_p S$. Then we get

$$\mathrm{d}N_p(w) = \frac{w_1}{a \cosh^2 x_0} \partial_1 - \frac{w_2}{a \cosh^2 x_0} \partial_2.$$

In particular, $\mathrm{d}N_p(T_p S) \subseteq T_p S$, once again.

Example 4.4.13. Let $S \subset \mathbb{R}^3$ be a surface of revolution, oriented by the Gauss map $N: S \to S^2$ we computed in Example 4.3.9. Then

$$\mathrm{d}N_p\left(\left.\frac{\partial}{\partial t}\right|_p\right) = \frac{\beta'\alpha'' - \alpha'\beta''}{\left((\alpha')^2 + (\beta')^2\right)^{3/2}} \left.\frac{\partial}{\partial t}\right|_p ,$$

$$\mathrm{d}N_p\left(\left.\frac{\partial}{\partial \theta}\right|_p\right) = \frac{-\beta'/\alpha}{\sqrt{(\alpha')^2 + (\beta')^2}} \left.\frac{\partial}{\partial \theta}\right|_p ,$$

and again $\mathrm{d}N_p(T_pS) \subseteq T_pS$ for all $p \in S$.

In all previous examples the differential of the Gauss map maps the tangent plane of the surface in itself; this is not a coincidence. By definition, $\mathrm{d}N_p$ maps T_pS in $T_{N(p)}S^2$. But, as already remarked (Example 3.3.13), the tangent plane to the sphere in a point is orthogonal to that point; so $T_{N(p)}S^2$ is orthogonal to $N(p)$, and thus it coincides with T_pS. Summing up, *we may consider the differential of the Gauss map at a point $p \in S$ as an endomorphism of T_pS*. And it is not just any endomorphism: it is symmetric. To prove this, we need a result from Differential Calculus (see [5, Theorem 3.3, p. 92]):

Theorem 4.4.14 (Schwarz). *Let $\Omega \subseteq \mathbb{R}^n$ be an open set and $f \in C^2(\Omega)$. Then*

$$\forall i, j = 1, \ldots, n \qquad \frac{\partial^2 f}{\partial x_i \partial x_j} \equiv \frac{\partial^2 f}{\partial x_j \partial x_i} .$$

Hence:

Proposition 4.4.15. *Let $S \subset \mathbb{R}^3$ be an oriented surface with Gauss map $N: S \to S^2$. Then $\mathrm{d}N_p$ is an endomorphism of T_pS, symmetric with respect to the scalar product $\langle \cdot, \cdot \rangle_p$ for all $p \in S$.*

Proof. Choose a local parametrization φ centered at p, and let $\{\partial_1, \partial_2\}$ be the basis of T_pS induced by φ. It suffices (why?) to prove that $\mathrm{d}N_p$ is symmetric on the basis, that is, that

$$\langle \mathrm{d}N_p(\partial_1), \partial_2 \rangle_p = \langle \partial_1, \mathrm{d}N_p(\partial_2) \rangle_p . \tag{4.7}$$

Now, by definition $\langle N \circ \varphi, \partial_2 \rangle \equiv 0$. Differentiating with respect to x_1 and recalling Remark 3.4.26 we get

$$0 = \frac{\partial}{\partial x_1}\langle N \circ \varphi, \partial_2 \rangle(O) = \left\langle \frac{\partial(N \circ \varphi)}{\partial x_1}(O), \frac{\partial \varphi}{\partial x_2}(O) \right\rangle + \left\langle N(p), \frac{\partial^2 \varphi}{\partial x_1 \partial x_2}(O) \right\rangle$$

$$= \langle \mathrm{d}N_p(\partial_1), \partial_2 \rangle_p + \left\langle N(p), \frac{\partial^2 \varphi}{\partial x_1 \partial x_2}(O) \right\rangle .$$

Analogously, by differentiating $\langle N \circ \varphi, \partial_1 \rangle \equiv 0$ with respect to x_2 we get

$$0 = \langle \mathrm{d}N_p(\partial_2), \partial_1 \rangle_p + \left\langle N(p), \frac{\partial^2 \varphi}{\partial x_1 \partial x_2}(O) \right\rangle ,$$

and (4.7) follows from Theorem 4.4.14. $\qquad \square$

We have a scalar product and a symmetric endomorphism; Linear Algebra suggests us to mix them together.

Definition 4.4.16. Let $S \subset \mathbb{R}^3$ be an oriented surface, with Gauss map $N: S \to S^2$. The *second fundamental form* of S is the quadratic form $Q_p: T_pS \to \mathbb{R}$ given by

$$\forall v \in T_pS \qquad Q_p(v) = -\langle dN_p(v), v \rangle_p .$$

Remark 4.4.17. The minus sign in the previous definition will be necessary for equation (4.8) to hold.

Remark 4.4.18. By changing the orientation of S the Gauss map changes sign, and so the second fundamental form changes sign too.

Example 4.4.19. Of course, the second fundamental form of a plane is zero everywhere.

Example 4.4.20. The second fundamental form of a cylinder oriented by the Gauss map given in Example 4.4.9 is $Q_p(v) = -v_1^2 - v_2^2$.

Example 4.4.21. The second fundamental form of the sphere oriented by the Gauss map of Example 4.4.8 is the opposite of the first fundamental form: $Q_p = -I_p$.

Example 4.4.22. Let $\Gamma_h \subset \mathbb{R}^3$ be the graph of a function $h: U \to \mathbb{R}$, where $U \subseteq \mathbb{R}^2$ is open, oriented by the Gauss map of Example 4.4.10. Recalling the Example 4.1.8 we find

$$Q_p(v) = -\langle dN_p(\partial_1), \partial_1 \rangle_p v_1^2 - 2\langle dN_p(\partial_1), \partial_2 \rangle_p v_1 v_2 - \langle dN_p(\partial_2), \partial_2 \rangle_p v_2^2$$
$$= \frac{1}{\sqrt{1 + \|\nabla h(x)\|^2}} \left[\frac{\partial^2 h}{\partial x_1^2}(x) v_1^2 + 2\frac{\partial^2 h}{\partial x_1 \partial x_2}(x) v_1 v_2 + \frac{\partial^2 h}{\partial x_2^2}(x) v_2^2 \right]$$

for all $p = (x, h(x)) \in \Gamma_h$ and all $v = v_1\partial_1 + v_2\partial_2 \in T_p\Gamma_h$. In other words, the matrix representing the second fundamental form with respect to the basis $\{\partial_1, \partial_2\}$ is $(1 + \|\nabla h\|^2)^{-1/2}\text{Hess}(h)$, where $\text{Hess}(h)$ is the Hessian matrix of h.

Example 4.4.23. Let $S \subset \mathbb{R}^3$ be a helicoid, oriented by the Gauss map of Example 4.4.11. Then, recalling Problem 3.2 and Example 4.1.11, we get

$$Q_p(v) = \frac{a}{(a^2 + y_0^2)^{1/2}} \left[F(x_0, y_0) \left(v_1^2 + \frac{v_2^2}{a^2 + y_0^2} \right) + 2G(x_0, y_0)v_1 v_2 \right]$$
$$= \frac{2a}{(a^2 + y_0^2)^{1/2}} v_1 v_2$$

for all $p = \varphi(x_0, y_0) \in S$ and $v = v_1\partial_1 + v_2\partial_2 \in T_pS$.

Example 4.4.24. Let $S \subset \mathbb{R}^3$ be a catenoid, oriented by the Gauss map of Example 4.4.12. Then

$$Q_p(w) = -\frac{E(x_0, y_0)}{a \cosh^2 x_0} w_1^2 + \frac{G(x_0, y_0)}{a \cosh^2 x_0} w_2^2 = -a w_1^2 + a w_2^2$$

for all $p = \psi(x_0, y_0) \in S$ and $w = w_1 \partial_1 + w_2 \partial_2 \in T_p S$.

Example 4.4.25. Let $S \subset \mathbb{R}^3$ be a surface of revolution, oriented by the Gauss map of Example 4.4.13. Then

$$Q_p(v) = \frac{\alpha' \beta'' - \alpha'' \beta'}{\sqrt{(\alpha')^2 + (\beta')^2}} v_1^2 + \frac{\alpha \beta'}{\sqrt{(\alpha')^2 + (\beta')^2}} v_2^2$$

for all $p = \big(\alpha(t) \cos \theta, \alpha(t) \sin \theta, \beta(t)\big) \in S$ and $v = v_1 \partial/\partial t + v_2 \partial/\partial \theta \in T_p S$.

The second fundamental form, just like the normal curvature, allows us to associate a number with each tangent versor to a surface; moreover, the second fundamental form, like the normal curvature, has to do with how much a surface curves. The second fundamental form, however, has an obvious advantage: as we have seen in the previous examples, it is very easy to compute starting from a local parametrization. This is important because we shall now show that *the normal curvature coincides with the second fundamental form*. To prove this, take an arbitrary curve $\sigma \colon (-\varepsilon, \varepsilon) \to S$ in S, parametrized by arc length, and set $\sigma(0) = p \in S$ and $\dot{\sigma}(0) = v \in T_p S$. Set $N(s) = N\big(\sigma(s)\big)$; clearly, $\langle \dot{\sigma}(s), N(s) \rangle \equiv 0$. By differentiating, we find

$$\langle \ddot{\sigma}(s), N(s) \rangle \equiv -\langle \dot{\sigma}(s), \dot{N}(s) \rangle \ .$$

But $\dot{N}(0) = \mathrm{d}N_p(v)$; so

$$Q_p(v) = -\langle \mathrm{d}N_p(v), \dot{\sigma}(0) \rangle = \langle \ddot{\sigma}(0), N(p) \rangle \ . \tag{4.8}$$

Moreover, if σ is biregular we have $\ddot{\sigma} = \kappa \mathbf{n}$, where κ is the curvature of σ, and $\mathbf{n}$ is the normal versor of σ, and so we have

$$Q_p(v) = \kappa(0) \langle \mathbf{n}(0), N(p) \rangle \ .$$

This formulas suggest the following:

Definition 4.4.26. Let $\sigma \colon I \to S$ be a curve parametrized by arc length contained in an oriented surface S. The *normal curvature* of σ is the function $\kappa_n \colon I \to \mathbb{R}$ given by

$$\kappa_n = \langle \ddot{\sigma}, N \circ \sigma \rangle = \kappa \langle \mathbf{n}, N \circ \sigma \rangle \ ,$$

where the second equality holds when σ is biregular. In other words, the normal curvature of σ is the (signed) length of the projection of the acceleration vector $\ddot{\sigma}$ along the direction orthogonal to the surface. Moreover, by (4.8) we know that

$$\kappa_n(s) = Q_{\sigma(s)}\big(\dot{\sigma}(s)\big) \ . \tag{4.9}$$

Remark 4.4.27. If the orientation of S is inverted, the normal curvature function changes sign.

If σ is the normal section of S at p along v, its normal versor at p is (why?) exactly $N(p)$, so the normal curvature of S at p along v is the normal curvature of σ at p. Hence we are at last able to prove that the second fundamental form gives the normal curvature of the surface:

Proposition 4.4.28 (Meusnier). *Let $S \subset \mathbb{R}^3$ be an oriented surface with Gauss map $N\colon S \to S^2$, and $p \in S$. Then:*

(i) *two curves in S passing through p tangent to the same direction have the same normal curvature at p;*

(ii) *the normal curvature of S at p along a vector $v \in T_pS$ of length 1 is given by $Q_p(v)$.*

Proof. (i) Indeed, if σ_1 and σ_2 are curves in S with $\sigma_1(0) = \sigma_2(0) = p$ and $\dot{\sigma}_1(0) = \dot{\sigma}_2(0) = v$ then (4.9) tells us that the normal curvature at 0 of both curves is given by $Q_p(v)$.

(ii) If σ is the normal section of S at p along v we have already remarked that $\ddot{\sigma}(0) = \tilde{\kappa}(0)N(p)$, where $\tilde{\kappa}$ is the oriented curvature of σ, and the assertion follows from (4.8). $\qquad\square$

4.5 Principal, Gaussian and mean curvatures

We have now proved that the normal curvatures of a surface are exactly the values of the second fundamental form on the tangent versors. This suggests that a more in-depth study of normal curvatures by using the properties of the differential of the Gauss map should be possible (and useful). As we shall see, the basic fact is that $\mathrm{d}N_p$ is a symmetric endomorphism, and so (by the spectral theorem) it is diagonalizable.

Definition 4.5.1. Let $S \subset \mathbb{R}^3$ be an oriented surface with Gauss map $N\colon S \to S^2$, and $p \in S$. A *principal direction* of S at p is an eigenvector of $\mathrm{d}N_p$ of length one, and the corresponding eigenvalue with the sign changed is a *principal curvature*.

If $v \in T_pS$ is a principal direction with principal curvature k, we have

$$Q_p(v) = -\langle \mathrm{d}N_p(v), v\rangle_p = -\langle -kv, v\rangle_p = k \ ,$$

and so the principal curvatures are normal curvatures. To be precise, they are the smallest and the largest normal curvatures at the point:

Proposition 4.5.2. *Let $S \subset \mathbb{R}^3$ be an oriented surface with Gauss map $N\colon S \to S^2$, and $p \in S$. Then we may find principal directions v_1, $v_2 \in T_pS$ with corresponding principal curvatures k_1, $k_2 \in \mathbb{R}$, with $k_1 \leq k_2$ and such that:*

(i) $\{v_1, v_2\}$ *is an orthonormal basis of* T_pS;
(ii) *given a versor* $v \in T_pS$, *let* $\theta \in (-\pi, \pi]$ *be a determination of the angle between* v_1 *and* v, *so that* $\cos\theta = \langle v_1, v \rangle_p$ *and* $\sin\theta = \langle v_2, v \rangle_p$. *Then*

$$Q_p(v) = k_1 \cos^2\theta + k_2 \sin^2\theta \qquad (4.10)$$

(Euler's formula);
(iii) k_1 *is the smallest normal curvature at* p, *and* k_2 *is the largest normal curvature at* p. *More precisely, the set of possible normal curvatures of* S *at* p *is the interval* $[k_1, k_2]$, *that is,*

$$\{Q_p(v) \mid v \in T_pS, \; I_p(v) = 1\} = [k_1, k_2] \,.$$

Proof. Since dN_p is a symmetric endomorphism of T_pS, the spectral theorem (see [1, Theorem 13.5.5, p. 100], or [21, Theorem 22.2, p. 311]) provides us with an orthonormal basis consisting of eigenvectors $\{v_1, v_2\}$ that satisfies (i).

Given $v \in T_pS$ of length one, we may write $v = \cos\theta\, v_1 + \sin\theta\, v_2$, and so we get

$$Q_p(v) = -\langle dN_p(v), v \rangle_p = \langle k_1 \cos\theta\, v_1 + k_2 \sin\theta\, v_2, \cos\theta\, v_1 + \sin\theta\, v_2 \rangle_p$$
$$= k_1 \cos^2\theta + k_2 \sin^2\theta \,.$$

Finally, if $k_1 = k_2$ then dN_p is a multiple of the identity, all normal curvatures are equal and (iii) is trivial. If, on the other hand, $k_1 < k_2$ then (4.10) tells us that

$$Q_p(v) = k_1 + (k_2 - k_1)\sin^2\theta \,.$$

So the normal curvature has a maximum (respectively, a minimum) for $\theta = \pm\pi/2$ (respectively, $\theta = 0, \pi$), that is, for $v = \pm v_2$ (respectively, $v = \pm v_1$), and this maximum (respectively, minimum) is exactly k_2 (respectively, k_1). Moreover, for $\theta \in (-\pi, \pi]$ the normal curvature takes all possible values between k_1 and k_2, and (iii) is proved. $\qquad\qquad\square$

When you learned about linear endomorphisms you certainly saw that two fundamental quantities for describing their behavior are the trace (given by the sum of the eigenvalues) and the determinant (given by the product of the eigenvalues). You will then not be surprised in learning that the trace and (even more so) the determinant of dN_p are going to play a crucial role when studying surfaces.

Definition 4.5.3. Let $S \subset \mathbb{R}^3$ be an oriented surface with Gauss map $N: S \to S^2$. The *Gaussian curvature* of S is the function $K: S \to \mathbb{R}$ given by

$$\forall p \in S \qquad\qquad K(p) = \det(dN_p) \,,$$

while the *mean curvature* of S is the function $H: S \to \mathbb{R}$ given by

$$\forall p \in S \qquad\qquad H(p) = -\frac{1}{2}\mathrm{tr}(dN_p) \,.$$

Remark 4.5.4. If k_1 and k_2 are the principal curvatures of S at $p \in S$, then $K(p) = k_1 k_2$ and $H(p) = (k_1 + k_2)/2$.

Remark 4.5.5. If we change the orientation on S the Gauss map N changes sign, and so both the principal curvatures and the mean curvature change sign; the Gaussian curvature K, on the other hand, does not change. So we may define the Gaussian curvature for non-orientable surfaces too: if p is a point of an arbitrary surface S, the Gaussian curvature of S at p is the Gaussian curvature at p of the image of an arbitrary local parametrization of S centered at p (remember Remark 4.4.6 too). Analogously, the absolute value of the mean curvature is well defined on non-orientable surfaces too.

Remark 4.5.6. The Gaussian curvature admits an interesting interpretation in terms of ratios of areas. Let $\varphi: U \to \mathbb{R}^3$ be a local parametrization of a surface $S \subset \mathbb{R}^3$ centered at $p \in S$, and denote by $B_\delta \subset \mathbb{R}^2$ the open disk with center in the origin and radius $\delta > 0$. Then if $K(p) \neq 0$ we have

$$|K(p)| = \lim_{\delta \to 0} \frac{\mathrm{Area}\big(N \circ \varphi(B_\delta)\big)}{\mathrm{Area}\big(\varphi(B_\delta)\big)} \, ;$$

see Fig 4.5.

To prove this, note first that $K(p) = \det dN_p \neq 0$ implies that $N \circ \varphi|_{B_\delta}$ is a local parametrization of the sphere for $\delta > 0$ small enough. Then Theorem 4.2.6 and Lemma 4.2.7 imply

$$\mathrm{Area}\big(N \circ \varphi(B_\delta)\big) = \int_{B_\delta} \left\| \frac{\partial(N \circ \varphi)}{\partial x_1} \wedge \frac{\partial(N \circ \varphi)}{\partial x_2} \right\| \, dx_1 \, dx_2$$

$$= \int_{B_\delta} |K| \, \|\partial_1 \wedge \partial_2\| \, dx_1 \, dx_2 \,,$$

and

$$\mathrm{Area}\big(\varphi(B_\delta)\big) = \int_{B_\delta} \|\partial_1 \wedge \partial_2\| \, dx_1 \, dx_2 \,.$$

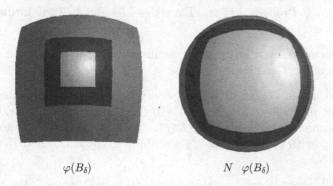

$\varphi(B_\delta)$ $N \; \varphi(B_\delta)$

Fig. 4.5.

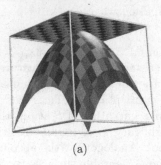 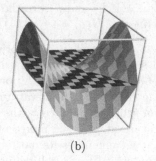

(a)   (b)

Fig. 4.6.

Hence,

$$\lim_{\delta \to 0} \frac{\text{Area}\big(N \circ \varphi(B_\delta)\big)}{\text{Area}\big(\varphi(B_\delta)\big)} = \frac{\lim\limits_{\delta \to 0}(\pi\delta^2)^{-1}\int_{B_\delta}|K|\,\|\partial_1 \wedge \partial_2\|\,\mathrm{d}x_1\,\mathrm{d}x_2}{\lim\limits_{\delta \to 0}(\pi\delta^2)^{-1}\int_{B_\delta}\|\partial_1 \wedge \partial_2\|\,\mathrm{d}x_1\,\mathrm{d}x_2}$$

$$= \frac{K(p)\|\partial_1|_p \wedge \partial_2|_p\|}{\|\partial_1|_p \wedge \partial_2|_p\|} = K(p)\,,$$

using the mean value theorem for multiple integrals (see [5, Problem 6, p. 190] for a sketch of the proof).

Remark 4.5.7. The sign of the Gaussian curvature may give an idea of how a surface looks like. If $p \in S$ is a point with $K(p) > 0$, all normal curvatures at p have the same sign. Intuitively, this means that all normal sections of S at p curve on the same side (why?) with respect to T_pS, and so in proximity of p the surface lies all on a single side of the tangent plane: see Fig. 4.6.(a). On the other hand, if $K(p) < 0$, we have normal curvatures of both signs at p; this means that the normal sections may curve on opposite sides with respect to T_pS, and so in a neighborhood of p the surface has sections on both sides of the tangent plane: see Fig. 4.6.(b). Nothing can be said a priori when $K(p) = 0$. Problem 4.17 and Exercises 4.54 and 4.49 will formalize these intuitive ideas.

The previous remark suggests a classification of the points of S according to the sign of the Gaussian curvature.

Definition 4.5.8. Let $S \subset \mathbb{R}^3$ be an oriented surface with Gauss map $N\colon S \to S^2$. A point $p \in S$ is *elliptic* if $K(p) > 0$ (and so all normal curvatures at p have the same sign); *hyperbolic* if $K(p) < 0$ (and so there are normal curvatures at p with opposite signs); *parabolic* if $K(p) = 0$ but $\mathrm{d}N_p \neq O$; and *planar* if $\mathrm{d}N_p = O$.

We shall systematically use this classification in Chapter 7; the rest of this section will be devoted to finding an effective procedure for computing the various kind of curvatures (principal, Gaussian and mean) we introduced. Let us begin by studying how to express the second fundamental form in local coordinates.

Fix a local parametrization $\varphi\colon U \to S$ at $p \in S$ of an oriented surface $S \subset \mathbb{R}^3$ with Gauss map $N\colon S \to S^2$. If $v = v_1\partial_1 + v_2\partial_2 \in T_pS$, then

$$Q_p(v) = Q_p(\partial_1)v_1^2 - 2\langle dN_p(\partial_1), \partial_2\rangle_p v_1 v_2 + Q_p(\partial_2)v_2^2 \,. \tag{4.11}$$

So it is natural to give the following:

Definition 4.5.9. Let $\varphi\colon U \to S$ be a local parametrization of a surface S. The *form coefficients* of S with respect to φ are the three functions e, f, and $g\colon U \to \mathbb{R}$ defined by

$$
\begin{aligned}
e(x) &= Q_{\varphi(x)}(\partial_1) = -\langle dN_{\varphi(x)}(\partial_1), \partial_1\rangle_{\varphi(x)} \,, \\
f(x) &= -\langle dN_{\varphi(x)}(\partial_1), \partial_2\rangle_{\varphi(x)} \,, \\
g(x) &= Q_{\varphi(x)}(\partial_2) = -\langle dN_{\varphi(x)}(\partial_2), \partial_2\rangle_{\varphi(x)}
\end{aligned}
\tag{4.12}
$$

for all $x \in U$, where $N = \partial_1 \wedge \partial_2/\|\partial_1 \wedge \partial_2\|$, as usual.

Remark 4.5.10. Again, this is Gauss' notation. We shall sometimes also use the more modern notation $e = h_{11}$, $f = h_{12} = h_{21}$, and $g = h_{22}$.

Remark 4.5.11. By differentiating the identities $\langle N \circ \varphi, \partial_j\rangle \equiv 0$ for $j = 1, 2$, it is straightforward to get the following expressions for form coefficients:

$$e = \left\langle N \circ \varphi, \frac{\partial^2\varphi}{\partial x_1^2}\right\rangle \,, \quad f = \left\langle N \circ \varphi, \frac{\partial^2\varphi}{\partial x_1\partial x_2}\right\rangle \,, \quad g = \left\langle N \circ \varphi, \frac{\partial^2\varphi}{\partial x_2^2}\right\rangle \,. \tag{4.13}$$

Remark 4.5.12. We have introduced e, f and g as functions defined on U. However, it will sometimes be more convenient to consider them as functions defined on $\varphi(U)$, that is, to replace them with $e \circ \varphi^{-1}$, $f \circ \varphi^{-1}$ and $g \circ \varphi^{-1}$, respectively. Finally, form coefficients also significantly depend on the local parametrization we have chosen, as it is easy to verify (see Example 4.5.18).

Remark 4.5.13. Metric and form coefficients depend on the chosen local parametrization, whereas the Gaussian curvature and the absolute value of the mean curvature do not, since they are defined directly from the Gauss map, without using local parametrizations.

Clearly, the form coefficients are (why?) functions of class C^∞ on U that completely determine the second fundamental form: indeed, from (4.11) we get

$$Q_p(v_1\partial_1 + v_2\partial_2) = e(x)v_1^2 + 2f(x)v_1v_2 + g(x)v_2^2$$

for all $p = \varphi(x) \in \varphi(U)$ and $v_1\partial_1 + v_2\partial_2 \in T_pS$.

Furthermore, (4.13) can be used to explicitly compute the form coefficients (as we shall momentarily verify on our usual examples). So, to get an effective way for computing principal, Gaussian and mean curvatures it will suffice to express them in terms of metric and form coefficients. Remember that principal, Gaussian and mean curvatures are defined from the eigenvalues of dN_p; so it may be helpful to try and write the matrix $A \in M_{2,2}(\mathbb{R})$ representing dN_p with respect to the basis $\{\partial_1, \partial_2\}$ using the functions E, F, G, e, f and g. Now, for all $v = v_1\partial_1 + v_2\partial_2$, $w = w_1\partial_1 + w_2\partial_2 \in T_pS$, we have

$$|w_1 \quad w_2| \begin{vmatrix} e & f \\ f & g \end{vmatrix} \begin{vmatrix} v_1 \\ v_2 \end{vmatrix} = -\langle dN_p(v), w \rangle_p = -|w_1 \quad w_2| \begin{vmatrix} E & F \\ F & G \end{vmatrix} A \begin{vmatrix} v_1 \\ v_2 \end{vmatrix} ;$$

from this it follows (why?) that

$$\begin{vmatrix} e & f \\ f & g \end{vmatrix} = - \begin{vmatrix} E & F \\ F & G \end{vmatrix} A .$$

Now, $\begin{vmatrix} E & F \\ F & G \end{vmatrix}$ is the matrix that represents a positive definite scalar product with respect to a basis; in particular, it is invertible and has positive determinant $EG - F^2$. So we have proved the following:

Proposition 4.5.14. *Let $\varphi : U \to S$ be a local parametrization of a surface $S \subset \mathbb{R}^3$, and set $N = \partial_1 \wedge \partial_2 / \|\partial_1 \wedge \partial_2\|$. Then the matrix $A \in M_{2,2}(\mathbb{R})$ representing the endomorphism dN with respect to the basis $\{\partial_1, \partial_2\}$ is given by*

$$A = \begin{vmatrix} a_{11} & a_{12} \\ a_{21} & a_{22} \end{vmatrix} = - \begin{vmatrix} E & F \\ F & G \end{vmatrix}^{-1} \begin{vmatrix} e & f \\ f & g \end{vmatrix}$$

$$= -\frac{1}{EG - F^2} \begin{vmatrix} eG - fF & fG - gF \\ fE - eF & gE - fF \end{vmatrix} . \qquad (4.14)$$

In particular, the Gaussian curvature is given by

$$K = \det(A) = \frac{eg - f^2}{EG - F^2} , \qquad (4.15)$$

the mean curvature is given by

$$H = -\frac{1}{2}\mathrm{tr}(A) = \frac{1}{2}\frac{eG - 2fF + gE}{EG - F^2} , \qquad (4.16)$$

and the principal curvatures by

$$k_{1,2} = H \pm \sqrt{H^2 - K} . \qquad (4.17)$$

Remark 4.5.15. If $\varphi : U \to S$ is a local parametrization with $F \equiv f \equiv 0$ the previous formulas become simpler:

$$K = \frac{eg}{EG} , \quad H = \frac{1}{2}\left(\frac{e}{E} + \frac{g}{G}\right) , \quad k_1 = \frac{e}{E} , \quad k_2 = \frac{g}{G} .$$

We may now compute the various curvatures for our usual examples.

Example 4.5.16. In the plane we have $e \equiv f \equiv g \equiv 0$, no matter which parametrization we are using, since the second fundamental form is zero everywhere. In particular, the principal, Gaussian and mean curvatures are all zero everywhere.

Example 4.5.17. For the right circular cylinder with the parametrization of Example 4.1.9 we have $e \equiv -1$ and $f \equiv g \equiv 0$, so $K \equiv 0$, $H \equiv -1/2$, $k_1 \equiv -1$, and $k_2 \equiv 0$.

Example 4.5.18. We have seen in Example 4.4.21 that on the sphere oriented as in Example 4.4.8 we have $Q_p = -I_p$. This means that for any parametrization the form coefficients have the same absolute value and opposite sign as the corresponding metric coefficients. In particular, $K \equiv 1$, $H \equiv -1$ and $k_1 \equiv k_2 \equiv -1$.

Example 4.5.19. Let $U \subseteq \mathbb{R}^2$ be an open set, $h \in C^\infty(U)$, and $\varphi \colon U \to \mathbb{R}^3$ the local parametrization of the graph Γ_h given by $\varphi(x) = \big(x, h(x)\big)$. Recalling Examples 4.1.8 and 4.4.10 we get

$$e = \frac{1}{\sqrt{1 + \|\nabla h\|^2}} \frac{\partial^2 h}{\partial x_1^2} \,, \quad f = \frac{1}{\sqrt{1 + \|\nabla h\|^2}} \frac{\partial^2 h}{\partial x_1 \partial x_2} \,, \quad g = \frac{1}{\sqrt{1 + \|\nabla h\|^2}} \frac{\partial^2 h}{\partial x_2^2} \,,$$

hence,

$$K = \frac{1}{(1 + \|\nabla h\|^2)^2} \det \mathrm{Hess}(h) \,,$$

$$H = \frac{1}{2(1 + \|\nabla h\|^2)^{3/2}} \left[\frac{\partial^2 h}{\partial x_1^2} \left(1 + \left| \frac{\partial h}{\partial x_2} \right|^2 \right) + \frac{\partial^2 h}{\partial x_2^2} \left(1 + \left| \frac{\partial h}{\partial x_1} \right|^2 \right) \right.$$
$$\left. -2 \frac{\partial^2 h}{\partial x_1 \partial x_2} \frac{\partial h}{\partial x_1} \frac{\partial h}{\partial x_2} \right] \,.$$

Example 4.5.20. For a helicoid parametrized as in Example 4.1.11, we find $f = a/\sqrt{a^2 + y^2}$ and $e \equiv g \equiv 0$, so

$$K = -\frac{a^2}{(a^2 + y^2)^2} \,, \qquad H \equiv 0 \,, \qquad k_{1,2} = \pm \frac{a}{a^2 + y^2} \,.$$

Example 4.5.21. For a catenoid parametrized as in Example 4.1.12, we find $e \equiv -a$, $f \equiv 0$ and $g \equiv a$, so

$$K = -\frac{1}{a^2 \cosh^4 x} \,, \qquad H \equiv 0 \,, \qquad k_{1,2} = \pm \frac{1}{a \cosh^2 x} \,.$$

Example 4.5.22. Let S be a surface of revolution, parametrized as in Example 4.1.13. The form coefficients are then given by

$$e = \frac{\alpha'\beta'' - \beta'\alpha''}{\sqrt{(\alpha')^2 + (\beta')^2}}, \quad f \equiv 0, \quad g = \frac{\alpha\beta'}{\sqrt{(\alpha')^2 + (\beta')^2}} \, .$$

Recalling Remark 4.5.15, we get

$$K = \frac{\beta'(\alpha'\beta'' - \beta'\alpha'')}{\alpha\big((\alpha')^2 + (\beta')^2\big)^2}, \quad H = \frac{\alpha(\alpha'\beta'' - \beta'\alpha'') + \beta'\big((\alpha')^2 + (\beta')^2\big)}{2\alpha\big((\alpha')^2 + (\beta')^2\big)^{3/2}},$$

$$k_1 = \frac{\alpha'\beta'' - \beta'\alpha''}{\big((\alpha')^2 + (\beta')^2\big)^{3/2}}, \quad k_2 = \frac{\beta'}{\alpha\big((\alpha')^2 + (\beta')^2\big)^{1/2}} \, .$$

If the generatrix of S is parametrized by arc length, these formulas become quite simpler: by differentiating $\dot{\alpha}^2 + \dot{\beta}^2 \equiv 1$ we get $\dot{\alpha}\ddot{\alpha} + \dot{\beta}\ddot{\beta} \equiv 0$, and so

$$K = -\frac{\ddot{\alpha}}{\alpha}, \quad H = \frac{\dot{\beta} + \alpha(\dot{\alpha}\ddot{\beta} - \dot{\beta}\ddot{\alpha})}{2\alpha}, \quad k_1 = \dot{\alpha}\ddot{\beta} - \dot{\beta}\ddot{\alpha}, \quad k_2 = \frac{\dot{\beta}}{\alpha} \, .$$

Example 4.5.23. Let $\sigma \colon (\pi/2, \pi) \to \mathbb{R}^3$ be the upper half of the tractrix given by

$$\sigma(t) = \left(\sin t, 0, \cos t + \log \tan \frac{t}{2} \right) ;$$

see Problem 1.3. The surface of revolution S obtained by rotating the tractrix around the z-axis is called *pseudosphere*; see Fig. 4.7 (and Exercise 3.18). By using the previous example, it is easy to find (see also Problem 4.8) that the pseudosphere has constant Gaussian curvature equal to -1.

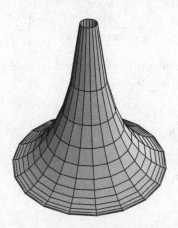

Fig. 4.7. The pseudosphere

Remark 4.5.24. The plane is an example of surface with constant Gaussian curvature equal to zero, and spheres are examples of surfaces with positive constant Gaussian curvature (Exercise 4.28). Other examples of surfaces with zero constant Gaussian curvature are cylinders (Exercise 4.27). The pseudo-sphere, on the other hand, is an example of a surface with negative constant Gaussian curvature but, unlike planes, cylinders, and spheres, it is not a closed surface in $\mathbb{R}^3$. This is not a coincidence: in Chapter 7 we shall prove that closed surfaces in $\mathbb{R}^3$ with negative constant Gaussian curvature *do not exist* (Theorem 7.3.6). Moreover, we shall also prove that spheres are the only closed surfaces with positive constant Gaussian curvature (Theorem 7.1.2 and Remark 7.1.4), and that planes and cylinder are the only closed surfaces with zero constant Gaussian curvature (Theorem 7.2.6).

4.6 Gauss' Theorema egregium

The goal of this section is to prove that the Gaussian curvature is an intrinsic property of a surface: it only depends on the first fundamental form, and not on the way the surface is immersed in $\mathbb{R}^3$. As you can imagine, it is an highly unexpected result; the definition of K directly involves the Gauss map, which is very strongly related to the embedding of the surface in $\mathbb{R}^3$. Nevertheless, the Gaussian curvature can be measured staying within the surface, forgetting the ambient space. In particular, two isometric surfaces have the same Gaussian curvature; and this will give us a necessary condition a surface has to satisfy for the existence of a similitude with an open subset of the plane.

The road to get to this result is almost as important as the result itself. The idea is to proceed as we did to get Frenet-Serret formulas for curves. The Frenet frame allows us to associate with each point of the curve a basis of $\mathbb{R}^3$; hence it is possible to express the derivatives of the Frenet frame as a linear combination of the frame itself, and the coefficients turn out to be fundamental geometric quantities for studying the curve.

Let us see how to adapt such an argument to surfaces. Let $\varphi\colon U \to S$ be a local parametrization of a surface $S \subset \mathbb{R}^3$, and let $N\colon \varphi(U) \to S^2$ be the Gauss map of $\varphi(U)$ given by $N = \partial_1 \wedge \partial_2/\|\partial_1 \wedge \partial_2\|$, as usual. The triple $\{\partial_1, \partial_2, N\}$ is a basis of $\mathbb{R}^3$ everywhere, and so we may express any vector of $\mathbb{R}^3$ as a linear combination of these vectors. In particular, there must exist functions $\Gamma_{ij}^h, h_{ij}, a_{ij} \in C^\infty(U)$ such that

$$\frac{\partial^2 \varphi}{\partial x_i \partial x_j} = \Gamma_{ij}^1 \partial_1 + \Gamma_{ij}^2 \partial_2 + h_{ij} N \,, \tag{4.18}$$

$$\frac{\partial(N \circ \varphi)}{\partial x_j} = a_{1j}\partial_1 + a_{2j}\partial_2 \,, \tag{4.19}$$

for $i, j = 1, 2$, where in last formula there are no terms proportional to N because $\|N\| \equiv 1$ implies that all partial derivatives of $N \circ \varphi$ are orthogonal to N. Note further that, by Theorem 4.4.14, the terms Γ_{ij}^r and h_{ij} are symmetric with respect to their lower indices, that is, $\Gamma_{ji}^r = \Gamma_{ij}^r$ and $h_{ji} = h_{ij}$ for all i, j, $r = 1, 2$.

We already know some of the functions appearing in (4.19). For instance, since $\partial(N \circ \varphi)/\partial x_i = \mathrm{d}N_p(\partial_i)$, the terms a_{ij} are just the components of the matrix A that represents $\mathrm{d}N_p$ with respect to the basis $\{\partial_1, \partial_2\}$, and so they are given by (4.14). The terms h_{ij} are known too: by (4.13) we know that they are exactly the form coefficients (thus the notation is consistent with Remark 4.5.10). So the only quantities that are still unknown are the coefficients Γ_{ij}^r.

Definition 4.6.1. The functions Γ_{ij}^r are the *Christoffel symbols* of the local parametrization φ.

We proceed now to compute Christoffel symbols. Taking the scalar product of (4.18) with ∂_1 and ∂_2 ($i = j = 1$) yields

$$\begin{cases} E\Gamma_{11}^1 + F\Gamma_{11}^2 = \left\langle \dfrac{\partial^2 \varphi}{\partial x_1^2}, \partial_1 \right\rangle = \dfrac{1}{2} \dfrac{\partial}{\partial x_1} \langle \partial_1, \partial_1 \rangle = \dfrac{1}{2} \dfrac{\partial E}{\partial x_1}, \\[2mm] F\Gamma_{11}^1 + G\Gamma_{11}^2 = \left\langle \dfrac{\partial^2 \varphi}{\partial x_1^2}, \partial_2 \right\rangle = \dfrac{\partial}{\partial x_1} \langle \partial_1, \partial_2 \rangle - \left\langle \partial_1, \dfrac{\partial^2 \varphi}{\partial x_1 \partial x_2} \right\rangle = \dfrac{\partial F}{\partial x_1} - \dfrac{1}{2} \dfrac{\partial E}{\partial x_2}. \end{cases}$$
$$(4.20)$$

Analogously, we find

$$\begin{cases} E\Gamma_{12}^1 + F\Gamma_{12}^2 = \dfrac{1}{2} \dfrac{\partial E}{\partial x_2}, \\[2mm] F\Gamma_{12}^1 + G\Gamma_{12}^2 = \dfrac{1}{2} \dfrac{\partial G}{\partial x_1}, \end{cases} \qquad (4.21)$$

and

$$\begin{cases} E\Gamma_{22}^1 + F\Gamma_{22}^2 = \dfrac{\partial F}{\partial x_2} - \dfrac{1}{2} \dfrac{\partial G}{\partial x_1}, \\[2mm] F\Gamma_{22}^1 + G\Gamma_{22}^2 = \dfrac{1}{2} \dfrac{\partial G}{\partial x_2}. \end{cases} \qquad (4.22)$$

These are three square linear systems whose matrix of coefficients has determinant $EG - F^2$, which is always positive; so they have a unique solution, and it can be expressed in terms of metric coefficients and of their derivatives (see Exercise 4.58).

Remark 4.6.2. Note that, in particular, *the Christoffel symbols only depend on the first fundamental form of S*, and so they are intrinsic. As a consequence, *any quantity that can be written in terms of Christoffel symbols is intrinsic*: it only depends on the metric structure of the surface, and not on the way the surface is immersed in $\mathbb{R}^3$.

Remark 4.6.3. We explicitly remark, since it will be useful later on, that if the local parametrization is orthogonal (that is, if $F \equiv 0$) the Christoffel symbols have a particularly simple expression:

$$\begin{cases} \Gamma_{11}^1 = \dfrac{1}{2E}\dfrac{\partial E}{\partial x_1}, & \Gamma_{12}^1 = \Gamma_{21}^1 = \dfrac{1}{2E}\dfrac{\partial E}{\partial x_2}, & \Gamma_{22}^1 = -\dfrac{1}{2E}\dfrac{\partial G}{\partial x_1}, \\[2mm] \Gamma_{11}^2 = -\dfrac{1}{2G}\dfrac{\partial E}{\partial x_2}, & \Gamma_{12}^2 = \Gamma_{21}^2 = \dfrac{1}{2G}\dfrac{\partial G}{\partial x_1}, & \Gamma_{22}^2 = \dfrac{1}{2G}\dfrac{\partial G}{\partial x_2}. \end{cases} \quad (4.23)$$

Let us see now the value of the Christoffel symbols in our canonical examples.

Example 4.6.4. By Example 4.1.7, we know that the Christoffel symbols of the plane are zero everywhere.

Example 4.6.5. The Christoffel symbols of the right circular cylinder parametrized as in Example 4.1.9 are identically zero too.

Example 4.6.6. The Christoffel symbols of the local parametrization of the sphere $\varphi(x,y) = (x, y, \sqrt{1 - x^2 - y^2})$ are

$$\begin{cases} \Gamma_{11}^1 = \dfrac{x}{1 - x^2 - y^2}, & \Gamma_{12}^1 = \Gamma_{21}^1 = \dfrac{x^2 y}{1 - x^2 - y^2}, \\[2mm] \Gamma_{22}^1 = \dfrac{x(1 - x^2)}{1 - x^2 - y^2}, & \Gamma_{11}^2 = \dfrac{y(1 - y^2)}{1 - x^2 - y^2}, \\[2mm] \Gamma_{12}^2 = \Gamma_{21}^2 = \dfrac{x y^2}{1 - x^2 - y^2}, & \Gamma_{22}^2 = \dfrac{y(1 - x^2)}{1 - x^2 - y^2}. \end{cases}$$

On the other hand, the Christoffel symbols of the other local parametrization of the sphere $\psi(\theta, \phi) = (\sin\theta\cos\phi, \sin\theta\sin\phi, \cos\theta)$ given in Example 4.1.10 are

$$\begin{cases} \Gamma_{11}^1 \equiv 0, & \Gamma_{12}^1 = \Gamma_{21}^1 \equiv 0, & \Gamma_{22}^1 = -\sin\theta\cos\theta, \\[2mm] \Gamma_{11}^2 \equiv 0, & \Gamma_{12}^2 = \Gamma_{21}^2 = \dfrac{\cos\theta}{\sin\theta}, & \Gamma_{22}^2 \equiv 0. \end{cases}$$

Example 4.6.7. Let $U \subseteq \mathbb{R}^2$ be an open set, $h \in C^\infty(U)$, and $\varphi : U \to \mathbb{R}^3$ the local parametrization of the graph Γ_h given by $\varphi(x) = (x, h(x))$. Recalling Example 4.1.8 we get

$$\begin{cases} \Gamma_{11}^1 = \dfrac{(\partial h/\partial x_1)(\partial^2 h/\partial x_1^2)}{1 + \|\nabla h\|^2}, & \Gamma_{12}^1 = \Gamma_{21}^1 = \dfrac{(\partial h/\partial x_1)(\partial^2 h/\partial x_1 \partial x_2)}{1 + \|\nabla h\|^2}, \\[2mm] \Gamma_{22}^1 = \dfrac{(\partial h/\partial x_1)(\partial^2 h/\partial x_2^2)}{1 + \|\nabla h\|^2}, & \Gamma_{11}^2 = \dfrac{(\partial h/\partial x_2)(\partial^2 h/\partial x_1^2)}{1 + \|\nabla h\|^2}, \\[2mm] \Gamma_{12}^2 = \Gamma_{21}^2 = \dfrac{(\partial h/\partial x_2)(\partial^2 h/\partial x_1 \partial x_2)}{1 + \|\nabla h\|^2}, & \Gamma_{22}^2 = \dfrac{(\partial h/\partial x_2)(\partial^2 h/\partial x_2^2)}{1 + \|\nabla h\|^2}. \end{cases}$$

Example 4.6.8. The Christoffel symbols of the helicoid parametrized as in Example 4.1.11 are

$$\begin{cases} \Gamma_{11}^1 \equiv 0, & \Gamma_{12}^1 = \Gamma_{21}^1 = \dfrac{y}{a^2 + y^2}, & \Gamma_{22}^1 \equiv 0, \\[2mm] \Gamma_{11}^2 = -y, & \Gamma_{12}^2 = \Gamma_{21}^2 \equiv 0, & \Gamma_{22}^2 \equiv 0. \end{cases}$$

Example 4.6.9. The Christoffel symbols of the catenoid parametrized as in Example 4.1.12 are

$$\begin{cases} \Gamma_{11}^1 = \dfrac{\sinh x}{\cosh x}\,, & \Gamma_{12}^1 = \Gamma_{21}^1 \equiv 0\,, & \Gamma_{22}^1 = -\dfrac{\sinh x}{\cosh x}\,, \\ \Gamma_{11}^2 \equiv 0\,, & \Gamma_{12}^2 = \Gamma_{21}^2 = \dfrac{\sinh x}{\cosh x}\,, & \Gamma_{22}^2 \equiv 0\,. \end{cases}$$

Example 4.6.10. We conclude with the Christoffel symbols of a surface of revolution parametrized as in Example 4.1.13:

$$\begin{cases} \Gamma_{11}^1 = \dfrac{\alpha'\alpha'' + \beta'\beta''}{(\alpha')^2 + (\beta')^2}\,, & \Gamma_{12}^1 = \Gamma_{21}^1 \equiv 0\,, & \Gamma_{22}^1 = -\dfrac{\alpha\alpha'}{(\alpha')^2 + (\beta')^2}\,, \\ \Gamma_{11}^2 \equiv 0\,, & \Gamma_{12}^2 = \Gamma_{21}^2 = \dfrac{\alpha'}{\alpha}\,, & \Gamma_{22}^2 \equiv 0\,. \end{cases} \tag{4.24}$$

Now, unlike what happened for curvature and torsion, the Christoffel symbols cannot be chosen arbitrarily; they must satisfy some compatibility conditions. To find them, let us compute the third derivatives of the parametrization.

As for the second derivatives, there exist functions A_{ijk}^r, $B_{ijk} \in C^\infty(U)$ such that

$$\frac{\partial^3 \varphi}{\partial x_i \partial x_j \partial x_k} = A_{ijk}^1 \partial_1 + A_{ijk}^2 \partial_2 + B_{ijk} N\,.$$

Again by Theorem 4.4.14 we are sure that the functions A_{ijk}^r and B_{ijk} are symmetric in the lower indices. In particular,

$$A_{ijk}^r = A_{jik}^r = A_{ikj}^r \qquad \text{and} \qquad B_{ijk} = B_{jik} = B_{ikj} \tag{4.25}$$

for all i, j, k, $r = 1, 2$.

To compute the expression of A_{ijk}^r and B_{ijk}, we differentiate (4.18) and then insert (4.18) and (4.19) in what we find. We get

$$A_{ijk}^r = \frac{\partial \Gamma_{jk}^r}{\partial x_i} + \Gamma_{jk}^1 \Gamma_{i1}^r + \Gamma_{jk}^2 \Gamma_{i2}^r + h_{jk} a_{ri}\,,$$

$$B_{ijk} = \Gamma_{jk}^1 h_{i1} + \Gamma_{jk}^2 h_{i2} + \frac{\partial h_{jk}}{\partial x_i}\,.$$

Recalling that $A_{ijk}^r - A_{jik}^r = 0$, we find, for all i, j, k, $r = 1, 2$ the fundamental *Gauss' equations:*

$$\frac{\partial \Gamma_{jk}^r}{\partial x_i} - \frac{\partial \Gamma_{ik}^r}{\partial x_j} + \sum_{s=1}^{2}\left(\Gamma_{jk}^s \Gamma_{is}^r - \Gamma_{ik}^s \Gamma_{js}^r\right) = -(h_{jk} a_{ri} - h_{ik} a_{rj})\,. \tag{4.26}$$

Before examining what can be deduced from the symmetry of B_{ijk}, note an important consequence of Gauss' equations. If we write (4.26) for $i = r = 1$

and $j = k = 2$ (see Exercise 4.59 for the other cases), we get

$$\frac{\partial \Gamma^1_{22}}{\partial x_1} - \frac{\partial \Gamma^1_{12}}{\partial x_2} + \sum_{s=1}^{2}(\Gamma^s_{22}\Gamma^1_{1s} - \Gamma^s_{12}\Gamma^1_{2s}) = -(h_{22}a_{11} - h_{12}a_{12})$$

$$= \frac{(eg - f^2)G}{EG - F^2} = GK \ .$$

Since, as already remarked, the Christoffel symbols only depend on the first fundamental form, we have proved the very famous *Gauss' Theorema Egregium*:

Theorem 4.6.11 (Gauss' Theorema Egregium). *The Gaussian curvature K of a surface is given by the formula*

$$K = \frac{1}{G}\left[\frac{\partial \Gamma^1_{22}}{\partial x_1} - \frac{\partial \Gamma^1_{12}}{\partial x_2} + \sum_{s=1}^{2}(\Gamma^s_{22}\Gamma^1_{1s} - \Gamma^s_{12}\Gamma^1_{2s})\right] \ . \qquad (4.27)$$

In particular, the Gaussian curvature of a surface is an intrinsic property, that is, it only depends on the first fundamental form.

As a consequence, two locally isometric surfaces must have the same Gaussian curvature:

Corollary 4.6.12. *Let $F\colon S \to \tilde{S}$ be a local isometry between two surfaces. Then $\tilde{K} \circ F = K$, where K is the Gaussian curvature of S and $\tilde{K}$ is the Gaussian curvature of $\tilde{S}$. More generally, if F is a similitude with scale factor $r > 0$ then $\tilde{K} \circ F = r^{-2}K$.*

Proof. It immediately follows from Theorem 4.6.11, Proposition 4.1.20, the definition of similitude, and Exercise 4.11. $\qquad \square$

Remark 4.6.13. Warning: there exist maps $F\colon S \to \tilde{S}$ that satisfy $\tilde{K} \circ F = K$ but are not local isometries; see Exercise 4.42. In the supplementary material of the next chapter we shall discuss necessary and sufficient conditions to determine when a map is a local isometry (Proposition 5.5.1 and Corollary 5.5.6).

As a consequence of Corollary 4.6.12, *if a surface S is locally isometric to (or, more in general, has a similitude with) a portion of a plane, then the Gaussian curvature of S is zero everywhere.* Hence, there is no local isometry between a portion of a sphere and a portion of a plane, because the sphere has Gaussian curvature positive everywhere while the plane has zero Gaussian curvature: anguished cartographers have to accept that it is not possible to draw a geographical map that preserves distances, not even scaled by some factor.

One last consequence of Theorem 4.6.11 is another explicit formula for computing the Gaussian curvature:

Lemma 4.6.14. *Let* $\varphi\colon U \to S$ *be an orthogonal local parametrization of a surface* S. *Then*

$$K = -\frac{1}{2\sqrt{EG}}\left\{\frac{\partial}{\partial x_2}\left(\frac{1}{\sqrt{EG}}\frac{\partial E}{\partial x_2}\right) + \frac{\partial}{\partial x_1}\left(\frac{1}{\sqrt{EG}}\frac{\partial G}{\partial x_1}\right)\right\}.$$

Proof. If we substitute (4.23) in (4.27), we get

$$K = \frac{1}{G}\left[-\frac{\partial}{\partial x_1}\left(\frac{1}{2E}\frac{\partial G}{\partial x_1}\right) - \frac{\partial}{\partial x_2}\left(\frac{1}{2E}\frac{\partial E}{\partial x_2}\right) - \frac{1}{4E^2}\frac{\partial G}{\partial x_1}\frac{\partial E}{\partial x_1}\right.$$

$$\left. + \frac{1}{4EG}\frac{\partial G}{\partial x_2}\frac{\partial E}{\partial x_2} - \frac{1}{4E^2}\left(\frac{\partial E}{\partial x_2}\right)^2 + \frac{1}{4EG}\left(\frac{\partial G}{\partial x_1}\right)^2\right]$$

$$= \frac{1}{4E^2G^2}\left(E\frac{\partial G}{\partial x_2} + G\frac{\partial E}{\partial x_2}\right)\frac{\partial E}{\partial x_2} - \frac{1}{2EG}\frac{\partial^2 E}{\partial x_2^2}$$

$$+ \frac{1}{4E^2G^2}\left(G\frac{\partial E}{\partial x_1} + E\frac{\partial G}{\partial x_1}\right)\frac{\partial G}{\partial x_1} - \frac{1}{2EG}\frac{\partial^2 G}{\partial x_1^2}$$

$$= -\frac{1}{2\sqrt{EG}}\left\{\frac{\partial}{\partial x_2}\left(\frac{1}{\sqrt{EG}}\frac{\partial E}{\partial x_2}\right) + \frac{\partial}{\partial x_1}\left(\frac{1}{\sqrt{EG}}\frac{\partial G}{\partial x_1}\right)\right\}.$$

$\square$

We close this chapter by completing the discussion of (4.25). The condition $B_{ijk} - B_{jik} = 0$ yields, for all i, j, $k = 1, 2$ the *Codazzi-Mainardi equations*:

$$\sum_{s=1}^{2}(\Gamma_{jk}^s h_{is} - \Gamma_{ik}^s h_{js}) = \frac{\partial h_{ik}}{\partial x_j} - \frac{\partial h_{jk}}{\partial x_i}. \tag{4.28}$$

Though less important than Gauss' equations, the Codazzi-Mainardi equations are nonetheless very useful when studying surfaces, as we shall see in Chapter 7.

Summing up, if φ is a local parametrization of a regular surface the coordinates of φ have to satisfy the systems of partial differential equations (4.18)–(4.19), whose coefficients depend on the metric and form coefficients E, F, G, e, f and g, which in turn satisfy the compatibility conditions (4.26) and (4.28). Conversely, in the supplementary material of this chapter we shall prove the *fundamental theorem of the local theory of surfaces* (also known as *Bonnet's theorem*), which basically says that functions E, F, G, e, f and g with E, G, $EG - F^2 > 0$ and satisfying (4.26) and (4.28) are locally the metric and form coefficients of a regular surface, unique up to a rigid motion of $\mathbb{R}^3$; see Theorem 4.9.4.

We conclude with two definitions which will be useful in Chapter 7 (and in the exercises of this chapter).

Definition 4.6.15. Let $S \subset \mathbb{R}^3$ be an oriented surface with Gauss map $N\colon S \to S^2$. A *line of curvature* of the surface S is a curve σ in S such that $\dot\sigma$ is always a principal direction.

Definition 4.6.16. Let $S \subset \mathbb{R}^3$ be an oriented surface with Gauss map $N \colon S \to S^2$. An *asymptotic direction* at $p \in S$ is a versor $v \in T_pS$ such that $Q_p(v) = 0$. An *asymptotic curve* of the surface S is a curve σ in S such that $\dot\sigma$ is always an asymptotic direction.

Remark 4.6.17. Since by exchanging orientations the second fundamental form just changes sign, and since each surface is locally orientable, the notions of principal direction, asymptotic direction, line of curvature, and asymptotic curve are well defined for every surface, not just orientable ones.

Guided problems

Notation. From this section onwards, we shall use the following convention for writing partial derivatives: if $\varphi \colon U \to \mathbb{R}$ is a function of class C^k (with $k \geq 2$) defined in an open set $U \subset \mathbb{R}^2$ with coordinates (u, v), we shall denote the partial derivatives of φ by

$$\begin{cases} \varphi_u = \dfrac{\partial \varphi}{\partial u} \,, \quad \varphi_v = \dfrac{\partial \varphi}{\partial v} \,, & \text{for first-order derivatives;} \\[2mm] \varphi_{uu} = \dfrac{\partial^2 \varphi}{\partial u^2} \,, \quad \varphi_{uv} = \dfrac{\partial^2 \varphi}{\partial u \partial v} \,, \quad \varphi_{vv} = \dfrac{\partial^2 \varphi}{\partial v^2} \,, & \text{for second-order derivatives.} \end{cases}$$

An analogous notation will sometimes be used for partial derivatives of functions of more than 2 variables, or for higher-order derivatives.

Definition 4.P.1. If $\varphi \colon U \to S$ is a local parametrization of a surface S, and we denote by (u, v) the coordinates in U, then a *u-curve* (respectively, a *v-curve*) is a coordinate curve of the form $u \mapsto \varphi(u, v_0)$ (respectively, $v \mapsto \varphi(u_0, v)$).

Problem 4.1. *Let $S \subset \mathbb{R}^3$ be the surface of equation $z = xy^2$.*

(i) *Determine the first fundamental form of S and its metric coefficients.*
(ii) *Determine the second fundamental form Q of S.*
(iii) *Prove that $K \leq 0$ everywhere, and that $K = 0$ only for the points of S with $y = 0$.*
(iv) *Prove that $(0, 0, 0)$ is a planar point of S.*
(v) *Determine the principal directions in the points of S with zero Gaussian curvature.*
(vi) *Prove that the curves σ_1, $\sigma_2 \colon \mathbb{R} \to S$ given by*

$$\sigma_1(t) = (x_0 + t, y_0, z_0 + t y_0^2) \quad \text{and} \quad \sigma_2(t) = (\mathrm{e}^t x_0, \mathrm{e}^{-2t} y_0, \mathrm{e}^{-3t} z_0)$$

are asymptotic curves passing through $(x_0, y_0, z_0) \in S$ for all x_0, $y_0 \in \mathbb{R}$.

Solution. (i) Let $\varphi \colon \mathbb{R}^2 \to S$ be the usual parametrization $\varphi(u, v) = (u, v, uv^2)$ of S seen as a graph. Then

$$\partial_1 = \varphi_u = (1, 0, v^2) \,, \qquad \partial_2 = \varphi_v = (0, 1, 2uv) \,,$$

and so the metric coefficients are given by

$$E = \langle \partial_1, \partial_1 \rangle = 1 + v^4 , \quad F = \langle \partial_1, \partial_2 \rangle = 2uv^3 , \quad G = \langle \partial_2, \partial_2 \rangle = 1 + 4u^2 v^2 .$$

In particular, $EG - F^2 = 1 + v^4 + 4u^2 v^2$; moreover, the first fundamental form is

$$I_{\varphi(u,v)}(v_1 \partial_1 + v_2 \partial_2) = E v_1^2 + 2F v_1 v_2 + G v_2^2$$
$$= (1 + v^4) v_1^2 + 4uv^3 v_1 v_2 + (1 + 4u^2 v^2) v_2^2 .$$

(ii) To determine the form coefficients e, f and g, we shall use (4.13). First of all, note that

$$\begin{cases} \varphi_u \wedge \varphi_v = (-v^2, -2uv, 1) , \\ N = \dfrac{\varphi_u \wedge \varphi_v}{\|\varphi_u \wedge \varphi_v\|} = \dfrac{1}{\sqrt{1 + v^4 + 4u^2 v^2}} (-v^2, -2uv, 1) ; \end{cases}$$

moreover, the second-order partial derivatives are

$$\varphi_{uu} = (0,0,0) , \quad \varphi_{uv} = (0,0,2v) , \quad \varphi_{vv} = (0,0,2u) .$$

Then,

$$e = \langle N, \varphi_{uu} \rangle = 0 ,$$
$$f = \langle N, \varphi_{uv} \rangle = \frac{2v}{\sqrt{1 + v^4 + 4u^2 v^2}} ,$$
$$g = \langle N, \varphi_{vv} \rangle = \frac{2u}{\sqrt{1 + v^4 + 4u^2 v^2}} .$$

In particular, $eg - f^2 = -4v^2/(1 + v^4 + 4u^2 v^2)$, and the second fundamental form is given by

$$Q_{\varphi(u,v)}(v_1 \partial_1 + v_2 \partial_2) = e v_1^2 + 2f v_1 v_2 + g v_2^2$$
$$= \frac{4v}{\sqrt{1 + v^4 + 4u^2 v^2}} v_1 v_2 + \frac{2u}{\sqrt{1 + v^4 + 4u^2 v^2}} v_2^2 .$$

(iii) The previous computations yield

$$K = \frac{eg - f^2}{EG - F^2} = \frac{-4v^2}{(1 + v^4 + 4u^2 v^2)^2} .$$

So K is always nonpositive, and is zero if and only if $v = 0$, which is equivalent to $y = 0$.

(iv) Since $(0,0,0) = \varphi(0,0)$, and $e = f = g = 0$ in the origin, it follows that $(0,0,0)$ is a planar point.

(v) Recall that the matrix A representing the differential of the Gauss map with respect to the basis $\{\varphi_u, \varphi_v\}$ is given by

$$A = \frac{-1}{EG - F^2} \begin{vmatrix} G & -F \\ -F & E \end{vmatrix} \begin{vmatrix} e & f \\ f & g \end{vmatrix}.$$

So in this case we have

$$A = \frac{-2}{(1 + v^4 + 4u^2 v^2)^{3/2}} \begin{vmatrix} -2uv^4 & v + 2u^2 v^3 \\ v + v^5 & u - uv^4 \end{vmatrix}.$$

In particular, when $v = y = 0$ we get

$$A = \begin{vmatrix} 0 & 0 \\ 0 & -2u \end{vmatrix},$$

and so the principal directions coincide with the coordinate directions.

(vi) First of all,

$$\sigma_1(t) = \varphi(x_0 + t, y_0) \quad \text{and} \quad \sigma_2(t) = \varphi(e^t x_0, e^{-2t} y_0),$$

so they actually are curves in S. Differentiating, we get

$$\sigma_1'(t) = (1, 0, y_0^2) = \varphi_u(x_0 + t, y_0),$$
$$\sigma_2'(t) = (e^t x_0, -2e^{-2t} y_0, -3e^{-3t} z_0)$$
$$= e^t x_0 \varphi_u(e^t x_0, e^{-2t} y_0) - 2e^{-2t} y_0 \varphi_v(e^t x_0, e^{-2t} y_0).$$

Recalling the expression we found for the second fundamental form, we obtain $Q(\sigma_1'(t)) \equiv 0$ and $Q(\sigma_2'(t)) \equiv 0$, and so σ_1 and σ_2 are asymptotic curves. $\square$

Problem 4.2. Let $S \subset \mathbb{R}^3$ be the regular surface with global parametrization $\varphi \colon \mathbb{R}^2 \to \mathbb{R}^3$ given by $\varphi(u, v) = (u, v, u^2 - v^2)$.

(i) Determine the metric coefficients of S with respect to φ.
(ii) Determine a Gauss map for S.
(iii) Compute the second fundamental form and the Gaussian curvature of S.
(iv) Let $\sigma \colon I \to S$ be a curve with $\sigma(0) = O \in S$. Prove that the normal curvature of σ at the origin belongs to the interval $[-2, 2]$.

Solution. (i) Differentiating we find

$$\partial_1 = \varphi_u = (1, 0, 2u), \qquad \partial_2 = \varphi_v = (0, 1, -2v);$$

so the metric coefficients of φ are given by

$$E = 1 + 4u^2, \quad F = -4uv, \quad G = 1 + 4v^2.$$

(ii) It is enough to consider

$$N = \frac{\varphi_u \wedge \varphi_v}{\|\varphi_u \wedge \varphi_v\|} = \frac{1}{\sqrt{4u^2 + 4v^2 + 1}} (-2u, 2v, 1).$$

(iii) The second-order partial derivatives of φ are

$$\varphi_{uu} = (0,0,2) , \quad \varphi_{uv} = (0,0,0) \quad \text{and} \quad \varphi_{vv} = (0,0,-2) ;$$

so the form coefficients of φ are

$$e = \frac{2}{\sqrt{4u^2 + 4v^2 + 1}} , \quad f \equiv 0 , \quad g = \frac{-2}{\sqrt{4u^2 + 4v^2 + 1}} ,$$

and the second fundamental form is given by

$$Q_{\varphi(u,v)}(v_1\partial_1 + v_2\partial_2) = e\, v_1^2 + 2f\, v_1 v_2 + g\, v_2^2 = \frac{2(v_1^2 - v_2^2)}{\sqrt{4u^2 + 4v^2 + 1}} .$$

Moreover,

$$K = \frac{eg - f^2}{EG - F^2} = \frac{-4}{(4u^2 + 4v^2 + 1)^2}$$

is always negative, and so all the points of S are hyperbolic.

(iv) We know that the normal curvature of σ is given by the second fundamental form computed in the tangent versor of σ, and that the second fundamental form of S at the origin $O = \varphi(0,0)$ is given by $Q_O(v_1\partial_1 + v_2\partial_2) = 2(v_1^2 - v_2^2)$. Moreover, if $\dot\sigma(0) = v_1\partial_1 + v_2\partial_2$ then

$$1 = \|\dot\sigma(0)\|^2 = E(0,0)v_1^2 + 2F(0,0)v_1 v_2 + G(0,0)v_2^2 = v_1^2 + v_2^2 .$$

In particular, we may write $v_1 = \cos\theta$ and $v_2 = \sin\theta$ for a suitable $\theta \in \mathbb{R}$; hence,

$$\kappa_n(0) = Q_O\big(\dot\sigma(0)\big) = 2(\cos^2\theta - \sin^2\theta) = 2\cos(2\theta) \in [-2,2] ,$$

as claimed.   □

Definition 4.P.2. A point p of a surface S is called *umbilical* if dN_p is a multiple of the identity map on T_pS. In other words, p is umbilical if the two principal curvatures in p coincide.

Problem 4.3. *Prove that an oriented surface S consisting entirely of umbilical points is necessarily contained in a sphere or in a plane (and these are surfaces consisting only of umbilical points; see Examples 4.4.7 and 4.4.8).*

Solution. By assumption, there exists a function $\lambda\colon S \to \mathbb{R}$ such that we have $dN_p(v) = \lambda(p)v$ for all $v \in T_pS$ and $p \in S$, where $N\colon S \to S^2$ is the Gauss map of S. In particular, if φ is a local parametrization we have

$$\frac{\partial(N \circ \varphi)}{\partial x_1} = dN(\partial_1) = (\lambda \circ \varphi)\partial_1 , \quad \text{and} \quad \frac{\partial(N \circ \varphi)}{\partial x_2} = dN(\partial_2) = (\lambda \circ \varphi)\partial_2 .$$

Differentiating again we get

$$\frac{\partial^2(N \circ \varphi)}{\partial x_2 \partial x_1} = \frac{\partial(\lambda \circ \varphi)}{\partial x_2}\partial_1 + (\lambda \circ \varphi)\frac{\partial^2 \varphi}{\partial x_2 \partial x_1},$$

$$\frac{\partial^2(N \circ \varphi)}{\partial x_1 \partial x_2} = \frac{\partial(\lambda \circ \varphi)}{\partial x_1}\partial_2 + (\lambda \circ \varphi)\frac{\partial^2 \varphi}{\partial x_1 \partial x_2},$$

and so

$$\frac{\partial(\lambda \circ \varphi)}{\partial x_2}\partial_1 - \frac{\partial(\lambda \circ \varphi)}{\partial x_1}\partial_2 \equiv O.$$

But ∂_1 and ∂_2 are linearly independent; therefore this implies

$$\frac{\partial(\lambda \circ \varphi)}{\partial x_2} \equiv \frac{\partial(\lambda \circ \varphi)}{\partial x_1} \equiv 0,$$

that is $\lambda \circ \varphi$ is constant.

So we have proved that λ is locally constant: being S is connected, λ is constant on all S. Indeed, choose $p_0 \in S$ and put $R = \{p \in S \mid \lambda(p) = \lambda(p_0)\}$. This set is not empty ($p_0 \in R$), it is closed since λ is continuous, and is open because λ is locally constant; so by the connectedness of S we have $R = S$, that is, λ is globally constant.

If $\lambda \equiv 0$, the differential of the Gauss map is zero everywhere, that is, N is everywhere equal to a vector $N_0 \in S^2$. Choose $p_0 \in S$, and define $h: S \to \mathbb{R}$ by setting $h(q) = \langle q - p_0, N_0 \rangle$. If $\varphi: U \to S$ is an arbitrary local parametrization of S, we have

$$\frac{\partial(h \circ \varphi)}{\partial x_j} = \langle \partial_j, N_0 \rangle \equiv 0$$

for $j = 1, 2$. It follows that h is locally constant, and so it is constant by the same argument as above. Since $h(p_0) = 0$, we get $h \equiv 0$, which means exactly that S is contained in the plane through p_0 and orthogonal to N_0.

If instead $\lambda \equiv \lambda_0 \neq 0$, let $q: S \to \mathbb{R}^3$ be given by $q(p) = p - \lambda_0^{-1}N(p)$. Then

$$dq_p = \mathrm{id} - \frac{1}{\lambda_0}dN_p = \mathrm{id} - \frac{1}{\lambda_0}\lambda_0\,\mathrm{id} \equiv O,$$

therefore q is (locally constant and thus) constant; denote by q_0 the value of q, that is, $q \equiv q_0$. Hence $p - q_0 \equiv \lambda_0^{-1}N(p)$, and so

$$\forall p \in S \qquad \qquad \|p - q_0\|^2 = \frac{1}{\lambda_0^2}.$$

In other words, S is contained in the sphere of center q_0 and radius $1/|\lambda_0|$, and we are done. $\qquad \qquad \square$

Problem 4.4. When are the coordinate lines lines of curvature? *Let $\varphi: U \to S \subset \mathbb{R}^3$ be a local parametrization of a regular surface S, and assume that no point of $\varphi(U)$ is umbilical. Prove that all the coordinate curves are lines of curvature if and only if $F \equiv f \equiv 0$.*

Solution. Saying that the coordinate curves are always lines of curvature is equivalent to saying that the coordinate directions are always principal directions, and this in turn is equivalent to saying that the matrix A representing the differential of the Gauss map in the basis $\{\varphi_u, \varphi_v\}$ is always diagonal.

Now, recalling (4.14), we immediately see that if $F \equiv f \equiv 0$ then A is diagonal; so, in this case, the coordinate curves are always curvature lines (even when there are umbilical points).

Conversely, assume that the coordinate lines are lines of curvature. This means that the vectors φ_u and φ_v are principal directions; in particular, since no point is umbilical, φ_u and φ_v are orthogonal, and so $F \equiv 0$. Now, the off-diagonal entries of A are $-f/G$ and $-f/E$; since they have to be zero, we get $f \equiv 0$, as claimed. Notice that in umbilical points all directions are principal, and so coordinate curves are always lines of curvature at umbilical points.   □

Problem 4.5. *Let S be an oriented surface, and $N\colon S \to S^2$ its Gauss map. Prove that a curve $\sigma\colon I \to S$ is a line of curvature if and only if, having set $N(t) = N(\sigma(t))$, we have $N'(t) = \lambda(t)\sigma'(t)$ for a suitable function $\lambda\colon I \to \mathbb{R}$ of class C^∞. In this case, $-\lambda(t)$ is the (principal) curvature of S along $\sigma'(t)$.*

Solution. It suffices to remark that

$$\mathrm{d}N_{\sigma(t)}\big(\sigma'(t)\big) = \frac{\mathrm{d}(N \circ \sigma)}{\mathrm{d}t}(t) = N'(t)\,,$$

so $\sigma'(t)$ is an eigenvector of $\mathrm{d}N_{\sigma(t)}$ if and only if $N'(t) = \lambda(t)\sigma'(t)$ for some $\lambda(t) \in \mathbb{R}$.   □

Problem 4.6. Characterization of the lines of curvature. *Let $S \subset \mathbb{R}^3$ be an oriented surface, $\varphi\colon U \to S$ a local parametrization, and let $\sigma\colon I \to \varphi(U)$ be a regular curve with support contained in $\varphi(U)$, so that we can write $\sigma(t) = \varphi\big(u(t), v(t)\big)$. Prove that σ is a line of curvature if and only if*

$$(fE - eF)(u')^2 + (gE - eG)u'v' + (gF - fG)(v')^2 \equiv 0\,.$$

Solution. By definition, we know that σ is a line of curvature if and only if $\mathrm{d}N_{\sigma(t)}\big(\sigma'(t)\big) = \lambda(t)\sigma'(t)$ for a suitable function λ of class C^∞. Now it suffices to use Proposition 4.5.14 for expressing $\mathrm{d}N_{\sigma(t)}\big(\sigma'(t)\big)$, and eliminate λ from the system of equations given by $\mathrm{d}N_{\sigma(t)}\big(\sigma'(t)\big) = \lambda(t)\sigma'(t)$, recalling that $\sigma'(t) \neq O$ always because σ is regular.   □

Problem 4.7. Characterization of asymptotic curves. *Let $\varphi\colon U \to S$ be a local parametrization of an oriented surface, and let $\sigma\colon I \to \varphi(U)$ be a regular curve with support contained in $\varphi(U)$, so we can write $\sigma(t) = \varphi\big(u(t), v(t)\big)$. Prove that σ is an asymptotic curve if and only if*

$$e(u')^2 + 2fu'v' + g(v')^2 \equiv 0\,.$$

In particular, deduce that the coordinate curves are asymptotic curves (necessarily in a neighborhood of a hyperbolic point) if and only if $e = g = 0$.

Solution. By definition, a curve σ is an asymptotic curve if and only if $Q_{\sigma(t)}\big(\sigma'(t)\big) \equiv 0$, where Q_p is the second fundamental form at $p \in S$. Since $\sigma'(t) = u'\,\varphi_u + v'\,\varphi_v$, where $\varphi_u = \partial_1$ and $\varphi_v = \partial_2$ are computed in $\sigma(t)$, the assertions immediately follows recalling that the form coefficients e, f, g represent the second fundamental form in the basis $\{\partial_1, \partial_2\}$.   □

Problem 4.8. *Let $S \subset \mathbb{R}^3$ the (upper half of the) pseudosphere obtained by rotating around the z-axis the (upper half of the) tractrix $\sigma\colon(\pi/2,\pi) \to \mathbb{R}^3$ given by*

$$\sigma(t) = \big(\sin t, 0, \cos t + \log\tan(t/2)\big) ;$$

see Exercise 3.18 and Example 4.5.23. In particular, S is the support of the immersed surface $\varphi\colon(\pi/2,\pi) \times \mathbb{R} \to \mathbb{R}^3$ given by

$$\varphi(t,\theta) = \left(\sin t \cos\theta, \sin t \sin\theta, \cos t + \log\tan\frac{t}{2}\right).$$

(i)   *Determine the Gauss map $N\colon S \to S^2$ induced by φ.*
(ii) *Determine the differential $\mathrm{d}N$ of the Gauss map and the Gaussian curvature of S.*
(iii) *Determine the mean curvature of S.*

Solution. (i) To determine the Gauss map of S we compute, as usual, the partial derivatives of the parametrization:

$$\partial_1 = \varphi_t = \cos t(\cos\theta, \sin\theta, \cot t) , \quad \partial_2 = \varphi_\theta = \sin t(-\sin\theta, \cos\theta, 0) .$$

Hence, $\varphi_t \wedge \varphi_\theta = \cos t(-\cos t \cos\theta, -\cos t \sin\theta, \sin t)$ and $\|\varphi_t \wedge \varphi_\theta\| = \cos t$, and so

$$N\big(\varphi(t,\theta)\big) = (-\cos t \cos\theta, -\cos t \sin\theta, \sin t) .$$

(ii) To determine the differential $\mathrm{d}N_p$ at $p = \varphi(t,\theta) \in S$, we use the fact that $\mathrm{d}N_p(\varphi_t) = \partial(N \circ \varphi)/\partial t$ and $\mathrm{d}N_p(\varphi_\theta) = \partial(N \circ \varphi)/\partial\theta$. We find that

$$\mathrm{d}N_p(\varphi_t) = \sin t\,(\cos\theta, \sin\theta, \cot t) = (\tan t)\,\varphi_t , \;\; \mathrm{d}N_p(\varphi_\theta) = -(\cot t)\varphi_\theta .$$

So the matrix representing $\mathrm{d}N_p$ with respect to the basis $\{\varphi_t, \varphi_\theta\}$ is

$$A = \begin{vmatrix} \tan t & 0 \\ 0 & -\cot t \end{vmatrix} ;$$

in particular, $K = \det(A) \equiv -1$, as claimed in Example 4.5.23. Incidentally, S is called "pseudosphere" exactly because it has constant — even if negative — Gaussian curvature, like the usual sphere.

(iii) It suffices to notice that the mean curvature is given by

$$H = -\frac{1}{2}\mathrm{tr}(A) = -\frac{1}{2}(\tan t + \cot t) = -\frac{1}{\sin 2t} .$$

□

Problem 4.9. *Compute the Gaussian curvature and the mean curvature of the ellipsoid* $S = \{(x, y, z) \in \mathbb{R}^3 \mid x^2 + 4y^2 + 9z^2 = 1\}$ *without using local parametrizations.*

Solution. Since S is the vanishing locus of the function $h \colon \mathbb{R}^3 \to \mathbb{R}$ given by $h(x, y, z) = x^2 + 4y^2 + 9z^2 - 1$, Corollary 4.3.12 tells us that a Gauss map $N \colon S \to S^2$ of S is

$$N(x, y, z) = \alpha(x, y, z)(x, 4y, 9z) \,,$$

where $\alpha \colon S \to \mathbb{R}$ is the function $\alpha(x, y, z) = (x^2 + 16y^2 + 81z^2)^{-1/2}$. Moreover, the tangent plane at $p = (x_0, y_0, z_0)$ is given by

$$T_p S = \{(v_1, v_2, v_3) \in \mathbb{R}^3 \mid x_0 v_1 + 4y_0 v_2 + 9z_0 v_3 = 0\} \,.$$

The Jacobian matrix of N seen as a map from $\mathbb{R}^3 \setminus \{O\}$ to $\mathbb{R}^3$ is

$$J = \begin{vmatrix} \alpha + x\alpha_x & x\alpha_y & x\alpha_z \\ 4y\alpha_x & 4\alpha + 4y\alpha_y & 4y\alpha_z \\ 9z\alpha_x & 9z\alpha_y & 9\alpha + 9z\alpha_z \end{vmatrix} \,;$$

so

$$\mathrm{d}N_p(v_1, v_2, v_3) = J \begin{vmatrix} v_1 \\ v_2 \\ v_3 \end{vmatrix} = \alpha(p) \begin{vmatrix} v_1 \\ 4v_2 \\ 9v_3 \end{vmatrix} - \alpha^3 (x_0 v_1 + 16 y_0 v_2 + 81 z_0 v_3) \begin{vmatrix} x_0 \\ 4y_0 \\ 9z_0 \end{vmatrix} \,.$$

Suppose that $x_0 \neq 0$. A basis $\mathcal{B} = \{w_1, w_2\}$ of the tangent vectors to S at p is then given by

$$w_1 = (-9z_0, 0, x_0) \quad \text{and} \quad w_2 = (-4y_0, x_0, 0) \,.$$

Computing explicitly $\mathrm{d}N_p(w_1)$ and $\mathrm{d}N_p(w_2)$, and writing them as linear combinations with respect to the basis $\mathcal{B}$, we get that the matrix representing $\mathrm{d}N_p$ with respect to $\mathcal{B}$ is

$$A = \alpha \begin{vmatrix} 9(1 - 72\alpha^2 z_0^2) & -108\alpha^2 y_0 z_0 \\ -288\alpha^2 y_0 z_0 & 4(1 - 12\alpha^2 y_0^2) \end{vmatrix} \,.$$

So, if $x_0 \neq 0$ the Gaussian curvature and the mean curvature are

$$K = \frac{36}{(x_0^2 + 16y_0^2 + 81z_0^2)^2} \,, \quad H = \frac{36(x_0^2 + y_0^2 + z_0^2 - 1) - 13}{2(x_0^2 + 16y_0^2 + 81z_0^2)^{3/2}} \,;$$

note that K is always positive.

We have now found the the Gaussian and mean curvatures in all points $p = (x, y, z) \in S$ with $x \neq 0$. But $S \cap \{x = 0\}$ is an ellipse C, and $S \setminus C$ is an open set dense in S. Since the Gaussian and mean curvatures are continuous, and the expressions we have found are defined and continuous on all S, they give the values of K and H on all S. □

Problem 4.10. *Let $S \subset \mathbb{R}^3$ be a regular surface with a global parametriza-tion $\varphi \colon \mathbb{R} \times \mathbb{R}^+ \to S$ whose metric coefficients satisfy $E(u,v) = G(u,v) = v$ e $F(u,v) \equiv 0$. Prove that S is not locally isometric to a sphere.*

Solution. Since the parametrization is orthogonal, we may use (4.23) to com-pute the Christoffel symbols of φ. We obtain $\Gamma_{11}^1 = 0$, $\Gamma_{11}^2 = -\frac{1}{2v}$, $\Gamma_{12}^1 = \frac{1}{2v}$, $\Gamma_{12}^2 = 0$, $\Gamma_{22}^1 = 0$, $\Gamma_{22}^2 = \frac{1}{2v}$. Gauss' Theorema Egregium 4.6.11 then implies

$$K = \frac{1}{G} \left[\frac{\partial \Gamma_{22}^1}{\partial u} - \frac{\partial \Gamma_{12}^1}{\partial v} + \sum_{s=1}^{2} (\Gamma_{22}^s \Gamma_{1s}^1 - \Gamma_{12}^s \Gamma_{2s}^1) \right] = \frac{1}{2v^3} .$$

Hence, K is not constant in any open set of S, and so (Corollary 4.6.12) S cannot be locally isometric to a sphere. $\qquad\square$

Problem 4.11. *Let $\sigma \colon (a,b) \to R^3$ be a biregular curve whose support is con-tained in the sphere S^2 with radius 1 and center in the origin of $\mathbb{R}^3$. Show that if the curvature of σ is constant then the support of σ is contained in a circle.*

Solution. We may assume that σ is parametrized by arc length; moreover, remember that if we orient S^2 as in Example 4.1.10 we have $\sigma = N \circ \sigma$. Since the support of σ is contained in S^2, the derivative $\dot{\sigma}$ is tangent to S^2, and so it is orthogonal to σ. Further, by Proposition 4.4.28 (Meusnier) and Exam-ple 4.4.21, the normal curvature of σ is equal to -1 everywhere, once more because σ takes values in S^2. Hence, (4.8) and (4.9) imply $\langle \sigma, \mathbf{n} \rangle \equiv -1/\kappa$.

Now, $1 = \|\sigma\|^2 = |\langle \sigma, \dot{\sigma} \rangle|^2 + |\langle \sigma, \mathbf{n} \rangle|^2 + |\langle \sigma, \mathbf{b} \rangle|^2$; since $\langle \sigma, \dot{\sigma} \rangle \equiv 0$ and $\langle \sigma, \mathbf{n} \rangle$ is a non zero constant, we deduce that $\langle \sigma, \mathbf{b} \rangle$ is a constant too. Hence,

$$0 = \frac{\mathrm{d}}{\mathrm{d}t} \langle \sigma, \mathbf{b} \rangle = -\tau \langle \sigma, \mathbf{n} \rangle = \frac{\tau}{\kappa} ,$$

so $\tau \equiv 0$ and σ is plane. But the support of a plane regular curve with con-stant curvature is contained in a circle, and we are done. $\qquad\square$

Problem 4.12. *Put $U = (0,1) \times (0,\pi)$ and let $\varphi \colon U \to \mathbb{R}^3$ be the map defined by $\varphi(u,v) = \bigl(u \cos v, u \sin v, \phi(v)\bigr)$, where $\phi \in C^\infty\bigl((0,\pi)\bigr)$ is a homeomor-phism with its image.*

(i)   *Show that the image S of φ is a regular surface.*
(ii) *Compute the Gaussian curvature in every point of S and check whether there exists an open subset of S that is locally isometric to a plane.*
(iii) *Give conditions for a point of S to be an umbilical point.*

Solution. (i) Note that

$$\partial_1 = \varphi_u = (\cos v, \sin v, 0) \quad \partial_2 = \varphi_v = \bigl(-u \sin v, u \cos v, \phi'(v)\bigr) ;$$

hence,

$$\varphi_u \wedge \varphi_v = (\phi'(v) \sin v, -\phi'(v) \cos v, u)$$

is never zero, because its third component is never zero. So the differential of φ is injective in every point. Moreover, φ is injective, and $\psi \colon S \to U$ given by $\psi(x, y, z) = (\sqrt{x^2 + y^2}, \phi^{-1}(z))$ is a continuous inverse of φ.

(ii) The metric coefficients are $E = 1$, $F = 0$ and $G = u^2 + \phi'(v)^2$, while $\|\varphi_u \wedge \varphi_v\| = \sqrt{u^2 + \phi'(v)^2}$. To determine the form coefficients, we compute

$$\varphi_{uu} = (0, 0, 0)\,, \quad \varphi_{uv} = (-\sin v, \cos v, 0)\,, \quad \varphi_{vv} = (-u\cos v, -u\sin v, \phi''(v))\,,$$

and so

$$e \equiv 0\,, \quad f = \frac{-\phi'(v)}{\sqrt{u^2 + \phi'(v)^2}}\,, \quad g = \frac{u\phi''(v)}{\sqrt{u^2 + \phi'(v)^2}}\,.$$

In particular,

$$K = -\frac{\phi'(v)^2}{(u^2 + \phi'(v)^2)^2}\,.$$

Since ϕ is injective, ϕ' cannot be zero on an interval, so K cannot be zero in an open set. Consequently, no open subset of S can be locally isometric to a plane.

(iii) Recalling (4.17), which gives the principal curvatures in terms of the mean and Gaussian curvatures, the relation that characterizes the umbilical points is $H^2 - K = 0$. Using (4.16), we find that the mean curvature of S is given by

$$H = \frac{1}{2}\frac{u\phi''(v)}{(u^2 + \phi'(v)^2)^{3/2}}\,,$$

and so

$$H^2 - K = \frac{1}{4}\frac{u^2\phi''(v)^2}{(u^2 + \phi'(v)^2)^3} + \frac{\phi'(v)^2}{(u^2 + \phi'(v)^2)^2}\,.$$

Hence we have $H^2 - K = 0$ if and only if $\phi'(v) = \phi''(v) = 0$, and so the umbilical points of S are exactly the points of the form $(u\cos v_0, u\sin v_0, \phi(v_0))$, where $v_0 \in (0, \pi)$ satisfies $\phi'(v_0) = \phi''(v_0) = 0$. $\square$

Problem 4.13. *Let* $\Sigma = \{(x, y, z) \in \mathbb{R}^3 \,|\, xyz = 1\}$.

(i) *Determine the largest subset* S *of* Σ *such that* S *is a regular surface.*
(ii) *Prove that the points* $(x, y, z) \in S$ *such that* $|x| = |y| = |z| = 1$ *are umbilical points of* S.

Solution. (i) Consider the function $f \colon \mathbb{R}^3 \to \mathbb{R}$ given by $f(x, y, z) = xyz$. Since $\nabla f = (yz, xz, xy)$, we find that 1 is a regular value for f, and so $\Sigma = S$ is a regular surface.

(ii) Let $p = (x, y, z) \in \Sigma$. In a neighborhood of p, the surface Σ is the graph of the function $g \colon \mathbb{R}^* \times \mathbb{R}^* \to \mathbb{R}$ given by $g(x, y) = 1/xy$. So we may take as parametrization of Σ near p the parametrization of the graph of g

given by $\varphi(u, v) = \big(u, v, g(u, v)\big)$. Then, proceeding in the usual way, we find

$$\partial_1 = \left(1, 0, -\frac{1}{u^2 v}\right), \ \partial_2 = \left(0, 1, -\frac{1}{uv^2}\right), \ N = \frac{1}{\sqrt{u^2 + v^2 + u^4 v^4}}(v, u, u^2 v^2),$$

$$E = 1 + \frac{1}{u^4 v^2}, \ F = \frac{1}{u^3 v^3}, \ G = 1 + \frac{1}{u^2 v^4},$$

$$e = \frac{2v}{u\sqrt{u^2 + v^2 + u^4 v^4}}, \ f = \frac{1}{\sqrt{u^2 + v^2 + u^4 v^4}}, \ g = \frac{2u}{v\sqrt{u^2 + v^2 + u^4 v^4}},$$

$$K = \frac{3u^4 v^4}{(u^2 + v^2 + u^4 v^4)^2}, \ H = \frac{uv(1 + u^2 v^4 + u^4 v^2)}{(u^2 + v^2 + u^4 v^4)^{3/2}},$$

$$H^2 - K = \frac{u^2 v^2 (1 + u^8 v^4 + u^4 v^8 - u^2 v^4 - u^6 v^6 - u^4 v^2)}{(u^2 + v^2 + u^4 v^4)^3}.$$

In particular, all points of the form $\varphi(u, v)$ with $|u| = |v| = 1$, that is, all points $p \in S$ with $|x| = |y| = |z| = 1$, are umbilical points.   □

Problem 4.14. *Let S be an oriented surface in $\mathbb{R}^3$ and $\sigma: \mathbb{R} \to S$ a biregular curve of class C^∞ that is an asymptotic curve of S. Prove that $T_{\sigma(s)}S$ is the osculating plane to σ at $\sigma(s)$ for all $s \in \mathbb{R}$.*

Solution. We may assume that σ is parametrized by arc length. Using the usual notation, we have to show that the versors $\mathbf{t}(s)$ and $\mathbf{n}(s)$ span the plane $T_{\sigma(s)}S$ tangent to S at $\sigma(s)$; in other words, we have to prove that $\mathbf{t}(s)$ and $\mathbf{n}(s)$ are orthogonal to the normal versor $N\big(\sigma(s)\big)$. Since $\mathbf{t}(s) \in T_{\sigma(s)}S$, by definition of a tangent plane to a surface, it suffices to show that $\mathbf{n}(s)$ and $N\big(\sigma(s)\big)$ are orthogonal. But by the biregularity of σ we know that

$$\big\langle \mathbf{n}(s), N\big(\sigma(s)\big) \big\rangle = \frac{1}{\kappa(s)} Q_{\sigma(s)}\big(\dot\sigma(s)\big) = 0,$$

since σ is an asymptotic curve.   □

Definition 4.P.3. Let $S \subset \mathbb{R}^3$ be a surface oriented by an atlas $\mathcal{A}$. Then the atlas $\mathcal{A}^-$ obtained by exchanging coordinates in all the parametrizations of $\mathcal{A}$, that is, $\varphi \in \mathcal{A}^-$ if and only if $\varphi \circ \chi \in \mathcal{A}$ where $\chi(x, y) = (y, x)$, is called *opposite* of $\mathcal{A}$.

Problem 4.15. *Let S be a surface oriented by an atlas $\mathcal{A}$, and take another local parametrization $\varphi: U \to S$ of S, with U connected. Prove that either φ has the same orientation as all local parametrizations of $\mathcal{A}$, or has the same orientation as all local parametrizations of $\mathcal{A}^-$.*

Solution. Let N be the normal versor field determining the given orientation, and $\{\partial_1, \partial_2\}$ the basis induced by φ. Exactly as in the proof of Proposition 4.3.7, we find that $\partial_1 \wedge \partial_2 / \|\partial_1 \wedge \partial_2\| \equiv \pm N$ on $\varphi(U)$, with constant sign since U is connected. So (4.6) implies that if the sign is positive then φ determines the same orientation of all elements of $\mathcal{A}$, while if the sign is negative it determines the opposite orientation.   □

Problem 4.16. *Let $S \subset \mathbb{R}^3$ be a surface in which the absolute value of the mean curvature is never zero. Prove that S is orientable.*

Solution. Let $\mathcal{A} = \{\varphi_\alpha\}$ be an atlas on S such that the domain U_α of each φ_α is connected. Using the usual Gauss map N_α induced by φ_α, we may define a mean curvature on $\varphi_\alpha(U_\alpha)$ with a well defined sign, since its absolute value is never zero and U_α is connected. Up to exchanging coordinates in U_α, we may then assume that the mean curvature induced by N_α is always positive.

Define now $N: S \to S^2$ by setting $N(p) = N_\alpha(p)$ for all $p \in \varphi_\alpha(U_\alpha)$. To conclude, it suffices to verify that N is well defined, that is it does not depend on α. Take $p \in \varphi_\alpha(U_\alpha) \cap \varphi_\beta(U_\beta)$. If we had $N_\alpha(p) = -N_\beta(p)$, then we would have $N_\alpha \equiv -N_\beta$ in a whole neighborhood of p; so the mean curvature induced by N_α and the mean curvature induced by N_β would have opposite sign in a neighborhood of p, against our assumptions.   □

Problem 4.17. *Let $p \in S$ be a point of a surface $S \subset \mathbb{R}^3$. Prove that if p is elliptic then there exists a neighborhood V of p in S such that $V \setminus \{p\}$ is contained in one of the two open half-spaces bounded by the affine tangent plane $p + T_pS$. Prove that if, on the other hand, p is hyperbolic then every neighborhood of p in S intersects both the open half-spaces bounded by the plane $p + T_pS$.*

Solution. Let $\varphi: U \to S$ be a local parametrization centered at p, and define the function $d: U \to \mathbb{R}$ by setting $d(x) = \langle \varphi(x) - p, N(p) \rangle$, where N is the Gauss map induced by φ. Clearly, $\varphi(x) \in p + T_pS$ if and only if $d(x) = 0$, and $\varphi(x)$ belongs to one or the other of the half-spaces bounded by $p + T_pS$ depending on the sign of $d(x)$. Expanding d as a Taylor series around the origin, we get

$$
\begin{aligned}
d(x) &= d(O) + \sum_{j=1}^{2} \frac{\partial d}{\partial x_j}(O)x_j + \frac{1}{2} \sum_{i,j=1}^{2} \frac{\partial^2 d}{\partial x_i \partial x_j}(O)x_i x_j + o(\|x\|^2) \\
&= e(p)x_1^2 + 2f(p)x_1 x_2 + g(p)x_2^2 + o(\|x\|^2) \qquad\qquad (4.29) \\
&= Q_p(x_1 \partial_1 + x_2 \partial_2) + o(\|x\|^2) \, .
\end{aligned}
$$

Now, if p is elliptic then the two principal curvatures at p have the same sign and are different from zero; in particular, Q_p is positive (or negative) definite. But then (4.29) implies that $d(x)$ has constant sign in a punctured neighborhood of the origin, and so there exists a neighborhood $V \subset S$ of p such that all points of $V \setminus \{p\}$ belong to one of the two open half-spaces bounded by $p + T_pS$.

If, on the other hand, p is hyperbolic, the two principal curvatures in p have opposite signs and are different from zero; in particular, Q_p is indefinite. Hence $d(x)$ changes sign in every neighborhood of the origin, and so every neighborhood of p in S intersects both the open half-spaces bounded by $p + T_pS$.   □

Problem 4.18. Osculating quadric to a level surface. *Given a function* $f \in C^\infty(\Omega)$ *admitting* 0 *as regular value, where* $\Omega \subset \mathbb{R}^3$ *is an open set, let* $p_0 = (x_1^o, x_2^o, x_3^o) \in S = f^{-1}(0)$ *be a point of the level surface of* f.

(i) *Determine a quadric* Q *passing through* p_0 *and such that* S *and* Q *have the same tangent plane at* p_0 *and the same second fundamental form. The quadric* Q *is called* osculating *quadric.*

(ii) *Show that* p_0 *is elliptic, hyperbolic or parabolic for* S *if and only if it is for* Q.

(iii) *Let* S *be the surface of* R^3 *of equation* $x_1 + x_1^3 + x_2^2 + x_3^3 = 0$. *Using the osculating quadric, show that the point* $p_0 = (-1, 1, 1)$ *is hyperbolic.*

Solution. Developing f in Taylor series around p_0, we find

$$f(x) = \sum_{j=1}^{3} \frac{\partial f}{\partial x_j}(p_0)(x_j - x_j^o)$$

$$+ \frac{1}{2} \sum_{i,j=1}^{3} \frac{\partial^2 f}{\partial x_i \partial x_j}(p_0)(x_i - x_i^o)(x_j - x_j^o) + o(\|x\|^2).$$

Choose as Q the quadric determined by the polynomial

$$P(x) = \sum_{j=1}^{3} \frac{\partial f}{\partial x_j}(p_0)(x_j - x_j^o) + \frac{1}{2} \sum_{i,j=1}^{3} \frac{\partial^2 f}{\partial x_i \partial x_j}(p_0)(x_i - x_i^o)(x_j - x_j^o).$$

So f and P have the same first and second derivatives in p_0. Since the tangent plane to S (respectively, to Q) at p_0 is orthogonal to the gradient of f (respectively, P) at p_0, and $\nabla f(p_0) = \nabla P(p_0)$, we immediately find that $T_{p_0}S = T_{p_0}Q$. Moreover, the differential of the Gauss map of S at p_0 only depends on the first derivatives of ∇f at p_0, that is, on the second derivatives of f at p_0; since P has the same (first and) second derivatives at p_0 as f, it follows that the differential of the Gauss map for S acts on $T_{p_0}S = T_{p_0}Q$ like the differential of the Gauss map for Q, and as a consequence S and Q have the same second fundamental form at p_0, and p_0 is elliptic (hyperbolic, parabolic) for S if and only if it is for Q.

In case (iii), the polynomial P is

$$P(x) = 4(x_1 + 1) + 2(x_2 - 1) + 3(x_3 - 1) - 3(x_1 + 1)^2 + (x_2 - 1)^2 + 3(x_3 - 1)^2.$$

The theorem of metric classification for quadrics (see [1, Vol. II, p. 163]) tells us that the quadric Q is obtained by a rigid motion from a one-sheeted hyperboloid. Since all points of Q are hyperbolic (see Exercise 4.55), p_0 is hyperbolic for S too.   □

Definition 4.P.4. A surface $S \subset \mathbb{R}$ is *ruled* if there exists a family $\{r_\lambda\}_{\lambda \in \mathbb{R}}$ of disjoint open line segments (or whole straight lines) whose union is S. The lines r_λ are called *generators* (or *rulings*) of S. A *cone* is a ruled surface whose generators all pass through a common point.

Problem 4.19. *Let S be a ruled (regular) surface. Show that S does not contain elliptic points, and as a consequence $K \leq 0$ in each point of S.*

Solution. By definition, for each point $p \in S$, there is a line segment contained in S and passing through p. A line segment within a surface always has zero normal curvature; so every point $p \in S$ has an asymptotic direction, which necessarily implies $K(p) \leq 0$. $\qquad\square$

Problem 4.20. Tangent surface to a curve. *Let $\sigma: I \to \mathbb{R}^3$ be a regular curve of class C^∞, with $I \subseteq \mathbb{R}$ an open interval. The map $\tilde{\varphi}: I \times \mathbb{R} \to \mathbb{R}^3$, defined by $\tilde{\varphi}(t, v) = \sigma(t) + v\,\sigma'(t)$, is called* tangent surface *to σ. Every affine tangent line to σ is called a* generator *of the tangent surface.*

(i) *Show that $\tilde{\varphi}$ is not an immersed surface.*
(ii) *Show that if the curvature κ of σ is nowhere zero then the restriction $\varphi = \tilde{\varphi}|_U: U \to \mathbb{R}^3$ of $\tilde{\varphi}$ to the subset $U = \{(t, v) \in I \times \mathbb{R} \mid v > 0\}$ is an immersed surface.*
(iii) *Show that the tangent plane along a generator of the tangent surface is constant in $S = \varphi(U)$.*

Solution. (i) It suffices to prove that the differential of $\tilde{\varphi}$ is not injective somewhere. Since $\varphi_t = \sigma' + v\,\sigma''$ and $\varphi_v = \sigma'$, we have $\varphi_t \wedge \varphi_v = v\,\sigma'' \wedge \sigma'$. So, using the expression (1.13) of the curvature of a curve in an arbitrary parameter, we find

$$\|\varphi_t \wedge \varphi_v\| = |v|\,\|\sigma'\|^3 \kappa\,.$$

In particular, the differential of φ is not injective when $v = 0$.

(ii) More precisely, we have proved that the differential of φ is injective in (t, v) if and only if $v \neq 0$ and $\kappa(t) \neq 0$, and so φ is an immersed surface.

(iii) It is sufficient to remark that the direction of the vector $\varphi_t \wedge \varphi_v$ is orthogonal to the tangent plane to S at the required point. Since this direction does not depend on v, the tangent plane is constant along a generator of S. $\qquad\square$

Exercises

FIRST FUNDAMENTAL FORM

4.1. Determine the metric coefficients and the first fundamental form for the regular surface with global parametrization $\varphi(u, v) = (u, v, u^4 + v^4)$.

4.2. Let $S \subset \mathbb{R}^3$ be the surface with global parametrization $\varphi: \mathbb{R}^+ \times \mathbb{R}^+ \to \mathbb{R}^3$ given by $\varphi(u, v) = (u\cos v, u\sin v, u)$. Prove that the coordinate curves of φ are orthogonal to each other in every point.

4.3. Let $S \subset \mathbb{R}^3$ be the catenoid, parametrized as in Problem 3.1. Given $r \in \mathbb{R}$, let $\sigma \colon \mathbb{R} \to S$ be the curve contained in the catenoid defined by $\sigma(t) = \varphi(t, rt)$. Compute the length of σ between $t = 0$ and $t = t_0$, using the first fundamental form of S.

4.4. Let $\varphi \colon \mathbb{R}^+ \times (0, 2\pi) \to \mathbb{R}^3$ be the local parametrization of the one-sheeted cone $S \subset \mathbb{R}^3$ given by $\varphi(u, v) = (u \cos v, u \sin v, u)$. Given $\beta \in \mathbb{R}$, determine the length of the curve $\sigma \colon [0, \pi] \to S$ expressed by $\sigma(t) = \varphi(e^{t \cotan(\beta)/\sqrt{2}}, t)$.

4.5. Let $S \subset \mathbb{R}^3$ be a regular surface with local parametrization $\varphi(u, v)$ whose metric coefficients satisfy $E \equiv 1$ and $F \equiv 0$. Show that the coordinate v-curves cut each u-curve in segments of equal length.

4.6. Determine the metric coefficients of the unit sphere $S^2 \subset \mathbb{R}^3$ with respect to the parametrization found by using the stereographic projection (see Exercise 3.4).

4.7. Determine the first fundamental form of the xy-plane minus the origin, parametrized by polar coordinates.

ISOMETRIES AND SIMILITUDES

4.8. Find two surfaces S_1 and S_2 such that S_1 is locally isometric to S_2 but S_2 is not locally isometric to S_1.

4.9. Determine for which values of $a, b \in \mathbb{R}$ the surface

$$S_{a,b} = \{(x, y, z) \in \mathbb{R}^3 \mid z = a\,x^2 + b\,y^2\}$$

is locally isometric to a plane.

4.10. Let $\sigma = (\sigma_1, \sigma_2) \colon \mathbb{R} \to \mathbb{R}^2$ be a regular plane curve parametrized by arc length. Let $S \subset \mathbb{R}^3$ be the right cylinder on σ parametrized by $\varphi(u, v) = (\sigma_1(u), \sigma_2(u), v)$. Prove that S is locally isometric to the cylinder of equation $x^2 + y^2 + 2x = 0$.

4.11. Let $H \colon S \to \tilde{S}$ be a similitude with scale factor $r > 0$. Given a local parametrization $\varphi \colon U \to S$, put $\tilde{\varphi} = H \circ \varphi$ and let E, F, G (respectively, $\tilde{E}$, $\tilde{F}$, $\tilde{G}$) be the metric coefficients with respect to φ (respectively, $\tilde{\varphi}$). Prove that $\tilde{E} = r^2 E$, $\tilde{F} = r^2 F$ and $\tilde{G} = r^2 G$.

ORIENTABLE SURFACES

4.12. Let $\sigma, \tau \colon \mathbb{R} \to \mathbb{R}^3$ be the trajectories, parametrized by arc length, of two points that are moving subject to the following conditions:

(a) σ starts at $\sigma(0) = (0,0,0)$, and moves along the x-axis in the positive direction;

(b) τ starts at $\tau(0) = (0,a,0)$, where $a \neq 0$, and moves parallel to the positive direction of the z-axis.

Denote by $S \subset \mathbb{R}^3$ the union of the straight lines passing through $\sigma(t)$ and $\tau(t)$ as t varies in $\mathbb{R}$.

(i)   Prove that S is a regular surface.
(ii) Find, for every point $p \in S$, a basis of the tangent plane T_pS.
(iii) Prove that S is orientable.

4.13. Let $S \subset \mathbb{R}^3$ be a surface oriented by an atlas $\mathcal{A} = \{\varphi_\alpha\}$. Given $p \in S$ and a basis $\{v_1, v_2\}$ of T_pS, prove that $\{v_1, v_2\}$ is a positive basis of T_pS if and only if it determines on T_pS the same orientation as the basis $\{\partial_{1,\alpha}|_p, \partial_{2,\alpha}|_p\}$ for all $\varphi_\alpha \in \mathcal{A}$ such that p belongs to the image of φ_α.

4.14. How many orientations does an orientable surface admit?

4.15. Determine a normal versor field for the surface S in $\mathbb{R}^3$ with global parametrization $\varphi: \mathbb{R}^2 \to \mathbb{R}^3$ given by $\varphi(u,v) = (e^u, u+v, u)$, and compute the angle between the coordinate curves.

4.16. Determine a normal versor field for the surface S in $\mathbb{R}^3$ of equation $z = e^{xy}$. Find for which values of $\lambda, \mu \in \mathbb{R}$ the vector $(\lambda, 0, \mu)$ is tangent to S at $p_0 = (0,0,1)$.

4.17. Let $S \subset \mathbb{R}^3$ be a surface oriented by an atlas $\mathcal{A}$, and let $\mathcal{A}^-$ be the opposite of $\mathcal{A}$. Prove that $\mathcal{A}^-$ is also oriented, and that all $\varphi \in \mathcal{A}$ and $\varphi^- \in \mathcal{A}^-$ with intersecting images determine opposite orientations.

SECOND FUNDAMENTAL FORM

4.18. Prove that if S is an oriented surface with $dN \equiv O$ then S is contained in a plane.

4.19. Let S be a regular level surface defined by $F(x,y,z) = 0$, with $F \in C^\infty(U)$ and $U \subset \mathbb{R}^3$ open. Show that, for all $p \in S$, the second fundamental form Q_p is the restriction to T_pS of the quadratic form on $\mathbb{R}^3$ induced by the Hessian matrix $\mathrm{Hess}(F)(p)$.

4.20. Consider the surface in $\mathbb{R}^3$ parametrized by $\varphi(u,v) = (u, v, u^2 + v^2)$. Determine the normal curvature of the curve $t \mapsto \varphi(t^2, t)$ contained in it.

4.21. Let ℓ be a line tangent to a regular surface in $\mathbb{R}^3$ at a point p, along a non-asymptotic direction. Show that the osculating circles (see Example 1.4.3) at p of all the curves on S passing through p and tangent to ℓ at p are contained in a sphere.

4.22. Determine the normal curvature of a regular curve σ whose support is contained in a sphere of radius 3.

PRINCIPAL, GAUSSIAN AND MEAN CURVATURES

4.23. Let $\sigma: I \to \mathbb{R}^3$ be a biregular curve parametrized by arc length, and assume there is $M > 0$ such that $\kappa(s) \le M$ for all $s \in I$. For all $\varepsilon > 0$ let $\varphi^\varepsilon: I \times (0, 2\pi) \to \mathbb{R}^3$ be given by

$$\varphi^\varepsilon(s, \theta) = \sigma(s) + \varepsilon \cos\theta\, \mathbf{n}(s) + \varepsilon \sin\theta\, \mathbf{b}(s).$$

(i) Prove that if $\varepsilon < 1/M$ then $\mathrm{d}\varphi^\varepsilon_x$ is injective for all $x \in I \times (0, 2\pi)$.

(ii) Assume that there exists $\varepsilon > 0$ such that φ^ε is globally injective and a homeomorphism with its image, so that it is a local parametrization of a surface $S^\varepsilon = \varphi^\varepsilon\big(I \times (0, 2\pi)\big)$. Find a normal versor field on S^ε, and compute the Gaussian and mean curvatures of S^ε.

(iii) Prove that for any interval $[a, b] \subset I$ there exists $\varepsilon > 0$ such that the restriction $\varphi^\varepsilon|_{(a,b) \times (0,2\pi)}$ is globally injective and a homeomorphism with its image.

4.24. Let $\rho: \mathbb{R} \to \mathbb{R}$ be a C^∞ function, and let $\varphi: \mathbb{R} \times (0, 2\pi) \to \mathbb{R}^3$ be given by

$$\varphi(z, \theta) = \big(\rho(z)\cos\theta, \rho(z)\sin\theta, z\big).$$

(i) Prove that φ parametrizes a regular surface S if and only if ρ is nowhere zero.

(ii) When S is a surface, write using ρ the first fundamental form with respect to the parametrization φ, and compute the Gaussian curvature of S.

4.25. Let $S \subset \mathbb{R}^3$ be the paraboloid of revolution of equation $z = x^2 + y^2$.

(i) Compute the Gaussian and mean curvatures of S at each point.

(ii) Compute the principal directions of S at the points of the support of the curve $\sigma: \mathbb{R} \to S$ given by

$$\sigma(t) = (2\cos t, 2\sin t, 4)\,.$$

4.26. Prove that $H^2 \ge K$ always on an orientable surface S. For which points $p \in S$ does equality hold?

4.27. Prove that cylinders have Gaussian curvature equal to zero everywhere.

4.28. Prove that the Gaussian curvature of a sphere with radius $R > 0$ is $K \equiv 1/R^2$, while its mean curvature (with respect to the usual orientation) is $H \equiv -1/R$.

4.29. Let $\sigma = (\sigma_1, \sigma_2): \mathbb{R} \to \mathbb{R}^2$ be a regular plane curve parametrized by arc length. Let $S \subset \mathbb{R}^3$ be the right cylinder on σ parametrized by $\varphi(u, v) = \big(\sigma_1(u), \sigma_2(u), v\big)$. Find the curvatures and the principal directions of S as functions of the curvature κ of σ.

4.30. Denote by $S \subset \mathbb{R}^3$ the subset

$$S = \{(x, y, z) \in \mathbb{R}^3 \mid (1 + |x|)^2 - y^2 - z^2 = 0\} .$$

(i)   Prove that $T = S \cap \{(x, y, z) \in \mathbb{R}^3 \mid x > 0\}$ is a regular surface.
(ii) Prove that S is not a regular surface.
(iii) Compute the Gaussian curvature and the mean curvature of T.

4.31. Let $S \subset \mathbb{R}^3$ be a surface, and $H \subset \mathbb{R}^3$ a plane such that $C = H \cap S$ is the support of a regular curve. Assume moreover that H is tangent to S at every point of C. Prove that the Gaussian curvature of S is zero at each point of C.

4.32. Let $S \subset \mathbb{R}^3$ be an orientable surface, and let N be a normal versor field on S. Consider the map $F \colon S \times \mathbb{R} \to \mathbb{R}^3$ defined by $F(p, t) = p + tN_p$.

(i) Show that F is smooth.
(ii) Show that the differential dF is singular at the point (p, t) if and only if $-1/t$ is one of the principal curvatures of S at p.

4.33. Prove that a surface with Gaussian curvature positive everywhere is necessarily orientable.

4.34. Let $\varphi \colon \mathbb{R}^2 \to \mathbb{R}^3$ be given by

$$\varphi(u, v) = (e^u \cos v, e^v \cos u, v) .$$

(i) Find the largest $c > 0$ such that φ restricted to $\mathbb{R} \times (-c, c)$ is a local parametrization of a regular surface $S \subset \mathbb{R}^3$.
(ii) Prove that the Gaussian curvature of S is nowhere positive.

(*Hint:* use the well-known formula $K = (eg - f^2)/(EG - F^2)$, without explicitly computing $eg - f^2$.)

4.35. Let $\varphi \colon U \to S$ be a local parametrization of a surface $S \subset \mathbb{R}^3$, and let N be the Gauss map induced by φ. Show that we have $N_u \wedge N_v = K(\varphi_u \wedge \varphi_v)$, where K is the Gaussian curvature.

4.36. Let $\sigma \colon [a, b] \to \mathbb{R}^3$ be a regular closed curve of class C^∞. Assume that the support of σ is contained in the ball with center in the origin and radius r. Show that there exists at least one point where σ has curvature at least $1/r$.

4.37. Let $\sigma \colon \mathbb{R} \to \mathbb{R}^3$ be a regular curve of class C^∞. Assume that the curvature of σ is greater than $1/r$ at every point. Is it true that the support of σ is contained in a ball of radius r?

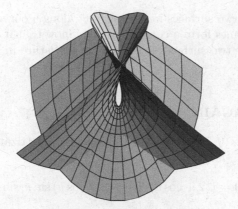

Fig. 4.8. Enneper's surface

LINES OF CURVATURE

4.38. Let $\varphi\colon \mathbb{R}^2 \to \mathbb{R}^3$ be the immersed surface (*Enneper's surface*; see Fig. 4.8) given by

$$\varphi(u,v) = \left(u - \frac{u^3}{3} + uv^2, v - \frac{v^3}{3} + vu^2, u^2 - v^2\right).$$

(i)   Prove that a connected component S of $\varphi(\mathbb{R}^2) \setminus (\{x = 0\} \cup \{y = 0\})$ is a regular surface.

(ii)   Show that the metric coefficients of S are $F \equiv 0$, $E = G = (1 + u^2 + v^2)^2$.

(iii) Prove that the form coefficients of S are $e = 2$, $g = -2$, $f = 0$.

(iv)   Compute the principal curvatures of S at each point.

(v)   Determine the lines of curvature of S.

4.39. Let $S \subset \mathbb{R}^3$ be an oriented surface with Gauss map $N\colon S \to S^2$, and take $p \in S$.

(i)   Prove that a vector $v \in T_p S$ is a principal direction if and only if

$$\langle dN_p(v) \wedge v, N(p) \rangle = 0.$$

(ii)   If $S = f^{-1}(a)$ is a level surface at a regular value $a \in \mathbb{R}$ for some function $f \in C^\infty(\mathbb{R}^3)$ prove that a vector $v \in T_p S$ is a principal direction if and only if

$$\det \begin{vmatrix} \dfrac{\partial f}{\partial x_1}(p) & \displaystyle\sum_{i=1}^{3} v_i \dfrac{\partial^2 f}{\partial x_i \partial x_1}(p) & v^1 \\[2ex] \dfrac{\partial f}{\partial x_2}(p) & \displaystyle\sum_{i=1}^{3} v_i \dfrac{\partial^2 f}{\partial x_i \partial x_2}(p) & v_2 \\[2ex] \dfrac{\partial f}{\partial x_3}(p) & \displaystyle\sum_{i=1}^{3} v_i \dfrac{\partial^2 f}{\partial x_i \partial x_3}(p) & v_3 \end{vmatrix} = 0.$$

4.40. Assume that two surfaces in $\mathbb{R}^3$ intersect along a curve σ in such a way that the tangent planes form a constant angle. Show that if σ is a line of curvature in one of the two surfaces it is a line of curvature in the other surface too.

ISOMETRIES, AGAIN

4.41. Let $\varphi \colon (0, 2\pi) \times (0, 2\pi) \to \mathbb{R}^3$ be the global parametrization of the surface $S \subset \mathbb{R}^3$ given by

$$\varphi(u, v) = \big((2 + \cos u) \cos v, (2 + \cos u) \sin v, \sin u\big) \, .$$

(i)   Determine the metric and form coefficients.
(ii) Determine the principal curvatures and the lines of curvature.
(iii) Determine whether S is locally isometric to a plane.

4.42. Let $\varphi, \tilde{\varphi} \colon \mathbb{R}^+ \times (0, 2\pi) \to \mathbb{R}^3$ be given by

$$\varphi(u, v) = (u \cos v, u \sin v, \log u), \qquad \tilde{\varphi}(u, v) = (u \cos v, u \sin v, v) \, ;$$

the image S of φ is the surface of revolution generated by the curve $(t, \log t)$, while the image $\tilde{S}$ of $\tilde{\varphi}$ is a portion of an helicoid. Prove that $K \circ \varphi \equiv \tilde{K} \circ \tilde{\varphi}$, where K (respectively, $\tilde{K}$) is the Gaussian curvature of S (respectively, $\tilde{S}$), but that $\tilde{\varphi} \circ \varphi^{-1}$ is not an isometry. Prove next that S and $\tilde{S}$ are not locally isometric. (*Hint:* in these parametrizations K depends on a single parameter, which can be determined in the same way in both surfaces. Assuming the existence of a local isometry, write the action on the coefficients of the first fundamental form: since the conditions imposed by the equality on these coefficients cannot be satisfied, the local isometry cannot exists.)

ASYMPTOTIC CURVES

4.43. Let $S \subset \mathbb{R}^3$ be the surface with global parametrization $\varphi \colon \mathbb{R}^2 \to \mathbb{R}^3$ given by $\varphi(u, v) = (u, v, uv)$.

(i) Determine the asymptotic curves of S.
(ii) Determine the values the curvature of the normal sections of S takes at the origin.

4.44. Let σ be a regular curve of class C^∞ on a surface S in $\mathbb{R}^3$. Show that if σ is an asymptotic curve then the normal to σ is always tangent to S.

4.45. Let S be a regular surface in $\mathbb{R}^3$.

(i) Show that if ℓ is a line segment contained in S then ℓ is an asymptotic curve for S.

(ii) Show that if S contains three distinct line segments passing through a given point $p \in S$ then the second fundamental form of S at p is zero, that is, p is a planar point.

4.46. Determine the asymptotic curves of the regular surface (see also Exercise 4.1) with global parametrization $\varphi(u,v) = (u,v,u^4 + v^4)$.

4.47. Let p be a point of a regular surface $S \subset \mathbb{R}^3$. Assume that at p there are exactly two distinct asymptotic directions. Show that there exist a neighborhood U of p in S and two maps $X, Y : U \to \mathbb{R}^3$ of class C^∞ such that for all $q \in U$ the vectors $X(q)$ and $Y(q)$ are linearly independent and asymptotic tangent vectors to S at q.

ELLIPTIC, HYPERBOLIC, PARABOLIC, PLANAR AND UMBILICAL POINTS

4.48. Characterization of umbilical points. Show that a point p of a regular surface S is umbilical if and only if, in a local parametrization of S, the first and the second fundamental form are equal. In particular, show that, when the form coefficients are different from zero, the point p is umbilical if and only if

$$\frac{E}{e} \equiv \frac{F}{f} \equiv \frac{G}{g} \, ,$$

and that in this case the normal curvature equals $\kappa_n = E/e$.

4.49. Let S be the graph of the function $f(x,y) = x^4 + y^4$. Prove that the point $O \in S$ is planar and that S lies within one of the two closed half-spaces bounded by the plane $T_O S$.

4.50. Find the umbilical points of the two-sheeted hyperboloid of equation

$$\frac{x^2}{a^2} + \frac{y^2}{b^2} - \frac{z^2}{c^2} = -1.$$

4.51. Find the umbilical points of the ellipsoid

$$\frac{x^2}{a^2} + \frac{y^2}{b^2} + \frac{z^2}{c^2} = 1 \, .$$

4.52. Let S be a connected regular surface in which every point is planar. Show that S is contained in a plane.

4.53. Let S be a closed, connected regular surface in $\mathbb{R}^3$. Show that S is a plane if and only if through every point p of S (at least) three distinct straight lines lying entirely in S pass.

4.54. Let S be the graph of the function $f(x,y) = x^3 - 3y^2x$ (this surface is sometimes called *monkey saddle*). Show that the point $O \in S$ is planar and that every neighborhood of O in S intersects both the open half-spaces bounded by the plane $T_O S$.

4.55. Let Q be a quadric in $\mathbb{R}^3$ such that Q is a regular surface but not a plane (see Problem 3.4).

(i)   Show that Q has no parabolic points.
(ii) Show that if Q has a hyperbolic point then all points of Q are hyperbolic.
(iii) Conclude that if Q has an elliptic point, then all points are elliptic.
(iv) Determine which quadrics have only hyperbolic points and which quadrics have only elliptic points.

4.56. Determine whether the origin O is an elliptic, hyperbolic, parabolic or planar point in the surface of equation:

(i) $z - xy = 0$;
(ii) $z - y^2 - x^4 = 0$;
(iii) $x + y + z - x^2 - y^2 - z^3 = 0$.

GAUSS' THEOREMA EGREGIUM

4.57. Compute the Christoffel symbols for the polar coordinates of the plane.

4.58. Prove that the Christoffel symbols can be computed using the following formula:

$$\Gamma_{ij}^k = \frac{1}{2} \sum_{l=1}^{2} g^{kl} \left(\frac{\partial g_{il}}{\partial x_j} + \frac{\partial g_{lj}}{\partial x_i} - \frac{\partial g_{ij}}{\partial x_l} \right) ,$$

where $g_{11} = E$, $g_{12} = g_{21} = F$, $g_{22} = G$, and (g^{ij}) is the inverse matrix of matrix (g_{ij}).

4.59. Check that the equations (4.26) written for the other possible values of i, j, k and r are either trivially satisfied, or a consequence of the symmetry of the Christoffel symbols, or equivalent to (4.27).

4.60. Let E be the ellipsoid of equation

$$\frac{1}{4}x^2 + y^2 + \frac{1}{9}z^2 = 3 .$$

(i) Compute the Gaussian curvature K and the principal directions of E at the point $p = (2,1,3) \in E$.
(ii) Compute the integral of the Gaussian curvature K on the intersection of E with the octant

$$Q = \{(x,y,z) \mid x \geq 0, y \geq 0, z \geq 0\} .$$

4.61. Check that the compatibility conditions that are a consequence of the identity $\partial^2(N \circ \varphi)/\partial x_i \partial x_j \equiv \partial^2(N \circ \varphi)/\partial x_j \partial x_i$ are either always satisfied or equivalent to (4.28).

CONFORMAL MAPS

Definition 4.E.1. A map $H: S_1 \to S_2$ of class C^∞ between two surfaces in $\mathbb{R}^3$ is *conformal* if there exists a function $\lambda: S_1 \to \mathbb{R}^*$ of class C^∞ nowhere vanishing such that

$$\langle dH_p(v_1), dH_p(v_2) \rangle_{H(p)} = \lambda^2(p) \langle v_1, v_2 \rangle_p$$

for all $p \in S_1$ and all $v_1, v_2 \in T_p S_1$. The map H is *locally conformal at* p if there are neighborhoods U_1 of p in S_1 and U_2 of $H(p)$ in S_2 such that the restriction of $H|_{U_1}: U_1 \to U_2$ is conformal. Two surfaces S_1 and S_2 are *conformally equivalent* if there exists a conformal diffeomorphism $H: S_1 \to S_2$. Finally, S_1 is *locally conformal* to S_2 if for all $p \in S_1$ there exist a point $q \in S_2$ and a conformal diffeomorphism between a neighborhood of p in S_1 and a neighborhood of q in S_2.

4.62. Show that the stereographic projection (see Exercise 3.4) is a conformal map.

4.63. Prove an analogue of Proposition 4.1.20 for conformal maps: Let $S, \tilde{S} \subset \mathbb{R}^3$ be two surfaces. Then S is locally conformal to $\tilde{S}$ if and only if for every point $p \in S$ there exist a point $\tilde{p} \in \tilde{S}$, an open set $U \subseteq \mathbb{R}^2$, a function $\lambda \in C^\infty(U)$ nowhere zero, a local parametrization $\varphi: U \to S$ of S centered at p, and a local parametrization $\tilde{\varphi}: U \to \tilde{S}$ of $\tilde{S}$ centered at $\tilde{p}$ such that $\tilde{E} \equiv \lambda^2 E$, $\tilde{F} \equiv \lambda^2 F$ and $\tilde{G} \equiv \lambda^2 G$, where E, F, G (respectively, $\tilde{E}$, $\tilde{F}$, $\tilde{G}$) are the metric coefficients of S with respect to φ (respectively, of $\tilde{S}$ with respect to $\tilde{\varphi}$).

Definition 4.E.2. A local parametrization of a surface S is called *isothermal* if $E \equiv G$ e $F \equiv 0$.

4.64. Prove that two surfaces both having an atlas consisting of isothermal local parametrizations are locally conformal. (*Remark*: It is possible to prove that every regular surface admits an atlas consisting of isothermal local parametrizations; as a consequence, two regular surfaces are always locally conformal. See, for instance, [3].)

4.65. Let $\varphi: U \to S$ be a isothermal local parametrization. Prove that

$$\varphi_{uu} + \varphi_{vv} = 2EHN , \quad \text{and that} \quad K = -\frac{\Delta \log G}{G} ,$$

where Δ denotes the Laplacian.

RULED SURFACES

Definition 4.E.3. A *conoid* in $\mathbb{R}^3$ is a ruled surface in $\mathbb{R}^3$ whose rulings are parallel to a plane H and intersect a straight line ℓ. The conoid is said to be *right* if the line ℓ is orthogonal to the plane H. The line ℓ is called *axis* of the conoid.

4.66. Show that the right helicoid parametrized as in Problem 3.2 is a right conoid.

4.67. Let $S \subset \mathbb{R}^3$ be a right conoid having rulings parallel to the plane $z = 0$ and the z-axis as its axis. Prove that it is the image of a map $\varphi: \mathbb{R}^2 \to \mathbb{R}^3$ of the form

$$\varphi(t, v) = \big(v \cos f(t), v \sin f(t), t\big) ,$$

where $f: \mathbb{R} \to \mathbb{R}$ is such that $f(t)$ is a determination of the angle between the ruling contained in $z = t$ and the plane $y = 0$. Prove that the map φ is an immersed surface if f is of class C^∞.

4.68. Prove that the cylinders introduced in Definition 3.P.3 are ruled surfaces.

4.69. Given a regular curve $\sigma: I \to \mathbb{R}^3$ of class C^∞, and a curve $\mathbf{v}: I \to S^2$ of class C^∞ on the sphere, let $\varphi: I \times \mathbb{R}^* \to \mathbb{R}^3$ be defined by

$$\varphi(t, v) = \sigma(t) + v\,\mathbf{v}(t) . \tag{4.30}$$

Prove that φ is an immersed surface if and only if $\mathbf{v}$ and $\sigma' + v\mathbf{v}'$ are everywhere linearly independent. In this case, φ is called a *parametrization in ruled form* of its support S, the curve σ is called *base curve* or *directrix*, and the lines $v \mapsto \varphi(t_0, v)$ are called *(rectilinear) generators* of S.

4.70. Let $S \subset \mathbb{R}^3$ be the hyperbolic paraboloid of equation $z = x^2 - y^2$.

(i) Find two parametric representations in ruled form (see Exercise 4.69) of S, corresponding to two different systems of generators.
(ii) Determine the generators of the two systems passing through the point $p = (1, 1, 0)$.

4.71. Prove that the tangent plane at the points of a generator of the (non-singular part of the) tangent surface (see Problem 4.20) to a biregular curve C coincides with the osculating plane to the curve C at the intersection point with the generator.

4.72. Let $\sigma: I \to \mathbb{R}^3$ be a regular plane curve of class C^∞, parametrized by arc length, with curvature $0 < \kappa < 1$. Let $\varphi: I \times (0, 2\pi) \to \mathbb{R}^3$ be the immersed surface given by $\varphi(t, v) = \sigma(t) + \cos v\,\mathbf{n}(t) + \mathbf{b}(t)$.

(i) Determine the Gaussian curvature and the mean curvature at every point of the support S of φ.

(ii) Determine the lines of curvature at each point of S.

4.73. Let $\sigma: I \to \mathbb{R}^3$ be a biregular curve of class C^∞, parametrized by arc length, and let $\varphi: I \times (-\varepsilon, \varepsilon) \to \mathbb{R}^3$ be the map given by $\varphi(s, \lambda) = \sigma(s) + \lambda \mathbf{n}(s)$.

(i) Show that, when ε is small enough, φ is a global parametrization of a surface S, called *normal surface* of σ.

(ii) Show that the tangent plane to S at a point of σ is the osculating plane to σ.

MINIMAL SURFACES

Definition 4.E.4. A surface $S \subset \mathbb{R}^3$ is *minimal* if its mean curvature vanishes everywhere.

4.74. Prove that there are no compact minimal surfaces.

4.75. Let $\varphi: U \to S$ be a global parametrization of a surface S. Given $h \in C^\infty(U)$, the *normal variation* of φ along h is the map $\varphi^h: U \times (-\varepsilon, \varepsilon) \to \mathbb{R}^3$ defined by

$$\varphi^h(x, t) = \varphi(x) + t\, h(x) N\big(\varphi(x)\big) \,,$$

where $N: \varphi(U) \to S^2$ is the Gauss map induced by φ.

(i) Prove that for every open set $U_0 \subset U$ with compact closure in U there exists an $\varepsilon > 0$ such that $\varphi^h|_{U_0 \times (-\varepsilon, \varepsilon)}$ is an immersed surface.

(ii) Let $R \subset U$ be a regular region, and $A_R^h: (-\varepsilon, \varepsilon) \to \mathbb{R}$ the function defined by $A_R^h(t) = \text{Area}\big(\varphi^h(R)\big)$. Prove that A_R^h is differentiable at zero and that

$$\frac{\mathrm{d}A_R^h}{\mathrm{d}t}(0) = -\int_{\varphi^h(R)} 2hH \, \mathrm{d}\nu \,.$$

(iii) Prove that $\varphi(U)$ is minimal if and only if

$$\frac{\mathrm{d}A_R^h}{\mathrm{d}t}(0) = 0$$

for every $h \in C^\infty(U)$ and every regular region $R \subset U$.

4.76. Prove that the catenoid is a minimal surface, and that no other surface of revolution is minimal.

4.77. Prove that the helicoid is a minimal surface. Conversely, prove that if $S \subset \mathbb{R}^3$ is a minimal ruled surface whose planar points are isolated then S is a helicoid. (*Hint:* Exercise 1.62 can help.)

4.78. Prove that Enneper's surface (see Exercise 4.38) is minimal where it is regular.

4.79. Let $S \subset \mathbb{R}^3$ be an oriented surface without umbilical points. Prove that S is a minimal surface if and only if the Gauss map $N: S \to S^2$ is a conformal map. Use this result to construct isothermal local parametrizations on minimal surfaces without umbilical points.

Supplementary material

4.7 Transversality

In this section we want to show that every closed surface in $\mathbb{R}^3$ is orientable, by using an argument due to Samelson (see [20]). To this aim, we shall also give a short introduction to an important notion in differential geometry: transversality.

Definition 4.7.1. A map $F: \Omega \to \mathbb{R}^3$ of class C^∞ defined on an open set $\Omega \subseteq \mathbb{R}^n$ is *transversal to a surface* $S \subset \mathbb{R}^3$ *at a point* $x \in \Omega$ if $F(x) \notin S$ or $F(x) \in S$ and $dF_x(\mathbb{R}^n) + T_{F(x)}S = \mathbb{R}^3$; and we shall say that F is *transversal* to S if it is so at all points.

Example 4.7.2. A curve $\sigma: I \to \mathbb{R}^3$ of class C^∞ is transversal to S at $t \in I$ if $\sigma(t) \notin S$ or $\sigma(t) \in S$ and $\sigma'(t) \notin T_{\sigma(t)}S$; see Fig. 4.8.(a). Analogously, a map $F: U \to \mathbb{R}^3$ of class C^∞ defined on an open set $U \subseteq \mathbb{R}^2$ of $\mathbb{R}^2$ is transversal to S at $x \in U$ if $F(x) \notin S$ or $F(x) \in S$ and $dF_x(\mathbb{R}^2)$ is not contained in $T_{F(x)}S$; see Fig. 4.9.(b).

Remark 4.7.3. It is clear (check it!) that if a map $F: \Omega \to \mathbb{R}^3$ is transversal to a surface S in a point, then it is so in all nearby points too. As a result, the set of points of transversality is always open.

One of the reasons why transversal maps are useful is given by the following:

Lemma 4.7.4. *Let* $F: U \to \mathbb{R}^3$, *where* U *is an open subset of* $\mathbb{R}^2$, *be a map of class* C^∞ *transversal to a surface* $S \subset \mathbb{R}^3$. *Then the connected components of* $F^{-1}(S)$ *are 1-submanifolds of* $\mathbb{R}^2$.

Proof. Let $x_0 \in U$ be such that $p_0 = F(x_0) \in S$. Without loss of generality, we may find a neighborhood $W \subseteq \mathbb{R}^3$ of p_0 such that $S \cap W$ is the graph of a function $f: U \to \mathbb{R}$ with respect to the xy-plane. Let $U_0 \subseteq U$ be a neighborhood of x_0 such that $F(U_0) \subset W$. Then $x \in U_0$ is such that $F(x) \in S$ if and only if

$$f\big(F_1(x), F_2(x)\big) - F_3(t) = 0 \,.$$

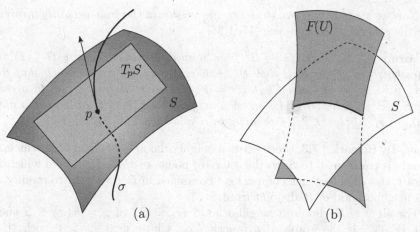

Fig. 4.9.

Define $g: U_0 \to \mathbb{R}$ with $g = f \circ (F_1, F_2) - F_3$; so x belongs to $U_0 \cap F^{-1}(S)$ if and only if $g(x) = 0$. Now, $g(x_0) = 0$ and

$$\frac{\partial g}{\partial x_j}(x_0) = \frac{\partial F_1}{\partial x_j}(x_0)\frac{\partial f}{\partial x_1}\big(F_1(x_0), F_2(x_0)\big) + \frac{\partial F_2}{\partial x_j}(x_0)\frac{\partial f}{\partial x_2}\big(F_1(x_0), F_2(x_0)\big)$$
$$- \frac{\partial F_3}{\partial x_j}(x_0).$$

If it were $\nabla g(x_0) = O$, both $\partial F/\partial x_1(x_0)$ and $\partial F/\partial x_2(x_0)$ would belong to $T_{p_0}S$, by Example 3.3.14, against the hypothesis that F is transversal to S. So $\nabla g(x_0) \neq O$, and by Proposition 1.1.18 we have that $U_0 \cap F^{-1}(S)$ is a 1-submanifold. Since $x_0 \in F^{-1}(S)$ was arbitrary, it follows that every component of $F^{-1}(S)$ is a 1-submanifold. $\qquad\square$

Another important property of trasversality is that if a map is transversal to a surface at some point then we can always modify it to obtain a map transversal everywhere. We shall just need a weaker version of this result, but still representative of the general case. We start with the following

Lemma 4.7.5. *For all open $U \subseteq \mathbb{R}^2$, $f \in C^\infty(U)$ and $\delta > 0$ there exist $a \in \mathbb{R}^2$ and $b \in \mathbb{R}$ with $\|a\|$, $|b| < \delta$ such that 0 is a regular value of the function $g: U \to \mathbb{R}$ given by $g(x) = f(x) - \langle a, x \rangle - b$.*

Proof. Choose a sequence $\{a_j\} \subset \mathbb{R}^2$ converging at the origin, and set $f_j(x) = f(x) - \langle a_j, x \rangle$. By Sard's Theorem 3.5.2, the set $S_j \subset \mathbb{R}$ of singular values of f_j has measure zero in $\mathbb{R}$; hences, $S = \bigcup_j S_j$ has measure zero in $\mathbb{R}$ too. In particular, we may find a sequence $\{b_h\} \subset \mathbb{R}$ converging to 0 such that each b_h is a regular value for all f_j. It follows that if we choose j and h so that $\|a_h\|$, $b_j < \delta$ then 0 is a regular value of $f(x) - \langle a_j, x \rangle - b_h$, as required. $\qquad\square$

Now we are able to prove the following version of the *transversality theorem* (for the general statement, see [15, 1.35]):

Theorem 4.7.6. *Let $F: \Omega \to \mathbb{R}^3$ be a map of class C^∞, where $\Omega \subseteq \mathbb{R}^2$ is a neighborhood of the unit disk $\overline{B^2}$ with center in the origin, such that the curve $\sigma: [0, 2\pi] \to \mathbb{R}^3$ given by $\sigma(t) = F(\cos t, \sin t)$ is transversal to a surface $S \subset \mathbb{R}^3$. Then there exist an open neighborhood $\Omega_0 \subseteq \Omega$ of $\overline{B^2}$ and a map $\tilde{F}: \Omega_0 \to \mathbb{R}^3$ of class C^∞ transversal to S and such that $\tilde{F}|_{S^1} \equiv F|_{S^1}$.*

Proof. By Remark 4.7.3, there exists a neighborhood V of S^1 in Ω in which F already is transversal to S; so the set K of points of $\Omega_0 = \overline{B^2} \cup V$ in which F is not transversal to S is a compact set contained in B^2; we want to modify F in a neighborhood of K disjoint from S^1.

For all $x \in K$ choose a neighborhood $W_x \subseteq \mathbb{R}^3$ of $p = F(x) \in S$ such that $W_x \cap S$ is the graph of a function g_x. Chose next $\varepsilon_x > 0$ such that $\overline{B}(x, 3\varepsilon_x) \subseteq B^2 \cap F^{-1}(W_x)$. Since K is compact, we may find $x_1, \ldots, x_r \in K$ such that $\{B(x_1, \varepsilon_1), \ldots, B(x_r, \varepsilon_r)\}$ is an open cover of K (where we have written, for the sake of simplicity, $\varepsilon_j = \varepsilon_{x_j}$). Set $W_j = W_{x_j}$ and $g_j = g_{x_j}$ for $j = 1, \ldots, r$. We shall modify F on each $B(x_j, 2\varepsilon_j)$ until we get the required map.

Let $U = \bigcup_{j=1}^r B(x_j, 2\varepsilon_j)$. Set $F_0 = F$ and, given $k = 1, \ldots, r$, suppose we have constructed a $F_{k-1}: \Omega_0 \to \mathbb{R}^3$ that coincides with F on $\Omega_0 \setminus U$, is transversal to S in all points of $\bigcup_{j=1}^{k-1} \overline{B}(x_j, \varepsilon_j)$, and maps each $\overline{B}(x_j, 2\varepsilon_j)$ to the corresponding W_j; we want to construct F_k with analogous properties.

Let $\chi: \mathbb{R}^2 \to [0, 1]$ be the function given by Corollary 1.5.3 with $p = x_k$ and $r = 2\varepsilon_k$, so that $\chi|_{\overline{B}(x_k, \varepsilon_k)} \equiv 1$ and $\chi|_{\mathbb{R}^2 \setminus B(x_k, 2\varepsilon_k)} \equiv 0$. Assume, without loss of generality, that $W_k \cap S$ is a graph on the xy-plane, and for all $a \in \mathbb{R}^2$ and $b \in \mathbb{R}$ define $F_{a,b}: \overline{B^2} \to \mathbb{R}^3$ by setting

$$F_{a,b}(x) = F_{k-1}(x) + \chi(x)\big[\langle a, x \rangle + b\big]\mathbf{e}_3 \ ;$$

we want to choose a and b in a suitable way. First of all, it is clear that $F_{a,b}$ coincides with F on $\Omega_0 \setminus U$ for all a and b. Next, we may find $\delta_1 > 0$ such that $F_{a,b}\big(B(x_j, 2\varepsilon_j)\big) \subset W_j$ for all $j = 1, \ldots, r$ as soon as $\|a\|, |b| < \delta_1$. We may also find $\delta_2 > 0$ such that if $\|a\|, |b| < \delta_2 \le \delta_1$ then $F_{a,b}$ is transversal to S in all points of $\bigcup_{j=1}^{k-1} \overline{B}(x_j, \varepsilon_j)$. Indeed, clearly it already is so in $\bigcup_{j=1}^{k-1} \overline{B}(x_j, \varepsilon_j) \setminus B(x_k, 2\varepsilon_k)$. Set $H = \big(\bigcup_{j=1}^{k-1} \overline{B}(x_j, \varepsilon_j)\big) \cap \overline{B}(x_k, 2\varepsilon_k)$, and let $\Phi: H \times \mathbb{R}^2 \times \mathbb{R} \to \mathbb{R} \times \mathbb{R}^2$ be given by

$$\Phi(x, a, b) = \big(h_k \circ F_{a,b}(x), \nabla(h_k \circ F_{a,b})(x)\big) \ ,$$

where $h_k: W_k \to \mathbb{R}$ is given by $h_k(p) = g_k(p_1, p_2) - p_3$, so that $h_k(p) = 0$ if and only if $p \in S$. In particular, $F_{a,b}$ is not transversal to S in $x \in \overline{B}(x_k, 2\varepsilon_k)$ if and only if $\Phi(x, a, b) = (0, O)$, by (why?) Proposition 3.3.11. By the assumption on $F_{k-1} = F_{O,0}$, we know that $\Phi\big(H \times \{(O, 0)\}\big) \subseteq \mathbb{R}^3 \setminus \{O\}$, which is an open set;

since H is compact, we find $0 < \delta_2 \leq \delta_1$ such that $\Phi(H \times \{(a,b)\}) \subseteq \mathbb{R}^3 \setminus \{O\}$ as soon as $\|a\|, |b| < \delta_2$.

Finally, we may apply Lemma 4.7.5 to find $a \in \mathbb{R}^2$ and $b \in \mathbb{R}$ with $\|a\|$, $|b| < \delta_2$ such that 0 is a regular value for $h_k \circ F_{a,b}$ on $\overline{B_k(x_k, \varepsilon_k)}$; this means that $\Phi(x, a, b) \neq (0, O)$ for all $x \in \overline{B_k(x_k, \varepsilon_k)}$, and so $F_k = F_{a,b}$ is transversal to S in all points of $\bigcup_{j=1}^k \overline{B(x_j, \varepsilon_j)}$.

Proceeding up to $k = r$ we get that $\tilde{F} = F_r$ is as required. $\qquad\square$

Corollary 4.7.7. *Let $\sigma: [0, 2\pi] \to \mathbb{R}^3$ be a closed curve of class C^∞, transversal to a surface $S \subset \mathbb{R}^3$. Then there exist a neighborhood $\Omega_0 \subset \mathbb{R}^2$ of $\overline{B^2}$ in the plane and a map $F: \Omega_0 \to \mathbb{R}^3$ of class C^∞ transversal to S and such that $F(\cos t, \sin t) = \sigma(t)$ for all $t \in [0, 2\pi]$.*

Proof. First of all, let $\beta: \mathbb{R} \to [0, 1]$ be the function given by Corollary 1.5.2 with $a = 0$ and $b = 2\pi$, and set $\alpha = 1 - \beta$; hence, $\alpha: \mathbb{R} \to [0, 1]$ is a C^∞ function such that $\alpha^{-1}(0) = (-\infty, 0]$ and $\alpha^{-1}(1) = [1, +\infty)$. Let then $\hat{F}: \mathbb{R}^2 \to \mathbb{R}^3$ be defined by

$$\hat{F}(r \cos t, r \sin t) = \alpha(r^2)\sigma(t) \; ;$$

it is straightforward to verify (exercise) that $\hat{F}$ is of class C^∞, and obviously satisfies $\hat{F}(\cos t, \sin t) = \sigma(t)$. So it is sufficient to apply the previous theorem to $\hat{F}$. $\qquad\square$

We shall need these results about transversality to prove the following:

Proposition 4.7.8. *Let $S \subset \mathbb{R}^3$ be a closed surface, and $\sigma: [0, 2\pi] \to \mathbb{R}^3$ a closed curve of class C^∞ transversal to S. Then either the support of σ is disjoint from S or it intersects S in at least two points.*

Proof. By contradiction, let $\sigma: [0, 1] \to \mathbb{R}^3$ be a closed curve of class C^∞, transversal to S and with support intersecting S in exactly one point, and let $F: \Omega_0 \to \mathbb{R}^3$ be given by Corollary 4.7.7. Since S is closed in $\mathbb{R}^3$, the set $F^{-1}(S) \cap B^2$ must be closed in B^2. In particular, the 1-submanifolds (Lemma 4.7.4) that make up $F^{-1}(S) \cap B^2$ are either compact sets contained in B^2 or homeomorphic to intervals whose closure intersects S^1. Since σ intersects S, there is at least one component C of $F^{-1}(S) \cap B^2$ whose closure intersects S^1. Since $F(S^1)$ is the support of σ, the intersection $\overline{C} \cap S^1$ consists of a single point. But by Lemma 4.7.4 (applied to $U = \Omega_0$) and Theorem 1.6.8 we know that $\overline{C}$, being forced to intersect S^1 in one point only, is the support of a closed curve, necessarily (why?) tangent to S^1 in the point of intersection. But then σ in that point should be tangent to S (as $F(C) \subset S$), against the transversality of σ. $\qquad\square$

In what follows we shall often have to construct curves of class C^∞ joining specific points. To do so, we shall use the following:

Lemma 4.7.9. *Let $\sigma\colon [0,1] \to \mathbb{R}^3$ and $\tau\colon [0,1] \to \mathbb{R}^3$ be two curves of class C^∞ with $\sigma(1) = \tau(0)$. Then there exists a curve $\sigma \star \tau\colon [0,1] \to \mathbb{R}^3$ of class C^∞ such that $\sigma \star \tau|_{[0,1/2]}$ is a new parametrization of σ, and $\sigma \star \tau|_{[1/2,1]}$ is a new parametrization of τ. Moreover, if $\tau(1) = \sigma(0)$, then $\sigma \star \tau$ is of class C^∞ as a closed curve as well.*

Proof. If $\beta\colon \mathbb{R} \to [0,1]$ is the C^∞ function given by Corollary 1.5.2 with $a = 0$ and $b = 1$, then $\sigma \circ \beta$ is a reparametrization of σ whose derivatives are all zero in 0 and 1. In particular, the curve $\sigma \star \tau\colon [0,1] \to \mathbb{R}^3$ defined by

$$\sigma \star \tau(t) = \begin{cases} \sigma\big(\beta(2t)\big) & \text{if } 0 \le t \le 1/2\,, \\ \tau\big(\beta(2t-1)\big) & \text{if } 1/2 \le t \le 1\,, \end{cases}$$

is a curve of class C^∞ (in the joining point all derivatives are zero, and thus they attach continuously) with the required properties. $\qquad\square$

As a first application of this lemma we prove that two points of a surface (or of a connected open set of $\mathbb{R}^3$) can always be joined by a C^∞ curve. We need a definition:

Definition 4.7.10. A subset $X \subseteq \mathbb{R}^3$ is *locally connected by C^∞ arcs* if every $x \in X$ has a neighborhood W in $\mathbb{R}^3$ such that for all $y \in W \cap X$ there exists a curve $\sigma\colon [0,1] \to \mathbb{R}^3$ of class C^∞ with support contained in X and such that $\sigma(0) = x$ and $\sigma(1) = y$.

For instance, connected surfaces and connected open subsets in $\mathbb{R}^3$ are locally connected by C^∞ arcs (exercise). Then:

Corollary 4.7.11. *Let $X \subseteq \mathbb{R}^3$ be a subset of $\mathbb{R}^3$ connected and locally connected by C^∞ arcs. Then for all x, $y \in X$ there exists a curve $\sigma\colon [0,1] \to \mathbb{R}^3$ of class C^∞ with support is contained in X such that $\sigma(0) = x$ and $\sigma(1) = y$.*

Proof. Introduce on X the equivalence relation $\sim$ defined by saying that $x \sim y$ if and only if there exists a curve $\sigma\colon [0,1] \to \mathbb{R}^3$ of class C^∞ with support contained in X such that $\sigma(0) = x$ and $\sigma(1) = y$. By Lemma 4.7.9, this actually is an equivalence relation. Moreover, local connectivity by C^∞ arcs (together, again, with Lemma 4.7.9) tells us that equivalence classes are open. So the equivalence classes form a partition in open sets of X; since X is connected, there is only one equivalence class, that is every pair of points in X can be joined by a curve of class C^∞, as required. $\qquad\square$

At last, we may begin to reap the benefits of our work, and see what all this implies for closed surfaces in $\mathbb{R}^3$. The first result is the first half of the *Jordan-Brouwer theorem for surfaces* (see Theorem 4.8.4 for the complete statement):

Proposition 4.7.12. *Let $S \subset \mathbb{R}^3$ be a closed surface. Then $\mathbb{R}^3 \setminus S$ is disconnected.*

Proof. Suppose to the contrary that $\mathbb{R}^3 \setminus S$ is connected. Choose $p \in S$, and a versor $N(p) \in \mathbb{R}^3$ orthogonal to T_pS. It is easy to see (exercise) that there exists $\varepsilon > 0$ such that $p + tN(p) \notin S$ if $0 < |t| \le \varepsilon$. Corollary 4.7.11 provides us with a curve τ of class C^∞ with support is contained in $\mathbb{R}^3 \setminus S$ and joining $p + \varepsilon N(p)$ and $p - \varepsilon N(p)$. Lemma 4.7.9 allows us to smoothly join the curve τ and the line segment $t \mapsto p + tN(p)$; but in this way we get a closed curve of class C^∞, transversal to S, and intersecting the surface in a single point, against Proposition 4.7.8. □

To show that every closed surface in $\mathbb{R}^3$ is orientable, we shall also need a new characterization of non-orientable surfaces.

Definition 4.7.13. Let $\sigma \colon I \to S$ be a curve of class C^∞ with support contained in a surface $S \subset \mathbb{R}^3$. A *normal field along σ* is a map $\tilde{N} \colon I \to S^2$ of class C^∞ such that $\tilde{N}(t)$ is a versor orthogonal to $T_{\sigma(t)}S$ for all $t \in I$.

Lemma 4.7.14. *Let $S \subset \mathbb{R}^3$ be a surface. Then S is not orientable if and only if there exist a closed curve $\sigma \colon [0,1] \to S$ of class C^∞ and a normal field $\tilde{N} \colon [0,1] \to S^2$ along σ such that $\tilde{N}(0) = -\tilde{N}(1)$.*

Proof. Suppose S is orientable. Given a closed curve $\sigma \colon [0,1] \to S$ of class C^∞ and a normal field $\tilde{N} \colon [0,1] \to S^2$ along σ, choose the Gauss map $N \colon S \to S^2$ such that $N(\sigma(0)) = \tilde{N}(0)$. Then, by continuity, we have $\tilde{N} \equiv N \circ \sigma$, and so in particular $\tilde{N}(1) = N(\sigma(1)) = N(\sigma(0)) = \tilde{N}(0)$.

Conversely, suppose that for every closed curve $\sigma \colon [0,1] \to S$ of class C^∞ and every normal field $\tilde{N} \colon [0,1] \to \mathbb{R}^3$ along σ we have $\tilde{N}(1) = \tilde{N}(0)$; we want to define a Gauss map $N \colon S \to S^2$.

Fix a $p_0 \in S$ and a versor $N(p_0)$ orthogonal to $T_{p_0}S$. Let $p \in S$, and choose a curve $\sigma \colon [0,1] \to S$ of class C^∞ such that $\sigma(0) = p_0$ and $\sigma(1) = p$ (it exists by Corollary 4.7.11). Now, arguing as usual using the Lebesgue number, we find a partition $0 = t_0 < \cdots < t_k = 1$ of $[0,1]$ such that $\sigma([t_{j-1}, t_j])$ is contained in the image of a local parametrization φ_j for $j = 1, \ldots, k$. Without loss of generality, we may suppose that the Gauss map N_1 induced by φ_1 coincides with $N(p_0)$ in p_0; define then $\tilde{N} \colon [0, t_1] \to S^2$ by setting $\tilde{N}(t) = N_1(\sigma(t))$. Analogously, we may suppose that the Gauss map N_2 induced by φ_2 coincides with $\tilde{N}(t_1)$ in $\sigma(t_1)$; so we may use N_2 to extend $\tilde{N}$ in a C^∞ way to $[0, t_2]$. Proceeding in this way we get a $\tilde{N} \colon [0,1] \to S^2$ of class C^∞ (more precisely, it is the restriction to the support of σ of a C^∞ map defined in a neighborhood) such that $\tilde{N}(t)$ is a versor orthogonal to $T_{\sigma(t)}S$ for all $t \in [0,1]$; set $N(p) = \tilde{N}(1)$. To conclude the proof it suffices (why?) to remark that $N(p)$ does not depend on the curve we have chosen. Indeed, let $\tau \colon [0,1] \to S$ be another curve of class C^∞ from p_0 to p, such that the normal field $\tilde{N}_\tau \colon [0,1] \to S^2$ along τ obtained with the previous argument satisfies $\tilde{N}_\tau(1) = -\tilde{N}(1)$. Set $\hat{\tau}(t) = \tau(1-t)$ and

$\hat{N}(t) = -\tilde{N}_\tau(1-t)$. Then $\sigma \star \hat{\tau}$ is a closed curve of class C^∞ such that the map $\tilde{N} \star \hat{N} \colon [0,1] \to S^2$, constructed as $\sigma \star \hat{\tau}$, is a normal field along $\sigma \star \hat{\tau}$ with $\tilde{N} \star \hat{N}(1) = -\tilde{N} \star \hat{N}(0)$, against the initial hypothesis. □

We are now finally able to prove the *Brouwer-Samelson theorem*:

Theorem 4.7.15 (Brouwer-Samelson). *Every closed surface $S \subset \mathbb{R}^3$ in $\mathbb{R}^3$ is orientable.*

Proof. Assume by contradiction that S is not orientable. Then the previous lemma provides us with a closed curve $\tau \colon [0,1] \to S$ of class C^∞ and a normal field $\tilde{N} \colon [0,1] \to S^2$ along τ with $\tilde{N}(0) = -\tilde{N}(1)$.

It is easy to verify (exercise) that for all $t_0 \in [0,1]$ there exist $\delta, \varepsilon > 0$ such that $\tau(t) + s\tilde{N}(t) \notin S$ for all $t \in (t_0 - \delta, t_0 + \delta)$ and $s \in (0, \varepsilon]$. Since $[0,1]$ is compact, we may find $\varepsilon_0 > 0$ such that $\tau(t) + \varepsilon_0 \tilde{N}(t) \notin S$ for all $t \in [0,1]$. Let then $\sigma \colon [0, 2\pi] \to \mathbb{R}^3$ be a closed curve of class C^∞ such that $\sigma(t) = \tau(t) + \varepsilon_0 \tilde{N}(t)$ for $t \in [0,1]$ and such that $\sigma|_{[1,2\pi]}$ describes the segment from $\tau(1) + \varepsilon_0 \tilde{N}(1)$ to $\tau(0) + \varepsilon_0 \tilde{N}(0)$ forming a C^∞ closed curve; such a curve exists by Lemma 4.7.9. But then σ is transversal to S and its support intersects S in a single point, contradicting Proposition 4.7.8. □

Exercises

Definition 4.E.5. Two surfaces S_1, $S_2 \subset \mathbb{R}^3$ are *transversal* (or *intersects transversally*) *at a point* $p \in S_1 \cap S_2$ if $T_p S_1 + T_p S_2 = \mathbb{R}^3$; and they are *transversal* if they are so at each intersection point. Finally, the *angle* between S_1 and S_2 at $p \in S_1 \cap S_2$ is the angle between a normal versor to S_1 at p and a normal versor to S_2 at p.

4.80. Prove that if two surfaces S_1, $S_2 \subset \mathbb{R}^3$ are transversal then $S_1 \cap S_2$ is a 1-submanifold of $\mathbb{R}^3$ (if not empty).

4.81. Show that if a sphere or a plane intersects transversally a surface forming a non-zero angle, then the intersection is a line of curvature of the surface.

4.82. Let C be the intersection between the surface $S_1 \subset \mathbb{R}^3$ of equation $x_1^2 + x_2^2 + x_3^2 = 4$ and the surface $S_2 \subset \mathbb{R}^3$ of equation $(x_1 - 1)^2 + x_2^2 = 1$. Determine which subset of C is a 1-submanifold, and draw C and the two surfaces.

4.83. Let C be the intersection between the surface $S_1 \subset \mathbb{R}^3$ of equation $x_1^2 + x_2^2 + x_3^2 = 1$ and the surface $S_2 \subset \mathbb{R}^3$ of equation $(x_1 - 1)^2 + x_2^2 = 1$. Prove that C is a (non-empty) 1-submanifold, and draw C and the two surfaces.

4.84. Let $S \subset \mathbb{R}^3$ be a regular surface. The *affine normal line* of S at $p \in S$ is the straight line through p and parallel to a versor normal to S at p. Show that if all affine normal lines of S properly intersect a fixed line ℓ then S is a surface of revolution (*Hint:* if $p \in S \setminus \ell$, the plane H_p generated by p and ℓ contains all the normal lines through the points of $H_p \cap S$. Note that the plane $H_p^\perp$ through p and orthogonal to ℓ intersects S transversally.)

4.85. Let C be a 1-manifold defined implicitly as the intersection of two transversal surfaces S_1 and S_2. At a point p of C, denote by α the angle between the normal lines through p to the surfaces, by κ the curvature of C at p, and (for $j = 1, 2$) by κ_j the normal curvature of S_j at p in the direction of the tangent versor to C. Prove that

$$\kappa^2 \sin^2 \alpha = \kappa_1^2 + \kappa_2^2 - 2\kappa_1 \kappa_2 \cos \alpha .$$

4.86. Prove that the asymptotic curves on the surface $S \subset \mathbb{R}^3$ of equation $z = (x/a)^4 - (y/b)^4$ are the curves along which the surface intersects the two families of cylinders

$$\frac{x^2}{a^2} + \frac{y^2}{b^2} = \text{constant} , \qquad \frac{x^2}{a^2} - \frac{y^2}{b^2} = \text{constant} .$$

Are these intersections transversal?

4.87. Check whether the intersection between the unit sphere $S^2 \subset \mathbb{R}^3$ and the surface $S_2 = \{(x, y, z) \in \mathbb{R}^3 \mid e^x - z = 0\}$ is a regular curve σ and, if so, determine the equations of the tangent line and of the normal plane at all points of σ.

4.8 Tubular neighborhoods

In this section we intend to prove the Jordan-Brouwer theorem for closed surfaces in $\mathbb{R}^3$, saying that the complement of a closed surface has exactly two connected components (just like the complement of the support of a Jordan curve in the plane). As a consequence, we shall prove that every closed surface in $\mathbb{R}^3$ is the level surface of a suitable C^∞ function; more in general, we shall see when the converse of Corollary 4.3.12 holds. The main tool to do all of this is the tubular neighborhood of a surface.

Definition 4.8.1. Let $S \subset \mathbb{R}^3$ be a surface. Given $\varepsilon > 0$ and $p \in S$, denote by $I_S(p, \varepsilon)$ the line segment $p + (-\varepsilon, \varepsilon)N(p)$ of length 2ε centered in p and orthogonal to T_pS, where $N(p) \in S^2$ is a versor orthogonal to T_pS. If $\varepsilon : S \to \mathbb{R}^+$ is a continuous function positive everywhere, and $R \subseteq S$, we shall denote by $N_R(\varepsilon)$ the union of segments $I_S(p, \varepsilon(p))$, obtained for $p \in R$. The set $N_S(\varepsilon)$ is called *tubular neighborhood* of S if $I_S(p_1, \varepsilon(p_1)) \cap I_S(p_2, \varepsilon(p_2)) = \varnothing$ for all $p_1 \neq p_2 \in S$.

Theorem 4.8.2. *Every surface $S \subset \mathbb{R}^3$ has a continuous function $\varepsilon\colon S \to \mathbb{R}^+$ positive everywhere such that $N_S(\varepsilon)$ is a tubular neighborhood of S. Moreover:*

(i) *$N_S(\varepsilon)$ is a connected open neighborhood of S in $\mathbb{R}^3$;*
(ii) *if $\Omega \subseteq \mathbb{R}^3$ is an open neighborhood of S such that S is closed in Ω then we may choose ε in such a way that $N_S(\varepsilon) \subseteq \Omega$;*
(iii) *there exists a map $\pi\colon N_S(\varepsilon) \to S$ of class C^∞ such that $\pi|_S \equiv \mathrm{id}_S$ and $y \in I_S\big(\pi(y), \varepsilon(\pi(y))\big)$ for all $y \in N_S(\varepsilon)$;*
(iv) *if S is orientable then there exists a function $h\colon N_S(\varepsilon) \to \mathbb{R}$ of class C^∞ such that $y = \pi(y) + h(y)N\big(\pi(y)\big)$ for all $y \in N_S(\varepsilon)$, where $N\colon S \to S^2$ is the Gauss map of S. In particular, $S = h^{-1}(0)$;*
(v) *$N_S(\varepsilon) \setminus S$ has at most two connected components, and if S is orientable, then $N_S(\varepsilon) \setminus S$ has exactly two connected components.*

Proof. Let $\varphi\colon U \to S$ be a local parametrization of S, compatible with the given orientation if S is orientable, and set $N = \partial_1 \wedge \partial_2 / \|\partial_1 \wedge \partial_2\|$ as usual.

Let $F\colon U \times \mathbb{R} \to \mathbb{R}^3$ be defined by

$$F(x,t) = \varphi(x) + tN\big(\varphi(x)\big) \,.$$

Then F is of class C^∞; moreover,

$$\det \mathrm{Jac}(F)(x,0) = \left| \begin{matrix} \dfrac{\partial \varphi}{\partial x_1}(x) & \dfrac{\partial \varphi}{\partial x_2}(x) & N\big(\varphi(x)\big) \end{matrix} \right| \neq 0 \,.$$

So, for all $p \in \varphi(U)$ we may find a connected neighborhood $U_0 \subseteq U$ of $x = \varphi^{-1}(p)$, an $\varepsilon_0 > 0$ and a connected neighborhood $W_0 \subseteq \Omega$ of p such that $F|_{U_0 \times (-\varepsilon_0, \varepsilon_0)}$ is a diffeomorphism between $U_0 \times (-\varepsilon_0, \varepsilon_0)$ and W_0. In other words, $W_0 = N_{\varphi(U_0)}(\varepsilon_0)$ is a tubular neighborhood of $\varphi(U_0)$. Moreover, if we set $F^{-1} = G = (G_1, G_2, G_3)$, let $\pi = \varphi \circ (G_1, G_2)\colon W_0 \to S$ and $h = G_3\colon W_0 \to \mathbb{R}$; then π and h are of class C^∞, and $y = \pi(y) + h(y)N\big(\pi(y)\big)$ for all $y \in W_0$. Moreover, $S \cap W_0 = h^{-1}(0)$, and $W_0 \setminus S$ has exactly two components, one where h is positive and one where h is negative. Finally, if $\tilde{\varphi}\colon \tilde{U} \to S$ is another local parametrization of S with $p \in \varphi(U) \cap \tilde{\varphi}(\tilde{U})$ and we denote with the tilde the entities obtained with this construction starting from $\tilde{\varphi}$, then $\tilde{\pi}|_{W_0 \cap \tilde{W}_0} \equiv \pi|_{W_0 \cap \tilde{W}_0}$ and $\tilde{h}|_{W_0 \cap \tilde{W}_0} \equiv \pm h|_{W_0 \cap \tilde{W}_0}$, with the plus sign if φ and $\tilde{\varphi}$ have the same orientation, the minus sign otherwise.

So we have proved the theorem locally, in a neighborhood of each point of S; now we have to globalize this construction. Set $\partial_{\mathbb{R}^3} S = \overline{S} \setminus S$, where $\overline{S}$ is the closure of S in $\mathbb{R}^3$ (so $\partial_{\mathbb{R}^3} S \cap \Omega = \varnothing$), and for all $k \in \mathbb{N}^*$ set

$$S_k = \left\{ p \in S \;\middle|\; \|p\| + \frac{1}{d(p, \partial_{\mathbb{R}^3} S)} < k \right\} \,.$$

Each S_k is open in S, has compact closure contained in S_{k+1} and every point of S is contained in some S_k. Arguing as in the proof of the existence of the tubular neighborhood for curves (Theorem 2.2.5), for all $k \in \mathbb{N}^*$ we find an

$\varepsilon'_k > 0$ such that $N_{S_k}(\varepsilon'_k)$ is a tubular neighborhood of S_k contained in Ω and for which properties (i)–(v) hold. Moreover, $\overline{S_{k-1}}$ and $S \setminus S_k$ are a compact subset and a closed subset of S that are disjoint, and so they have a positive distance; hence we may find a $\delta_k > 0$ such that $N_{\overline{S_{k-1}}}(\delta_k)$ and $N_{S \setminus S_k}(\delta_k)$ are disjoint. Finally, we may also assume that $\varepsilon'_k \leq \varepsilon'_{k-1}$ and $\delta_k \leq \delta_{k-1}$.

Set $\varepsilon_0 = 1$. For $k \geq 1$, let $\varepsilon_k = \min\{\varepsilon'_{k+1}, \varepsilon_{k-1}, \delta_{k+1}\} > 0$, and define

$$V = \bigcup_{k \in \mathbb{N}^*} N_{S_k \setminus S_{k-1}}(\varepsilon_k) \,,$$

where $S_0 = \varnothing$. It is straightforward to see (check it!) that V is an open neighborhood of S contained in Ω. Moreover, for all $y \in V$ there exists a unique $p \in S$ such that $y \in I_S(p, \delta)$, where $\delta > 0$ is such that $I_S(p, \delta) \subset V$. Indeed, suppose that there exist $p_1, p_2 \in S$ and $t_1, t_2 \in \mathbb{R}$ such that $p_1 + t_1 N(p_1) = p_2 + t_2 N(p_2) = y_0 \in V$ and $I_S(p_j, |t_j|) \subset V$ for $j = 1, 2$, where $N(p_j)$ is an arbitrary versor orthogonal to $T_{p_j} S$. Let $k_0 \geq 1$ be the least k such that $p_1, p_2 \in S_k$; without loss of generality we may assume that $p_1 \in S_{k_0} \setminus S_{k_0 - 1}$ and so, in particular, $|t_1| \leq \varepsilon_{k_0} \leq \varepsilon'_{k_0}$. We have two possibilities:

- $p_2 \in S_{k_0} \setminus S_{k_0 - 2}$. In this case, $|t_2| < \varepsilon_{k_0 - 1} \leq \varepsilon'_{k_0}$ and, since $N_{S_{k_0}}(\varepsilon'_{k_0})$ is a tubular neighborhood of S_{k_0}, from this follows that $p_1 = p_2$ and $|t_1| = |t_2|$, as required;
- $p_2 \in S_{k_0 - 2}$. In this case, $|t_2| < \varepsilon_{k_0 - 2} \leq \delta_{k_0 - 1}$, so $y_0 \in N_{S_{k_0 - 2}}(\delta_{k_0 - 1})$; but $y_0 \in N_{S_{k_0} \setminus S_{k_0 - 1}}(\varepsilon_{k_0}) \subset N_{S \setminus S_{k_0 - 1}}(\delta_{k_0 \ 1})$, and thus $N_{S_{k_0 - 2}}(\delta_{k_0 - 1})$ and $N_{S \setminus S_{k_0 - 1}}(\delta_{k_0 - 1})$ would not be disjoint, a contradiction.

Using this property we may define a map $\pi \colon V \to S$ so that for all $y \in V$ there exists $h(y) \in \mathbb{R}$ such that $y = \pi(y) + h(y) N\big(\pi(y)\big)$, and π is of class C^∞ as it coincides with the restriction to V of the maps π defined in each $N_{S_k}(\varepsilon_k)$. Moreover, if S is orientable the function $h \in C^\infty(V)$ is also well defined, and is such that $h^{-1}(0) = S$.

Let $\chi \in C^\infty(\mathbb{R})$ be a positive non-increasing function such that $\chi(k) = \varepsilon_{k+1}$ for all $k \in \mathbb{N}^*$, and define $\varepsilon \colon S \to \mathbb{R}^+$ by setting

$$\varepsilon(p) = \chi \left(\|p\| + \frac{1}{d(p, \partial_{\mathbb{R}^3} S)} \right) \,.$$

In particular, if $p \in S_k \setminus S_{k-1}$ then $\varepsilon(p) \leq \varepsilon_k$; so $N_S(\varepsilon) \subseteq V$, and (ii) and (iii) hold. Clearly, $N_S(\varepsilon)$ is an open neighborhood of S; let us prove that it is connected. Indeed, if $y_1, y_2 \in N_S(\varepsilon)$, we may find a curve in $N_S(\varepsilon)$ that joins y_1 to y_2 going first down from y_1 to $\pi(y_1)$ along $I_S\big(\pi(y_1), \varepsilon(\pi(y_1))\big)$, then along a curve in S to $\pi(y_2)$, and finally going up to y_2 along $I_S\big(\pi(y_2), \varepsilon(\pi(y_2))\big)$.

We still have to prove (v). Since S is closed in $N_S(\varepsilon)$, the boundary in $N_S(\varepsilon)$ of each connected component of $N_S(\varepsilon) \setminus S$ coincides with S. But we have seen that every $p \in S$ has a connected neighborhood $W \subseteq N_S(\varepsilon)$ such that $W \setminus S$ has exactly two components; since every component of $N_S(\varepsilon) \setminus S$ must contain a component of $W \setminus S$, then $N_S(\varepsilon) \setminus S$ has at most two components.

Finally, let S be orientable. By construction, the function h takes both positive and negative values, and $S = h^{-1}(0)$; then $N_S(\varepsilon) \setminus S$ is the disjoint union of the open sets Ω^+ and Ω^-, where $\Omega^{\pm} = \{y \in N_S(\varepsilon) \mid \pm h(y) > 0\}$. To finish the proof we just need to show that each $\Omega^{\pm}$ is connected. Take $y_1, y_2 \in \Omega^+$, and choose a curve $\sigma \colon [0,1] \to S$ that joins $\pi(y_1)$ and $\pi(y_2)$. Let $\tau \colon [0,1] \to \Omega^+$ be given by

$$\tau(t) = \sigma(t) + \frac{\varepsilon\big(\sigma(t)\big)}{\varepsilon\big(\pi(y_1)\big)} h(y_1) N\big(\sigma(t)\big) \ .$$

Then we may go from y_1 to y_2 in Ω^+ by first following τ and then moving along $I_S(\pi(y_2), \varepsilon(\pi(y_2)))$. The connectedness of Ω^- can be shown analogously, and we are done. □

As a first consequence, we obtain the following:

Corollary 4.8.3. *Let $S \subset \Omega \subseteq \mathbb{R}^3$ be a surface, closed in the open subset Ω of $\mathbb{R}^3$. Then $\Omega \setminus S$ has at most two components, whose boundary (in Ω) coincides with S.*

Proof. Choose $\varepsilon \colon S \to \mathbb{R}^+$ such that $N_S(\varepsilon)$ is a tubular neighborhood of S contained in Ω. Since S is closed in Ω, the boundary (in Ω) of each component of $\Omega \setminus S$ coincides with S. In particular, each component of $\Omega \setminus S$ has to contain one of the components of $N_S(\varepsilon) \setminus S$. As a result, $\Omega \setminus S$ has at most two components. □

We then have proved the *Jordan-Brouwer theorem* for surfaces:

Theorem 4.8.4 (Jordan-Brouwer). *Let $S \subset \mathbb{R}^3$ be a closed surface. Then $\mathbb{R}^3 \setminus S$ has exactly two components, each having S as its boundary.*

Proof. Follows from Proposition 4.7.12 and from Corollary 4.8.3. □

We may also give a first converse of Corollary 4.3.12:

Corollary 4.8.5. *Let $S \subset \Omega \subseteq \mathbb{R}^3$ be an orientable surface, closed in the open subset Ω of $\mathbb{R}^3$. Then there exist an open neighborhood $V \subseteq \Omega$ of S and a function $h \in C^{\infty}(V)$ with 0 as a regular value such that $S = f^{-1}(0)$.*

Proof. Let $\varepsilon \colon S \to \mathbb{R}^+$ be a continuous function, positive everywhere, such that $N_S(\varepsilon) \subseteq \Omega$ is a tubular neighborhood of S, and set $V = N_S(\varepsilon)$. Let $h \in C^{\infty}(V)$ be the function given by Theorem 4.8.2.(iv) with respect to a Gauss map $N \colon S \to S^2$ of S. Clearly, $S = h^{-1}(0)$; moreover,

$$\mathrm{d}h_p\big(N(p)\big) = \lim_{t \to 0} \frac{h\big(p + tN(p)\big)}{t} = 1$$

for all $p \in S$, and so 0 is a regular value of h. □

For f to be defined in all Ω, we have to add a topological hypothesis:

Theorem 4.8.6. *Let $S \subset \Omega \subseteq \mathbb{R}^3$ be a surface, closed in the open subset Ω of $\mathbb{R}^3$. Then there exists a function $f \in C^\infty(\Omega)$ with 0 as a regular value such that $S = f^{-1}(0)$ if and only if S is orientable and $\Omega \setminus S$ is disconnected.*

Proof. Assume that there exists a function $f \in C^\infty(\Omega)$ with 0 as a regular value and such that $S = f^{-1}(0)$; in particular, S is orientable by Corollary 4.3.12. Choose $p_0 \in S$, and set $g(t) = f(p_0 + t\nabla f(p_0))$. Since Ω is open, there is an $\varepsilon > 0$ such that g is defined and of class C^∞ on $(-\varepsilon, \varepsilon)$. Moreover, $g(0) = 0$ and $g'(0) = \|\nabla f(p_0)\|^2 > 0$; so $g(t)$ is strictly positive (respectively, negative) for small $t > 0$ (respectively, $t < 0$). In particular, f changes sign in Ω. Set then $\Omega^\pm = \{x \in \Omega \mid \pm f(x) > 0\}$; it follows that $\Omega \setminus S = \Omega^+ \cup \Omega^-$ is a union of two disjoint non-empty open sets, and so it is disconnected.

Conversely, suppose that S is orientable and $\Omega \setminus S$ is disconnected (so it has exactly two components, by Corollary 4.8.3). Choose $\varepsilon \colon S \to \mathbb{R}^+$ such that $N_S(\varepsilon)$ is a tubular neighborhood of S contained in Ω, and let $h \in C^\infty(N_S(\varepsilon))$ be the function given by Theorem 4.8.2.(iv) with respect to a Gauss map $N \colon S \to S^2$ of S. Set $N_S^\pm(\varepsilon) = N_S(\varepsilon) \cap \{\pm h > 0\}$, so that $N_S(\varepsilon) \setminus S$ is the disjoint union of the connected open sets $N_S^+(\varepsilon)$ and $N_S^-(\varepsilon)$. Denote by $\Omega^\pm$ the component of $\Omega \setminus S$ that contains $N_S^\pm(\varepsilon)$. Then $\{N_S(\varepsilon), \Omega^+ \setminus \overline{N_S^+(\varepsilon/2)},$ $\Omega^- \setminus \overline{N_S^-(\varepsilon/2)}\}$ is an open cover of Ω. Let $\{\rho_0, \rho_+, \rho_-\}$ be a partition of unity subordinate to this cover (Theorem 3.6.6), and define $f \colon \Omega \to \mathbb{R}$ by setting

$$f(x) = \rho_0(x)h(x) + \rho_+(x) - \rho_-(x) \ .$$

Note that $f \in C^\infty(\Omega)$ because $\operatorname{supp}(\rho_0) \subset N_S(\varepsilon)$. Next, $\rho_0 \equiv 1$ and $\rho_\pm \equiv 0$ in a neighborhood of S; so $f \equiv h$ in a neighborhood of S and, in particular, $S \subseteq f^{-1}(0)$. Now, if $x \in \Omega^+$ then $\rho_-(x) = 0$; so $f(x) > 0$ since either $\rho_+(x) > 0$ (and $\rho_0(x)h(x) \geq 0$) or $\rho_+(x) = 0$, which implies $x \in N_S^+(\varepsilon)$ and $\rho_0(x)h(x) = h(x) > 0$. Analogously, it can be verified that $f(x) < 0$ if $x \in \Omega^-$, and thus $S = f^{-1}(0)$. Since f coincides with h in a neighborhood of S, it can be proved as in the previous corollary that 0 is a regular value for f, and we are done. $\qquad\square$

Putting all together, we have proved that every closed surface of $\mathbb{R}^3$ is a level surface:

Corollary 4.8.7. *Let $S \subset \mathbb{R}^3$ be a surface. Then S is closed in $\mathbb{R}^3$ if and only if there is a $h \in C^\infty(\mathbb{R}^3)$ with 0 as a regular value such that $S = h^{-1}(0)$.*

Proof. An implication is trivial. Conversely, if S is closed, the assertion follows from Theorems 4.7.15 and 4.8.6 and from Proposition 4.7.12. $\qquad\square$

4.9 The fundamental theorem of the local theory of surfaces

In Chapter 1 we saw that curvature and torsion determine uniquely (up to rigid motions) a curve in space and, conversely, that we may always find a regular curve with assigned curvature and torsion. In this section we want to study an analogous problem for surfaces: whether metric and form coefficients uniquely identify a regular surface. As you will see, the answer will be similar but with some significant differences.

The proof of the fundamental theorem of the local theory of curves (Theorem 1.3.37) relied heavily on Theorem 1.3.36 of existence and uniqueness for the solutions of a system of ordinary differential equations; in the study of surface we shall meet, instead, partial differential equations. Since the result we need does not appear in standard calculus textbooks, we give here both the statement and a proof, starting from the classical theorem of existence and uniqueness for solutions of systems of ordinary differential equations depending on a parameter (see [24, pp. 150–157] and [8, pp. 65–86]):

Theorem 4.9.1. *Given two open sets $\Omega_1 \subseteq \mathbb{R}^n$ and $V_0 \subseteq \mathbb{R}^l$, an interval $I \subseteq \mathbb{R}$, and two maps $a\colon V_0 \times I \times \Omega_1 \to \mathbb{R}^n$ and $b\colon V_0 \to \Omega_1$ of class C^k, with $k \in \mathbb{N}^* \cup \{\infty\}$, consider the following Cauchy problem:*

$$\begin{cases} \dfrac{\mathrm{d}\sigma}{\mathrm{d}t}(z,t) = a\big(z,t,\sigma(t)\big)\,, \\ \sigma(z,t_0) = b(z)\,, \end{cases} \tag{4.31}$$

where $z = (z_1,\ldots,z_l)$ are coordinates in $\mathbb{R}^l$. Then:

(i) *for all $t_0 \in I$ and $z_0 \in V_0$ there exist $\delta > 0$, an open neighborhood $V \subseteq V_0$ of z_0 and a map $\sigma\colon V \times (t_0 - \delta, t_0 + \delta) \to \Omega$ of class C^k satisfying (4.31);*
(ii) *two solutions of (4.31) always coincide in the intersection of their domains.*

Then:

Theorem 4.9.2. *Let $\Omega_0 \subseteq \mathbb{R}^n$, $\Omega_1 \subseteq \mathbb{R}^m$ and $V_0 \subseteq \mathbb{R}^l$ be open sets, and $G\colon V_0 \times \Omega_0 \times \Omega_1 \to M_{m,n}(\mathbb{R})$ and $b\colon V_0 \to \Omega_1$ maps of class C^k, with $k \in \{2,3,\ldots,\infty\}$. Denote by $x = (x_1,\ldots,x_n)$ the coordinates in $\mathbb{R}^n$, by $y = (y_1,\ldots,y_m)$ the coordinates in $\mathbb{R}^m$, and by $z = (z_1,\ldots,z_l)$ the coordinates in $\mathbb{R}^l$. Suppose that G satisfies the following compatibility conditions:*

$$\frac{\partial G_{jr}}{\partial x_s} + \sum_{h=1}^{m} \frac{\partial G_{jr}}{\partial y_h} G_{hs} \equiv \frac{\partial G_{js}}{\partial x_r} + \sum_{h=1}^{m} \frac{\partial G_{js}}{\partial y_h} G_{hr} \tag{4.32}$$

for $j = 1,\ldots,m$ and $r,\ s = 1,\ldots,n$. Then, for all $x^o \in \Omega_0$ and all $z^o \in V_0$ there exist a neighborhood $\Omega \subseteq \Omega_0$ of x^o and a neighborhood $V \subseteq V_0$ of z^o

such that the system

$$
\begin{cases}
\dfrac{\partial F_j}{\partial x_r}(z,x) = G_{jr}\big(z,x,F(z,x)\big) \,, \quad j=1,\ldots,m,\ r=1,\ldots,n \,, \\[2mm]
F(z,x^o) = b(z) \,,
\end{cases}
\tag{4.33}
$$

admits a solution $F\colon V \times \Omega \to \Omega_1$ *of class* C^k. *Moreover, two such solutions always coincide in the intersection of their domains.*

Remark 4.9.3. If $F\colon V \times \Omega \to \Omega_1$ satisfies (4.33) and is of class at least C^2, Theorem 4.4.14 implies that

$$
\begin{aligned}
\frac{\partial}{\partial x_s} G_{jr}\big(z,x,F(z,x)\big) &= \frac{\partial^2 F_j}{\partial x_s \partial x_r}(z,x) \\
&\equiv \frac{\partial^2 F_j}{\partial x_r \partial x_s}(z,x) = \frac{\partial}{\partial x_r} G_{js}\big(z,x,F(z,x)\big) \,,
\end{aligned}
$$

which is exactly (4.32) restricted to the graph of F.

Proof (of Theorem 4.9.2). Suppose (4.32) are satisfied; we want to prove that (4.33) has a unique solution. We use induction on n. For $n=1$, the compatibility conditions (4.32) are automatically satisfied, and the assertion follows from Theorem 4.9.1.

Suppose next that the claim is true for $n-1$, and write $x = (x_1, \hat{x})$ with $\hat{x} = (x_2,\ldots,x_{n-1}) \in \mathbb{R}^{n-1}$. By Theorem 4.9.1, the Cauchy problem

$$
\begin{cases}
\dfrac{\mathrm{d}\sigma_j}{\mathrm{d}t}(z,t) = G_{j1}\big(z,t,\hat{x}^o,\sigma(z,t)\big) \,, \quad j=1,\ldots,m \,, \\[2mm]
\sigma(z,x_1^o) = b(z) \,,
\end{cases}
\tag{4.34}
$$

admits a unique solution $\sigma\colon V \times (x_1^o - \delta, x_1^o + \delta) \to \Omega_1$ of class C^k, for suitable $\delta > 0$ and $V \subseteq V_0$ neighborhood of z^o. Then for all $x_1 \in (x_1^o - \delta, x_1^o + \delta)$ we may consider the Cauchy problem

$$
\begin{cases}
\dfrac{\partial F_j}{\partial x_r}(z,x_1,\hat{x}) = G_{jr}\big(z,x_1,\hat{x},F(z,x_1,\hat{x})\big) \,, \quad j=1,\ldots,m,\ r=2,\ldots,n \,, \\[2mm]
F(z,x_1,\hat{x}^o) = \sigma(z,x_1) \,,
\end{cases}
\tag{4.35}
$$

which (up to taking a smaller V if necessary) by the induction hypothesis has a unique solution $F\colon V \times \Omega \to \Omega_1$ of class C^k, where $\Omega \subseteq \Omega_0$ is a neighborhood of x^o. In particular, $F(z,x^o) = \sigma(z,x_1^o) = b(z)$; so we want to prove that this F is a solution of (4.33). Note that, if (4.33) admits a solution F, then $(z,t) \mapsto F(z,t,\hat{x}^o)$ solves (4.34), and $(z,\hat{x}) \mapsto F(z,x_1,\hat{x})$ solves (4.35), so the solution of (4.33), if it exists, is necessarily unique.

By construction, we have

$$
\frac{\partial F_j}{\partial x_r}(z,x) = G_{jr}\big(z,x,F(z,x)\big)
\tag{4.36}
$$

for $j = 1, \ldots m$ and $r = 2, \ldots, n$; we are left to check the behavior of $\partial F_j / \partial x_1$. First of all, we have

$$\frac{\partial F_j}{\partial x_1}(z, x_1, \hat{x}^o) = \frac{\mathrm{d}\sigma_j}{\mathrm{d}t}(z, x_1) = G_{j1}\big(z, x_1, \hat{x}^o, F(z, x_1, \hat{x}^o)\big) \ .$$

Set

$$U_j(z, x) = \frac{\partial F_j}{\partial x_1}(z, x) - G_{j1}\big(z, x, F(z, x)\big) \ ;$$

then $U_j(z, x_1, \hat{x}^o) = 0$, and we want to prove that $U_j \equiv 0$ for $j = 1, \ldots, m$. Fix $2 \le r \le n$. Then:

$$\begin{aligned}
\frac{\partial U_j}{\partial x_r}(z, x) &= \frac{\partial^2 F_j}{\partial x_r \partial x_1}(z, x) - \frac{\partial G_{j1}}{\partial x_r}(z, x, F(z, x)) \\
&\quad - \sum_{h=1}^m \frac{\partial G_{j1}}{\partial y_h}(z, x, F(z, x)) G_{hr}(z, x, F(z, x)) \\
&= \frac{\partial^2 F_j}{\partial x_1 \partial x_r}(z, x) - \frac{\partial G_{j1}}{\partial x_r}(z, x, F(z, x)) \\
&\quad - \sum_{h=1}^m \frac{\partial G_{j1}}{\partial y_h}(z, x, F(z, x)) G_{hr}(z, x, F(z, x)) \\
&= \frac{\partial}{\partial x_1} G_{jr}(z, x, F(z, x)) - \frac{\partial G_{j1}}{\partial x_r}(z, x, F(z, x)) \\
&\quad - \sum_{h=1}^m \frac{\partial G_{j1}}{\partial y_h}(z, x, F(z, x)) G_{hr}(z, x, F(z, x)) \\
&= \frac{\partial G_{jr}}{\partial x_1}(z, x, F(z, x)) + \sum_{h=1}^m \frac{\partial G_{jr}}{\partial y_h}(z, x, F(z, x)) \frac{\partial F_h}{\partial x_1}(z, x) \\
&\quad - \frac{\partial G_{h1}}{\partial x_r}(z, x, F(z, x)) - \sum_{h=1}^m \frac{\partial G_{j1}}{\partial y_h}(z, x, F(z, x)) G_{hr}(z, x, F(z, x)) \\
&= \sum_{h=1}^m \frac{\partial G_{jr}}{\partial y_h}(z, x, F(z, x)) U_h(z, x) \ ,
\end{aligned}$$

by (4.36) and the compatibility equations (4.32). So, for every fixed x_1 the function $\hat{x} \mapsto U(z, x_1, \hat{x})$ is a solution of the system

$$\begin{cases}
\dfrac{\partial U_j}{\partial x_r} = \displaystyle\sum_{h=1}^m \frac{\partial G_{jr}}{\partial y_h}\big(z, x_1, \hat{x}, F(z, x_1, \hat{x})\big) U_h(z, \hat{x}) \ , \ j = 1, \ldots, m, \ r = 2, \ldots, n \ , \\
U(z, \hat{x}^o) \equiv O \ ,
\end{cases}$$

which satisfies (exercise) the compatibility conditions (4.32). As a consequence, the induction hypothesis tells us that the unique solution is $U \equiv O$, and we are done. $\square$

We are now ready to state and prove the *fundamental theorem of the local theory of surfaces*, also known as *Bonnet's theorem*:

Theorem 4.9.4 (Bonnet). *Let E, F, G, e, f, $g \in C^\infty(\Omega_0)$, where $\Omega_0 \subset \mathbb{R}^2$ is an open subset of the plane, be functions satisfying Gauss' equations (4.26) and the Codazzi-Mainardi equations (4.28), where a_{ij} are given by (4.14) and Γ_{ij}^k are given by (4.20)–(4.22), and such that E, G, and $EG - F^2$ are positive everywhere. Then for all $q \in \Omega_0$ there is a connected neighborhood $\Omega \subseteq \Omega_0$ of q and an immersed surface $\varphi\colon \Omega \to \varphi(\Omega) \subset \mathbb{R}^3$ such that $\varphi(\Omega)$ is a regular surface with E, F, G as its metric coefficients and e, f, g as its form coefficients. Moreover, if $\tilde\varphi\colon \Omega \to \mathbb{R}^3$ is another immersed surface satisfying the same conditions then there exist a rotation $\rho \in SO(3)$ and a vector $b \in \mathbb{R}^3$ such that $\tilde\varphi = \rho \circ \varphi + b$.*

Remark 4.9.5. Before the proof, note the two main differences between this result and Theorem 1.3.37: the given functions cannot be arbitrary, but have to satisfy some compatibility relations; and we obtain only the local existence of the surface. A moment's thought will show that these are the same differences existing between Theorem 1.3.36 and Theorem 4.9.2.

Proof (of Theorem 4.9.4). As in the case of curves, the main idea consists in studying the following system of partial differential equations

$$\begin{cases} \dfrac{\partial(\partial_j)}{\partial x_i} = \Gamma_{ij}^1 \partial_1 + \Gamma_{ij}^2 \partial_2 + h_{ij} N \,, & i, j = 1,\, 2 \,, \\[2mm] \dfrac{\partial N}{\partial x_j} = a_{1j}\partial_1 + a_{2j}\partial_2 \,, & j = 1,\, 2 \,, \end{cases} \tag{4.37}$$

in the unknowns ∂_1, ∂_2, $N\colon \Omega_0 \to \mathbb{R}^3$. It is a system of the form (4.33); moreover, we have seen in Section 4.5 that the compatibility conditions of this system are exactly Gauss' and the Codazzi-Mainardi equations. Given $q \in \Omega_0$, choose three vectors ∂_1^o, ∂_2^o, $N^o \in \mathbb{R}^3$ such that

$$\begin{aligned} \|\partial_1^o\|^2 = E(q) \,, \ \langle \partial_1^o, \partial_2^o \rangle = F(q) \,, \ \|\partial_2^o\|^2 = G(q) \,, \\ \|N^o\|^2 = 1 \,, \ \langle \partial_1^o, N^o \rangle = \langle \partial_2^o, N^o \rangle = 0 \,, \ \langle \partial_1^o \wedge \partial_2^o, N^o \rangle > 0 \,. \end{aligned} \tag{4.38}$$

Theorem 4.9.2 provides us then with a connected neighborhood $\Omega_1 \subseteq \Omega_0$ of q and maps ∂_1, ∂_2, $N\colon \Omega_1 \to \mathbb{R}^3$ of class C^∞ solving (4.37) with the initial conditions $\partial_1(q) = \partial_1^o$, $\partial_2(q) = \partial_2^o$ and $N(q) = N^o$.

Now, the functions $\langle \partial_i, \partial_j \rangle$, $\langle \partial_j, N \rangle$, $\langle N, N \rangle\colon \Omega_1 \to \mathbb{R}$, for $i, j = 1, 2$, are a solution of the system

$$\begin{cases} \dfrac{\partial f_{ij}}{\partial x_k} = \Gamma_{ki}^1 f_{1j} + \Gamma_{ki}^2 f_{2j} + \Gamma_{kj}^1 f_{1i} + \Gamma_{kj}^2 f_{2i} + h_{ki} f_{j3} + h_{kj} f_{i3}, & i, j, k = 1, 2, \\[2mm] \dfrac{\partial f_{j3}}{\partial x_k} = \Gamma_{kj}^1 f_{13} + \Gamma_{kj}^2 f_{23} + h_{kj} f_{33} + a_{1k} f_{j1} + a_{2k} f_{j2} \,, & j, k = 1,\, 2 \,, \\[2mm] \dfrac{\partial f_{33}}{\partial x_k} = 2a_{1k} f_{13} + 2a_{2k} f_{23} \,, & k = 1,\, 2 \,, \end{cases}$$

with the initial conditions (4.38). With a bit of patience, you may verify (exercise) that this system satisfies the compatibility conditions (4.32), and that (another exercise) $f_{11} \equiv E$, $f_{12} \equiv f_{21} \equiv F$, $f_{22} \equiv G$, $f_{13} \equiv f_{23} \equiv 0$, $f_{33} \equiv 1$ satisfy the same system with the same initial conditions: so, Theorem 4.9.2 tells us that

$$\|\partial_1\|^2 \equiv E' , \ \langle \partial_1, \partial_2 \rangle \equiv F , \ \|\partial_2\|^2 \equiv G , \ \|N\|^2 \equiv 1 , \ \langle \partial_1, N \rangle \equiv \langle \partial_2, N \rangle \equiv 0 ,$$

on all Ω_1. In particular, $\|\partial_1 \wedge \partial_2\|^2 \equiv EG - F^2 > 0$ on all Ω_1, and so ∂_1 and ∂_2 are always linearly independent. Moreover, $N \equiv \pm \partial_1 \wedge \partial_2 / \|\partial_1 \wedge \partial_2\|$; since the sign is positive in q, it is positive everywhere.

Having fixed $p^o \in \mathbb{R}^3$, consider now the system

$$\begin{cases} \dfrac{\partial \varphi}{\partial x_j} = \partial_j , \quad j = 1, 2 , \\ \varphi(q) = p^o . \end{cases} \tag{4.39}$$

By the symmetry of the Γ_{ij}^k's and of the h_{ij}'s, the compatibility conditions are satisfied; so we find a connected neighborhood $\Omega \subseteq \Omega_1$ of q and a map $\varphi \colon \Omega \to \mathbb{R}^3$ of class C^∞ that solves (4.39). We want to prove that, up to taking a smaller Ω if necessary, the map φ is as required.

Since ∂_1 and ∂_2 are always linearly independent, the differential of φ always has rank 2, and by Corollary 3.1.5 we know that, up to further restricting Ω, φ is a global parametrization of the regular surface $S = \varphi(\Omega)$ with metric coefficients E, F and G. Moreover, N is by construction the Gauss map of S, and so (4.37) tells us that e, f and g are the form coefficients of S, as required.

We are just left with uniqueness. Let $\tilde{\varphi} \colon \Omega \to \mathbb{R}^3$ be another immersed surface that is a homeomorphism with its image and has the same metric and form coefficients. Up to a translation and a rotation, we may suppose that $\tilde{\varphi}(q) = \varphi(q)$, that $\partial \tilde{\varphi}/\partial x_j(q) = \partial \varphi/\partial x_j(q)$ for $j = 1, 2$, and consequently $\tilde{N}(q) = N(q)$, where $\tilde{N}$ is the Gauss map of $\tilde{\varphi}(\Omega)$ induced by $\tilde{\varphi}$. In this case both $\partial \varphi/\partial x_j$ and N on one side, and $\partial \tilde{\varphi}/\partial x_j$ and $\tilde{N}$ on the other side satisfy the system (4.39) with the same initial conditions; so they coincide everywhere. Thus φ and $\tilde{\varphi}$ satisfy the system (4.39) with the same initial conditions; it follows that $\tilde{\varphi} \equiv \varphi$, and we are done. $\qquad \square$

5

Geodesics

In the study of the geometry of the plane, straight lines clearly play a fundamental role. The goal of this chapter is to introduces the curves playing on surfaces a role analogous to the one played by lines in the plane.

There are (at least) two distinct ways of characterizing straight lines (or, more in general, line segments) among all plane curves: one geometric and global, and another analytic and local. A line segment is the shortest curve between its endpoints (global geometric characterization); and it is a curve with constant tangent vector (local analytic characterization).

The global geometric characterization is hardly suitable to be transferred to surfaces: as we shall see, the shortest curve joining two points on a surface might not exist at all (Remark 5.2.10), or might not be unique (Remark 5.2.11). The local analytic characterization is more consistent with the approach we have been following to study surfaces, but seems to pose a problem: a curve with constant tangent vector is a straight line in space too. But let us see what "constant tangent vector" actually means. The tangent vector σ' to a curve $\sigma: I \to \mathbb{R}^2$ is constant if it does not vary; geometrically, the vector $\sigma'(t_1)$, seen as a bound vector based at $\sigma(t_1)$, is *parallel* to the vector $\sigma'(t_2)$ based at $\sigma(t_2)$ for all $t_1, t_2 \in I$. In other words, σ' is obtained by parallel translation of the same vector along the support of σ. Now, if $\sigma: I \to S$ is a curve in a surface S instead staying inside S we do not actually see the whole variation of the tangent vector σ': if the tangent vector varies in the direction orthogonal to the surface, it does not seem to change at all when seen from S. In other words, the variation of the tangent vector to a curve in a surface, from the viewpoint of the surface, is not measured by the derivative of the tangent vector, but by the *orthogonal projection* of the derivative on the tangent planes to the surfaces. This operation (orthogonally projecting the derivative) is called *covariant derivative*, and we shall see in Section 5.1 that it is completely intrinsic to the surface. Summing up, from the viewpoint of a surface the tangent vector to a curve is "constant" (or, using a more precise terminology suggested by the above, *parallel*) if it has covariant derivative

Abate M., Tovena F.: Curves and Surfaces.
DOI 10.1007/978-88-470-1941-6_5, © Springer-Verlag Italia 2012

equal to zero. A *geodesic* on a surface is a curve with parallel (in this sense) tangent vector.

In Section 5.1 we shall show that geodesics exist. More precisely, we shall show that for every point $p \in S$ of a surface S and every tangent vector $v \in T_pS$ there is a unique maximal geodesic $\sigma_v : I_v \to S$ passing through p and tangent to the direction v, that is, such that $\sigma_v(0) = p$ and $\sigma'_v(0) = v$. Moreover, we shall also define a measure, the *geodesic curvature*, of how much a curve is far from being a geodesic.

In the next two chapters we shall see how to effectively use geodesics to study the geometry of surfaces. In Section 5.2 we shall instead see how to retrieve from the local analytic characterization the global geometric characterization mentioned at the beginning. To be precise, we shall prove that the shortest curve between two points of a surface is always a geodesic and, conversely, that every geodesic is locally the shortest curve between points in its support (Theorem 5.2.8). However, you have to be careful, since a geodesic is not necessarily the shortest curve between its endpoints (Remark 5.2.9).

Section 5.3 of this chapter covers a slightly different topic: vector fields. A *vector field* on a surface S is a smooth map that associates each point $p \in S$ with a vector tangent to S at p. In a sense, we are prescribing the direction (and the speed) of the motion of points on the surface. The curves described by the points under the action of the vector field (the trajectories of this motion) are called *integral curves* of the field. Vector fields have countless applications outside mathematics too (for instance, they are widely used in physics); we shall just prove some elementary properties (existence and uniqueness of integral curves, for instance) and show how to use them to construct local parametrizations with specific properties (Theorem 5.3.19).

Finally, in the supplementary material of this chapter we shall return to geodesics, giving in Section 5.4 a necessary and sufficient condition for the geodesics of a surface S to be defined for all times (*Hopf-Rinow's Theorem* 5.4.8). Finally, in Section 5.5 we shall give a necessary and sufficient condition in terms of geodesics and Gaussian curvature for two surfaces to be locally isometric (Theorem 5.5.5).

5.1 Geodesics and geodesic curvature

Straight lines in the plane are obviously extremely important for plane geometry; in this section we shall introduce the class of curves that play for surfaces the role played by lines in the plane.

As described in the introduction to this chapter, we want to generalize to surfaces the local analytic characterization of straight lines as curves with constant tangent vector; we shall deduce the global geometric characterization in the next section, a characterization that will turn out to be slightly but significantly different from that of lines in plane.

Let us begin by defining the class of objects, among which tangent vectors to a curve are a typical example, we want to be able to differentiate intrinsically.

Definition 5.1.1. A *vector field along a curve* $\sigma: I \to S$ of class C^∞ is a map $\xi: I \to \mathbb{R}^3$ of class C^∞ such that $\xi(t) \in T_{\sigma(t)}S$ for all $t \in I$. More in general, if $\sigma: I \to S$ is a piecewise C^∞ curve, a *vector field along* σ is a *continuous* map $\xi: I \to \mathbb{R}^3$ such that $\xi(t) \in T_{\sigma(t)}S$ for all $t \in I$, smooth in every subinterval of I where σ is smooth. The vector space of vector fields along σ will be denoted by $\mathcal{T}(\sigma)$.

Example 5.1.2. The tangent vector $\sigma': I \to \mathbb{R}^3$ of a curve $\sigma: I \to S$ of class C^∞ is a typical example of a vector field along a curve.

We want to measure how much a vector field $\xi \in \mathcal{T}(\sigma)$ varies along a curve $\sigma: I \to S$ of class C^∞. From the viewpoint of $\mathbb{R}^3$, the variation of ξ is given by its derivative ξ'. But, as anticipated in the introduction to this chapter, from the viewpoint of the surface S this derivative has no meaning; only the component of ξ' tangent to S can be seen from inside the surface. So a geometrically more meaningful measure of the variation of a vector field along a curve contained in a surface is the following:

Definition 5.1.3. The *covariant derivative* of a vector field $\xi \in \mathcal{T}(\sigma)$ along a curve $\sigma: I \to S$ of class C^∞ in a surface S is the vector field $D\xi \in \mathcal{T}(\sigma)$ defined by

$$D\xi(t) = \pi_{\sigma(t)}\left(\frac{\mathrm{d}\xi}{\mathrm{d}t}(t)\right),$$

where $\pi_{\sigma(t)}: \mathbb{R}^3 \to T_{\sigma(t)}S$ is the orthogonal projection on the tangent plane to S at $\sigma(t)$.

Remark 5.1.4. If $\xi \in \mathcal{T}(\sigma)$ is a vector field along a smooth curve $\sigma: I \to S$ then we may write $\xi' = D\xi + w$, where $w: I \to \mathbb{R}^3$ is a C^∞ map such that $w(t)$ is orthogonal to $T_{\sigma(t)}S$ for all $t \in S$. In particular,

$$\langle \xi', \tilde{\xi} \rangle_\sigma \equiv \langle D\xi, \tilde{\xi} \rangle_\sigma$$

for every vector field $\tilde{\xi} \in \mathcal{T}(\sigma)$ along σ. From this immediately follows that

$$\frac{\mathrm{d}}{\mathrm{d}t}\langle \xi, \tilde{\xi} \rangle_\sigma = \langle D\xi, \tilde{\xi} \rangle_\sigma + \langle \xi, D\tilde{\xi} \rangle_\sigma \tag{5.1}$$

for every pair of vector fields $\xi, \tilde{\xi} \in \mathcal{T}(\sigma)$.

From the definition it might seem that the covariant derivative depends on the way the surface is immersed in $\mathbb{R}^3$. On the contrary, *the covariant derivative is a purely intrinsic notion*: it only depends on the first fundamental form of S. To see this, let us express $D\xi$ in local coordinates.

Let $\varphi\colon U \to S$ be a local parametrization whose image contains the support of a curve $\sigma\colon I \to S$, so that we may write $\sigma(t) = \varphi(\sigma_1(t), \sigma_2(t))$. A vector field ξ along σ can be written as $\xi(t) = \xi_1(t)\partial_1|_{\sigma(t)} + \xi_2(t)\partial_2|_{\sigma(t)}$; then

$$
\frac{d\xi}{dt} = \frac{d}{dt}\left(\xi_1 \frac{\partial\varphi}{\partial x_1} \circ \sigma\right) + \frac{d}{dt}\left(\xi_2 \frac{\partial\varphi}{\partial x_2} \circ \sigma\right)
$$
$$
= \sum_{j=1}^{2}\left[\frac{d\xi_j}{dt}\partial_j|_\sigma + \xi_j\left(\sigma_1' \frac{\partial^2\varphi}{\partial x_1 \partial x_j} \circ \sigma + \sigma_2' \frac{\partial^2\varphi}{\partial x_2 \partial x_j} \circ \sigma\right)\right],
$$

and so, recalling (4.18), we find

$$
D\xi = \sum_{k=1}^{2}\left[\frac{d\xi_k}{dt} + \sum_{i,j=1}^{2}(\Gamma_{ij}^k \circ \sigma)\sigma_i'\xi_j\right]\partial_k|_\sigma . \tag{5.2}
$$

Thus D is expressed in terms of the Christoffel symbols, and as a consequence it only depends on the first fundamental form, as claimed.

A constant (or parallel) vector field along a curve is a field that does not vary. The following definition formalizes this idea:

Definition 5.1.5. Let $\sigma\colon I \to S$ be a curve of class C^∞ in a surface S. A vector field $\xi \in \mathcal{T}(\sigma)$ along σ is *parallel* if $D\xi \equiv O$. More in general, if $\sigma\colon I \to S$ is a piecewise C^∞ curve, a vector field $\xi \in \mathcal{T}(\sigma)$ is *parallel* if it is so when restricted to each subinterval of I where σ is C^∞.

The next result not only shows that parallel vector fields actually exist (and lots of them!), but also that they behave with respect of the first fundamental form in a way consistent with the intuitive idea of parallel field:

Proposition 5.1.6. *Let $\sigma\colon I \to S$ be a piecewise C^∞ curve in a surface S. Then:*

(i) *given $t_0 \in I$ and $v \in T_{\sigma(t_0)}S$, there is a unique parallel vector field $\xi \in \mathcal{T}(\sigma)$ such that $\xi(t_0) = v$;*
(ii) *if $\xi, \tilde{\xi} \in \mathcal{T}(\sigma)$ are parallel vector fields along σ, the inner product $\langle \xi, \tilde{\xi} \rangle_\sigma$ is constant. In particular, the norm of a parallel field is constant.*

Proof. (i) Suppose for now that σ is of class C^∞. Equality (5.2) says that $D\xi \equiv O$ locally is a system of two linear ordinary differential equations; so the claim follows as usual rom Theorem 1.3.36 of existence and uniqueness for the solution of a linear system of ordinary differential equations. Indeed, first of all, by Theorem 1.3.36 there is a unique solution defined on a subinterval $\tilde{I}$ of I containing t_0 and such that $\sigma(\tilde{I})$ is contained in the image of a local parametrization. Let now $I_0 \subset I$ be the maximal interval containing t_0 on which a parallel vector field ξ such that $\xi(t_0') = v$ is defined. If $I_0 \neq I$, let $t_1 \in I$ be an endpoint of I_0 and take a local parametrization $\psi\colon V \to S$

centered at $\sigma(t_1)$. Then there exists for sure a $t_2 \in I_0$ such that $\sigma(t_2) \in \psi(V)$, and Theorem 1.3.36 yields the existence of a unique vector field $\tilde{\xi}$ defined on $\sigma^{-1}(\psi(V))$ such that $\tilde{\xi}(t_2) = \xi(t_2)$; in particular, $\tilde{\xi}$ is also defined in t_1. But uniqueness tells us that $\tilde{\xi}$ and ξ coincide on the intersection of the domains, and so $\tilde{\xi}$ extends ξ to a neighborhood of t_1 too, against the hypothesis that t_1 is an endpoint of I_0. So $I_0 = I$, as claimed.

In general, if σ is piecewise C^∞, it suffices to apply the above argument to each subinterval of I on which σ is of class C^∞ (beginning with an interval containing t_0) to construct the required parallel field $\xi \in \mathcal{T}(\sigma)$.

(ii) Equality (5.1) implies

$$\frac{\mathrm{d}}{\mathrm{d}t}\langle \xi, \tilde{\xi}\rangle_\sigma = \langle D\xi, \tilde{\xi}\rangle_\sigma + \langle \xi, D\tilde{\xi}\rangle_\sigma \equiv 0\,,$$

since $D\xi$, $D\tilde{\xi} \equiv O$, and $\langle \xi, \tilde{\xi}\rangle_\sigma$ is constant.   □

We are now able to define the curves that play on surfaces a role analogous to the one played by straight lines in the plane: curves with parallel tangent vector.

Definition 5.1.7. A *geodesic* on a surface S is a curve $\sigma \colon I \to S$ of class C^∞ such that $\sigma' \in \mathcal{T}(\sigma)$ is parallel, that is, such that $D\sigma' \equiv O$.

Remark 5.1.8. Traditionally, geodesics are known to cartographers as the shortest routes between two points on Earth's surface. We shall see in the next section that (when they exist) the shortest curves joining two points on a surface are geodesics in our sense too; and, conversely, every geodesic as just defined is locally the shortest route joining two points of its support.

Obviously, the first thing to do is to show that geodesics exist. Let $\sigma \colon I \to S$ be a curve of class C^∞ whose support is contained in the image of a local parametrization $\varphi \colon U \to S$. Then (5.2) tells us that $\sigma = \varphi(\sigma_1, \sigma_2)$ is a geodesic if and only if it satisfies the following system of differential equations, called *geodesic equation*:

$$\sigma_j'' + \sum_{h,k=1}^{2} (\Gamma_{hk}^j \circ \sigma)\sigma_h'\sigma_k' = 0\,, \qquad j = 1, 2\,. \tag{5.3}$$

This is a system of second-order non-linear ordinary differential equations. We may transform it into a system of first-order ordinary differential equations by introducing auxiliary variables v_1, v_2 to represent the components of σ'. In other words, we consider instead the equivalent first-order system

$$\begin{cases} v_j' + \displaystyle\sum_{h,k=1}^{2} (\Gamma_{hk}^j \circ \sigma)\,v_h v_k = 0\,, & j = 1, 2\,, \\[2mm] \sigma_j' = v_j\,, & j = 1, 2\,. \end{cases} \tag{5.4}$$

So it is natural to use Theorem 4.9.1, which is repeated here in a simplified version (obtained by taking $V_0 = \Omega_1$ and $b = \mathrm{id}_{V_0}$), sufficient for our aims:

Theorem 5.1.9. *Consider the Cauchy problem for a curve* $\sigma \colon I \to U$ *given by*

$$
\begin{cases}
\dfrac{d\sigma_j}{dt}(t) = a_j\big(\sigma(t)\big), & j = 1, \ldots, n, \\[2mm]
\sigma(t_0) = x \in U,
\end{cases}
\tag{5.5}
$$

where $U \subseteq \mathbb{R}^n$ *is an open set and* $a_1, \ldots, a_n \in C^\infty(U)$. *Then:*

(i) *for all* $t_0 \in \mathbb{R}$ *and* $x_0 \in U$ *there exist* $\delta > 0$ *and an open neighborhood* $U_0 \subseteq U$ *of* x_0 *such that for all* $x \in U_0$ *there exists a curve* $\sigma_x \colon (t_0 - \delta, t_0 + \delta) \to U$ *of class* C^∞ *that is a solution of* (5.5);

(ii) *the map* $\Sigma \colon U_0 \times (t_0 - \delta, t_0 + \delta) \to U$ *given by* $\Sigma(x, t) = \sigma_x(t)$ *is of class* C^∞;

(iii) *two solutions of* (5.5) *always coincide in the intersection of their domains.*

As a consequence:

Proposition 5.1.10. *Let* $S \subset \mathbb{R}^3$ *be a surface. Then for all* $p \in S$ *and* $v \in T_pS$ *there exists a geodesic* $\sigma \colon I \to S$ *such that* $0 \in I$, $\sigma(0) = p$ *and* $\sigma'(0) = v$. *More-over, if* $\tilde{\sigma} \colon \tilde{I} \to S$ *is another geodesic satisfying the same conditions then* σ *and* $\tilde{\sigma}$ *coincide in* $I \cap \tilde{I}$. *In particular, for all* $p \in S$ *and* $v \in T_pS$ *there exists a maximal open interval* $I_v \subseteq \mathbb{R}$ *and a unique geodesic* $\sigma_v \colon I_v \to S$ *such that* $\sigma_v(0) = p$ *and* $\sigma_v'(0) = v$.

Proof. Theorem 5.1.9 applied to (5.4) tells us that there are $\varepsilon > 0$ and a curve $\sigma \colon (-\varepsilon, \varepsilon) \to U \subset S$ that is a solution of (5.3) with initial conditions $\sigma(0) = p$ and $\sigma'(0) = v$. Moreover, if $\tilde{\sigma}$ is another geodesic that satisfies the same initial conditions, then σ and $\tilde{\sigma}$ coincide in some neighborhood of 0. Let I_0 be the maximum interval contained in $I \cap \tilde{I}$ on which σ and $\tilde{\sigma}$ coincide. If I_0 were strictly contained in $I \cap \tilde{I}$, there would exist an endpoint t_0 of I_0 contained in $I \cap \tilde{I}$, and we could apply once more Theorem 5.1.9 with initial conditions $\sigma(t_0)$ and $\sigma'(t_0)$. But in this case σ and $\tilde{\sigma}$ would coincide in a neighborhood of t_0 too, against the definition of I_0. Hence, $I_0 = I \cap \tilde{I}$, and so there exists a unique maximal geodesic for a given point and tangent to a given direction. $\square$

A remark to be made immediately is that since the notion of covariant derivative (and consequently of parallelism) is intrinsic, local isometries map geodesics to geodesics:

Proposition 5.1.11. *Let* $H \colon S \to \tilde{S}$ *be a local isometry between surfaces, and* $\sigma \colon I \to S_1$ *a piecewise* C^∞ *curve. Then, for all* $\xi \in \mathcal{T}(\sigma)$ *we have* $\mathrm{d}H_\sigma(\xi) \in \mathcal{T}(H \circ \sigma)$ *and* $D\big(\mathrm{d}H_\sigma(\xi)\big) = \mathrm{d}H_\sigma(D\xi)$. *In particular,* ξ *is paral-lel along* σ *if and only if* $\mathrm{d}H_\sigma(\xi)$ *is parallel along* $H \circ \sigma$, *and* σ *is a geodesic of* S *if and only if* $H \circ \sigma$ *is a geodesic of* $\tilde{S}$.

Proof. Choose $t_0 \in I$ and a local parametrization $\varphi \colon U \to S$ centered at $\sigma(t_0)$ such that $H|_{\varphi(U)}$ is an isometry with its image. In particular, $\tilde{\varphi} = H \circ \varphi$ is a local parametrization of $\tilde{S}$ and, if we denote by Γ_{ij}^k the Christoffel symbols of φ and by $\tilde{\Gamma}_{ij}^k$ the Christoffel symbols of $\tilde{\varphi}$, we have $\tilde{\Gamma}_{ij}^k \circ H = \Gamma_{ij}^k$ for all i, j, $k = 1$, 2. Moreover, (3.7) implies that $\tilde{\partial}_j|_{H \circ \sigma} = \mathrm{d}H_\sigma(\partial_j|_\sigma)$ for $j = 1$, 2, where $\{\tilde{\partial}_1, \tilde{\partial}_2\}$ is the basis induced by $\tilde{\varphi}$, so if we write $\sigma = \varphi(\sigma_1, \sigma_2)$ and $\xi = \xi_1 \partial_1|_\sigma + \xi_2 \partial_2|_\sigma$ we obtain $H \circ \varphi = \tilde{\varphi}(\sigma_1, \sigma_2)$ and $\mathrm{d}H_\sigma(\xi) = \xi_1 \tilde{\partial}_1|_{H \circ \sigma} + \xi_2 \tilde{\partial}_2|_{H \circ \sigma}$. So (5.2) gives

$$
\mathrm{d}H_{\sigma(t_0)}\big(D\xi(t_0)\big) = \sum_{k=1}^{2} \left[\frac{\mathrm{d}\xi_k}{\mathrm{d}t}(t_0) + \sum_{i,j=1}^{2} \Gamma_{ij}^k(\sigma(t_0))\sigma_i'(t_0)\xi_j(t_0) \right] \tilde{\partial}_k|_{H(\sigma(t_0))}
$$

$$
= \sum_{k=1}^{2} \left[\frac{\mathrm{d}\xi_k}{\mathrm{d}t}(t_0) + \sum_{i,j=1}^{2} \tilde{\Gamma}_{ij}^k\big(H(\sigma(t_0))\big)\sigma_i'(t_0)\xi_j(t_0) \right] \tilde{\partial}_k|_{H(\sigma(t_0))}
$$

$$
= D\big(\mathrm{d}H_{\sigma(t_0)}\big(\xi(t_0)\big)\big) \,,
$$

as required. In particular, since $\mathrm{d}H$ is injective, we have $D\xi \equiv O$ if and only if $D\big(\mathrm{d}H_\sigma(\xi)\big) = O$, and $D\big((II \circ \sigma)'\big) \equiv O$ if and only if $D\sigma' \equiv O$, and we are done. $\qquad\square$

The next proposition contains some elementary properties of geodesics, which are useful in the (rare) cases in which they can be explicitly determined:

Proposition 5.1.12. *Let $\sigma \colon I \to S$ be a curve of class C^∞ in a surface S. Then:*

(i) *σ is a geodesic if and only if σ'' is always orthogonal to the surface, that is, if and only if $\sigma''(t) \perp T_{\sigma(t)}S$ for all $t \in I$;*

(ii) *if σ is a geodesic then it is parametrized by a multiple of arc length, that is, $\|\sigma'\|$ is constant.*

Proof. (i) Follows from the definition of $D\sigma'$ as orthogonal projection of σ'' on the tangent planes to S.

(ii) It follows immediately from Proposition 5.1.6.(ii). $\qquad\square$

Let us see now some examples of geodesics.

Example 5.1.13. If σ is a regular curve whose support is contained in a plane H, the vector σ'' may be orthogonal to H only if it is zero (why?). So the geodesics of a plane are the curves with second derivative equal to zero, that is, straight lines (parametrized by a multiple of arc length). This can also be deduced from the geodesic equation by noting that the Christoffel symbols of a plane are zero everywhere.

More in general, a straight line (parametrized by a multiple of arc length) contained in a surface S is always a geodesics, since it has second derivative zero everywhere.

Example 5.1.14. We want to determine all the geodesics of a right circular cylinder S parametrized as in Example 4.1.9. First of all, by the previous example we know that the meridians of the cylinder (the vertical straight lines contained in the cylinder), parametrized by a multiple of arc length, are geodesics. Parallels (the circles obtained by intersecting the cylinder with a plane orthogonal to its axis) parametrized by a multiple of arc length are geodesics too: indeed, their second derivative is parallel to the radius, hence to the normal versor of the cylinder (see Example 4.4.9), and so we may apply Proposition 5.1.12.(i).

We want to determine now the geodesics through the point $p_0 = (1, 0, 0)$. The map $H: \mathbb{R}^2 \to S$ given by $H(x_1, x_2) = (\cos x_1, \sin x_1, x_2)$ is, by Example 4.1.17, a local isometry, and consequently (Proposition 5.1.11) maps geodesics to geodesics. The geodesics through the origin of the plane are the lines $t \mapsto (at, bt)$, con $(a, b) \neq (0, 0)$; so the curves

$$\sigma_{a,b}(t) = \big(\cos(at), \sin(at), bt\big)$$

are geodesics of the cylinder issuing from p_0. Since it is immediately verified that for all $v \in T_{p_0} S$ there is a pair $(a, b) \in \mathbb{R}^2$ such that $\sigma'_{a,b}(0) = v$, Proposition 5.1.10 tells us that we have found all geodesics issuing from p_0; in particular, if they are neither a parallel ($b = 0$) nor a meridian ($a = 0$), they are circular helices ($a, b \neq 0$).

Finally, by a translation and a rotation, which clearly are isometries, we may map p_0 to any other point p of the cylinder, and the geodesics issuing from p are obtained by rotation and translation from the geodesics issuing from p_0. Summing up, the geodesics of the cylinder are the meridians, the parallels and the circular helices contained in the cylinder.

Example 5.1.15. All the geodesics of the sphere are great circles — parametrized by a multiple of arc length — and vice versa all great circles are geodesics. Indeed, a great circle is given by the intersection of the sphere with a plane passing through the center of the sphere; hence its second derivative is parallel to the normal versor of the sphere (see Example 4.4.8), and we may apply again Proposition 5.1.12.(i). On the other hand, given a point and a tangent direction, there is always a great circle passing through that point and tangent to that direction, and so by Proposition 5.1.10 there are no other geodesics.

Example 5.1.16. To give an idea of the difficulty of explicitly computing geodesics even in simple cases, let us write the geodesic equation for the graph Γ_h of a function $h \in C^\infty(U)$ defined on an open set U of $\mathbb{R}^2$. Recalling Example 4.6.7, equations (5.3) become

$$\sigma_j'' + \frac{1}{1 + \|(\nabla h) \circ \sigma\|^2} \left(\frac{\partial h}{\partial x_j} \circ \sigma\right) \big\langle (\mathrm{Hess}(h) \circ \sigma) \cdot \sigma', \sigma' \big\rangle = 0 \,, \quad j = 1, 2 \,,$$

a system of differential equations almost impossible to solve explicitly except in very particular cases.

Example 5.1.17 (Geodesics of surfaces of revolution). We want to study the geodesics on a surface of revolution obtained by rotating around the z-axis the curve $\sigma(t) = (\alpha(t), 0, \beta(t))$ with $\alpha(t) > 0$ everywhere. Let $\varphi \colon \mathbb{R}^2 \to \mathbb{R}^3$ be the usual immersed surface with support S given by $\varphi(t, \theta) = (\alpha(t) \cos \theta, \alpha(t) \sin \theta, \beta(t))$, whose restrictions give local parametrizations for S as shown in Example 3.1.18. Using the Christoffel symbols computed in Example 4.6.10, we see that a curve $\tau(s) = \varphi(t(s), \theta(s))$ is a geodesic if and only if

$$
\begin{cases}
t'' + \left(\dfrac{\alpha' \alpha'' + \beta' \beta''}{(\alpha')^2 + (\beta')^2} \circ t \right) (t')^2 - \left(\dfrac{\alpha \alpha'}{(\alpha')^2 + (\beta')^2} \circ t \right) (\theta')^2 = 0 \, , \\
\theta'' + \left(\dfrac{2\alpha'}{\alpha} \circ t \right) \theta' t' = 0 \, .
\end{cases}
\tag{5.6}
$$

In particular, if the curve σ is parametrized by arc length, the geodesic equation for surfaces of revolution becomes

$$
\begin{cases}
t'' - \big((\alpha \dot{\alpha}) \circ t \big) (\theta')^2 = 0 \, , \\
\theta'' + \left(\dfrac{2\dot{\alpha}}{\alpha} \circ t \right) \theta' t' = 0 \, .
\end{cases}
\tag{5.7}
$$

Let us see which conclusions we may draw from these equations.

- The meridians, that is, the curves $\tau(s) = \varphi(t(s), \theta_0)$ where $\theta_0 \in \mathbb{R}$ is a constant, parametrized by a multiple of arc length, are geodesics. Indeed, the second equation of (5.7) is clearly satisfied. Moreover, saying that τ is parametrized by a multiple of arc length is equivalent to saying that $E(t')^2 \equiv k^2$ for some $k \in \mathbb{R}^+$, that is, by Example 4.1.13, $(t')^2 \equiv k^2$. Hence, $2t't'' \equiv 0$; from this we deduce $t'' \equiv 0$ (why?), and the first equation of (5.7) is satisfied too.
- Let us see when a parallel, that is a curve $\tau(s) = \varphi(t_0, \theta(s))$ where $t_0 \in \mathbb{R}$ is a constant, parametrized by a multiple of arc length, is a geodesic. The second equation of (5.7) tells us that θ' has to be constant; moreover, since τ is parametrized by a multiple of arc length, we have $G(\theta')^2 \equiv k^2$ for some $k \in \mathbb{R}^+$, and so Example 4.1.13 implies $|\theta'| \equiv k/\alpha(t_0)$. The first equation of (5.7) becomes then $\dot{\alpha}(t_0)/\alpha(t_0) = 0$, that is, $\dot{\alpha}(t_0) = 0$. So we have proved that τ is a geodesic if and only if θ' is a non-zero constant and t_0 is a critical point for α or, in other words, that *a parallel (parametrized by a multiple of arc length) is a geodesic if and only if it is obtained by rotating a point where the tangent vector of the generator is parallel to the axis of rotation of the surface*; see Fig. 5.1.
- Except for meridians and parallels, it is quite rare to be able to explicitly solve (5.7). However, it is possible to deduce from those equations much information about the qualitative behavior of geodesics of surfaces of revolution; let us see how. Since α is nowhere zero, we may multiply the second equation in (5.7) by $\alpha^2 \circ t$, obtaining $(\theta' \alpha^2 \circ t)' \equiv 0$; so $\alpha^2 \theta'$ is constant

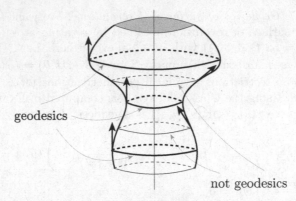

geodesics

not geodesics

Fig. 5.1.

along every geodesic. In other words, (5.7) is equivalent to

$$\begin{cases} t'' = c^2 \dfrac{\dot\alpha}{\alpha^3} \circ t, & c \in \mathbb{R}, \\[2mm] \theta' = \dfrac{c}{\alpha^2 \circ t}. \end{cases}$$

The constant c (which is different for each geodesic) has a geometric meaning. To find it, note first that $c = 0$ corresponds to θ being constant, that is, to meridians. On the other hand, $\alpha(t(s))$ is the radius of the parallel passing through $\tau(s)$, parametrized by $\theta \mapsto \varphi(t(s), \theta)$; so the angle $\psi(s)$ between the geodesic τ and the parallel passing through $\tau(s)$ is such that

$$\cos\psi = \frac{\langle \tau', \partial/\partial\theta \rangle_\tau}{\|\tau'\|\sqrt{G}} = \frac{1}{\|\tau'\|}\theta'(\alpha \circ t) = \frac{c}{\|\tau'\|\,\alpha \circ t},$$

with $\|\tau'\|$ constant, where we used $\tau' = t'(\partial/\partial t) + \theta'(\partial/\partial\theta)$ and Example 4.1.13. So

$$\alpha \cos\psi = \text{costant} \tag{5.8}$$

along the geodesic. We have obtained the important *Clairaut's relation*: *for every geodesic of a surface of revolution, the product $\alpha \cos\psi$ of the cosine of the the radius of the parallel with the angle between the parallel and the geodesic is constant.*

Clairaut's relation is very useful to study qualitatively the behavior of geodesics on surfaces of revolution. For instance, if $\tau\colon I \to S$ is a geodesic through $p_0 = \tau(s_0)$ in a direction forming an angle ψ_0 with the parallel through p_0, we immediately find that

$$\forall s \in I \qquad \alpha(t(s)) \geq \alpha(t(s))|\cos\psi(s)| = \alpha(p_0)|\cos\psi_0|,$$

or, in other words, *the geodesic τ never intersects parallels with radius less than $\alpha(p_0)|\cos\psi_0|$.* Other applications of Clairaut's relation are described in Problem 5.3 and in Exercise 5.1.

We now want to introduce now a measure of how much a curve is far from being a geodesic. We begin with:

Proposition 5.1.18. *Let $\sigma\colon I \to S$ be a regular curve in a surface S. Then σ is a geodesic if and only if it is parametrized by a multiple of arc length and its curvature κ coincides with the absolute value of its normal curvature $|\kappa_n|$.*

Proof. Since the claim is a local one, we may assume S to be orientable. Let N be a normal versor field on S; note that the absolute value of the normal curvature does not depend on the choice of N, and that the formula

$$\kappa_n = \langle \ddot{\sigma}, N \circ \sigma \rangle = \frac{\langle \sigma'', N \circ \sigma \rangle}{\|\sigma'\|^2} \, , \tag{5.9}$$

holds since

$$\ddot{\sigma} = \frac{1}{\|\sigma'\|^2} \left(\sigma'' - \left(\log \|\sigma'\| \right)' \sigma' \right) \tag{5.10}$$

and σ' is orthogonal to $N \circ \sigma$. Proposition 5.1.12.(i) tells us that the curve σ is a geodesic if and only if σ'' is parallel to $N \circ \sigma$. Moreover, (5.10) implies that $\|\sigma'\|$ is constant if and only if $\ddot{\sigma} = \sigma''/\|\sigma'\|^2$, and in this case we have $\kappa = \|\ddot{\sigma}\| = \|\sigma''\|/\|\sigma'\|^2$.

Suppose then that σ is a geodesic. Thus, σ'' is parallel to $N \circ \sigma$, and so $|\langle \sigma'', N \circ \sigma \rangle| = \|\sigma''\|$. Moreover, Proposition 5.1.12.(ii) says that $\|\sigma'\|$ is constant; hence, $\kappa = \|\sigma''\|/\|\sigma'\|^2$ and formula (5.9) implies

$$|\kappa_n| = \frac{|\langle \sigma'', N \circ \sigma \rangle|}{\|\sigma'\|^2} = \frac{\|\sigma''\|}{\|\sigma'\|^2} = \kappa \, .$$

Conversely, suppose that $\|\sigma'\|$ is constant and $|\kappa_n| \equiv \kappa$; then formula (5.9) yields $|\langle \sigma'', N \circ \sigma \rangle| \equiv \|\sigma''\|$, and this may happen only if σ'' is parallel to $N \circ \sigma$, that is, by Proposition 5.1.12.(i), only if σ is a geodesic. $\qquad\square$

As a result, we may determine how much a curve is far from being a geodesic by measuring how much its curvature differs from its normal curvature. Let us quantify this remark.

Definition 5.1.19. *Let $\sigma\colon I \to S$ be a regular curve on a surface S. A normal versor field along σ is a C^∞ map $N\colon I \to \mathbb{R}^3$ such that $\|N(t)\| = 1$ and $N(t) \perp T_{\sigma(t)}S$ for all $t \in I$.*

Remark 5.1.20. There always exist exactly two normal versor fields along a curve. Indeed, first of all assume that N, $\tilde{N}\colon I \to \mathbb{R}^3$ are two normal versor fields along the curve $\sigma\colon I \to S$. Then, by continuity, the function $t \mapsto \langle N(t), \tilde{N}(t) \rangle$ must be identically equal to either $+1$ or -1 — and consequently we have either $\tilde{N} \equiv N$ or $\tilde{N} \equiv -N$.

For the existence, if S is orientable then we can define a normal versor field N along σ simply by setting $N(t) = N(\sigma(t))$, where $N\colon S \to S^2$ is a

Gauss map for S. Since every surface is locally orientable, given $t_0 \in I$ we can always find a subinterval $I_0 \subseteq I$ containing t_0 and a normal versor field N along $\sigma|_{I_0}$. Assume that I_0 is the largest such interval; we claim that $I_0 = I$. If not, there is an endpoint t_1 of I_0 contained in I. Since S is locally orientable around $\sigma(t_1)$, we can find (how?) a normal versor field defined in a neighborhood I_1 of t_1 and coinciding with N on $I_0 \cap I_1$; but this means that we can extend N to a subinterval larger than I_0, contradiction.

Let $\sigma: I \to S$ be a regular curve on a surface S, and choose a normal versor field $N: I \to \mathbb{R}^3$ along σ. If $\xi \in \mathcal{T}(\sigma)$ is a field of *versors* along σ, by differentiating $\langle \xi, \xi \rangle_\sigma \equiv 1$ keeping in mind (5.1) we obtain

$$0 = \frac{\mathrm{d}}{\mathrm{d}t} \langle \xi, \xi \rangle_\sigma = 2 \langle D\xi, \xi \rangle_\sigma \;.$$

Hence $D\xi$ is orthogonal both to N and ξ; so there has to exist a function $\lambda: I \to \mathbb{R}$ such that $D\xi = \lambda N \wedge \xi$. To be precise, since $N \wedge \xi$ is a versor, we get $\lambda = \langle D\xi, N \wedge \xi \rangle_\sigma$, and in particular λ is of class C^∞.

Definition 5.1.21. Let $\sigma: I \to S$ be a regular curve parametrized by arc length on a surface S, and let $N: I \to \mathbb{R}^3$ be a normal versor field along σ. The *geodesic curvature* of σ is the function $\kappa_g: I \to \mathbb{R}$ given by

$$\kappa_g = \langle D\dot{\sigma}, N \wedge \dot{\sigma} \rangle = \langle \ddot{\sigma}, N \wedge \dot{\sigma} \rangle \;,$$

so that

$$D\dot{\sigma} = \kappa_g \, N \wedge \dot{\sigma} \;.$$

Remark 5.1.22. Replacing the normal vector field N along σ by $-N$ the geodesic curvature changes sign, and so the *absolute value* of the geodesic curvature depends only on the curve. Furthermore, the geodesic curvature also changes sign when inverting the orientation of the curve.

In particular, $\|D\dot{\sigma}\|^2 = \kappa_g^2$. Since $\ddot{\sigma} = D\dot{\sigma} + \langle \ddot{\sigma}, N \rangle N$ is an orthogonal decomposition, and we have $\|\ddot{\sigma}\|^2 = \kappa^2$ and $|\langle \ddot{\sigma}, N \rangle|^2 = |\kappa_n|^2$, it is immediate to conclude:

Corollary 5.1.23. *Let $\sigma: I \to S$ be a curve parametrized by arc length in a surface S. Then*

$$\kappa^2 = \kappa_n^2 + \kappa_g^2 \;. \tag{5.11}$$

In particular, a curve σ parametrized by arc length is a geodesic if and only if its geodesic curvature is zero everywhere.

We end this section with an explicit formula for computing the geodesic curvature we shall need later on. To obtain it, we begin with:

Lemma 5.1.24. *Let* $\varphi\colon U \to S$ *be an orthogonal local parametrization of a surface* S, *and set* $X_j = \partial_j/\|\partial_j\|$ *for* $j = 1,\ 2$. *Let next* $\sigma\colon I \to \varphi(U) \subseteq S$ *be a regular curve parametrized by arc length and write* $\sigma(s) = \varphi\big(\sigma_1(s), \sigma_2(s)\big)$ *and* $\xi_j = X_j \circ \sigma \in \mathcal{T}(\sigma)$. *Then*

$$\langle D\xi_1, \xi_2\rangle_\sigma = \frac{1}{2\sqrt{EG}}\left[\dot{\sigma}_2\frac{\partial G}{\partial x_1} - \dot{\sigma}_1\frac{\partial E}{\partial x_2}\right].$$

Proof. We have

$$\langle D\xi_1, \xi_2\rangle_\sigma = \left\langle \frac{\mathrm{d}(X_1 \circ \sigma)}{\mathrm{d}s}, X_2\right\rangle_\sigma = \left\langle \frac{\partial X_1}{\partial x_1}, X_2\right\rangle_\sigma \dot{\sigma}_1 + \left\langle \frac{\partial X_1}{\partial x_2}, X_2\right\rangle_\sigma \dot{\sigma}_2\ .$$

Differentiating $F = \langle \partial_1, \partial_2\rangle \equiv 0$ with respect to x_1 we find

$$\left\langle \frac{\partial^2\varphi}{\partial x_1^2}, \frac{\partial\varphi}{\partial x_2}\right\rangle = -\left\langle \frac{\partial\varphi}{\partial x_1}, \frac{\partial^2\varphi}{\partial x_1 \partial x_2}\right\rangle = -\frac{1}{2}\frac{\partial E}{\partial x_2}\ ,$$

and so

$$\left\langle \frac{\partial X_1}{\partial x_1}, X_2\right\rangle = \left\langle \frac{\partial}{\partial x_1}\left(\frac{1}{\sqrt{E}}\frac{\partial\varphi}{\partial x_1}\right), \frac{1}{\sqrt{G}}\frac{\partial\varphi}{\partial x_2}\right\rangle = -\frac{1}{2\sqrt{EG}}\frac{\partial E}{\partial x_2}\ .$$

Analogously, we find

$$\left\langle \frac{\partial X_1}{\partial x_2}, X_2\right\rangle = \left\langle \frac{\partial}{\partial x_2}\left(\frac{1}{\sqrt{E}}\frac{\partial\varphi}{\partial x_1}\right), \frac{1}{\sqrt{G}}\frac{\partial\varphi}{\partial x_2}\right\rangle = \frac{1}{2\sqrt{EG}}\frac{\partial G}{\partial x_1}\ ,$$

and we are done. $\qquad\square$

We need one more definition. Let $\sigma\colon [a,b] \to S$ be a curve parametrized by arc length in a surface $S \subset \mathbb{R}^3$, and choose a normal versor field $N\colon I \to \mathbb{R}^3$ along σ. Suppose we have two never vanishing vector fields $\xi_1,\ \xi_2 \in \mathcal{T}(\sigma)$ along σ; in particular, $\{\xi_1(s), N(s) \wedge \xi_1(s)\}$ is an orthogonal basis of $T_{\sigma(s)}S$ for all $s \in [a,b]$, and we may write

$$\frac{\xi_2}{\|\xi_2\|} = \frac{\langle\xi_1, \xi_2\rangle}{\|\xi_1\|\|\xi_2\|}\frac{\xi_1}{\|\xi_1\|} + \frac{\langle N \wedge \xi_1, \xi_2\rangle}{\|\xi_1\|\|\xi_2\|}\frac{N \wedge \xi_1}{\|\xi_1\|}\ .$$

Hence we may define a continuous map $\phi\colon [a,b] \to S^1$ by setting

$$\phi = \left(\frac{\langle\xi_1, \xi_2\rangle}{\|\xi_1\|\|\xi_2\|}, \frac{\langle N \wedge \xi_1, \xi_2\rangle}{\|\xi_1\|\|\xi_2\|}\right)\ . \tag{5.12}$$

Definition 5.1.25. Let $\sigma\colon [a,b] \to S$ be a curve parametrized by arc length in a surface $S \subset \mathbb{R}^3$, and choose a normal versor field $N\colon I \to \mathbb{R}^3$ along σ. Given two never vanishing vector fields $\xi_1,\ \xi_2 \in \mathcal{T}(\sigma)$ along σ, a *continuous* *determination of the angle between ξ_1 and ξ_2 is a lift $\theta: [a, b] \to \mathbb{R}$ of the function $\phi: [a, b] \to S^1$ given by (5.12); see also Proposition 2.1.4. In particular, by construction we have*

$$\frac{\xi_2}{\|\xi_2\|} = \cos\theta \frac{\xi_1}{\|\xi_1\|} + \sin\theta \frac{N \wedge \xi_1}{\|\xi_1\|} \, .$$

Replacing N by $-N$ the continuous determination of the angle clearly changes sign.

Remark 5.1.26. If $\sigma: [a, b] \to \varphi(U) \subseteq S$ is a curve with support contained in the image of a local parametrization $\varphi: U \to S$ then as normal versor field along σ we shall always take $N \circ \sigma$, where $N = \partial_1 \wedge \partial_2 / \|\partial_1 \wedge \partial_2\|$ is the Gauss map on $\varphi(U)$ induced by the local parametrization φ.

Remark 5.1.27. Let $\varphi: U \to S$ be an *orthogonal* local parametrization of a surface S, and $\sigma: [a, b] \to \varphi(U) \subseteq S$ a curve parametrized by arc length. If we take as usual $N = \partial_1 \wedge \partial_2 / \|\partial_1 \wedge \partial_2\|$, then $N \wedge (\partial_1/\|\partial_1\|) = \partial_2/\|\partial_2\|$. So every continuous determination $\theta: [a, b] \to \mathbb{R}$ of the angle between $\partial_1|_\sigma$ and $\dot\sigma$ is such that

$$\dot\sigma = \cos\theta \frac{\partial_1|_\sigma}{\|\partial_1|_\sigma\|} + \sin\theta \frac{\partial_2|_\sigma}{\|\partial_2|_\sigma\|} \, . \tag{5.13}$$

We can then close this section with the following:

Proposition 5.1.28. *Let $\varphi: U \to S$ be an orthogonal local parametrization of a surface S, and $\sigma: I \to \varphi(U) \subseteq S$ a curve parametrized by arc length. Write $\sigma(s) = \varphi(\sigma_1(s), \sigma_2(s))$; then the geodesic curvature of σ is given by*

$$\kappa_g = \frac{1}{2\sqrt{EG}} \left[\dot\sigma_2 \frac{\partial G}{\partial x_1} - \dot\sigma_1 \frac{\partial E}{\partial x_2} \right] + \frac{d\theta}{ds} \, ,$$

where $\theta: I \to \mathbb{R}$ is a continuous determination of the angle between $\partial_1|_\sigma$ and $\dot\sigma$.

Proof. Set $\xi_1 = \partial_1|_\sigma / \|\partial_1|_\sigma\|$, $\xi_2 = \partial_2|_\sigma / \|\partial_2|_\sigma\|$, and $N = \xi_1 \wedge \xi_2$; in particular, $N \wedge \xi_1 = \xi_2$ and $N \wedge \xi_2 = -\xi_1$. Recalling (5.13) we obtain

$$N \wedge \dot\sigma = -(\sin\theta)\xi_1 + (\cos\theta)\xi_2 \, ,$$

$$\ddot\sigma = -(\sin\theta)\dot\theta\xi_1 + (\cos\theta)\dot\xi_1 + (\cos\theta)\dot\theta\xi_2 + (\sin\theta)\dot\xi_2 \, ,$$

and

$$D\dot\sigma = -(\sin\theta)\dot\theta\xi_1 + (\cos\theta)D\xi_1 + (\cos\theta)\dot\theta\xi_2 + (\sin\theta)D\xi_2 \, .$$

Differentiating $\langle \xi_1, \xi_1 \rangle_\sigma \equiv \langle \xi_2, \xi_2 \rangle_\sigma \equiv 1$ and $\langle \xi_1, \xi_2 \rangle_\sigma \equiv 0$ we find

$$\langle D\xi_1, \xi_1 \rangle_\sigma = \langle D\xi_2, \xi_2 \rangle_\sigma \equiv 0 \quad \text{and} \quad \langle D\xi_1, \xi_2 \rangle_\sigma = -\langle \xi_1, D\xi_2 \rangle_\sigma \, ,$$

so that

$$\begin{aligned}
\kappa_g &= \langle D\dot\sigma, N \wedge \dot\sigma \rangle_\sigma \\
&= (\sin\theta)^2\dot\theta + (\cos\theta)^2\langle D\xi_1, \xi_2 \rangle_\sigma + (\cos\theta)^2\dot\theta - (\sin\theta)^2\langle \xi_1, D\xi_2 \rangle_\sigma \\
&= \langle D\xi_1, \xi_2 \rangle_\sigma + \dot\theta \, .
\end{aligned}$$

The assertion follows now from Lemma 5.1.24.   □

5.2 Minimizing properties of geodesics

The goal of this section is to keep a promise: we shall prove that the shortest curves between two points of a surface are geodesics, and that the converse holds locally.

Proposition 5.1.10 tells us that for all $p \in S$ and $v \in T_pS$ there exists a unique maximal geodesic $\sigma_v: I_v \to S$ with $\sigma_v(0) = p$ and $\sigma_v'(0) = v$. A consequence of next lemma is that the support of σ_v only depends on the direction of v:

Lemma 5.2.1. *Let $S \subset \mathbb{R}^3$ be a surface, $p \in M$ and $v \in T_pM$. Then:*

(i) *for all $c, t \in \mathbb{R}$ we have*

$$\sigma_{cv}(t) = \sigma_v(ct) \tag{5.14}$$

as soon as one of the terms is defined;

(ii) *if σ_v is defined at $t \in \mathbb{R}$ and $s \in \mathbb{R}$, then σ_v is defined at $s + t$ if and only if $\sigma_{\sigma_v'(t)}$ is defined at s, and in this case*

$$\sigma_{\sigma_v'(t)}(s) = \sigma_v(t + s) . \tag{5.15}$$

Remark 5.2.2. The phrase "σ_v is defined at $t \in \mathbb{R}$" means that there exists an open interval containing both the origin and t where σ_v is defined.

Proof (of Lemma 5.2.1). (i) If $c = 0$ there is nothing to prove. If $c \neq 0$ we begin by proving that (5.14) holds as soon as $\sigma_v(ct)$ exists. Set $\tilde{\sigma}(t) = \sigma_v(ct)$; clearly, $\tilde{\sigma}(0) = p$ and $\tilde{\sigma}'(0) = cv$, so it suffices to prove that $\tilde{\sigma}$ is a geodesic. Indeed

$$D\tilde{\sigma}'(t) = \pi_{\tilde{\sigma}(t)}\left(\frac{\mathrm{d}^2\tilde{\sigma}}{\mathrm{d}t^2}(t)\right) = c^2\pi_{\sigma_v(ct)}\left(\frac{\mathrm{d}^2\sigma_v}{\mathrm{d}t^2}(ct)\right) = c^2 D\sigma_v'(ct) = O ,$$

and so we are done.

Conversely, suppose that $\sigma_{cv}(t)$ exists, and set $v' = cv$ and $s = ct$. Then $\sigma_{cv}(t) = \sigma_{v'}(c^{-1}s)$ exists, so it is equal to $\sigma_{c^{-1}v'}(s) = \sigma_v(ct)$, and we are done.

(ii) Having set $I_v - t = \{s \in \mathbb{R} \mid s + t \in I_v\}$, the curve $\sigma: I_v - t \to S$ defined by $\sigma(s) = \sigma_v(s+t)$ is clearly a geodesic with $\sigma(0) = \sigma_v(t)$ and $\sigma'(0) = \sigma_v'(t)$; so $\sigma(s) = \sigma_{\sigma_v'(t)}(s)$ for all $s \in I_v - t$. This means that if σ_v is defined at $s+t$ then $\sigma_{\sigma_v'(t)}$ is defined at s (that is, $I_v - t \subseteq I_{\sigma_v'(t)}$), and (5.15) holds. In particular, $-t \in I_v - t \subseteq I_{\sigma_v'(t)}$, so $\sigma_{\sigma_v'(t)}(-t) = p$ and $\sigma_{\sigma_v'(t)}'(-t) = \sigma_v'(0) = v$. Applying the same argument to $(-t, \sigma_v(t))$ instead of (t, p), we get $I_{\sigma_v'(t)} + t \subseteq I_v$; therefore $I_{\sigma_v'(t)} + t = I_v$ (which exactly means that σ_v is defined at $s + t$ if and only if $\sigma_{\sigma_v'(t)}$ is defined at s), and (5.15) always holds. □

In particular, we get $\sigma_v(t) = \sigma_{tv}(1)$ as soon as one of the two terms is defined; so, studying the behavior of $\sigma_v(t)$ as a function of t is equivalent to studying the behavior of $\sigma_v(1)$ as a function of the length of v. This suggests that we study the map $v \mapsto \sigma_v(1)$ defined on a suitable subset of T_pS:

Definition 5.2.3. Let S be a surface, and $p \in S$. Denote by

$$\mathcal{E}_p = \{v \in T_pS \mid 1 \in I_v\}$$

the set of tangent vectors $v \in T_pS$ such that σ_v is defined in a neighborhood of $[0,1]$. Then the *exponential map* $\exp_p \colon \mathcal{E}_p \to S$ of S at p is defined by $\exp_p(v) = \sigma_v(1)$.

Remark 5.2.4. The reason of the appearance of the word "exponential" in this context is related to the differential geometry of higher-dimension manifolds (the multidimensional equivalent of surfaces). To be precise, it is also possible to define in a natural way a notion of geodesic for compact groups of matrices such as the orthogonal group $O(n)$, and it is possible to prove that the corresponding exponential map in the identity element of the group is given by the exponential of matrices.

We can immediately prove that the exponential map enjoys some nice properties:

Proposition 5.2.5. *Let $S \subset \mathbb{R}^3$ be a surface, and $p \in S$. Then:*

(i) *$\mathcal{E}_p$ is an open star-shaped neighborhood of the origin in T_pS;*
(ii) *for all $v \in T_pS$ and $t_0 \in \mathbb{R}$ such that σ_v is defined at t_0 there exist $\delta > 0$ and a local parametrization $\varphi \colon U \to S$ centered at p such that, having set $w_0 = (\mathrm{d}\varphi_O)^{-1}(v)$, the map $(x,w,t) \mapsto \sigma_{\mathrm{d}\varphi_x(w)}(t)$ is defined and of class C^∞ on $U \times B(w_0, \delta) \times (t_0 - \delta, t_0 + \delta) \subset \mathbb{R}^2 \times \mathbb{R}^2 \times \mathbb{R} = \mathbb{R}^5$;*
(iii) *the map $\exp_p \colon \mathcal{E}_p \to S$ is of class C^∞;*
(iv) *$\mathrm{d}(\exp_p)_O = \mathrm{id}$ and, in particular, $\exp_p$ is a diffeomorphism between a neighborhood of O in T_pS and a neighborhood of p in S.*

Proof. Lemma 5.2.1 immediately implies that $\mathcal{E}_p$ is star-shaped with respect to the origin. Define now the set

$$\mathcal{U} = \{(p,v,t_0) \mid p \in S, \ v \in T_pS, \ \sigma_v \text{ is defined at } t_0 \in \mathbb{R}\},$$

and the subset $\mathcal{W} \subseteq \mathcal{U}$ of the points $(p,v,t_0) \in \mathcal{U}$ for which a $\delta > 0$ and a local parametrization $\varphi \colon U \to S$ centered at p exist such that, having set $w_0 = (\mathrm{d}\varphi_O)^{-1}(v)$, the map $(x,w,t) \mapsto \sigma_{\mathrm{d}\varphi_x(w)}(t)$ is of class C^∞ in the open set $U \times B(w_0,\delta) \times (t_0 - \delta, t_0 + \delta) \subset \mathbb{R}^2 \times \mathbb{R}^2 \times \mathbb{R} = \mathbb{R}^5$.

Clearly, $\mathcal{E}_p = \pi\big(\mathcal{U} \cap (\{p\} \times T_pS \times \{1\})\big)$, where $\pi \colon S \times \mathbb{R}^3 \times \mathbb{R} \to \mathbb{R}^3$ is the projection on the central factor; on the other hand, $\mathcal{W}_p = \pi\big(\mathcal{W} \cap (\{p\} \times T_pS \times \{1\})\big)$ is by definition open in T_pS and $\exp_p$ is of class C^∞ in $\mathcal{W}_p$; so, to get (i)–(iii) it is enough to prove that $\mathcal{U} = \mathcal{W}$.

Note first that $(p,v,0) \in \mathcal{W}$ for all $p \in S$ and $v \in T_pS$, by Theorem 5.1.9 applied to the system (5.4).

Suppose by contradiction that there exists $(p_0,v_0,t_0) \in \mathcal{U} \backslash \mathcal{W}$; since $t_0 \neq 0$, we may assume $t_0 > 0$; the case $t_0 < 0$ is analogous.

Let $\hat{t} = \sup\{t \in \mathbb{R} \mid \{p_0\} \times \{v_0\} \times [0, t] \subset \mathcal{W}\}$; clearly, $0 < \hat{t} \le t_0$. On the other hand, since $(p_0, v_0, t_0) \in \mathcal{U}$, the geodesic σ_{v_0} is defined at $\hat{t}$; set $\hat{p} = \sigma_{v_0}(\hat{t})$ and $\hat{v} = \sigma'_{v_0}(\hat{t})$. Since $(\hat{p}, \hat{v}, 0) \in \mathcal{W}$, there exist a local parametrization $\hat{\varphi} \colon \hat{U} \to S$ centered at $\hat{p}$, a $\delta > 0$ and a neighborhood $\hat{W} \subseteq \mathbb{R}^2$ of $\hat{w} = (\mathrm{d}\hat{\varphi}_O)^{-1}(\hat{v})$ such that the map $(x, w, t) \mapsto \sigma_{\mathrm{d}\hat{\varphi}_x(w)}(t)$ is of class C^∞ in $\hat{U} \times \hat{W} \times (-\delta, \delta)$.

Choose now $t_1 < \hat{t}$ such that $t_1 + \delta > \hat{t}$, $\sigma_{v_0}(t_1) \in \hat{V} = \hat{\varphi}(\hat{U})$, and $\sigma'_{v_0}(t_1) \in \mathrm{d}\hat{\varphi}_{x_1}(\hat{W})$, where $x_1 = \hat{\varphi}^{-1}(\sigma_{v_0}(t_1))$. Since we know that $\{p_0\} \times \{v_0\} \times [0, t_1] \subset \mathcal{W}$, we may find a local parametrization $\varphi \colon U \to S$ centered at p_0, an $\varepsilon > 0$ and a neighborhood $W \subseteq \mathbb{R}^2$ of $w_0 = (\mathrm{d}\varphi_O)^{-1}(v_0)$ such that the map $(x, w, t) \mapsto \sigma_{\mathrm{d}\varphi_x(v)}(t)$ is of class C^∞ in $U \times W \times (-\varepsilon, t_1 + \varepsilon)$; moreover, up to shrinking U and W, we may also assume that $\sigma_{\mathrm{d}\varphi_x(w)}(t_1) \in \hat{V}$ and $\sigma'_{\mathrm{d}\varphi_x(w)}(t_1) \in \mathrm{d}\hat{\varphi}_{x'}(\hat{W})$ for all $(x, w) \in U \times W$, where $x' = \hat{\varphi}^{-1}(\sigma_{\mathrm{d}\varphi_x(w)}(t_1))$.

Now, if $(x, w) \in U \times W$ the point $\sigma_{\mathrm{d}\varphi_x(w)}(t_1)$ is well defined and depends in a C^∞ way on x and w. So $x' = \hat{\varphi}^{-1}(\sigma_{\mathrm{d}\varphi_x(w)}(t_1)) \in \hat{U}$ and the vector $w' = (\mathrm{d}\hat{\varphi}_{x'})^{-1}(\sigma'_{\mathrm{d}\varphi_x(w)}(t_1)) \in \hat{W}$ also depend in a C^∞ way on x and w. Hence for all $t \in (t_1 - \delta, t_1 + \delta)$ the point

$$\sigma_{\sigma'_{\mathrm{d}\varphi_x(w)}(t_1)}(t - t_1) = \sigma_{\mathrm{d}\hat{\varphi}_{x'}(w')}(t - t_1)$$

depends in a C^∞ way on x, w and t. But Lemma 5.2.1.(ii) implies

$$\sigma_{\sigma'_{\mathrm{d}\varphi_x(w)}(t_1)}(t - t_1) = \sigma_{\mathrm{d}\varphi_x(w)}(t) \ ;$$

so we have proved that the map $(x, w, t) \mapsto \sigma_{\mathrm{d}\varphi_x(w)}(t)$ is of class C^∞ in $U \times W \times (-\varepsilon, t_1 + \delta)$, and this contradicts the definition of $\hat{t}$.

Finally, let us compute the differential of $\exp_p$ in the origin. If $w \in T_pS$, we have by definition

$$\mathrm{d}(\exp_p)_O(w) = \left. \frac{\mathrm{d}}{\mathrm{d}t}(\exp_p \circ \tau)\right|_{t=0},$$

where τ is an arbitrary curve in T_pS with $\tau(0) = O$ and $\tau'(0) = w$. For instance, we may take $\tau(t) = tw$; so,

$$\mathrm{d}(\exp_p)_O(w) = \left. \frac{\mathrm{d}}{\mathrm{d}t}\exp_p(tw)\right|_{t=0} = \left. \frac{\mathrm{d}}{\mathrm{d}t}\sigma_{tw}(1)\right|_{t=0} = \left. \frac{\mathrm{d}}{\mathrm{d}t}\sigma_w(t)\right|_{t=0} = w \ ,$$

by (5.14). Hence, $\mathrm{d}(\exp_p)_O = \mathrm{id}$ is invertible, and the exponential map is a diffeomorphism between a neighborhood of O in T_pS and a neighborhood of p in S. $\qquad\square$

In particular, note that (5.14) implies that the geodesics issuing from a point $p \in S$ can be written in the form

$$\sigma_v(t) = \exp_p(tv) \ .$$

The latter proposition allows us to introduce some definitions:

Definition 5.2.6. If $p \in S$ and $\delta > 0$, let $B_p(O, \delta) = \{v \in T_pS \mid \|v\|_p < \delta\}$ be the ball with center in the origin and radius $\delta > 0$ in the tangent plane at p. The *injectivity radius* inj rad(p) of S at p is the largest $\delta > 0$ for which $\exp_p \colon B_p(O, \delta) \to S$ is a diffeomorphism with its image. If $0 < \delta \leq$ inj rad(p), we shall call the set $B_\delta(p) = \exp_p\big(B_p(O, \delta)\big)$ a *geodesic ball* with center p and radius δ. The geodesics issuing from p, that is, the curves of the form $t \mapsto \exp_p(tv)$, are called *radial geodesics*; the curves that are image by $\exp_p$ of the circles with center in the origin of T_pS and radius less than inj rad(p) are called *geodesic circles*. Finally, having set $B_\delta^*(p) = B_\delta(p) \setminus \{p\}$, the *radial field* $\partial/\partial r \colon B_\delta^*(p) \to \mathbb{R}^3$ is defined by

$$\frac{\partial}{\partial r}\bigg|_q = \dot{\sigma}_v(1) \in T_qS$$

for all $q = \exp_p(v) \in B_\delta^*(p)$.

In the plane, the radial geodesics from the origin are half-lines: they are orthogonal to the (geodesic) circles with center in the origin, and are tangent everywhere to the radial field. The important *Gauss lemma* we are going to prove claims that all these properties hold for every surface; as we shall see, this is the essential step needed to prove the minimizing properties of geodesics.

Lemma 5.2.7. *Given $p \in S$ and $0 < \delta \leq$ inj rad(p), let $B_\delta(p) \subset S$ be a geodesic ball with center p. Then:*

(i) *for all $q = \exp_p(v) \in B_\delta^*(p)$ we have*

$$\frac{\partial}{\partial r}\bigg|_q = \frac{\sigma_v'(1)}{\|v\|} = \dot{\sigma}_{v/\|v\|}(\|v\|) = \frac{1}{\|v\|}\mathrm{d}(\exp_p)_v(v) \ ;$$

in particular, $\|\partial/\partial r\| \equiv 1$ and $\|\mathrm{d}(\exp_p)_v(v)\| = \|v\|$;

(ii) *the radial geodesics through p, parametrized by arc length, are tangent to the radial field in $B_\delta^*(p)$;*

(iii) *(Gauss' lemma) the radial field is orthogonal to all geodesic circles lying in $B_\delta^*(p)$ and, in particular,*

$$\langle \mathrm{d}(\exp_p)_v(v), \mathrm{d}(\exp_p)_v(w) \rangle = \langle v, w \rangle \tag{5.16}$$

for all $w \in T_pS$.

Proof. (i) The first equality follows from $\|\sigma_v'(1)\| = \|\sigma_v'(0)\| = \|v\|$; the second one can be obtained by differentiating $\sigma_{v/\|v\|}(t) = \sigma_v(t/\|v\|)$ and setting $t = \|v\|$. Finally, a curve in T_pS issuing from v and tangent to v is $\sigma(t) = v + tv$; so

$$\mathrm{d}(\exp_p)_v(v) = \frac{\mathrm{d}}{\mathrm{d}t}\exp_p\big((1+t)v\big)\bigg|_{t=0} = \frac{\mathrm{d}}{\mathrm{d}t}\sigma_v(1+t)\bigg|_{t=0} = \sigma_v'(1) \ . \tag{5.17}$$

(ii) The radial geodesics parametrized by arc length issuing from p are the curves σ_v with $v \in T_pS$ having norm equal to 1. Since $\sigma_{sv}(t) = \sigma_v(ts)$, we find $\sigma'_{sv}(1) = s\dot\sigma_v(s)$, and by (i) we have:

$$\dot\sigma_v(s) = \frac{\sigma'_{sv}(1)}{\|sv\|} = \left.\frac{\partial}{\partial r}\right|_{\exp_p(sv)} = \left.\frac{\partial}{\partial r}\right|_{\sigma_v(s)},$$

and so σ_v is always tangent to $\partial/\partial r$.

(iii) Take $q \in B^*_\delta(p)$ and $v \in T_pS$ such that $q = \exp_p(v)$. Fix furthermore an orthonormal basis $\{E_1, E_2\}$ of T_pS such that $E_1 = v/\|v\|$. Then σ_{E_1} is the radial geodesic from p to $q = \sigma_{E_1}(\|v\|)$ parametrized by arc length, while the geodesic circle through q is parametrized by the curve $\tau(s) = \exp_p(\|v\|(\cos s) E_1 + \|v\|(\sin s)E_2)$. So our goal is to prove that $\dot\sigma_{E_1}(\|v\|)$ is orthogonal to $\tau'(0)$. To this aim, let us consider the map $\Sigma: (-\pi, \pi) \times (-\delta, \delta) \to S$ defined by

$$\Sigma(s, t) = \exp_p(t(\cos s) E_1 + t(\sin s)E_2) \ ;$$

since $\|t(\cos s)E_1 + t(\sin s)E_2\| = |t| < \delta$, the map Σ is well defined and of class C^∞. Now, $\sigma_{E_1}(t) = \Sigma(0, t)$ and $\tau(s) = \Sigma(s, \|v\|)$; so we have to prove that the scalar product

$$\left\langle \frac{\partial\Sigma}{\partial t}(0, \|v\|), \frac{\partial\Sigma}{\partial s}(0, \|v\|) \right\rangle$$

is zero.

Set $v_s = (\cos s)E_1 + (\sin s)E_2$, to get $\Sigma(s,t) = \sigma_{v_s}(t)$. In particular, every σ_{v_s} is a geodesic parametrized by arc length; so $D\dot\sigma_{v_s} \equiv O$ and $\|\dot\sigma_{v_s}(t)\| \equiv 1$. Hence

$$\frac{\partial}{\partial t}\left\langle \frac{\partial\Sigma}{\partial t}(s,t), \frac{\partial\Sigma}{\partial s}(s,t) \right\rangle = \left\langle \frac{\partial}{\partial t}\dot\sigma_{v_s}(t), \frac{\partial\Sigma}{\partial s}(s,t) \right\rangle + \left\langle \frac{\partial\Sigma}{\partial t}(s,t), \frac{\partial^2\Sigma}{\partial t\partial s}(s,t) \right\rangle$$

$$= \left\langle D\dot\sigma_{v_s}(t), \frac{\partial\Sigma}{\partial s}(s,t) \right\rangle + \left\langle \frac{\partial\Sigma}{\partial t}(s,t), \frac{\partial^2\Sigma}{\partial s\partial t}(s,t) \right\rangle$$

$$= \left\langle \frac{\partial\Sigma}{\partial t}(s,t), \frac{\partial^2\Sigma}{\partial s\partial t}(s,t) \right\rangle = \frac{1}{2}\frac{\partial}{\partial s}\left\| \frac{\partial\Sigma}{\partial t}(s,t) \right\|^2$$

$$= \frac{1}{2}\frac{\partial}{\partial s}\|\dot\sigma_{v_s}(t)\|^2$$

$$= 0 \ .$$

It follows that $\langle \partial\Sigma/\partial t, \partial\Sigma/\partial s \rangle$ does not depend on t, and so

$$\left\langle \frac{\partial\Sigma}{\partial t}(0, \|v\|), \frac{\partial\Sigma}{\partial s}(0, \|v\|) \right\rangle = \left\langle \frac{\partial\Sigma}{\partial t}(0,0), \frac{\partial\Sigma}{\partial s}(0,0) \right\rangle \ .$$

Since

$$\frac{\partial\Sigma}{\partial s}(0,t) = t\, \mathrm{d}(\exp_p)_{tE_1}(E_2) \ ,$$

we get $\partial\Sigma/\partial s(0,0) = O$, and we are done.

In particular, since $\dot{\sigma}_{E_1}(\|v\|) = \|v\|\mathrm{d}(\exp_p)_v(v)$ by (i), and

$$\tau'(0) = \|v\|\mathrm{d}(\exp_p)_v(E_2) ,$$

we have proved (5.16) for $w = v$ and for w orthogonal to v. Since every vector of T_pS can be written as a sum of a multiple of v and a vector orthogonal to v, we have the assertion.   □

At last, we are able to prove the characterization we promised, of geodesics as (locally) shortest curves between two points:

Theorem 5.2.8. *Let S be a surface, and $p \in S$.*

(i) *if $0 < \delta \le \mathrm{inj}\ \mathrm{rad}(p)$, then for all $q \in B_\delta(p)$ the radial geodesic from p to q is the unique (up to reparametrizations) shortest curve in S joining p and q;*

(ii) *if $\sigma\colon [0,1] \to S$ has the shortest length between all piecewise regular curve in S joining two points $p = \sigma(0)$ and $q = \sigma(1) \in S$ then σ is a geodesic (and in particular is of class C^∞);*

(iii) *let $\sigma\colon [a,b] \to S$ be a geodesic parametrized by arc length, $t_0 \in [a,b]$, and $\delta = \mathrm{inj}\ \mathrm{rad}(\sigma(t_0))$. Then σ is the shortest curve in S from $\sigma(t_0)$ to $\sigma(t_1)$ if $|t_0 - t_1| < \delta$.*

Proof. (i) Let $q_0 \in B_\delta(p)$; choose $v_0 \in T_pS$ such that $q_0 = \exp_p(v_0)$, and set $E_1 = v_0/\|v_0\|$, so that $\sigma_{E_1}\colon [0, \|v_0\|] \to S$ is the geodesic parametrized by arc length that joins p and q_0. We shall prove that if $\tau\colon [0,\ell] \to S$ is another piecewise regular curve parametrized by arc length in S from p to q_0 then $L(\tau) \ge L(\sigma_{E_1}) = \|v_0\|$, with equality holding if and only if $\tau = \sigma_{E_1}$.

If τ passes more than once through p, clearly we can remove a portion of τ finding a shorter curve from p to q_0; so we may assume that $\tau(s) \ne p$ for all $s > 0$.

Suppose now that the support of τ is contained in $B_\delta(p)$; we shall see later how to remove this assumption. In this case, for all $s \in (0,\ell]$ where τ is smooth there exist $a(s) \in \mathbb{R}$ and $w(s) \in T_{\tau(s)}S$ orthogonal to the radial field such that

$$\dot{\tau}(s) = a(s) \left.\frac{\partial}{\partial r}\right|_{\tau(s)} + w(s) ;$$

note that, by Gauss' Lemma 5.2.7.(iii), the vector $w(s)$ is tangent to the geodesic circle passing through $\tau(s)$.

Let $r\colon B_\delta^*(p) \to \mathbb{R}^+$ be given by

$$r(q) = \| \exp_p^{-1}(q)\|_p .$$

Clearly, r is of class C^∞, and the geodesic circles are the level curves of r; so (why?) we have $\mathrm{d}r_q(w) = 0$ for every vector $w \in T_qS$ tangent to the geodesic circle passing through $q \in B^*_\delta(p)$. Moreover, if $q = \exp_p(v)$ we have

$$
dr_q\left(\left.\frac{\partial}{\partial r}\right|_q\right) = dr_{\exp_p(v)}\left(\frac{\sigma'_v(1)}{\|v\|}\right)
$$

$$
= \frac{1}{\|v\|}\frac{d}{dt}(r \circ \exp_p)(tv)\Big|_{t=1} = \frac{1}{\|v\|}\frac{d}{dt}(t\|v\|)\Big|_{t=1} \equiv 1 \,,
$$

and so we obtain

$$
\frac{d(r \circ \tau)}{ds}(s) = dr_{\tau(s)}\big(\dot\tau(s)\big) = a(s) \,.
$$

We are now ready to estimate the length of τ. Let $0 = s_0 < \cdots < s_k = \ell$ be a partition of $[0, \ell]$ such that $\tau|_{[s_{j-1}, s_j]}$ is regular for $j = 1, \ldots, k$. Then

$$
L(\tau) = \sum_{j=1}^{k}\int_{s_{j-1}}^{s_j}\|\dot\tau(s)\|\,ds = \sum_{j=1}^{k}\int_{s_{j-1}}^{s_j}\sqrt{|a(s)|^2 + \|w(s)\|^2}\,ds
$$

$$
\geq \sum_{j=1}^{k}\int_{s_{j-1}}^{s_j}|a(s)|\,ds \geq \sum_{j=1}^{k}\int_{s_{j-1}}^{s_j}a(s)\,ds = \sum_{j=1}^{k}\int_{s_{j-1}}^{s_j}\frac{d(r \circ \tau)}{ds}(s)\,ds
$$

$$
= \sum_{j=1}^{k}\big[r\big(\tau(s_j)\big) - r\big(\tau(s_{j-1})\big)\big] = r(q_0) - r(p) = \|v_0\| = L(\sigma_{E_1}) \,.
$$

Hence, τ is at least as long as σ_{E_1}. Equality holds if and only if $w(s) = O$ and $a(s) \geq 0$ for all s; in this case, since $\dot\tau$ has length 1, we get $\dot\tau \equiv \partial/\partial r|_\tau$. So τ is a curve from p to q always tangent to the radial field, just like σ_{E_1}, by Lemma 5.2.7.(ii); so σ_{E_1} and τ solve the same system of ordinary differential equations with the same initial conditions, and by Theorem 5.1.9 we have $\tau \equiv \sigma_{E_1}$, as required.

Finally, if the image of τ is not contained in $B_p(\delta)$, let $t_1 > 0$ be the first value for which τ intersects $\partial B_p(\delta)$. Then the previous argument shows that

$$
L(\tau) \geq L(\tau|_{[0,t_1]}) \geq \delta > \|v_0\| \,,
$$

and we are done.

(ii) If σ is the shortest curve between p and q, it is also so between any two points of its support; if this were not the case, we might replace a piece of σ by a shorter piece, getting a piecewise regular curve from p and q shorter than σ. But we have just shown in (i) that locally the shortest curves are geodesic; since being a geodesic is a local property, σ has to be a geodesic.

(iii) Under these assumptions $\sigma(t_1)$ is in the geodesic ball with center $\sigma(t_0)$ and radius δ, and σ is the radial geodesic from $\sigma(t_0)$ to $\sigma(t_1)$. $\square$

Remark 5.2.9. In general, a geodesic is not necessarily the shortest curve between its endpoints. For instance, a segment of a great circle on the unit sphere

S^2 that is longer than π is not: the complementary segment of the same great circle is shorter.

Remark 5.2.10. A shortest curve joining two points of a surface does not always exist. For instance, if $S = \mathbb{R}^2 \setminus \{O\}$ and $p \in S$, then for any curve in S joining p and $-p$ there is always (why?) a shorter curve in S from p to $-p$. However, in the supplementary material of this chapter we shall show that *if $S \subset \mathbb{R}^3$ is a closed surface in $\mathbb{R}^3$, then every pair of points of S can be joined by a curve in S of minimum length, which is necessarily a geodesic* (Corollary 5.4.11).

Remark 5.2.11. Even when it exists, the shortest curve joining two points may well not be unique. For instance, two antipodal points on the sphere are joined by infinitely many great circles, all of the same length.

5.3 Vector fields

In this section we shall introduce the fundamental notion of tangent vector field on a surface. As a first application, we shall prove the existence of orthogonal local parametrizations.

Definition 5.3.1. Let $S \subset \mathbb{R}^3$ be a surface. A *(tangent) vector field* on S is a map $X \colon S \to \mathbb{R}^3$ of class C^∞ such that $X(p) \in T_p S$ for all $p \in S$. We shall denote by $\mathcal{T}(S)$ the vector space of tangent vector fields to S.

Example 5.3.2. If $\varphi \colon U \to S$ is a local parametrization of a surface S then ∂_1, $\partial_2 \colon \varphi(U) \to \mathbb{R}^3$ are vector fields defined on $\varphi(U)$, the *coordinate fields* induced by φ.

Example 5.3.3. Let $S \subset \mathbb{R}^3$ be the surface of revolution obtained by rotating around the z-axis a curve whose support lies in the xz-plane. Let $X \colon S \to \mathbb{R}^3$ be defined by

$$X(p) = \begin{vmatrix} -p_2 \\ p_1 \\ 0 \end{vmatrix}$$

for all $p = (p_1, p_2, p_3) \in S$. Clearly, X is of class C^∞; let us prove that it is a vector field by showing that $X(p) \in T_p S$ for all $p \in S$. Indeed, if $\varphi \colon \mathbb{R}^2 \to \mathbb{R}^3$ is the immersed surface having S as its support defined in Example 3.1.18, by Example 3.3.15 we know that

$$X(p) = \frac{\partial \varphi}{\partial \theta}(\varphi^{-1}(p)) = \left. \frac{\partial}{\partial \theta} \right|_p \in T_p S \, .$$

In other words, $X(p)$ is tangent to the parallel passing through p; see Fig. 5.2.

Fig. 5.2. A vector field on the torus

Remark 5.3.4. Let $X \in \mathcal{T}(S)$ be a vector field on a surface S, and $\varphi: U \to S$ a local parametrization of S. For all $x \in U$, the vector $X(\varphi(x))$ belongs to $T_{\varphi(x)}S$; so there exist $X_1(x)$, $X_2(x) \in \mathbb{R}$ such that

$$X(\varphi(x)) = X_1(x)\partial_1|_{\varphi(x)} + X_2(x)\partial_2|_{\varphi(x)} .$$

In this way we have defined two functions X_1, $X_2: U \to \mathbb{R}$ such that

$$X \circ \varphi \equiv X_1\partial_1 + X_2\partial_2 . \tag{5.18}$$

Note that X_1 and X_2 solve the linear system

$$\begin{cases} EX_1 + FX_2 = \langle X \circ \varphi, \partial_1 \rangle , \\ FX_1 + GX_2 = \langle X \circ \varphi, \partial_2 \rangle , \end{cases}$$

whose coefficients and right-hand terms are of class C^∞ and whose determinant is nowhere zero; hence, X_1 and X_2 are also of class C^∞. Conversely, it is easy to prove (see Problem 5.5) that a map $X: S \to \mathbb{R}^3$ is a vector field if and only if for all local parametrization $\varphi: U \to S$ there exist X_1, $X_2 \in C^\infty(U)$ such that (5.18) holds.

Remark 5.3.5. In Chapter 3 we have presented the vector tangents to a surface as tangent vectors to a curve and as derivations of germs of C^∞ functions. Similarly, vector fields have a dual nature too. We shall shortly discuss curves tangent to a vector field; here we want to discuss the other viewpoint. Let $X \in \mathcal{T}(S)$ be a tangent vector field on a surface S, and $f \in C^\infty(S)$. For all $p \in S$ the vector field X provides us with a derivation $X(p)$ of $C^\infty(p)$; and the function f gives us a germ $\mathbf{f}_p \in C^\infty(p)$ with (S, f) as a representative. So we may define a function $X(f): S \to \mathbb{R}$ by setting

$$\forall p \in S \qquad X(f)(p) = X(p)(\mathbf{f}_p) .$$

By (5.18) we know that if $\varphi \colon U \to S$ is a local parametrization then

$$X(f) \circ \varphi = X_1 \frac{\partial (f \circ \varphi)}{\partial x_1} + X_2 \frac{\partial (f \circ \varphi)}{\partial x_2} \in C^\infty(U) \, ;$$

hence $X(f)$ is a function of class C^∞ on S. Moreover, Remark 3.4.25 gives us the following important relation:

$$\forall X \in \mathcal{T}(S), \, \forall f \in C^\infty(S) \qquad X(f) = \mathrm{d}f(X) \, . \tag{5.19}$$

So we have seen that a tangent vector field X induces a mapping of $C^\infty(S)$ in itself, which we shall again denote by X. This map is obviously linear; further, it is a *derivation* of $C^\infty(S)$, that is,

$$X(fg) = fX(g) + gX(f)$$

for all $f, g \in C^\infty(S)$, as immediately follows from its definition. Conversely, it can be shown (Exercise 5.29) that every linear derivation $X \colon C^\infty(S) \to C^\infty(S)$ is induced by a vector field on S.

Thus, a vector field is a way of smoothly associating a tangent vector with each point of the surface. Since tangent vectors of the surface can be thought of as tangent vectors to curves on the surface, it is only natural to ask ourselves whether there are curves whose tangent vectors give the vector field:

Definition 5.3.6. An *integral curve* (or *trajectory*) of a vector field $X \in \mathcal{T}(S)$ on a surface S is a curve $\sigma \colon I \to S$ such that $\sigma'(t) = X(\sigma(t))$ for all $t \in I$. Fig. 5.3 shows some vector fields with their integral curves.

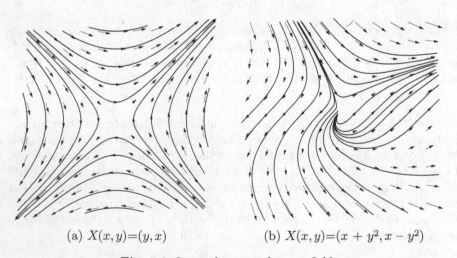

(a) $X(x,y)=(y,x)$ (b) $X(x,y)=(x + y^2, x - y^2)$

Fig. 5.3. Integral curves of vector fields

Let $X \in \mathcal{T}(S)$ be a vector field on a surface S, and $p \in S$; we want to see if there exists an integral curve of X passing through p. Take a local parametrization $\varphi\colon U \to S$ centered at p, and write $X \circ \varphi = X_1\partial_1 + X_2\partial_2$. Every curve $\sigma\colon(-\varepsilon,\varepsilon) \to \varphi(U) \subseteq S$ with $\sigma(0) = p$ will have the form $\sigma = \varphi\circ\sigma^o = \varphi(\sigma_1^o,\sigma_2^o)$ for a suitable curve $\sigma^o\colon(-\varepsilon,\varepsilon) \to U$ with $\sigma^o(0) = O$. So, σ is an integral curve of X if and only if

$$(\sigma_1^o)'\partial_1 + (\sigma_2^o)'\partial_2 = \sigma' = X \circ \sigma = (X_1 \circ \sigma^o)\partial_1 + (X_2 \circ \sigma^o)\partial_2\,,$$

that is, if and only if σ^o is a solution of the Cauchy problem

$$\begin{cases} (\sigma_1^o)' = X_1 \circ \sigma^o\,, \\ (\sigma_2^o)' = X_2 \circ \sigma^o\,, \\ \sigma^o(0) = O\,. \end{cases} \tag{5.20}$$

Thus, at least locally, to find the integral curves of a vector field we have to solve systems of ordinary differential equations. Hence it is natural to use Theorem 5.1.9, and we obtain the following result:

Theorem 5.3.7. *Let $X \in \mathcal{T}(S)$ be a vector field on a surface S. Then:*

(i) *for every $p_0 \in S$ there is an integral curve $\sigma\colon I \to S$ of X with $\sigma(0) = p_0$, and two such integral curves coincide in the intersection of their domains;*

(ii) *for every $p_0 \in S$ there exist $\varepsilon > 0$, a neighborhood V of p_0 in S and a map $\Sigma\colon V \times (-\varepsilon,\varepsilon) \to S$ of class C^∞ such that for all $p \in V$ the curve $\sigma_p = \Sigma(p,\cdot)$ is an integral curve of X with $\sigma_p(0) = p$.*

Proof. Given $p_0 \in S$ and a local parametrization $\varphi\colon U \to S$ centered at p_0, we have seen that finding the integral curves of X with support in $\varphi(U)$ is equivalent to solving the system (5.20). Hence, Theorem 5.1.9 directly yields (ii) and the first part of (i).

Let now $\sigma\colon I \to S$ and $\hat{\sigma}\colon \hat{I} \to S$ be two integral curves of X (with support not necessarily contained in the image of a local parametrization) starting from the same point $p_0 \in S$. By Theorem 5.1.9.(iii) we know that σ and $\hat{\sigma}$ coincide in an open interval I_0 containing 0 and contained in $I \cap \hat{I}$; we want to show that $I_0 = I \cap \hat{I}$. If I_0 were strictly smaller than $I \cap \hat{I}$ we might find an endpoint $t_0 \in \partial I_0 \cap (I \cap \hat{I})$. Since $t_0 \in I \cap \hat{I}$, both σ and $\hat{\sigma}$ are defined in a neighborhood of t_0. As $t_0 \in \partial I_0$, by continuity $\sigma(t_0) = \hat{\sigma}(t_0)$. Now, σ and $\hat{\sigma}$ are integral curves of X from the same point; so Theorem 5.1.9.(iii) implies that σ and $\hat{\sigma}$ coincide in a neighborhood of t_0, against the assumption that $t_0 \in \partial I_0$. Hence, $I_0 = I \cap \hat{I}$, and we are done. $\square$

Remark 5.3.8. Since two integral curves starting from the same point $p \in S$ coincide in the intersection of their domains, we can always find a *maximal* integral curve starting from p.

Definition 5.3.9. Let $X \in \mathcal{T}(S)$ be a vector field on a surface S, and $p_0 \in S$. The map $\Sigma\colon V \times (-\varepsilon,\varepsilon) \to S$ defined in Theorem 5.3.7.(ii) is the *local flow* of X near p_0.

Another consequence of Theorem 5.1.9 is that the supports of distinct maximal integral curves are necessarily disjoint (as the maximal integral curve from any point is unique). So the surface S is partitioned in the disjoint union of the supports of these curves (sometimes is said that the integral curves form a *foliation* of the surface S).

Example 5.3.10. Let $U \subseteq \mathbb{R}^2$ be an open set, and take $X \in \mathcal{T}(U)$ given by $X \equiv \partial/\partial x_1$. Then the integral curve of X starting from the point $p = (p_1, p_2)$ is the line segment $t \mapsto (t + p_1, p_2)$. In particular, the supports of the integral curves of X are the intersection with U of the lines $\{x_2 = \text{cost.}\}$. As you will see in the proof of Theorem 5.3.18, this very simple situation is a blueprint for the general case.

Example 5.3.11. Let $X \in \mathcal{T}(\mathbb{R}^2)$ be given by

$$X(x) = x_2 \frac{\partial}{\partial x_1} + x_1 \frac{\partial}{\partial x_2} \ ;$$

we want to prove that the integral curve of X are the ones shown in Fig. 5.3.(a). The integral curve σ of X passing through $p^o = (x_1^o, x_2^o)$ is the solution of the Cauchy problem

$$\begin{cases} \sigma_1'(t) = \sigma_2(t) \ , \\ \sigma_2'(t) = \sigma_1(t) \ , \\ \sigma(0) = p^o \ . \end{cases} \tag{5.21}$$

Note first that if p^o is the origin then the constant curve $\sigma(t) \equiv O$ is a solution (and consequently the only solution) of (5.21). This is true in general: if $p \in S$ is a point where a vector field $X \in \mathcal{T}(S)$ is zero, then the integral curve of S through p is the constant curve (check it, please).

Let now $p^0 \neq O$. A curve σ is a solution of (5.21) if and only if it solves

$$\begin{cases} \sigma_1''(t) = \sigma_1(t) \ , \\ \sigma_2(t) = \sigma_1'(t) \ , \\ \sigma(0) = p^o \ . \end{cases}$$

Hence the integral curve of X passing through p^o is

$$\sigma(t) = \left(\frac{x_1^o + x_2^o}{2} \, e^t + \frac{x_1^o - x_2^o}{2} \, e^{-t}, \frac{x_1^o + x_2^o}{2} \, e^t - \frac{x_1^o - x_2^o}{2} \, e^{-t} \right) \ .$$

In particular, if $\sigma = (\sigma_1, \sigma_2)$ is an arbitrary integral curve of X, then $\sigma_1^2 - \sigma_2^2$ is constant; so the supports of the integral curves of X are hyperbolas, the level curves of the function $f(x) = x_1^2 - x_2^2$.

Example 5.3.12. Coordinate curves are, essentially by definition, the integral curves of the coordinate fields induced by a local parametrization.

Example 5.3.13. The integral curves of the vector field $X \in \mathcal{T}(S)$ defined in Example 5.3.3 on the surface of revolution $S \subset \mathbb{R}^3$ are clearly the parallels. This means that the supports of the integral curves of X are the sets of the form $\varphi(\{t = \text{const.}\})$, where $\varphi \colon \mathbb{R}^2 \to \mathbb{R}^3$ is the usual immersed surface with support S. In particular, if $U \subset \mathbb{R}^2$ is an open set on which φ is invertible then the supports of the integral curves contained in $\varphi(U)$ are the level sets of the first coordinate of $(\varphi|_U)^{-1}$.

The previous examples seem to suggest that the integral curves of a vector field might be, at least locally, the level sets of suitable C^∞ functions. Let us then give a name to functions which are constant on the supports of the integral curves of a vector field:

Definition 5.3.14. Let $X \in \mathcal{T}(S)$ be a vector field on surface S, and $V \subseteq S$ an open set. A *first integral* of X in V is a function $f \in C^\infty(V)$ constant on the support of the integral curves of X contained in V. Moreover, f is a *proper first integral* if $\mathrm{d}f_q \neq O$ for all $q \in V$ (and so, in particular, the function f is not constant).

Remark 5.3.15. If f is a proper first integral of the vector field $X \in \mathcal{T}(S)$ in an open subset V of a surface S, and $p_0 \in S$ then the set

$$C_{p_0} = \{q \in V \mid f(q) = f(p_0)\}$$

is the support of a regular curve (why? recall Proposition 1.1.18). On the other hand, the support of the integral curve of X that passes through p_0 has to be contained in C_{p_0}, by definition of first integral. Hence C_{p_0} is the support of the integral curve of X passing through p_0, and thus determining a first integral allows one to find the supports of the integral curves of a vector field.

It is possible to give a characterization of first integrals that does not require knowing the integral curves in the first place:

Lemma 5.3.16. *Let $X \in \mathcal{T}(S)$ be a vector field on a surface S, and $V \subseteq S$ an open subset. Then a function $f \in C^\infty(V)$ is a first integral of X if and only if $\mathrm{d}f(X) \equiv 0$ if and only if $X(f) \equiv 0$.*

Proof. Let $\sigma \colon I \to U$ be an integral curve of X, and $t \in I$. Then Remarks 3.4.25 and 5.3.5 tell us that

$$X(f)\big(\sigma(t)\big) = \mathrm{d}f_{\sigma(t)}\big(X(\sigma(t))\big) = (f \circ \sigma)'(t)\,,$$

and so f is constant along all integral curves if and only if $\mathrm{d}f(X) \equiv 0$ if and only if $X(f) \equiv 0$. $\qquad\square$

Remark 5.3.17. Let $\varphi \colon U \to S$ be a local parametrization and $X \in \mathcal{T}(S)$. Then a function $f \in C^\infty\big(\varphi(U)\big)$ is a first integral of X on $\varphi(U)$ if and only if

$$X_1 \frac{\partial(f \circ \varphi)}{\partial x_1} + X_2 \frac{\partial(f \circ \varphi)}{\partial x_2} \equiv 0\,,$$

where we have written $X \circ \varphi = X_1 \partial_1 + X_2 \partial_2$, as usual. So, whereas to find integral curves it is necessary to solve a system of *non-linear ordinary* differential equations, to find first integrals (and consequently the supports of the integral curves) it is necessary to solve a *linear partial* differential equation.

We show next how to use Example 5.3.10 to construct proper first integrals in general:

Theorem 5.3.18. *Let $X \in \mathcal{T}(S)$ be a vector field on a surface S, and $p \in S$ such that $X(p) \neq O$. Then there exists a proper first integral f of X defined in a neighborhood V of p.*

Proof. Let $\varphi \colon U \to S$ be a local parametrization centered at p, and write $X(p) = a_1 \partial_1|_p + a_2 \partial_2|_p$. Let $A \in GL(2, \mathbb{R})$ be an invertible matrix such that

$$A \begin{vmatrix} a_1 \\ a_2 \end{vmatrix} = \begin{vmatrix} 1 \\ 0 \end{vmatrix} ;$$

then it is not difficult to verify (check it!) that $\tilde{\varphi} = \varphi \circ A^{-1}$ is a local parametrization of S centered at p and such that $X(p) = \tilde{\partial}_1|_p$, where $\tilde{\partial}_1 = \partial \tilde{\varphi} / \partial x_1$. So, without loss of generality, we may suppose we have a local parametrization $\varphi \colon U \to S$ centered at p and such that $\partial_1|_p = X(p)$.

Let $\Sigma \colon V \times (-\varepsilon, \varepsilon) \to S$ be the local flow of X near p, with $V \subseteq \varphi(U)$, and set $\hat{V} = \pi(\varphi^{-1}(V) \cap \{x_1 = 0\})$, where $\pi \colon \mathbb{R}^2 \to \mathbb{R}$ is the projection on the second coordinate. Define $\hat{\Sigma} \colon (-\varepsilon, \varepsilon) \times \hat{V} \to S$ by setting

$$\hat{\Sigma}(t, x_2) = \Sigma(\varphi(0, x_2), t) ;$$

in particular, $\hat{\Sigma}(0, x_2) = \varphi(0, x_2)$ for all $x_2 \in \hat{V}$.

Roughly speaking, $\hat{\Sigma}$ maps line segments $\{x_2 = \text{const.}\}$ to integral curves of X; so, if $\hat{\Sigma}$ is invertible the second coordinate of the inverse of $\hat{\Sigma}$ is constant on the integral curves, that is, it is a first integral; see Fig. 5.4.

Hence we want to prove that $\hat{\Sigma}$ is invertible in a neighborhood of the origin, by applying the inverse function theorem for surfaces (Corollary 3.4.28). Denote, as per Remark 3.4.17, by $\partial / \partial t$ and $\partial / \partial x_2$ the canonical basis of $T_O \mathbb{R}^2$, and by σ_p the maximal integral curve of X starting from p. By definition, we have

$$d\hat{\Sigma}_O \left(\frac{\partial}{\partial t} \right) = \frac{\partial}{\partial t} \hat{\Sigma}(t, 0) \bigg|_{t=0} = \frac{\partial}{\partial t} \Sigma(p, t) \bigg|_{t=0} = \sigma_p'(0) = X(p) = \frac{\partial}{\partial x_1} \bigg|_p ,$$

$$d\hat{\Sigma}_O \left(\frac{\partial}{\partial x_2} \right) = \frac{\partial}{\partial x_2} \hat{\Sigma}(0, x_2) \bigg|_{x_2=0} = \frac{\partial}{\partial x_2} \Sigma(\varphi(0, x_2), 0) \bigg|_{x_2=0} = \frac{\partial}{\partial x_2} \bigg|_p ,$$

and so $d\hat{\Sigma}_O$ is invertible, since it maps a basis of $T_O \mathbb{R}^2$ to a basis of $T_p S$. We can then apply Corollary 3.4.28; let $W \subseteq V$ be a neighborhood of p on which $\hat{\Sigma}^{-1} = (\hat{\Sigma}_1^{-1}, \hat{\Sigma}_2^{-1})$ exists. As already observed, $\hat{\Sigma}^{-1}$ maps the integral curves

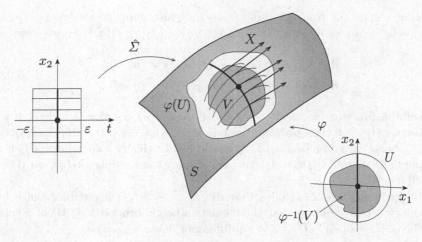

Fig. 5.4.

of X to the lines $\{x_2 = \text{const.}\}$, since $\hat{\Sigma}$ maps the lines $\{x_2 = \text{const.}\}$ to integral curves of X. Thus the function $f = \hat{\Sigma}_2^{-1}$ is of class C^∞ and constant on the integral curves of X, and so is a first integral of X. Moreover, the above computation tells us that

$$\mathrm{d}f_p\left(\left.\frac{\partial}{\partial x_2}\right|_p\right) = \mathrm{d}(\hat{\Sigma}_2^{-1})_p\left(\left.\frac{\partial}{\partial x_2}\right|_p\right) = \frac{\partial}{\partial x_2} \, ;$$

so $\mathrm{d}f_p \neq O$, and f is a proper first integral of X in a neighborhood of p.   □

We shall talk again in general about vector fields in Section 6.4; here we draw to a close by showing how it is possible to use vector fields and first integrals to construct local parametrizations with prescribed support for the coordinate curves.

Theorem 5.3.19. *Let X_1, $X_2 \in \mathcal{T}(S)$ be vector fields on a surface S such that $X_1(p) \wedge X_2(p) \neq O$ for some $p \in S$. Then there exists a local parametrization $\varphi\colon U \to S$ centered at p such that ∂_j is proportional to X_j for $j = 1, 2$. In particular, the supports of the coordinate curves of φ in $\varphi(U)$ coincide with the supports of the integral curves of X_1 and X_2 in $\varphi(U)$.*

Remark 5.3.20. Saying that $X_1(p) \wedge X_2(p) \neq O$ is equivalent to saying that $X_1(p)$ and $X_2(p)$ are linearly independent, or that $\{X_1(p), X_2(p)\}$ is a basis of T_pS. Note that, arguing as at the beginning of the proof of Theorem 5.3.18, it is easy to find a local parametrization $\varphi\colon U \to S$ centered at p such that $\partial_1|_p = X_1(p)$ and $\partial_2|_p = X_2(p)$; here the problem lies in obtaining something analogous in a neighborhood of p, and not just at the point.

Proof (of Theorem 5.3.19). We choose a neighborhood $W \subseteq S$ of p such that there exist proper first integrals $f_j \in C^\infty(W)$ of X_j, for $j = 1, 2$; up to subtracting a constant from f_1 and f_2, we may also suppose we have $f_1(p) = f_2(p) = 0$. Define $\psi: W \to \mathbb{R}^2$ by setting $\psi(q) = (f_2(q), f_1(q))$. So we have

$$\begin{aligned}
\mathrm{d}\psi(X_1) &= \big(\mathrm{d}f_2(X_1), \mathrm{d}f_1(X_1)\big) = (a_1, 0) , \\
\mathrm{d}\psi(X_2) &= \big(\mathrm{d}f_2(X_2), \mathrm{d}f_1(X_2)\big) = (0, a_2) ,
\end{aligned} \tag{5.22}$$

for suitable functions $a_1, a_2: W \to \mathbb{R}$. Note that $a_1(p), a_2(p) \neq 0$: Indeed, if we had, say, $a_1(p) = 0$ then we would also have $\mathrm{d}(f_2)_p(X_1) = 0 = \mathrm{d}(f_2)_p(X_2)$; so $\mathrm{d}(f_2)_p$, being zero on a basis of $T_p S$, would be identically zero, whereas f_2 is a proper first integral. Up to restricting W, we may then suppose $a_1(q), a_2(q) \neq 0$ for all $q \in W$.

In particular, (5.22) implies that $\mathrm{d}\psi_p: T_p S \to \mathbb{R}^2$ is invertible; hence by Corollary 3.4.28 we know that there exists a neighborhood $V \subseteq W$ of p such that $\psi|_V: V \to \psi(V) = U \subseteq \mathbb{R}^2$ is a diffeomorphism.

Set $\varphi = \psi^{-1}$; then φ is a local parametrization centered at p, and is the required parametrization. Indeed, (5.22) says that $\mathrm{d}\psi(X_j) = a_j \mathbf{e}_j$, where $\{\mathbf{e}_1, \mathbf{e}_2\}$ is the canonical basis of $\mathbb{R}^2$, and so

$$X_j = (\mathrm{d}\psi)^{-1}(a_j \mathbf{e}_j) = a_j \mathrm{d}\varphi(\mathbf{e}_j) = a_j \partial_j$$

for $j = 1, 2$, as required.

Finally, if σ is a coordinate curve of φ tangent to ∂_j, and b_j is a solution of the differential equation $b_j' = a_j \circ \sigma \circ b_j$, then b_j is invertible (since a_j is never zero) and the curve $\tau = \sigma \circ b_j$ is a reparametrization of σ such that

$$\tau' = b_j'(\sigma' \circ b_j) = (a_j \circ \tau)\partial_j|_\tau = X_j \circ \tau ,$$

that is, τ is an integral curve of X_j. Hence, integral curves of X_j and coordinate curves of φ have the same support, and we are done. $\square$

For instance, we may use this technique to prove that orthogonal local parametrizations centered at a given point always exist (a fact that, as already seen for instance in Remarks 4.5.15 and 4.6.3, is often quite useful to simplify computations):

Corollary 5.3.21. *Let $S \subset \mathbb{R}^3$ be a surface, and $p \in S$. Then there exists an orthogonal local parametrization centered at p.*

Proof. Let $\varphi: U \to S$ an arbitrary local parametrization centered at p. Set $X_1 = \partial_1$ and

$$X_2 = \partial_2 - \frac{\langle \partial_1, \partial_2 \rangle}{\langle \partial_1, \partial_1 \rangle}\partial_1 .$$

In particular, X_1 is always perpendicular to X_2, and so by applying Theorem 5.3.19 we obtain the orthogonal local parametrization we are looking for. $\square$

Remark 5.3.22. One might ask if there exist *orthonormal* local parametrizations, that is, such that $\|\partial_1\| \equiv \|\partial_2\| \equiv 1$ and $\langle \partial_1, \partial_2 \rangle \equiv 0$. The answer is that this happens if and only if the surface S is locally isometric to a plane. Indeed, in an orthonormal local parametrization we have $E \equiv G \equiv 1$ and $F \equiv 0$, and the assertion follows from Proposition 4.1.20. In particular, the Gaussian curvature of S has to be zero; so, if $K \not\equiv 0$ orthonormal local parametrizations cannot possibly exist.

Remark 5.3.23. A consequence of the previous remark is that, given two vector fields X_1, X_2 on a surface S such that $X_1(p) \wedge X_2(p) \neq O$, it is well possible that there is no local parametrization φ centered at p such that $\partial_1 \equiv X_1$ and $\partial_2 \equiv X_2$ in a neighborhood of p. Indeed, since it is always possible to find two orthonormal vector fields defined in a neighborhood of p (it suffices to apply Gram-Schmidt orthogonalization process to two arbitrary linearly independent vector fields at p), if such a parametrization existed we could always find an orthonormal local parametrization, but we have just seen that this is not the case in general. Exercise 5.35 describes a necessary and sufficient condition for two fields X_1 and X_2 to be the coordinate fields of a local parametrization.

We conclude by proving the existence of two more kinds of special local parametrizations, which will be useful in Chapter 7.

Corollary 5.3.24. *Let $S \subset \mathbb{R}^3$ be a surface, and $p \in S$. Then:*

(i) *if p is not an umbilical point, then there exists a local parametrization centered at p whose coordinate curves are lines of curvature;*
(ii) *if p is a hyperbolic point, then there exists a local parametrization centered at p whose coordinate curves are asymptotic curves.*

Proof. (i) If p is not umbilical there exists a neighborhood $V \subset S$ of p such that every point $q \in V$ has two distinct principal directions, which are the eigenvectors of dN_q, where N is an arbitrary Gauss map for S defined in a neighborhood of p. Since the eigenvectors can be expressed in terms of the matrix that represents dN_q with respect to the basis induced by any local parametrization centered at p, and so (by Proposition 4.5.14) in terms of the metric and form coefficients, it follows that, up to restricting V, we may define two vector fields X_1, $X_2 \in \mathcal{T}(V)$ such that $X_1(q)$ and $X_2(q)$ are distinct principal directions for every $q \in V$. Then the assertion immediately follows from Theorem 5.3.19.

(ii) Sylvester's law of inertia (see [1, Vol. II, Theorem 13.4.7, p. 98]) implies that at every hyperbolic point the second fundamental form admits two linearly independent asymptotic directions. Then arguing (see also Exercise 4.47) as in (i) we get two vector fields X_1, $X_2 \in \mathcal{T}(V)$ defined in a neighborhood $V \subset S$ of p such that $X_1(q)$ and $X_2(q)$ are linearly independent asymptotic directions for every $q \in V$. We conclude again using Theorem 5.3.19. $\square$

Guided problems

Problem 5.1. *Show that if a geodesic σ on an oriented surface is also a line of curvature then σ is a plane curve.*

Solution. We may assume that the curve σ is parametrized by a multiple of the arc length. As σ is a geodesic, $\ddot{\sigma}$ is always parallel to $N \circ \sigma$. Moreover, $\dot{\sigma}$ is always parallel to $\mathrm{d}N_\sigma(\dot{\sigma}) = \mathrm{d}(N \circ \sigma)/\mathrm{d}s$, since σ is a line of curvature. Hence setting $\mathbf{v} = \dot{\sigma} \wedge (N \circ \sigma)$ we get $\dot{\mathbf{v}} \equiv 0$, and so $\mathbf{v}$ is equal everywhere to a fixed vector $\mathbf{v}_0$, which is not zero as $\dot{\sigma}$ is orthogonal to $\ddot{\sigma}$ and thus to $N \circ \sigma$. But then the derivative of the function $\langle \sigma, \mathbf{v}_0 \rangle$ is zero everywhere; so $\langle \sigma, \mathbf{v}_0 \rangle$ is constant, and this means exactly that the support of σ lies in a plane orthogonal to $\mathbf{v}_0$. $\square$

Problem 5.2. *Let $S \subset \mathbb{R}^3$ be a surface of revolution support of the immersed surface $\varphi \colon \mathbb{R} \times \mathbb{R} \to \mathbb{R}^3$ given by $\varphi(t, \theta) = \big(\alpha(t) \cos\theta, \alpha(t) \sin\theta, \beta(t)\big)$. Let further $\sigma \colon I \to S$ be a geodesic parametrized by arc length; fix $s_0 \in I$ and write $\sigma(s) = \varphi\big(t(s), \theta(s)\big)$. If σ is not a parallel prove that there exist constants $b, c \in \mathbb{R}$ such that*

$$\theta(s) = c \int_{s_0}^{s} \frac{1}{\alpha} \sqrt{\frac{(\alpha')^2 + (\beta')^2}{\alpha^2 - c^2}} \, \mathrm{d}t + b \,. \tag{5.23}$$

Solution. If σ is a meridian given by $\theta = \theta_0$ it suffices to take $c = 0$ and $b = \theta_0$ to have (5.23) satisfied; so from now on we shall assume that σ is not a meridian either.

The fact that σ is parametrized by arc length translates to

$$1 \equiv \|\dot{\sigma}\|^2 = (t')^2[(\alpha')^2 + (\beta')^2] \circ t + (\alpha \circ t)^2 (\theta')^2 \,. \tag{5.24}$$

We have already remarked (in Example 5.1.17), that the second equation in (5.6) is equivalent to $(\alpha \circ t)^2 \theta' \equiv c \in \mathbb{R}$, with $c \neq 0$ because σ is not a meridian. So we get

$$(t')^2[(\alpha')^2 + (\beta')^2] \circ t = 1 - \frac{c^2}{(\alpha \circ t)^2}$$

and, differentiating with respect to s,

$$(t')^3[\alpha'\alpha'' + \beta'\beta''] \circ t + t't''[(\alpha')^2 + (\beta')^2] \circ t = \frac{c^2 \alpha' \circ t}{(\alpha \circ t)^3} t' \,.$$

Since σ is not a parallel, $t' \not\equiv 0$; more generally, the set $\{s \in I \mid t'(s) = 0\}$ must have empty interior, because otherwise σ would be a parallel (why?). So we can divide the latter formula by t' for a dense set of parameter values, and by continuity we recover the first equation in (5.6). In other words, we have proved that a curve parametrized by arc length that is not a meridian is a geodesic if and only if it is tangent to the parallels in a set of parameters with empty interior and satisfies the second equation in (5.6).

Now, $\theta'(s)$ can never be zero, since $c \neq 0$. So we may invert $s \mapsto \theta(s)$, to obtain s (and consequently t) as a function of θ. Multiplying (5.24) by $(ds/d\theta)^2 = (\alpha \circ t \circ s)^4/c^2$ we get

$$(\alpha \circ t)^2 = c^2 + c^2 \left(\frac{dt}{d\theta} \right)^2 \frac{(\alpha')^2 + (\beta')^2}{\alpha^2} \circ t \,,$$

where t is being considered as a function of θ. Since $\alpha \geq |c|$ by Clairaut's relation (see Example 5.1.17) , we find

$$\frac{dt}{d\theta} = \frac{\alpha \circ t}{c} \sqrt{\frac{\alpha^2 - c^2}{(\alpha')^2 + (\beta')^2}} \circ t \,.$$

Inverting and integrating we obtain (5.23).   □

Problem 5.3. *Let $S \subset \mathbb{R}^3$ be the elliptic paraboloid of revolution of equation $z = x^2 + y^2$, with local parametrization $\varphi \colon \mathbb{R}^+ \times (0, 2\pi) \to \mathbb{R}^3$ given by $\varphi(t, \theta) = (t \cos \theta, t \sin \theta, t^2)$.*

(i) *Determine the geodesic curvature of the parallels.*
(ii) *Show that the meridians are geodesics.*
(iii) *Show that a geodesic $\sigma \colon \mathbb{R} \to S$ of S that is not a meridian, when traced in the direction of increasing parallels' radii, intersects infinitely many times all the meridians.*

Solution. (i) Let us compute:

$$\varphi_t = (\cos \theta, \sin \theta, 2t) \,, \quad \varphi_\theta = (-t \sin \theta, t \cos \theta, 0) \,,$$
$$E = \langle \varphi_t, \varphi_t \rangle = 1 + 4t^2 \,, \quad F = \langle \varphi_t, \varphi_\theta \rangle = 0 \,, \quad G = \langle \varphi_\theta, \varphi_\theta \rangle = t^2 \,,$$
$$N = \frac{1}{\sqrt{1 + 4t^2}} (-2t \cos \theta, -2t \sin \theta, 1) \,.$$

So for the parallel corresponding to $t = t_0$ we find

$$\mathbf{t} = \frac{\varphi_\theta}{\|\varphi_\theta\|} = (-\sin \theta, \cos \theta, 0) \,, \quad \dot{\mathbf{t}} = \frac{1}{\|\varphi_\theta\|} \mathbf{t}' = \frac{1}{t_0}(-\cos \theta, -\sin \theta, 0) \,,$$
$$N \wedge \mathbf{t} = \frac{1}{\sqrt{1 + 4t_0^2}} (-\cos \theta, -\sin \theta, -2t_0) \,,$$

and so

$$\kappa_g = \langle \dot{\mathbf{t}}, N \wedge \mathbf{t} \rangle = \frac{1}{t_0 \sqrt{1 + 4t_0^2}} \,.$$

In particular, the geodesic curvature does not depend on θ and is never zero (in agreement with the discussion in Example 5.1.17 about the parallels that are geodesics in a surface of revolution).

(ii) For meridians we may proceed in an analogous way, or we may resort to Example 5.1.17, where it was proved that all the meridians of surfaces of revolution are geodesics. Still another option consists in using, since φ is an orthogonal local parametrization, Proposition 5.1.28 to directly compute the geodesic curvature of meridians, keeping in mind that in this case $\dot{\sigma}$ is parallel to $\partial_1|_\sigma = \varphi_r|_\sigma$ and thus the angle θ does not appear in the formula. We obtain

$$\kappa_g = -\frac{1}{2\sqrt{EG}}\frac{\partial E}{\partial \theta} \equiv 0 ,$$

and so meridians are geodesics.

(iii) Let p_0 be a point of the paraboloid and denote by $\mathbf{v}_0$ a tangent vector to the parallel, of radius t_0, passing through $p_0 = \sigma(s_0)$. Let $\sigma: \mathbb{R} \to S$ be a geodesic parametrized by arc length, passing through p_0, forming an angle $\psi_0 \in (0, \pi/2)$ with $\mathbf{v}_0$ and not a meridian. Write, as usual, $\sigma(s) = \varphi\big(t(s), \theta(s)\big)$, and denote by $\psi(s)$ the angle between $\dot{\sigma}(s)$ and the parallel through $\sigma(s)$. First of all, σ cannot be tangent to a meridian in any point, by Proposition 5.1.10 about the uniqueness of the geodesic for a point tangent to a given direction. By Clairaut's relation (5.8), we know that $t(s)\cos\psi(s) = t_0\cos\psi_0 = c$ is a positive constant; hence ψ increases as t increases. Moreover, (5.23) implies

$$\theta(s) = c\int_{s_0}^{s}\frac{1}{t}\sqrt{\frac{1+4t^2}{t^2-c^2}}\,\mathrm{d}t + b > c\int_{s_0}^{s}\frac{\mathrm{d}t}{t} + b \to +\infty$$

for $s \to +\infty$, and thus σ goes infinitely many times around the paraboloid, intersecting each meridian infinitely many times.   □

Problem 5.4. *Let $\sigma: [0,1] \to \mathbb{R}^3$ be a biregular curve of class C^∞ parametrized by arc length. Define $\varphi: [0,1] \times (-\varepsilon, \varepsilon) \to \mathbb{R}^3$ by setting*

$$\varphi(s,v) = \sigma(s) + v\mathbf{b}(s)$$

where $\mathbf{b}$ is the binormal versor of σ. Prove that φ is an immersed surface, and that the curve σ is a geodesic of the support of φ.

Solution. Since σ is parametrized by arc length, using Frenet-Serret formulas we find that

$$\partial_1 = \dot{\sigma} + v\dot{\mathbf{b}} = \mathbf{t} - v\tau\mathbf{n} , \quad \partial_2 = \mathbf{b} , \quad \partial_1 \wedge \partial_2 = -\mathbf{n} - v\tau\mathbf{t} .$$

In particular, the vector $\partial_1 \wedge \partial_2$ is non-zero for every (u,v), because $\mathbf{t}$ and $\mathbf{n}$ are linearly independent, and thus φ is an immersed surface.

In particular, for all $s_0 \in (0,1)$ there is a neighborhood $U \subset \mathbb{R}^2$ of $(s_0, 0)$ such that $\varphi|_U: U \to \mathbb{R}^3$ is a homeomorphism with its image; so $\varphi(U)$ is a regular surface containing the support of σ restricted to a neighborhood of s_0, and we may sensibly ask whether this restriction is a geodesic of $\varphi(U)$. Now, in the point $\sigma(s)$ the normal to the surface is given by $\partial_1 \wedge \partial_2(s, 0) = -\mathbf{n}(s)$; hence σ is a geodesic, because its normal versor is parallel to the normal to the surface.   □

Problem 5.5. *Let $X: S \to \mathbb{R}^3$ be a map. Prove that X is a vector field if and only if for every local parametrization $\varphi: U \to S$ of S there are X_1, $X_2 \in C^\infty(U)$ such that $X \circ \varphi = X_1 \partial_1 + X_2 \partial_2$, that is, $X \circ \varphi = \mathrm{d}\varphi(X_1, X_2)$.*

Solution. Suppose that X satisfies the assumptions. Since the fields ∂_1 and ∂_2 are of class C^∞, and $X_1, X_2 \in C^\infty(U)$, the composite function $X \circ \varphi$ is again of class C^∞ on U. Since this holds for every local parametrization, we have proved that X is of class C^∞ on S. Finally, the condition $X(p) \in T_p S$ for all $p \in S$ is clearly satisfied, since $X(p)$ is a linear combination of a basis of $T_p S$.

The converse has been proved in Remark 5.3.4.   □

Problem 5.6. *Let $X: \mathbb{R}^2 \to \mathbb{R}^2$ be the vector field defined by*

$$X(x_1, x_2) = x_1 \frac{\partial}{\partial x_1} + x_2^2 \frac{\partial}{\partial x_2}.$$

Determine the integral curves of X (see Fig. 5.5).

Solution. The field X vanishes only at the origin. A curve $\sigma = (\sigma_1, \sigma_2)$ is an integral curve for X if and only if it satisfies the following system of ordinary differential equations:

$$\begin{cases} \sigma_1'(t) = \sigma_1(t) \\ \sigma_2'(t) = \sigma_2(t)^2. \end{cases}$$

Let $p_0 = (x_1, x_2)$ be the starting point we are considering. If $p_0 = O$ then $\sigma \equiv (0,0)$ is the integral curve passing through p_0; suppose then $p_0 \neq O$. Firstly, we get $\sigma_1(t) = x_1 e^t$; hence if $x_2 = 0$ the integral curve starting in p_0 is given by $\sigma(t) = (x_1 e^t, 0)$.

If on the other hand $x_2 \neq 0$ we get $\sigma_2(t) = 1/(x_2^{-1} - t)$, and the integral curve starting in p_0 is given by

$$\sigma(t) = \left(x_1 e^t, \frac{1}{x_2^{-1} - t} \right).$$

Note that in this case σ is not defined on the whole real line but diverges as $t \to 1/x_2$.   □

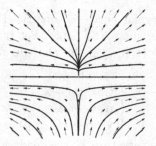

Fig. 5.5. $X(x, y) = (x, y^2)$

Exercises

GEODESICS AND GEODESIC CURVATURE

Definition 5.E.1. A geodesic $\sigma\colon\mathbb{R}\to S$ in a surface $S\subset\mathbb{R}^3$ is *closed* if σ is a closed periodic curve, that is, if there exists $a\in\mathbb{R}$ such that $\sigma(t)=\sigma(t+a)$ for all $t\in\mathbb{R}$.

5.1. If S is the hyperboloid in $\mathbb{R}^3$ of equation $x^2+y^2=1+z^2$, show that the only closed geodesic on S is the circle found by intersecting S with the plane $z=0$. (*Hint:* use Clairaut's relation.)

5.2. Let $S\subset\mathbb{R}^3$ be the one-sheeted hyperboloid, and $C\subset S$ the support of the central parallel, which we know to be a geodesic. Prove that through every point $p\in S\setminus C$ there are infinitely many geodesics whose support is disjoint from C.

5.3. Let $\varphi\colon\mathbb{R}^2\to\mathbb{R}^3$ be a global parametrization of a regular surface S, and suppose that for every curve $\sigma\colon I\to S$ we have $D\partial_1\equiv O$, where D is the covariant derivative along σ. Prove that the Gaussian curvature of S is zero everywhere.

5.4. Let $S\subset\mathbb{R}^3$ be the torus obtained as a surface of revolution by rotating the circle
$$(x-a)^2+z^2=r^2,\quad y=0,$$
around the z-axis, where $0<r<a$. Compute the geodesic curvature of the top parallel, generated by rotating the point $(a,0,r)$.

5.5. Let $H\subset\mathbb{R}^3$ be the plane passing through the x-axis and forming an angle θ (with $\theta\neq\pi/2,\ 3\pi/2$) with the xy-plane; let $S\subset\mathbb{R}^3$ be the cylinder of equation $x^2+y^2=1$, and set $C=S\cap H$.

(i)   Prove that the map $\psi\colon\mathbb{R}^2\to\mathbb{R}^3$ given by $\psi(s,t)=(s,t\cos\theta,t\sin\theta)$ is a global parametrization of the plane H.
(ii) Prove that C is an ellipse by computing its equation with respect to the coordinates (s,t) of H introduced in (i).
(iii) Compute the geodesic curvature of C (parametrized by arc length) in S.

5.6. Let $\sigma\colon I\to S$ be a regular curve parametrized by arc length in an oriented surface S with Gauss map $N\colon S\to S^2$. Set $\mathbf{u}(s)=N\big(\sigma(s)\big)\wedge\mathbf{t}(s)$ and $\mathbf{v}(s)=N\big(\sigma(s)\big)$; the triple $\{\mathbf{t},\mathbf{u},\mathbf{v}\}$ is the *Darboux frame* (or *trihedron*) of σ.

(i)   Prove that
$$\begin{cases} \dot{\mathbf{t}}=\kappa_g\mathbf{u}+\kappa_n\mathbf{v}\,,\\ \dot{\mathbf{u}}=-\kappa_g\mathbf{t}+\tau_g\mathbf{v}\,,\\ \dot{\mathbf{v}}=-\kappa_n\mathbf{t}-\tau_g\mathbf{u}\,, \end{cases}$$
where κ_n is the normal curvature, κ_g the geodesic curvature and $\tau_g\colon I\to\mathbb{R}$ is a function called *geodesic torsion*.

(ii) Prove that, just like for normal curvature, two curves in S parametrized by arc length passing through a point $p \in S$ and tangent in p to the same vector have the same geodesic torsion in p. So the geodesic torsion $\tau_g(v)$ of a unit vector $v \in T_pS$ is well defined.

(iii) Prove that if $v \in T_pS$ has unit length then

$$\tau_g(v) = -\langle dN_p(v), N \wedge v \rangle .$$

(iv) Let v_1, $v_2 \in T_pS$ be orthonormal principal directions, ordered in such a way that $v_1 \wedge v_2 = N(p)$, and let k_1, k_2 be the respective principal curvatures. Prove that for all $v \in T_pS$ of unit length we have

$$\tau_g(v) = \frac{k_2 - k_1}{2} \sin(2\theta) ,$$

where θ is the angle from v_1 to v.

(v) If ϕ is the angle between $\mathbf{n}$ and $\mathbf{v}$, prove that $\mathbf{u} = (\sin\phi)\mathbf{n} - (\cos\phi)\mathbf{b}$ and that

$$\tau_g = \tau + \frac{d\phi}{ds} .$$

Deduce that for all $v \in T_pS$ of unit length such that $\kappa_n(v) \neq 0$ the geodesic torsion $\tau_g(v)$ is the torsion at p of the geodesic passing through p and tangent to v. If instead $\kappa_n(v) = 0$ prove that $|\tau_g(v)| = \sqrt{-K(p)}$.

5.7. Set $p_n = (0,0,1)$, $S_1 = S^2 \setminus \{p_n\}$ and $S_2 = \{(x,y,z) \in \mathbb{R}^3 \mid z = 0\}$, and denote by $\pi \colon S_1 \to S_2$ the stereographic projection.

(i) Determine the points p of S_1 such that the differential $d\pi_p$ is an isometry from T_pS_1 to $T_{\pi(p)}S_2$.

(ii) For every point p of S_1 find a geodesic σ_p on S_1 passing through p and such that its image $\pi \circ \sigma$ is a straight line in S_2.

(iii) Is there a point $p \in S_1$ such that the geodesic σ_p is closed?

5.8. Given the one-sheeted hyperboloid

$$S_1 = \{(x,y,z) \mid x^2 + y^2 - z^2 - 1 = 0\}$$

and the cylinder $S_2 = \{(x,y,z) \mid x^2 + y^2 = 1\}$, let $h \colon S_1 \to S_2$ be the map obtained by associating with $p = (x,y,z) \in S_1$ the point $q \in S_2$ where the half-line $(0,0,z) + \mathbb{R}^+(x,y,0)$ intersects S_2.

(i) Using the standard local parametrizations of S_1 and S_2 seen as surfaces of revolution determine the expressions in coordinates of the first fundamental forms of S_1 and of S_2, and of the map h.

(ii) Find all the points $p \in S_1$ where dh_p is an isometry.

(iii) Find an isometry between S_1 and S_2, if it exists, or prove that no such isometry exists.

(iv) Find a curve on S_1 with constant, non-zero geodesic curvature, mapped by h in a geodesic of S_2.

5.9. Show that if a geodesic is a biregular plane curve then it is a line of curvature. Find an example of a regular (but not biregular) plane geodesic which is not a line of curvature.

5.10. Show that all geodesics in a surface are plane curves if and only if the surface is contained in a plane or in a sphere.

5.11. Let S be a surface in $\mathbb{R}^3$. Given a curve $\sigma: [-\delta, \delta] \to S$ of class C^∞ in S, and a field $\xi \in \mathcal{T}(\sigma)$ of tangent versors orthogonal to $\dot{\sigma}$, prove that there exists an $\varepsilon > 0$ such that the map $\Sigma: (-\varepsilon, \varepsilon) \times (-\varepsilon, \varepsilon) \to S$ given by

$$\Sigma(u, v) = \exp_{\sigma(v)}(u\xi(v))$$

is well defined. Moreover, show that:

(i) for ε small enough, Σ is the inverse of a local parametrization on a neighborhood of $\sigma(0)$;
(ii) the coordinate curves with respect to this local parametrization are orthogonal;
(iii) $E = 0$ in $\sigma(0)$.

5.12. Show that a regular surface in $\mathbb{R}^3$ has two mutually orthogonal families of geodesics if and only if its Gaussian curvature is zero everywhere.

5.13. Show that if two surfaces in $\mathbb{R}^3$ are tangent along a curve σ which is a geodesic for one of the two surfaces then σ is a geodesic for the other surface as well.

5.14. Let $\varphi: \mathbb{R}^2 \to \mathbb{R}^3$ be the map given by

$$\varphi(u, v) = (2\cos u, \sin u, 2v) .$$

(i) Find the largest $c > 0$ such that φ restricted to $(-c, c) \times \mathbb{R}$ is a global parametrization of a regular surface $S \subset \mathbb{R}^3$.
(ii) Prove that the Gaussian curvature of S is zero everywhere.
(iii) Prove that the coordinate directions φ_u and φ_v of this parametrization are always principal directions.
(iv) Show that if $\sigma(t) = \varphi\big(u(t), v(t)\big)$ is a geodesic in S then $v(t) = at + b$ for suitable $a, b \in \mathbb{R}$.

5.15. Let $T \subset \mathbb{R}^3$ be the torus obtained by rotating around the z-axis the circle in the yz-plane with center $(0, 2, 0)$ and radius 1.

(i) Prove that the map $\varphi: (-\pi/2, 3\pi/2) \times (0, 2\pi) \to \mathbb{R}^3$ given by

$$\varphi(u, v) = \big((2 + \cos u)\cos v, (2 + \cos u)\sin v, \sin u\big)$$

is a local parametrization of T.

(ii) Denote by $\sigma_u\colon (0, 2\pi) \to T$ the parallel

$$\sigma_u(t) = \big((2 + \cos u) \cos t, (2 + \cos u) \sin t, \sin u\big).$$

Compute the geodesic curvature of σ_u as a function of $u \in (-\pi/2, 3\pi/2)$.

5.16. Consider the regular surfaces

$$S = \{(x, y, z) \in \mathbb{R}^3 \mid x^2 - y^2 - z^2 = 0, x > 0\},$$
$$T = \{(x, y, z) \in \mathbb{R}^3 \mid y^2 + z^2 - 1 = 0, x > 0\},$$

and the map $\phi\colon S \to T$ given by

$$\phi(x, y, z) = \left(x, \frac{y}{\sqrt{y^2 + z^2}}, \frac{z}{\sqrt{y^2 + z^2}}\right).$$

(i) Prove that S and T are regular surfaces.
(ii) Prove that ϕ is a diffeomorphism between S and T.
(iii) Prove that for every point $p \in S$ there is a geodesic passing through p and mapped by ϕ in a geodesic of T.
(iv) Prove that ϕ is not an isometry by finding a closed curve σ in S which is not a geodesic and such that $\phi \circ \sigma$ is a geodesic in T.

5.17. Let S be a regular surface in $\mathbb{R}^3$, and $\varphi\colon U \to S$ an orthogonal local parametrization. Show that the geodesic curvature of the v-curves is given by

$$\frac{1}{\sqrt{EG}} \frac{\partial G}{\partial u} .$$

5.18. Let S be a regular surface in $\mathbb{R}^3$, and $\varphi\colon U \to S$ an orthogonal local parametrization such that $E \equiv 1$. Prove that the u-curves are geodesic.

5.19. Show that the parallels of a surface of revolution in $\mathbb{R}^3$ have constant geodesic curvature.

5.20. Let $\sigma = (\sigma_1, \sigma_2)\colon \mathbb{R} \to \mathbb{R}^2$ be a regular plane curve parametrized by arc length, and let S be the right cylinder on σ parametrized by the map $\varphi(u, v) = (\sigma_1(u), \sigma_2(u), v)$. Determine, in terms of the parametrization φ, the geodesics of S parametrized by arc length.

5.21. Let S be the surface of revolution obtained by rotating around the z-axis the curve $\sigma\colon \mathbb{R} \to \mathbb{R}^3$ given by

$$\sigma(t) = (3 + 2\cos t, 0, \sin t) .$$

Given $a \in [-1, 1]$, let σ_a the curve obtained by intersecting S with the plane $\{(x, y, z) \in \mathbb{R}^3 \mid z = a\}$.

(i)   Compute the geodesic and normal curvature of σ_a for all $a \in [-1, 1]$.
(ii) Find the values of a for which the curve σ_a is a geodesic, and compute the length of σ_a in these cases.
(iii) Find the values of a for which the curve σ_a has normal curvature zero everywhere, and compute the length of σ_a in these cases.

5.22. Let $S \in \mathbb{R}^3$ be a surface with a global parametrization $\varphi: U \to S$ such that $\partial G/\partial v \equiv 0$ and $\partial G/\partial u \equiv 2\partial F/\partial v$. Show that the v-curves are geodesics on S.

5.23. Find all the geodesics on the regular surface in $\mathbb{R}^3$ defined by the equation $x^2 + 2y^2 = 2$.

VECTOR FIELDS

5.24. Prove that for all $p \in S$ and $v \in T_pS$ there is a vector field $X \in \mathcal{T}(S)$ such that $X(p) = v$.

5.25. Let $X \in \mathcal{T}(S)$ be a vector field on a surface S, and $p \in S$ such that $X(p) \neq O$. Prove that there exists a local parametrization $\varphi: U \to S$ at p such that $X|_{\varphi(U)} \equiv \partial_1$.

5.26. Let $S \subset \mathbb{R}^3$ be a closed surface, and $X \in \mathcal{T}(S)$. Suppose that there exists $M > 0$ such that $\|X(p)\| \leq M$ for all $p \in S$. Prove that all integral curves of X are defined on $\mathbb{R}$.

5.27. Determine the integral curves of the vector fields $X, Y: \mathbb{R}^2 \to \mathbb{R}^2$ defined by

$$X(x,y) = -x\frac{\partial}{\partial x} + y\frac{\partial}{\partial y}, \quad Y(x,y) = -y\frac{\partial}{\partial x} + x\frac{\partial}{\partial y}.$$

5.28. Find a local parametrization of the hyperbolic paraboloid of equation $z = x^2 - y^2$ whose coordinate curves are asymptotic curves.

LIE BRACKET

5.29. Let $S \subset \mathbb{R}^3$ be a surface. Prove that every linear derivation of $C^\infty(S)$ is induced by a vector field on S.

5.30. Let $S \subset \mathbb{R}^3$ be a surface. Given two vector fields $X, Y \in \mathcal{T}(S)$, define a map $[X,Y]: C^\infty(S) \to C^\infty(S)$ by setting

$$[X,Y](f) = X\big(Y(f)\big) - Y\big(X(f)\big)$$

for all $f \in C^\infty(S)$. Prove that $[X, Y]$ is a linear derivation $C^\infty(S)$, and hence (Exercise 5.29) it is induced by a vector field on S, still denoted by $[X, Y]$.

Definition 5.E.2. Given two vector fields X, $Y \in \mathcal{T}(S)$ in a surface S, the vector field $[X,Y] \in \mathcal{T}(S)$ introduce in the previous exercise is the *Lie bracket* of X and Y.

5.31. Show that, if X, Y and Z are vector fields on a surface $S \subset \mathbb{R}^3$ and $f \in C^\infty(S)$, then the following properties hold for the Lie bracket as defined in Exercise 5.30:

(i) $[X,Y] = -[Y,X]$;
(ii) $[X+Y,Z] = [X,Y] + [Z,Y]$;
(iii) $[X,fY] = X(f)Y + f[X,Y]$;
(iv) $\big[X,[Y,Z]\big] + \big[Y,[Z,X]\big] + \big[Z,[X,Y]\big] \equiv O$ (*Jacobi identity*).

5.32. If X, $Y \in \mathcal{T}(S)$ are two vector fields on a surface $S \subset \mathbb{R}^3$, and $\varphi \colon U \to S$ is a local parametrization of S, write $X \circ \varphi = X_1 \partial_1 + X_2 \partial_2$ and $Y \circ \varphi = Y_1 \partial_1 + Y_2 \partial_2$. Prove that

$$[X,Y] \circ \varphi = \sum_{i,j=1}^{2} \left[X_i \frac{\partial Y_j}{\partial x_i} - Y_i \frac{\partial X_j}{\partial x_i} \right] \partial_j \ .$$

In particular, $[\partial_1, \partial_2] \equiv O$.

5.33. Let X, $Y \in \mathcal{T}(S)$ be two vector fields on a surface S, and $p \in S$. Let $\Theta \colon V \times (-\varepsilon, \varepsilon) \to S$ be the local flow of X near p, and for every $t \in (-\varepsilon, \varepsilon)$ define $\theta_t \colon V \to S$ by setting $\theta_t(q) = \Theta(q, t)$. Prove that

$$\lim_{t \to 0} \frac{\mathrm{d}(\theta_{-t})_{\theta_t(p)}(Y) - Y(p)}{t} = [X,Y](p) \ .$$

(*Hint:* given $f \in C^\infty(S)$, put $g(t,q) = f\big(\theta_{-t}(q)\big) - f(q)$ and prove that there exists $h \colon (-\varepsilon, \varepsilon) \times V \to \mathbb{R}$ so that $g \equiv th$.)

5.34. Let X, $Y \in \mathcal{T}(S)$ be two vector fields on a surface $S \subset \mathbb{R}^3$. Denote by Θ the local flow of X, and by Ψ the local flow of Y, and put $\theta_t = \Theta(\cdot, t)$ and $\psi_s = \Psi(\cdot, s)$. Prove that the following assertions are equivalent:

(i) $[X,Y] = O$;
(ii) $\mathrm{d}(\theta_{-t})_{\theta_t(p)}(Y) = Y(p)$ wherever $\theta_t(p)$ is defined;
(iii) $\mathrm{d}(\theta_{-s})_{\psi_s(p)}(X) = X(p)$ wherever $\psi_s(p)$ is defined;
(iv) $\psi_s \circ \theta_t = \theta_t \circ \psi_s$ wherever one of the two sides is defined.

5.35. Let X_1, $X_2 \in \mathcal{T}(S)$ be two vector fields on a surface $S \subset \mathbb{R}^3$, linearly independent in a neighborhood of a point $p \in S$. Prove that there is a local parametrization $\varphi \colon U \to S$ centered at p such that $X_j|_{\varphi(U)} = \partial_j$ for $j = 1$, 2 if and only if $[X_1, X_2] \equiv O$ in a neighborhood of p.

Supplementary material

5.4 The Hopf-Rinow theorem

The Euclidean distance between points in the plane (or in space) is strictly related to the length of curves, in the sense that the distance $\|p_2 - p_1\|$ between two points p_1, $p_2 \in \mathbb{R}^2$ is the minimum among the lengths of curves from p_1 to p_2; moreover, there is always a curve of length exactly equal to the distance between p_1 and p_2, the line segment joining them.

In this section we shall see how to use curves to define an intrinsic distance on any surface, and under which conditions this distance is actually realized by a curve on the surface (which, by Theorem 5.2.8, is necessarily a geodesic).

The basic idea is that the (intrinsic) distance between two points of a surface should be measured by taking the shortest curve joining the two points on the surface. Since, as we have already remarked, such a curve might not exist, rather than the minimum we have to use the infimum:

Definition 5.4.1. Let $S \subset \mathbb{R}^3$ be a surface. The function $d_S \colon S \times S \to \mathbb{R}^+$ given by

$$d_S(p, q) = \inf\{L(\sigma) \mid \sigma \colon [a, b] \to S \text{ is a piecewise } C^\infty \text{ curve from } p \text{ to } q\}$$

is the *intrinsic distance* on S.

Of course, for d_S to be useful we have to prove that it actually is a distance in the sense of metric spaces, and that the topology of (S, d_S) as a metric space coincides with the topology of S as a subset of $\mathbb{R}^3$:

Proposition 5.4.2. *Let $S \subset \mathbb{R}^3$ be a surface. Then the intrinsic distance d_S is actually a distance, and induces on S the usual topology of subspace of $\mathbb{R}^3$.*

Proof. It is clear from the definition that $d_S(p, q) = d_S(q, p) \geq 0$ and that $d_S(p, p) = 0$ for all p, $q \in S$. Furthermore, two points in S can always be joined by a smooth curve (Corollary 4.7.11), and so $d_S(p, q)$ is always finite. Take now p, q, $r \in S$; if σ is a piecewise C^∞ curve in S from p to q, and τ is a piecewise C^∞ curve S from q to r, the curve $\sigma \star \tau$ obtained by first going along σ and then along (a suitable reparametrization of) τ is a piecewise C^∞ curve from p to r with $L(\sigma \star \tau) = L(\sigma) + L(\tau)$. Considering the infima on σ and τ, and keeping in mind that $d_S(p, r)$ is the infimum of the lengths of all curves from p to r, we find

$$d_S(p, r) \leq d_S(p, q) + d_S(q, r) \, ,$$

that is, the triangle inequality. Finally, since the inequality

$$\|p - q\| \leq d_S(p, q) \tag{5.25}$$

clearly holds for all p, $q \in S$, we get $d_S(p, q) > 0$ if $p \neq q$, and so d_S is a distance.

It remains to prove that the topology induced by d_S on S is the usual one. Denote by $B_S(p, \varepsilon) \subset S$ the ball with center in $p \in S$ and radius $\varepsilon > 0$ for d_S, and by $B(p, \varepsilon) \subset \mathbb{R}^3$ the ball with center in p and radius ε with respect to the Euclidean distance in $\mathbb{R}^3$. Inequality (5.25) immediately implies

$$B_S(p, \varepsilon) \subseteq B(p, \varepsilon) \cap S ,$$

for all $p \in S$ and $\varepsilon > 0$, and so the open sets of the usual topology of S are open sets for the topology induced by d_S as well.

Conversely, we have to show that every ball $B_S(p, \varepsilon)$ contains the intersection of an Euclidean ball with S. Indeed, let $\varepsilon_1 = \min\{\text{inj rad}(p), \varepsilon\} > 0$. Since $\exp_p : B_p(O, \varepsilon_1) \to B_{\varepsilon_1}(p)$ is a diffeomorphism, the geodesic ball $B_{\varepsilon_1}(p)$ is an open set (for the usual topology) of S; hence there exists $\delta > 0$ such that $B(p, \delta) \cap S \subseteq B_{\varepsilon_1}(p)$. But, on the other hand, for all $q = \exp_p(v) \in B_{\varepsilon_1}(p)$ the radial geodesic $t \mapsto \exp_p(tv)$ is a C^∞ curve from p to q of length $\|v\| < \varepsilon_1$; hence $d_S(p, q) \leq \varepsilon_1$, and so

$$B(p, \delta) \cap S \subseteq B_{\varepsilon_1}(p) \subseteq B_S(p, \varepsilon) .$$

Thus the open sets for d_S are open sets for the usual topology as well, and we are done. $\qquad\square$

The curves realizing the distance clearly deserve a name.

Definition 5.4.3. A piecewise C^∞ curve $\sigma : [a, b] \to S$ on a surface S is called *minimizing* if it has length less or equal than every other piecewise C^∞ curve on S with the same endpoints, that is, if and only if $L(\sigma) = d_S(\sigma(a), \sigma(b))$. The curve σ is *locally minimizing* if for all $t \in [a, b]$ there is $\varepsilon > 0$ such that $\sigma|_{[t-\varepsilon, t+\varepsilon]}$ is minimizing (with the obvious modifications if $t = a$ or $t = b$).

Theorem 5.2.8 can be rephrased in this language, and implies that geodesic balls are actually balls for the intrinsic distance:

Corollary 5.4.4. Let $S \subset \mathbb{R}^3$ be a surface, $p \in M$ and $0 < \delta \leq \text{inj rad}(p)$. Then:

(i) if q belongs to the geodesic ball $B_\delta(p)$ with center in p then the radial geodesic from p to q is the unique (up to reparametrizations) minimizing curve from p to q;

(ii) the geodesic ball $B_\delta(p)$ coincides with the ball $B_S(p, \delta)$ with center p and radius δ with respect to the intrinsic distance of S, and $d_S(p, q) = \|\exp_p^{-1}(q)\|$ for all $q \in B_\delta(p)$;

(iii) every geodesic of S is locally minimizing and, conversely, every locally minimizing curve in S is a geodesic.

Proof. (i) This is exactly Theorem 5.2.8.(i).

(ii) If $q \in B_\delta(p)$ there is a unique $v \in B_p(O, \delta)$ such that $q = \exp_p(v)$, and the radial geodesic $t \mapsto \exp_p(tv)$ from p to q, which is minimizing, has length $d_S(p, q) = \|v\| < \delta$; hence $B_\delta(p) \subseteq B_S(p, \delta)$ and $d_S(p, q) = \|\exp_p^{-1}(q)\|$. Conversely, if $q \in B_S(p, \delta)$ there has to exist a curve σ from p to q of length less than δ; but we have seen in the proof of Theorem 5.2.8.(i) that every curve from p exiting $B_\delta(p)$ has length at least equal to δ. So $q \in B_\delta(p)$, and we are done.

(iii) It follows from Theorem 5.2.8.(ii) and (iii). $\square$

We want to give now a sufficient condition for the existence of minimizing curves. It is easy to find examples of situations where a minimizing curve does not exist: it is sufficient to take two points of a surface S that are joined by a unique minimizing curve σ (for instance, the center and another point of a geodesic ball), and remove from S a point of the support of σ. This example suggests that the existence of minimizing curves is related to the possibility of extending geodesics indefinitely; indeed, we shall shortly prove that if all geodesics of a surface have as their domain the whole real line then minimizing curves always exist (but the converse is not true: see Remark 5.4.10).

We start with a characterization of the possibility of extending geodesics indefinitely:

Lemma 5.4.5. *Let $S \subset \mathbb{R}^3$ be a surface, and $p \in S$. Then for all $v \in T_pS$ the geodesic $\sigma_v \colon I_v \to S$ is defined on the whole real line (that is, $I_v = \mathbb{R}$) if and only if the exponential map $\exp_p$ is defined on the whole tangent plane (that is, $\mathcal{E}_p = T_pS$).*

Proof. Suppose that $\exp_p$ is defined on all T_pS. Then for all $v \in T_pS$ the curve $\sigma(t) = \exp_p(tv)$ is defined on all $\mathbb{R}$, and Lemma 5.2.1.(i) tells us that $\sigma(t) = \sigma_{tv}(1) = \sigma_v(t)$, allowing us to conclude that σ_v is defined on all $\mathbb{R}$.

Conversely, if σ_v is defined on all $\mathbb{R}$ it is also defined at 1, and so $\exp_p(v)$ is well defined. $\square$

We shall also need a result about the injectivity radius. To be precise, we want to show that every point p of a surface S has a neighborhood $V \subset S$ such that the infimum of the injectivity radii of the points of V is positive.

Definition 5.4.6. The *injectivity radius* of a subset $C \subseteq S$ of a surface S is the number

$$\text{inj rad}(C) = \inf\{\text{inj rad}(q) \mid q \in C\} \geq 0 \ .$$

An open set $V \subseteq S$ is *uniformly normal* if its injectivity radius is positive. In other words, V is uniformly normal if and only if there exists $\delta > 0$ such that $\exp_q$ is a diffeomorphism of $B_q(O, \delta)$ with $B_\delta(q)$ for all $q \in V$.

Then:

Proposition 5.4.7. *Every point of a surface $S \subset \mathbb{R}^3$ has a uniformly normal neighborhood.*

Proof. Given $p \in S$, Proposition 5.2.5.(ii) yields $\delta > 0$ and a local parametrization $\varphi \colon U \to S$ centered at p such that the map $(x, w, t) \mapsto \sigma_{\mathrm{d}\varphi_x(w)}(t)$ is of class C^∞ in the open subset $U \times B(O, \delta) \times (1-\delta, 1+\delta)$ of $\mathbb{R}^2 \times \mathbb{R}^2 \times \mathbb{R} = \mathbb{R}^5$. In particular, the map $\tilde{E} \colon U \times B(O, \delta) \to S$ given by $\tilde{E}(x, w) = \exp_{\varphi(x)}(\mathrm{d}\varphi_x(w))$ is of class C^∞. Moreover, since $\tilde{E}(O, O) = p$, we may find a neighborhood $U_1 \subset U$ of O and a $0 < \delta_1 < \delta$ such that $\tilde{E}(U_1 \times B(O, \delta_1)) \subseteq \varphi(U)$. Define then $E \colon U_1 \times B(O, \delta_1) \to U \times U$ by setting

$$E(x, w) = \big(x, \varphi^{-1}(\tilde{E}(x, w))\big) \ ;$$

we want to show that E is invertible in a neighborhood of (O, O). As usual, it is sufficient to verify that $\mathrm{d}E_{(O,O)}$ is invertible. Denote by $x = (x_1, x_2)$ the coordinates of x, and by $w = (w_1, w_2)$ the coordinates of w. Then for $j = 1, 2$ we have

$$\mathrm{d}E_{(O,O)}\left(\frac{\partial}{\partial w_j}\right) = \frac{\mathrm{d}}{\mathrm{d}t} E(O, te_j)\Big|_{t=0} = \frac{\mathrm{d}}{\mathrm{d}t}\big(O, \varphi^{-1}(\exp_p(t\,\partial_j|_p))\big)\Big|_{t=0}$$
$$= \big(O, \mathrm{d}\varphi_p^{-1} \circ \mathrm{d}(\exp_p)_O(\partial_j|_p)\big) = \big(O, \mathrm{d}\varphi_p^{-1}(\partial_j|_p)\big)$$
$$= \left(O, \frac{\partial}{\partial x_j}\right),$$

since $\mathrm{d}(\exp_p)_O = \mathrm{id}$ by Proposition 5.2.5.(iv). On the other hand, we also have

$$\mathrm{d}E_{(O,O)}\left(\frac{\partial}{\partial x_j}\right) = \frac{\mathrm{d}}{\mathrm{d}t} E(te_j, O)\Big|_{t=0} = \frac{\mathrm{d}}{\mathrm{d}t}\big(te_j, \varphi^{-1}(\exp_{\varphi(te_j)}(O))\big)\Big|_{t=0}$$
$$= \frac{\mathrm{d}}{\mathrm{d}t}(te_j, te_j)\Big|_{t=0} = \left(\frac{\partial}{\partial x_j}, \frac{\partial}{\partial x_j}\right).$$

Hence $\mathrm{d}E_{O,O}$, mapping a basis of $\mathbb{R}^2 \times \mathbb{R}^2$ to a basis of $\mathbb{R}^2 \times \mathbb{R}^2$, is invertible, and so there exist a neighborhood $U_0 \subseteq U_1$ of O and a $0 < \delta_0 \le \delta_1$ such that $E|_{U_0 \times B(O, \delta_0)}$ is a diffeomorphism with its image. In particular, this implies that for all $x \in U_0$ the map $\tilde{e}_x \colon B(O, \delta_0) \to U$ given by $\tilde{e}_x(w) = \varphi^{-1}(\exp_{\varphi(x)}(\mathrm{d}\varphi_x(w)))$ is a diffeomorphism with its image.

Finally, up to restricting U_0, we may find (why?) an $\varepsilon > 0$ such that $B_{\varphi(x)}(O, \varepsilon) \subseteq \mathrm{d}\varphi_x(B(O, \delta_0))$ for all $x \in U_0$. Set $V_0 = \varphi(U_0)$; then, for all $q = \varphi(x) \in V_0$ the map $\exp_q \colon B_q(O, \varepsilon) \to S$ is a diffeomorphism with its image, as $\exp_q = \varphi \circ \tilde{e}_x \circ (\mathrm{d}\varphi_x)^{-1}$, and V_0 is a uniformly normal neighborhood of p. $\qquad\square$

We may now give, as promised, a necessary and sufficient condition for the extendability of geodesics, a condition which will also be sufficient for the existence of minimizing geodesics. An interesting feature of this result, the *Hopf-Rinow theorem*, is that this condition is expressed in terms of intrinsic distance:

Theorem 5.4.8 (Hopf-Rinow). *Let* $S \subset \mathbb{R}^3$ *be a surface. Then the following assertions are equivalent:*

(i) *the intrinsic distance d_S is complete;*
(ii) *for all $p \in S$ the exponential map $\exp_p$ is defined on all T_pS;*
(iii) *there is a point $p \in S$ for which the exponential map $\exp_p$ is defined on all T_pS;*
(iv) *every closed and bounded (with respect to d_S) subset of S is compact.*

Moreover, any of these conditions implies that:

(v) *every pair of points of S can be joined by a minimizing geodesic.*

Proof. (i) $\Longrightarrow$ (ii). By Lemma 5.4.5, it suffices to prove that for every $p \in S$ and $v \in T_pS$ the geodesic σ_v is defined on all $\mathbb{R}$. Set

$$t_0 = \sup\{t > 0 \mid \sigma_v \text{ is defined in } [0, t]\} \,,$$

and assume by contradiction that t_0 is finite. Since

$$d_S\big(\sigma_v(s), \sigma_v(t)\big) \le L(\sigma_v|_{[s,t]}) = \|v\| \, |s - t|$$

for every $0 \le s \le t < t_0$, if $\{t_k\} \subset [0, t_0)$ is increasing and tends to t_0, the sequence $\{\sigma_v(t_k)\}$ is Cauchy in S with respect to the distance d_S, and so it converges to a point $q \in S$ independent of (why?) the sequence we have chosen. Thus setting $\sigma_v(t_0) = q$, we get a continuous map from $[0, t_0]$ to S. Let V be a uniformly normal neighborhood of q, with injectivity radius $\varepsilon > 0$. For all sufficiently large k we have both $|t_k - t_0| < \delta/\|v\|$ and $\sigma_v(t_k) \in V$. In particular, the radial geodesics starting from $\sigma_v(t_k)$ can be prolonged for a length at least equal to δ; since $L(\sigma_v|_{[t_k, t_0]}) = |t_0 - t_k| \|v\| < \delta$, the geodesic σ_v can be prolonged beyond t_0, a contradiction. Hence, $t_0 = +\infty$; in the same way it can be proved that σ_v is defined on $(-\infty, 0]$, and we are done.

(ii) $\Longrightarrow$ (iii). Trivial.

Consider now the condition

(v$'$) *There exists a point $p \in S$ that can be joined to any other point of S by a minimizing geodesic.*

(iii) $\Longrightarrow$ (v$'$). Given $q \in S$, set $r = d_S(p, q)$, and let $B_{2\varepsilon}(p)$ be a geodesic ball with center p such that $q \notin \overline{B_\varepsilon(p)}$. Let $x_0 \in \partial B_\varepsilon(p)$ be a point of the geodesic circle $\partial B_\varepsilon(p)$ in which the continuous function $d_S(q, \cdot)$ admits a minimum. We may write $x_0 = \exp_p(\varepsilon v)$ for a suitable versor $v \in T_pS$; if we prove that $\sigma_v(r) = q$, we have found a minimizing geodesic from p to q.

Define

$$A = \{s \in [0, r] \mid d_S\big(\sigma_v(s), q\big) = d_S(p, q) - s\} \,.$$

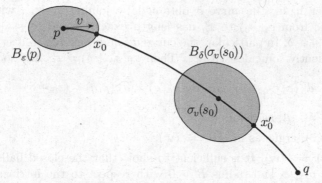

Fig. 5.6.

The set A is not empty ($0 \in A$), and is closed in $[0, r]$; if we prove that $\sup A = r$ we are done. Let $s_0 \in A$ be less than r; we only need to show that there is a sufficiently small $\delta > 0$ such that $s_0 + \delta \in A$. As $s_0 < r = d_S(p, q)$, the point $\sigma_v(s_0)$ cannot be q; choose $0 < \delta < \text{inj rad}(\sigma_v(s_0))$ in such a way that $q \notin B_\delta(\sigma_v(s_0))$. By construction,

$$d_S(p, \sigma_v(s_0)) \leq s_0 = d_S(p, q) - d_S(\sigma_v(s_0), q) \leq d_S(p, \sigma_v(s_0)) \,,$$

which is possible if and only if $d_S(p, \sigma_v(s_0)) = s_0$. Let $x_0' \in \partial B_\delta(\sigma_v(s_0))$ be a point of the geodesic circle $\partial B_\delta(\sigma_v(s_0))$ in which the continuous function $d_S(q, \cdot)$ has a minimum (see Fig. 5.6); in particular, if $s_0 = 0$, take $\delta = \varepsilon$ and $x_0' = x_0$. Then

$$d_S(p, q) - s_0 = d_S(\sigma_v(s_0), q) \leq d_S(\sigma_v(s_0), x_0') + d_S(x_0', q) = \delta + d_S(x_0', q) \,,$$
$$(5.26)$$

since by Corollary 5.4.4.(ii) the geodesic circle $\partial B_\delta(\sigma_v(s_0))$ consists exactly of the points of S at a distance δ from $\sigma_v(s_0)$. On the other hand, if $\tau : [a, b] \to S$ is a piecewise C^∞ curve from $\sigma_v(s_0)$ to q and $\tau(t_0)$ is the first intersection point between the support of τ and $\partial B_\delta(\sigma_v(s_0))$, we have

$$L(\tau) = L(\tau|_{[a, t_0]}) + L(\tau|_{[t_0, b]}) \geq \delta + \min_{x \in \partial B_\delta(\sigma_v(s_0))} d_S(x, q) = \delta + d_S(x_0', q) \,.$$

Taking the infimum with respect to τ, we get

$$d_S(\sigma_v(s_0), q) \geq \delta + d_S(x_0', q) \,;$$

then, recalling (5.26),

$$d_S(p, q) - s_0 = \delta + d_S(x_0', q) \,,$$

from which we deduce

$$d_S(p, x_0') \geq d_S(p, q) - d_S(q, x_0') = s_0 + \delta \,.$$

On the other hand, the curve $\tilde{\sigma}$ obtained by joining $\sigma_v|_{[0,s_0]}$ with the radial geodesic from $\sigma_v(s_0)$ to x_0' has length exactly equal to $s_0 + \delta$; hence, $d_S(p, x_0') = s_0 + \delta$. In particular, the curve $\tilde{\sigma}$ is minimizing; so it is a geodesic and consequently coincides with σ_v. Then $\sigma_v(s_0 + \delta) = x_0'$, and so

$$d_S\big(\sigma_v(s_0 + \delta), q\big) = d_S(x_0', q) = d_S(p, q) - (s_0 + \delta),$$

that is, $s_0 + \delta \in A$, as required.

(ii) $\Longrightarrow$ (v): argue as in (iii) $\Longrightarrow$ (v').

(iii)+(v') $\Longrightarrow$ (iv). It is sufficient to show that the closed balls $\overline{B_S(p, R)}$ with center $p \in S$ and radius $R > 0$ with respect to the intrinsic distance are compact. To do so, it suffices to prove that $\overline{B_S(p, R)} = \exp_p\big(\overline{B_p(O, R)}\big)$; indeed, $\exp_p$ is continuous, the closed and bounded sets of the plane $T_p S$ are compact, and the image of a compact set under a continuous map is compact.

If $v \in \overline{B_p(O, R)}$, then the radial geodesic $t \mapsto \exp_p(tv)$ from p to $q = \exp_p(v)$ has length $\|v\| \le R$; hence, $d_S(p, q) \le R$ and

$$\exp_p\big(\overline{B_p(O, R)}\big) \subseteq \overline{B_S(p, R)}.$$

Conversely, if $q \in \overline{B_S(p, R)}$, let σ be a minimizing geodesic from p to q. Since its length is at most R, it has the form $\sigma(t) = \exp_p(tv)$ with $\|v\| \le R$, and so $q = \exp_p(v) \in \exp_p\big(\overline{B_p(O, R)}\big)$, as required.

(iv) $\Longrightarrow$ (i). Let $\{p_n\} \subset S$ be a Cauchy sequence for d_S; in particular, it is a bounded subset of S, and for any $q \in S$ we can find $R > 0$ such that $\{p_n\} \subset \overline{B_S(q, R)}$. By assumption (iv), the closed ball $\overline{B_S(q, R)}$ is compact; so we can extract from the sequence $\{p_n\}$ a subsequence $\{p_{n_k}\}$ converging to a point $p \in \overline{B_S(q, R)}$. But when a subsequence converges to $p \in S$ the whole Cauchy sequence $\{p_n\}$ has to converge (why?) to the same point, and we have proved that S is complete. $\square$

Definition 5.4.9. A surface $S \subset \mathbb{R}^3$ whose intrinsic distance is complete is called *complete*.

Remark 5.4.10. Condition (v) in Theorem 5.4.8 is strictly weaker than the other ones: there are non-complete surfaces where each pair of points is joined by a minimizing geodesic. The simplest example is provided by a convex open subset Ω of the plane (different from the whole plane): every pair of points of Ω is joined by a minimizing geodesic in Ω (a line segment), but there exist geodesics of Ω that cannot be extended indefinitely (it suffices to consider a segment joining a point of Ω with a point of the boundary of Ω).

The main class of complete surfaces consists of the closed surfaces in $\mathbb{R}^3$:

Corollary 5.4.11. *Every closed surface S in $\mathbb{R}^3$ is complete. In particular, every pair of points of a closed surface is joined by a minimizing geodesic.*

Proof. The inequality (5.25) implies that every bounded (with respect to the intrinsic distance) subset of S is also bounded in $\mathbb{R}^3$; moreover, as S is closed, every closed subset of S is closed in $\mathbb{R}^3$. Hence, every closed and bounded subset of S, being closed and bounded in $\mathbb{R}^3$, is compact, and the assertion follows from the previous theorem. □

Remark 5.4.12. There are surfaces in $\mathbb{R}^3$ that are complete but not closed. Let $\sigma\colon \mathbb{R} \to \mathbb{R}^2$ be the curve $\sigma(t) = \big((1+\mathrm{e}^{-t})\cos t, (1+\mathrm{e}^{-t})\sin t\big)$; it is a spiral that approaches asymptotically the unit circle from outside without ever reaching it. Denoting by $C = \sigma(\mathbb{R})$ the support of σ, let $S = C \times \mathbb{R}$ be the right cylinder with basis C (see Fig. 5.7). Then S is clearly a non-closed surface in $\mathbb{R}^3$; we want to show that it is complete.

To do so it suffices to prove that for all $p_0 \in S$ and $R > 0$ the closed ball $\overline{B}_S(p_0, R) \subset S$ is closed in $\mathbb{R}^3$; indeed, in this case, since it is obviously bounded, it is compact, and the claim follows from Hopf-Rinow theorem. Choose a reparametrization by arc length $\sigma_0\colon \mathbb{R} \to \mathbb{R}^2$ of σ, and let s_0, $z_0 \in \mathbb{R}$ such that $p_0 = \big(\sigma_0(s_0), z_0\big)$. Take a piecewise C^∞ curve $\tau\colon [a,b] \to S$ with $\tau(a) = p_0$ and having length less than $2R$; it is sufficient (why?) to prove that the support of τ is contained in a compact subset of S depending only on p_0 and R but not on τ.

Since C is an open Jordan arc, we may find two piecewise C^∞ functions h, $z\colon [a,b] \to \mathbb{R}$ such that $\tau(t) = \big(\sigma_0(h(t)), z(t)\big)$ for all $t \in [a,b]$; in particular, $h(a) = s_0$ and $z(a) = z_0$. Now, for all $t_0 \in [a,b]$ we have

$$|z(t_0) - z_0| = \left| \int_a^{t_0} z'(t)\,\mathrm{d}t \right| \leq \int_a^{t_0} |z'(t)|\,\mathrm{d}t$$

$$\leq \int_a^{t_0} \|\tau'(t)\|\,\mathrm{d}t = L(\tau|_{[a,t_0]}) \leq 2R\,;$$

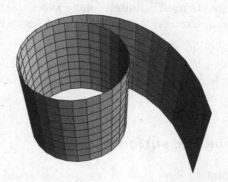

Fig. 5.7. A complete, non-closed surface

so the image of the function z is contained in the interval $[z_0 - 2R, z_0 + 2R]$. Analogously, since σ_0 is parametrized by arc length, we find

$$|h(t_0) - s_0| = \left| \int_a^{t_0} h'(t)\, dt \right| \leq \int_a^{t_0} |h'(t)|\, dt = \int_a^{t_0} \|(\sigma_0 \circ h)'(t)\|\, dt$$

$$\leq \int_a^{t_0} \|\tau'(t)\|\, dt = L(\tau|_{[a,t_0]}) \leq 2R \; ;$$

hence $h([a,b]) \subseteq [s_0 - 2R, s_0 + 2R]$. It follows that the support of τ is contained in $\sigma_0([s_0 - 2R, s_0 + 2R]) \times [z_0 - 2R, z_0 + 2R] \subset S$, and we are done.

Remark 5.4.13. Let $S \subset \mathbb{R}^3$ be a complete surface; then no surface $\tilde{S} \subset \mathbb{R}^3$ properly includes S (and then S is said to be *non-extendable*). Indeed, suppose by contradiction that there exists a surface $\tilde{S}$ such that $\tilde{S} \supset S$ properly, and let q_0 be a point of the boundary of S in $\tilde{S}$. Since S is a surface, it is open in $\tilde{S}$ (why?); hence, $q_0 \notin S$. Choose a geodesic ball $B_\delta(q_0) \subset \tilde{S}$ centered at q_0. Since q_0 belongs to the boundary of S, there is a point $p_0 \in B_\delta(q_0) \cap S$. Since p_0 is in the geodesic ball, it is joined to q_0 by a geodesic $\sigma\colon [0,1] \to \tilde{S}$. But, being p is in the interior of S, there is an $\varepsilon > 0$ such that $\sigma|_{[0,\varepsilon)}$ is a geodesic in S, and this geodesic is clearly not indefinitely extendable in S, as its support has to reach $q_0 \notin S$ in a finite time.

The converse of this claim is not true: there exist non-extendable surfaces that are not complete (see Exercise 5.36).

Exercises

5.36. Show that the one-sheeted cone with the origin removed is a non-extendable surface S that is not complete. Determine for every point $p \in S$ a geodesic on S passing through p that cannot be extended to all values of the parameter.

5.37. Prove that a geodesic in a surface of revolution cannot approach asymptotically a parallel that is not itself a geodesic.

5.38. Prove that every geodesic of the elliptic paraboloid of revolution of equation $z = x^2 + y^2$ intersects itself infinitely many times unless it is a meridian. (*Hint:* use Problem 5.3, Exercise 5.37 and Clairaut's relation.)

5.39. A *divergent curve* in a surface S is a curve $\sigma\colon \mathbb{R}^+ \to S$ such that for every compact set $K \subset S$ there is a $t_0 > 0$ such that $\sigma(t) \notin K$ for all $t \geq t_0$. Prove that S is complete if and only if every divergent curve in S has infinite length.

5.5 *Locally isometric surfaces*

In this section we want to show how to use geodesics and the Gaussian curvature to discover when two surfaces are locally isometric.

A first result in this context is very easy to prove:

Proposition 5.5.1. *Let $H: S \to \tilde{S}$ be a C^∞ map between surfaces. Then H is a local isometry if and only if it maps geodesics parametrized by arc length of S to geodesics parametrized by arc length of $\tilde{S}$.*

Proof. One implication is Proposition 5.1.11. Conversely, suppose H maps geodesics parametrized by arc length to geodesics parametrized by arc length. This means that for all $p \in S$ and $v \in T_p S$ of unit length the curve $H \circ \sigma_v$ is a geodesic of $\tilde{S}$ parametrized by arc length. In particular,

$$\|v\| = 1 = \|(H \circ \sigma_v)'(0)\| = \|\mathrm{d}H_p(\sigma_v'(0))\| = \|\mathrm{d}H_p(v)\| \,,$$

and so $\mathrm{d}H_p$ is an isometry. Since $p \in S$ is arbitrary, we have proved that H is a local isometry. $\qquad\square$

It is more interesting to attempt to directly determine when two surfaces are locally isometric in a point. To simplify our statements, let us introduce the following:

Definition 5.5.2. *Let S, $\tilde{S} \subset \mathbb{R}^3$ be two surfaces, $p \in S$ and $\tilde{p} \in \tilde{S}$. We shall say that (S, p) is locally isometric to $(\tilde{S}, \tilde{p})$ if there exists an isometry between a neighborhood of p in S and a neighborhood of $\tilde{p}$ in $\tilde{S}$.*

If (S, p) is locally isometric to $(\tilde{S}, \tilde{p})$, let $H: V \to \tilde{V}$ be an isometry between a neighborhood of p and a neighborhood of $\tilde{p}$. Since H has to map geodesics into geodesics, we have $H \circ \sigma_v = \sigma_{\mathrm{d}H_p(v)}$ for all $v \in T_p S$. Moreover, Gauss' Theorema Egregium (or, more precisely, Corollary 4.6.12) tells us that $\tilde{K} \circ H \equiv K$, where K is the Gaussian curvature of S and $\tilde{K}$ that of $\tilde{S}$. Putting all together, we find that if (S, p) is locally isometric to $(\tilde{S}, \tilde{p})$ then

$$\forall v \in T_p S \qquad\qquad \tilde{K} \circ \sigma_{I(v)} \equiv K \circ \sigma_v \,, \tag{5.27}$$

where $I = \mathrm{d}H_p: T_p S \to T_{\tilde{p}} \tilde{S}$ is a local isometry.

Our goal is to prove that (5.27) is also a sufficient condition for (S, p) to be locally isometric to $(\tilde{S}, \tilde{p})$. We begin with two quite technical but important lemmas.

Lemma 5.5.3. *Let S be a surface, $p \in S$, and fix a Gauss map $N: V \to S^2$ defined in a neighborhood $V \subseteq S$ of p. Then for every orthogonal basis $\{v_1, v_2\}$ of $T_p S$ we have*

$$\langle v_1, \mathrm{d}N_p(v_2) \rangle \mathrm{d}N_p(v_1) - \langle v_1, \mathrm{d}N_p(v_1) \rangle \mathrm{d}N_p(v_2) = -\|v_1\|^2 K(p) v_2 \,. \tag{5.28}$$

Proof. Note first that the left-hand side of (5.28) does not depend on the Gauss map we have chosen, so it is not necessary to assume S orientable.

Let $\{w_1, w_2\}$ be an orthonormal basis of $T_p S$ consisting of principal directions, with respective principal curvatures k_1 and k_2 (Proposition 4.5.2); in particular, $K(p) = k_1 k_2$. Since $\{v_1, v_2\}$ is an orthogonal basis of $T_p S$, there exists $\theta \in \mathbb{R}$ such that

$$v_1 = \|v_1\|(\cos\theta\, w_1 + \sin\theta\, w_2) \quad \text{and} \quad v_2 = \pm\|v_2\|(-\sin\theta\, w_1 + \cos\theta\, w_2) \,.$$

From this it follows that $dN_p(v_1) = -\|v_1\|(k_1\cos\theta\, w_1 + k_2\sin\theta\, w_2)$ and $dN_p(v_2) = \pm\|v_2\|(k_1\sin\theta\, w_1 - k_2\cos\theta\, w_2)$, and so:

$$\langle v_1, dN_p(v_2)\rangle\, dN_p(v_1) - \langle v_1, dN_p(v_1)\rangle\, dN_p(v_2)$$
$$= \mp\|v_1\|^2\|v_2\|(k_1 - k_2)\cos\theta\sin\theta(k_1\cos\theta\, w_1 + k_2\sin\theta\, w_2)$$
$$\pm\|v_1\|^2\|v_2\|(k_1\cos^2\theta + k_2\sin^2\theta)(k_1\sin\theta\, w_1 - k_2\cos\theta\, w_2)$$
$$= \pm\|v_1\|^2\|v_2\|k_1 k_2(\sin\theta\, w_1 - \cos\theta\, w_2) = -\|v_1\|^2 K(p)v_2 \,.$$

$\square$

Lemma 5.5.4. *Let $S \subset \mathbb{R}^3$ be a surface, and let $p \in S$ and $0 < \delta < \mathrm{inj\ rad}(p)$. Given $v \in B_p(O, \delta)$, denote by $\sigma_v\colon I_v \to S$ the geodesic through p tangent to v. Given $w \in T_p S$ orthogonal to v and $\varepsilon > 0$ small enough, define $\Sigma\colon (-\varepsilon, \varepsilon) \times [0, 1] \to B_\delta(p)$ by setting*

$$\Sigma(s, t) = \exp_p\big(t(v + sw)\big) \,;$$

in particular, $\Sigma(0, t) = \sigma_v(t)$. Let $J\colon [0, 1] \to \mathbb{R}^3$ be given by

$$J(t) = \frac{\partial\Sigma}{\partial s}(0, t) = d(\exp_p)_{tv}(tw) \in T_{\sigma_v(t)}S \,.$$

Then:

(i) $J(0) = O$ *and* $J(1) = d(\exp_p)_v(w)$;
(ii) $DJ(0) = w$;
(iii) $J(t)$ *is orthogonal to* $\sigma_v'(t)$ *for all* $t \in [0, 1]$;
(iv) $D^2 J \equiv -\|v\|^2(K \circ \sigma_v)J$.

Proof. (i) Obvious.

(ii) We have

$$\frac{\partial J}{\partial t}(t) = \frac{\partial^2\Sigma}{\partial t\partial s}(0, t) = d(\exp_p)_{tv}(w) + t\frac{\partial}{\partial t}\big(d(\exp_p)_{tv}(w)\big) \,;$$

so recalling that $d(\exp_p)_O = \mathrm{id}$ we get $\partial J/\partial t(0) = w$. In particular, since $w \in T_p S$, the covariant derivative of J along σ_v at 0 is equal to $\partial J/\partial t(0)$, and we are done.

(iii) Differentiating $\sigma_v(t) = \exp_p(tv)$, we get $\sigma_v'(t) = d(\exp_p)_{tv}(v)$. Since $J(t) = d(\exp_p)_{tv}(tw)$ and $\langle v, w\rangle = 0$, by Lemma 5.2.7.(iii) we know that $J(t)$ is orthogonal to $\sigma_v'(t)$ for all $t \in (0, 1]$. Since $J(0) = O$, we are done.

(iv) By construction, for all $s \in (-\varepsilon, \varepsilon)$ the curve $\sigma_s = \Sigma(s, \cdot)$ is a geodesic; hence, $D\sigma'_s \equiv O$, where D is the covariant derivative along σ_s. By the definition of covariant derivative, this means that

$$\frac{\partial^2 \Sigma}{\partial t^2} - \left\langle \frac{\partial^2 \Sigma}{\partial t^2}, N \circ \Sigma \right\rangle N \circ \Sigma \equiv O \,,$$

where $N: B_\delta(p) \to S^2$ is a Gauss map (which always exists: why?). Differentiating with respect to s and projecting again on the tangent plane we obtain

$$\begin{aligned} O &\equiv \frac{\partial^3 \Sigma}{\partial s \partial t^2} - \left\langle \frac{\partial^3 \Sigma}{\partial s \partial t^2}, N \circ \Sigma \right\rangle N \circ \Sigma - \left\langle \frac{\partial^2 \Sigma}{\partial t^2}, N \circ \Sigma \right\rangle \frac{\partial(N \circ \Sigma)}{\partial s} \\ &\equiv \frac{\partial^3 \Sigma}{\partial t^2 \partial s} - \left\langle \frac{\partial^3 \Sigma}{\partial t^2 \partial s}, N \circ \Sigma \right\rangle N \circ \Sigma + \left\langle \frac{\partial \Sigma}{\partial t}, \frac{\partial(N \circ \Sigma)}{\partial t} \right\rangle \frac{\partial(N \circ \Sigma)}{\partial s} \,, \end{aligned}$$

since $\langle \partial \Sigma / \partial t, N \circ \Sigma \rangle \equiv 0$.

Now, $\partial \Sigma / \partial s$ is also a vector field along σ_s, and we have

$$\begin{aligned} D^2 \frac{\partial \Sigma}{\partial s} &= D\left(\frac{\partial^2 \Sigma}{\partial t \partial s} - \left\langle \frac{\partial^2 \Sigma}{\partial t \partial s}, N \circ \Sigma \right\rangle N \circ \Sigma \right) \\ &= \frac{\partial^3 \Sigma}{\partial t^2 \partial s} - \left\langle \frac{\partial^3 \Sigma}{\partial^2 t \partial s}, N \circ \Sigma \right\rangle N \circ \Sigma + \left\langle \frac{\partial \Sigma}{\partial t}, \frac{\partial(N \circ \Sigma)}{\partial s} \right\rangle \frac{\partial(N \circ \Sigma)}{\partial t} \,, \end{aligned}$$

where we have used again $\langle \partial \Sigma / \partial t, N \circ \Sigma \rangle \equiv 0$. Taking now $s = 0$ we get

$$\begin{aligned} D^2 J &= \langle \sigma'_v, \mathrm{d}N_{\sigma_v}(J) \rangle \mathrm{d}N_{\sigma_v}(\sigma'_v) - \langle \sigma'_v, \mathrm{d}N_{\sigma_v}(\sigma'_v) \rangle \mathrm{d}N_{\sigma_v}(J) \\ &= -\|v\|^2 (K \circ \sigma_v) J \,, \end{aligned}$$

by Lemma 5.5.3, which can be applied because J is orthogonal to σ'_v. $\qquad \square$

We are now ready to prove the main result of this section:

Theorem 5.5.5. *Let $S, \tilde{S} \subset \mathbb{R}^3$ be two surfaces, and suppose we have a point $p \in S$, a point $\tilde{p} \in \tilde{S}$, a $0 < \delta < \mathrm{inj\ rad}(p)$, and a linear isometry $I: T_p S \to T_{\tilde{p}}\tilde{S}$, such that $\tilde{K} \circ \sigma_{I(v)} \equiv K \circ \sigma_v$ on $[0, 1]$ for all $v \in T_p S$ with $\|v\| < \delta$, where K is the Gaussian curvature of S and $\tilde{K}$ the Gaussian curvature of $\tilde{S}$. Then $H = \exp_{\tilde{p}} \circ I \circ \exp_p^{-1}: B_\delta(p) \to B_\delta(\tilde{p})$ is a local isometry. Moreover, if δ is less than $\mathrm{inj\ rad}(\tilde{p})$ too then H is an isometry.*

Proof. Since $\mathrm{d}H_p = I$, the map H is an isometry in p. Let $q = \exp_p(v) \in B_\delta^*(p)$ be arbitrary; we have to prove that

$$\forall \hat{w} \in T_q S \qquad \|\mathrm{d}(\exp_{\tilde{p}} \circ I \circ \exp_p^{-1})_q(\hat{w})\| = \|\hat{w}\| \,.$$

Now, $\mathrm{d}(\exp_p^{-1})_q = (\mathrm{d}(\exp_p)_v)^{-1}$ and

$$\mathrm{d}(\exp_{\tilde{p}} \circ I \circ \exp_p^{-1})_q = \mathrm{d}(\exp_{\tilde{p}})_{I(v)} \circ I \circ (\mathrm{d}(\exp_p)_v)^{-1} \,;$$

here, having set $w = (\mathrm{d}(\exp_p)_v)^{-1}(\hat{w}) \in T_pS$, it suffices to prove that

$$\forall w \in T_pS \qquad \left\|\mathrm{d}(\exp_{\tilde{p}})_{I(v)}\big(I(w)\big)\right\| = \left\|\mathrm{d}(\exp_p)_v(w)\right\| . \tag{5.29}$$

When $w = v$, formula (5.17) yields $\|\mathrm{d}(\exp_p)_v(v)\| = \|v\|$. Analogously, $\left\|\mathrm{d}(\exp_{\tilde{p}})_{I(v)}\big(I(v)\big)\right\| = \|I(v)\|$; since I is an isometry, we have proved (5.29) for $w = v$.

Suppose now that w is orthogonal to v. Define then $J: [0,1] \to \mathbb{R}^3$ as in the previous lemma, and introduce analogously $\tilde{J}: [0,1] \to \mathbb{R}^3$ on $\tilde{S}$ using $I(v)$ and $I(w)$ rather than v and w. By Lemma 5.5.4.(i), it is enough to prove that $\|J(1)\| = \|\tilde{J}(1)\|$.

Set $v_1 = v/\|v\|$ and $v_2 = w/\|w\|$, so that $\{v_1, v_2\}$ is an orthonormal basis of T_pS. Let $\xi_1, \xi_2 \in \mathcal{T}(\sigma_v)$ be the only (Proposition 5.1.6) parallel fields along σ_v such that $\xi_j(0) = v_j$. In particular, $\{\xi_1(t), \xi_2(t)\}$ is an orthonormal basis of $T_{\sigma_v(t)}S$ for all $t \in [0,1]$; so we may find two functions $J_1, J_2: [0,1] \to \mathbb{R}$ of class C^∞ such that $J \equiv J_1\xi_1 + J_2\xi_2$. In particular, $D^2J = J_1''\xi_1 + J_2''\xi_2$; hence Lemma 5.5.4.(iv) implies that J_1 and J_2 solve the linear system of ordinary differential equations

$$\begin{cases} \dfrac{\mathrm{d}^2 J_i}{\mathrm{d}t^2} = -\|v\|^2 (K \circ \sigma_v)J_i , & i = 1,\, 2 , \\[2mm] J_i(0) = 0, \quad \dfrac{\mathrm{d}J_i}{\mathrm{d}t}(0) = \|w\|\delta_{i2} , & i = 1,\, 2 . \end{cases}$$

Proceeding analogously in $\tilde{S}$, we find two functions $\tilde{J}_1, \tilde{J}_2: [0,1] \to \mathbb{R}$ and two parallel vector fields $\tilde{\xi}_1, \tilde{\xi}_2 \in \mathcal{T}(\sigma_{I(v)})$ such that $\tilde{J} = \tilde{J}_1\tilde{\xi}_1 + \tilde{J}_2\tilde{\xi}_2$; moreover, $\tilde{J}_1$ and $\tilde{J}_2$ satisfy the linear system of ordinary differential equations

$$\begin{cases} \dfrac{\mathrm{d}^2 \tilde{J}_i}{\mathrm{d}t^2} = -\|I(v)\|^2 (\tilde{K} \circ \sigma_{I(v)})\tilde{J}_i , & i = 1,\, 2 , \\[2mm] \tilde{J}_i(0) = 0, \quad \dfrac{\mathrm{d}\tilde{J}_i}{\mathrm{d}t}(0) = \|I(w)\|\delta_{i2} , & i = 1,\, 2 . \end{cases}$$

But I is an isometry, and $K \circ \sigma_v \equiv \tilde{K} \circ \sigma_{I(v)}$ by assumption; hence Theorem 1.3.36 implies $\tilde{J}_i \equiv J_i$ for $i = 1,\, 2$, and so

$$\|\tilde{J}(1)\|^2 = \tilde{J}_1(1)^2 + \tilde{J}_2(1)^2 = J_1(1)^2 + J_2(1)^2 = \|J(1)\|^2 .$$

Finally, for $w \in T_pS$ generic we may write $w = cv + w^\perp$, with $w^\perp$ orthogonal to v. We have remarked that $\mathrm{d}(\exp_p)_v(w^\perp)$ is orthogonal to $\sigma_v'(1) = \mathrm{d}(\exp_p)_v(v)$; hence

$$\|\mathrm{d}(\exp_p)_v(w)\|^2 = c^2\|\mathrm{d}(\exp_p)_v(v)\|^2 + \|\mathrm{d}(\exp_p)_v(w^\perp)\|^2 .$$

Analogously, we may write $I(w) = cI(v) + I(w^\perp)$, with $I(w^\perp)$ orthogonal to $I(v)$; so

$$\left\|\mathrm{d}(\exp_{\tilde{p}})_{I(v)}\big(I(w)\big)\right\|^2 = c^2\left\|\mathrm{d}(\exp_{\tilde{p}})_{I(v)}\big(I(v)\big)\right\|^2 + \left\|\mathrm{d}(\exp_{\tilde{p}})_{I(v)}\big(I(w^\perp)\big)\right\|^2 ,$$

and (5.29) follows from the the previous cases.

Finally, if $\delta < \operatorname{inj} \operatorname{rad}(\tilde{p})$ also holds then H is a diffeomorphism, and hence a global isometry. □

Consequently:

Corollary 5.5.6. *Let S, $\tilde{S} \subset \mathbb{R}^3$ be two surfaces, $p \in S$ and $\tilde{p} \in \tilde{S}$. Then (S, p) is locally isometric to $(\tilde{S}, \tilde{p})$ if and only if there exists a linear isometry $I : T_p S \to T_{\tilde{p}} \tilde{S}$ such that (5.27) holds. In particular, two surfaces with equal constant Gaussian curvature are always locally isometric.*

Proof. One implication has already been proved deriving (5.27). Conversely, if (5.27) holds then the previous theorem provides us with an isometry between a neighborhood of p and a neighborhood of $\tilde{p}$, as required.

Finally, if both S and $\tilde{S}$ have constant Gaussian curvature K_0 (5.27) is automatically satisfied, and we are done. □

Roughly speaking, we may summarize this result by saying that *the behavior of Gaussian curvature along the geodesics completely characterizes the surface.*

Remark 5.5.7. Exercise 4.42 shows that it is possible to find two surfaces S, $\tilde{S} \subset \mathbb{R}^3$, two points $p \in S$ and $\tilde{p} \in \tilde{S}$ and a diffeomorphism $F : S \to \tilde{S}$ such that $F(p) = \tilde{p}$ and $\tilde{K} \circ F \equiv K$, even if (S, p) is not locally isometric to $(\tilde{S}, \tilde{p})$.

6

The Gauss-Bonnet theorem

The purpose of this chapter is to give a proof of the Gauss-Bonnet theorem, undoubtedly one of the most important (if not simply *the* most important) results in the differential geometry of surfaces. The Gauss-Bonnet theorem uncovers an unexpected and deep relation between purely local notions, defined in differential terms, such as Gaussian and geodesic curvatures, and the global topology of a surface.

As we shall see, the Gauss-Bonnet theorem has a local version as well as a global version. The local version (which we shall prove in Section 6.1) is a statement that applies to simple (that is, homeomorphic to a closed disk), small (that is, lying within the image of an orthogonal local parametrization) regular regions. To obtain a version valid for arbitrary regular regions, we need a way to cut up a regular region into many simple small regular regions. This is always possible, by using the triangulations we shall introduce in Section 6.2 (even if the proof of the existence of triangulations is postponed to Section 6.5 of the supplementary material to this chapter). In particular, by using triangulations we shall also introduce the Euler-Poincaré characteristic, a fundamental topological invariant of regular regions.

In Section 6.3 we shall be able to prove the global Gauss-Bonnet theorem, and we shall give several of its applications, the most famous among them perhaps is the fact that the integral of the Gaussian curvature on a compact orientable surface S is always equal to 2π times the Euler-Poincaré characteristic of S (Corollary 6.3.10). Section 6.4 discusses the proof of the Poincaré-Hopf theorem, a noteworthy application of Gauss-Bonnet theorem to vector fields. Finally, the supplementary material to this chapter contains complete proofs of the results about triangulations stated in Section 6.2.

Abate M., Tovena F.: Curves and Surfaces.
DOI 10.1007/978-88-470-1941-6_6, © Springer-Verlag Italia 2012

6.1 The local Gauss-Bonnet theorem

The local version of Gauss-Bonnet theorem is a result about regular regions (see Definition 4.2.2) contained in the image of an orthogonal local parametrization and is, in a sense, a generalization of Hopf's turning tangents Theorem 2.4.7. So we start by defining the rotation index for curvilinear polygons on surfaces.

Remark 6.1.1. In this chapter, with a slight abuse of language, we shall sometimes use the phrase "curvilinear polygon" to mean what, strictly speaking, is just the support of a curvilinear polygon.

Definition 6.1.2. A *small* curvilinear polygon in a surface S is a curvilinear polygon whose support is contained in the image of a local parametrization.

Let $\sigma: [a, b] \rightarrow \varphi(U) \subseteq S$ be a small curvilinear polygon with support contained in the image of the local parametrization $\varphi: U \rightarrow S$. Choose a partition $a = s_0 < s_1 < \cdots < s_k = b$ of $[a, b]$ such that $\sigma|_{[s_{j-1}, s_j]}$ is regular for $j = 1, \ldots, k$. The *external angle* of σ at s_j is the angle $\varepsilon_j \in (-\pi, \pi)$ from $\dot{\sigma}(s_j^-)$ to $\dot{\sigma}(s_j^+)$, taken with the positive sign if $\{\dot{\sigma}(s_j^-), \dot{\sigma}(s_j^+)\}$ is a positive basis of $T_{\sigma(s_j)}S$, with the negative sign otherwise (and we are giving $T_{\sigma(s_j)}S$ the orientation induced by φ); the external angle is well defined because σ has no cusps. Define next the function *rotation angle* $\theta: [a, b] \rightarrow \mathbb{R}$ as follows: let $\theta: [a, t_1) \rightarrow \mathbb{R}$ be the continuous determination (see Remark 5.1.27) of the angle from $\partial_1|_\sigma$ to $\dot{\sigma}$, chosen in such a way that $\theta(a) \in (-\pi, \pi]$. Set next

$$\theta(s_1) = \lim_{s \rightarrow s_1^-} \theta(s) + \varepsilon_1 \,,$$

where ε_1 is the external angle at s_1; see Fig 6.1. Note that, by construction, $\theta(s_1)$ is a determination of the angle between $\partial_1|_{\sigma(s_1)}$ and $\dot{\sigma}(s_1^+)$, whereas $\lim_{s \rightarrow s_1^-} \theta(s)$ is a determination of the angle between $\partial_1|_{\sigma(s_1)}$ and $\dot{\sigma}(s_1^-)$.

Define now $\theta: [s_1, s_2) \rightarrow \mathbb{R}$ as the continuous determination of the angle from $\partial_1|_\sigma$ to $\dot{\sigma}$ with initial value $\theta(s_1)$, and set $\theta(s_2) = \lim_{s \rightarrow s_2^-} \theta(s) + \varepsilon_2$. Proceeding like this we define θ on the whole interval $[a, b)$ so that it is continuous from the right, and we conclude by setting

$$\theta(b) = \lim_{s \rightarrow b^-} \theta(s) + \varepsilon_k \,,$$

where ε_k is the external angle at $b = s_k$; clearly, $\theta(b)$ is a determination of the angle between $\partial_1|_{\sigma(b)}$ and $\dot{\sigma}(s_k^+)$. Then the *rotation index* $\rho(\sigma)$ of the curvilinear polygon σ is the number

$$\rho(\sigma) = \frac{1}{2\pi} \big(\theta(b) - \theta(a)\big) \,.$$

Since $\dot{\sigma}(s_k^+) = \dot{\sigma}(s_0^+)$, the rotation index is necessarily an integer number. Moreover, note that the choice of a different determination $\theta(a)$ of the angle

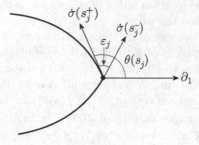

Fig. 6.1.

from $\partial_1|_{\sigma(a)}$ to $\dot{\sigma}(a)$ modifies by an additive constant the rotation angle, but does not change the rotation index.

By the turning tangent Theorem 2.4.7, the rotation index of a curvilinear polygon in the plane is ± 1. It is not difficult to verify that the same holds for small curvilinear polygons:

Proposition 6.1.3. *Let $\sigma\colon [a,b] \to S$ be a curvilinear polygon contained in the image of a local parametrization $\varphi\colon U \to S$ of a surface S, and set $\sigma_o = \varphi^{-1} \circ \sigma\colon [a,b] \to \mathbb{R}^2$. Then the rotation index of σ coincides with the rotation index of σ_o. In particular, $\rho(\sigma) = \pm 1$.*

Proof. The basic idea consists in comparing the way we compute the rotation angle and the external angles for σ and for σ_o. The rotation angle for σ_o is obtained by computing the angle between the constant direction $\partial/\partial x_1$ and the tangent vector σ_o' by using the canonical scalar product $\langle \cdot\,,\cdot \rangle^0$ of $\mathbb{R}^2$. On the other hand, the rotation angle for σ is obtained by computing the angle between the varying direction $\partial_1|_{\sigma(s)} = \mathrm{d}\varphi_{\sigma_o(s)}(\partial/\partial x_1)$ and the tangent vector $\sigma'(s) = \mathrm{d}\varphi_{\sigma_o(s)}(\sigma_o'(s))$ by using the scalar product in $T_{\sigma(s)}S$ given by the first fundamental form. An analogous remark also holds for the computation of external angles. This suggests that we may obtain the rotation angle and the external angles of σ working just with $\partial/\partial x_1$ and the tangent vectors of σ_o if to compute the angle of bound vectors at the point $\sigma_o(s)$ we use the scalar product

$$\forall v, w \in \mathbb{R}^2 \qquad \langle v, w \rangle_s^1 = \big\langle \mathrm{d}\varphi_{\sigma_o(s)}(v), \mathrm{d}\varphi_{\sigma_o(s)}(w) \big\rangle_{\sigma(s)} .$$

But we could also measure rotation and external angles at the point $\sigma_o(s)$ using, more in general, the scalar product on $\mathbb{R}^2$ given by

$$\langle \cdot\,,\cdot \rangle_s^\lambda = (1-\lambda)\langle \cdot\,,\cdot \rangle^0 + \lambda \langle \cdot\,,\cdot \rangle_s^1$$

with $\lambda \in [0,1]$ In this way we would get for every $\lambda \in [0,1]$ a rotation index $\rho^\lambda(\sigma)$ that should depend continuously on λ. But, on the other hand, $\rho^\lambda(\sigma)$ is always an integer, because $\dot{\sigma}_o(b^+) = \dot{\sigma}_o(a^+)$; hence $\rho^\lambda(\sigma)$ should be constant. Since ρ^0 is the rotation index of σ_o, and ρ^1 is the rotation index of σ we would have completed the proof.

| Let us make this argument more formal. Given $s \in [a, b]$ and $\lambda \in [0, 1]$, let $\{X_1^\lambda(s), X_2^\lambda(s)\}$ be the basis of $\mathbb{R}^2$ obtained by applying the Gram-Schmidt orthogonalization process to $\{\partial/\partial x_1, \partial/\partial x_2\}$ with respect to the scalar product $\langle \cdot, \cdot \rangle_s^\lambda$; clearly, $X_1^\lambda(s)$ and $X_2^\lambda(s)$ depend continuously on λ and s. For $s \in [a, s_1)$ and $j = 1, 2$, set $a_j^\lambda(s) = \langle \sigma_o'(s), X_j^\lambda(s) \rangle_s^\lambda / \|\sigma_o'(s)\|_s^\lambda$ so that $\alpha(\lambda, s) = \left(a_1^\lambda(s), a_2^\lambda(s) \right)$ is a continuous map from $[0, 1] \times [a, s_1)$ to S^1, extendable with continuity to all $[0, 1] \times [a, s_1]$ using $\sigma_o'(s_1^-)$. Now, Proposition 2.1.13 provides us with a unique lift $\Theta: [0, 1] \times [a, s_1] \to \mathbb{R}$ such that $\Theta(0, a) \in (-\pi, \pi]$. By construction, $\Theta(0, s)$ is the rotation angle of σ_o in $[a, s_1)$, while $\Theta(1, s)$ is, up to an additive constant, the rotation angle of σ in $[a, s_1)$.

We may define analogously a continuous function $\varepsilon_1: [0, 1] \to \mathbb{R}$ such that $\varepsilon_1(\lambda)$ is a determination of the angle from $\sigma_o'(s_1^-)$ to $\sigma_o'(s_1^+)$, measured using the scalar product $\langle \cdot, \cdot \rangle_{s_1}^\lambda$. So we may go on and define Θ on $[0, 1] \times [s_1, s_2)$ using $\Theta(0, s_1) + \varepsilon_1(0)$ as initial value. In this way we get a map $\Theta: [0, 1] \times [a, b] \to \mathbb{R}$ with $\Theta(\cdot, s)$ continuous with respect to λ for all $s \in [a, b]$; hence $\rho^\lambda(\sigma) = \left(\Theta(\lambda, b) - \Theta(\lambda, a) \right)/2\pi$ is continuous with respect to λ. But, as already remarked, $\rho^\lambda(\sigma)$ is integer-valued; hence, it is constant, and $\rho^0(\sigma) = \rho^1(\sigma)$, as required.

Finally, the last claim follows from the turning tangent Theorem 2.4.7.

$\square$

Definition 6.1.4. A small curvilinear polygon $\sigma: [a, b] \to S$ in a surface S is *positively oriented* (with respect to a given local parametrization containing the support of σ in its image) if its rotation index is $+1$.

We have seen (Definition 4.2.2) that a regular region is a compact connected subset of a surface obtained as the closure of its interior and with a boundary consisting of finitely many disjoint curvilinear polygons. Whereas the (global) Gauss-Bonnet theorem will hold for arbitrary regions, the local Gauss-Bonnet theorem applies only to regular regions of a particular kind.

Definition 6.1.5. A regular region $R \subseteq S$ of a surface S is *simple* if it is homeomorphic to a closed disk.

Remark 6.1.6. The boundary of a simple regular region is a single curvilinear polygon. Conversely, Schönflies' theorem mentioned in Remark 2.3.7 (and proved in Section 2.8 of the supplementary material of Chapter 2) implies that a regular region R whose boundary consists of a single curvilinear polygon and contained in the image of a local parametrization is necessarily simple. This is *not* true if R is not contained in the image of a local parametrization. For instance, a small curvilinear polygon in a torus can be the boundary of two regular regions: a simple one R_i, lying in the image of a local parametrization, and another one, R_e, not simple (and not contained in the image of any local parametrization); see Fig. 6.2.

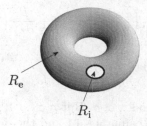

R_e

R_i

Fig. 6.2.

Finally, keeping in mind the definition of integral on a regular region contained in the image of a local parametrization (Definition 4.2.9), we may now prove the local version of the Gauss-Bonnet theorem:

Theorem 6.1.7 (local Gauss-Bonnet). *Let $R \subset S$ be a simple regular region contained in the image of an orthogonal local parametrization $\varphi: U \to S$. Denote by $\sigma: [a, b] \to S$ a parametrization by arc length of the boundary of R, positively oriented with respect to φ, with external angles $\varepsilon_1, \ldots, \varepsilon_k \in (-\pi, \pi)$. Finally, orient $\varphi(U)$ choosing $N = \partial_1 \wedge \partial_2 / \|\partial_1 \wedge \partial_2\|$ as normal versor field, and denote by κ_g the corresponding geodesic curvature of σ (defined outside the vertices). Then*

$$\int_R K \, d\nu + \int_a^b \kappa_g \, ds + \sum_{j=1}^k \varepsilon_j = 2\pi \,, \tag{6.1}$$

where K is the Gaussian curvature of S.

Proof. Write $\sigma = \varphi(\sigma_1, \sigma_2)$, as usual. Proposition 5.1.28 tells us that in the point where σ is regular we have

$$\kappa_g = \frac{1}{2\sqrt{EG}} \left[\dot{\sigma}_2 \frac{\partial G}{\partial x_1} - \dot{\sigma}_1 \frac{\partial E}{\partial x_2} \right] + \frac{d\theta}{ds} \,,$$

where $\theta: [a, b] \to \mathbb{R}$ is the rotation angle of σ. Hence, if $a = s_0 < \cdots < s_k = b$ is a partition of $[a, b]$ such that $\sigma|_{[s_{j-1}, s_j]}$ is regular for $j = 1, \ldots, k$, recalling the classical Gauss-Green theorem (Theorem 2.7.1) we obtain

$$\int_a^b \kappa_g \, ds = \sum_{j=1}^k \int_{s_{j-1}}^{s_j} \kappa_g(s) \, ds$$

$$= \int_a^b \left[\frac{1}{2\sqrt{EG}} \frac{\partial G}{\partial x_1} \dot{\sigma}_2 - \frac{1}{2\sqrt{EG}} \frac{\partial E}{\partial x_2} \dot{\sigma}_1 \right] ds$$

$$+ \sum_{j=1}^k \left[\theta(s_j) - \theta(s_{j-1}) - \varepsilon_j \right]$$

$$= \int_{\varphi^{-1}(R)} \left[\frac{\partial}{\partial x_1} \left(\frac{1}{2\sqrt{EG}} \frac{\partial G}{\partial x_1} \right) + \frac{\partial}{\partial x_2} \left(\frac{1}{2\sqrt{EG}} \frac{\partial E}{\partial x_2} \right) \right] dx_1 \, dx_2$$

$$+ 2\pi \rho(\sigma) - \sum_{j=1}^{k} \varepsilon_j \, ,$$

where $\rho(\sigma)$ is the rotation index of σ. But $\rho(\sigma) = 1$, as σ is positively oriented with respect to φ; so, by Lemma 4.6.14 and the definition of integral on R, we obtain

$$\int_a^b \kappa_g \, ds = - \int_{\varphi^{-1}(R)} K\sqrt{EG} \, dx_1 \, dx_2 + 2\pi - \sum_{j=1}^{k} \varepsilon_j = - \int_R K \, d\nu + 2\pi - \sum_{j=1}^{k} \varepsilon_j \, ,$$

and we are done. $\qquad\qquad\qquad\qquad\qquad\qquad\qquad\qquad\qquad\qquad\qquad\square$

Remark 6.1.8. The hypothesis that the regular region R is contained in the image of an orthogonal local parametrization is only needed to simplify the proof; as we shall see, (6.1) holds even without this hypothesis.

So the local Gauss-Bonnet theorem says that on simple regular regions the Gaussian curvature on the region, the geodesic curvature of the boundary and the external angles always add up to 2π. In other words, any change to one of these elements necessarily affects the other ones so that the sum remains constant.

The global Gauss-Bonnet theorem 6.3.9 (for whose proof we shall need the local version we have just proved) will give us the value of the left-hand side of (6.1) for *arbitrary* regular regions, not necessarily simple ones, in terms of the topology of the region. In particular, it will turn out that the right-hand side of (6.1) should be written as $2\pi \cdot 1$, where 1 is actually a *topological* invariant of simple regions. The next section provides an aside in topology of surfaces whose goal is precisely to introduce this invariant; we shall return to the Gauss-Bonnet theorem in section 6.3, where we shall also describe several applications of this powerful result.

6.2 Triangulations

The crucial idea is that to obtain a global version of the Gauss-Bonnet theorem we have to cut the regular region in small pieces on which we shall apply the local version, and then add the results. So our first goal is to describe formally the procedure of cutting a regular region in smaller pieces.

Definition 6.2.1. A (*smooth*) *triangle* in a surface is a simple regular region T with three points of the boundary, called *vertices*, singled out, with the only condition that all vertices of the boundary ∂T of T as a curvilinear polygon are vertices in the former sense too. In other words, ∂T has at most

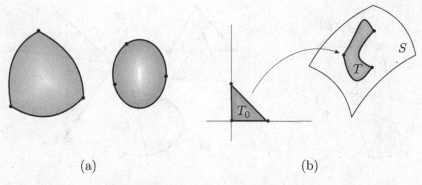

Fig. 6.3. (a) triangles; (b) a 2-simplex

three vertices as a curvilinear polygon, and all of them are vertices of T as a triangle, but some of the vertices of T as a triangle may not be vertices of ∂T as a curvilinear polygon; see Fig. 6.3.(a). The vertices partition the boundary of the triangle into three parts, called (of course) *sides*.

More in general, a *2-simplex* in a topological space X is a subset of X homeomorphic to the *standard triangle*

$$T_0 = \{(t_1, t_2) \in \mathbb{R}^2 \mid t_1, t_2 \geq 0, \ t_1 + t_2 \leq 1\} \subset \mathbb{R}^2 .$$

The *vertices* of a 2-simplex T are the points corresponding to the vertices $(0,0)$, $(1,0)$ and $(0,1)$ of the standard triangle; the *sides* of T are the subsets corresponding to the sides of T_0; and the *interior* consists of the image of the interior of T_0 in $\mathbb{R}^2$; see Fig. 6.3.(b).

Definition 6.2.2. Let $R \subseteq S$ be a regular region ($R = S$ with S compact is allowed). A *(regular) triangulation* of R is a finite family $\mathbf{T} = \{T_1, \ldots, T_r\}$ of triangles (the *faces* of the triangulation) such that:

(a) $R = \bigcup_{j=1}^{r} T_j$;
(b) the intersection of two distinct faces is either empty, or consists of vertices (common to both faces), or is a whole side (common to both faces);
(c) the intersection of a face with the border of R, if it is not empty, consists of vertices or whole sides;
(d) every vertex of the boundary of R is a vertex of (at least) one face of the triangulation.

We shall denote by $v(\mathbf{T})$ the total number of distinct vertices of the triangles of $\mathbf{T}$, by $l(\mathbf{T})$ the total number of distinct sides of the triangles of $\mathbf{T}$, and by $f(\mathbf{T}) = r$ the number of faces of $\mathbf{T}$. Finally, a *topological triangulation* of R (or, more in general, of a topological space homeomorphic to a regular region of a surface) is defined in the same way but using 2-simplices instead of smooth triangles.

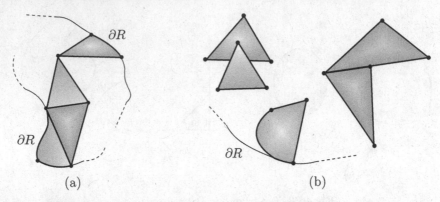

Fig. 6.4. (a) allowed intersections; (b) forbidden intersections

Remark 6.2.3. In other words, the interiors of the faces of a triangulation are always pairwise disjoint, and if two faces (or a face and the boundary of a region) intersect in part of a side, both of them have to contain the whole side. Fig. 6.4 shows allowed and forbidden intersections between the faces of a triangulation (and the boundary of the region).

Remark 6.2.4. If a side of a topological triangulation $\mathbf{T}$ of a regular region $R \subseteq S$ is contained in a single face of R, then it is necessarily (why?) a subset of the boundary of R. On the other hand, no side of $\mathbf{T}$ can be contained in three distinct faces. Indeed, suppose to the contrary that the side ℓ is contained in the three faces T_1, T_2 and T_3. Let two points $p_1, p_2 \in \ell$ be such that the whole line segment ℓ_0 of ℓ from p_1 to p_2 is included in the image of a local parametrization $\varphi: U \to S$. Then we may join p_1 and p_2 both with a continuous simple curve contained in the intersection of $\varphi(U)$ and the interior of T_1 and with a continuous simple curve contained in the intersection of $\varphi(U)$ and the interior of T_2; in this way we get a closed continuous simple curve in $\varphi(U)$ whose support intersects ℓ only in p_1 and p_2 (see Fig. 6.5). Applying φ^{-1}, we have found a Jordan curve in U whose support C intersects $\varphi^{-1}(\ell)$ only in $\varphi^{-1}(p_1)$ and $\varphi^{-1}(p_2)$, and whose interior is a neighborhood of $\varphi^{-1}(\ell_0)$ (endpoints excluded) contained in $\varphi^{-1}(T_1 \cup T_2)$. But then $\varphi^{-1}(T_3)$ should be contained in the unbounded component of $\mathbb{R}^2 \setminus C$, whereas it is adherent to ℓ_0, a contradiction.

Summing up, *each side of a topological triangulation of a regular region R belongs to exactly one face of the triangulation if it is in ∂R, and to exactly two faces of the triangulation otherwise.*

To use triangulations, we must be sure, at the very least, that they do exist. The first theorem of this section, whose (difficult) proof is postponed to Section 6.5 of the supplementary material of this chapter, has precisely this purpose:

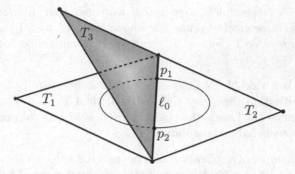

Fig. 6.5.

Theorem 6.2.5. *Let $R \subseteq S$ be a regular region on a surface S, and $\mathcal{U}$ an open cover of R. Then there exists a triangulation $\mathbf{T}$ of R such that for all $T \in \mathbf{T}$ there exists $U \in \mathcal{U}$ with $T \subset U$.*

So triangulations exist, and we may choose them with arbitrarily small triangles. However, they are all but unique; so we must find a way to ensure that the results we shall find are independent of the particular triangulation chosen to prove them. In other words, we have to find a way to compare different triangulations.

A first idea might consist in subdividing the faces of one of the two triangulations to be compared into triangles so small that the new triangulation refines both triangulations (in the sense that the triangles of the large triangulations can all be obtained as union of triangles of the small one), and then using this third triangulation as a link between the two original triangulations. Unfortunately, reality is somewhat more complicated, and this naive idea does not work; fortunately, it is possible to fix it so to reach the goal we set ourselves (even if we shall need, as you will see, three intermediate triangulations rather than just one). Let us begin by giving a name to the small triangulation:

Definition 6.2.6. A triangulation $\mathbf{T}'$ is a *refinement* of a triangulation $\mathbf{T}$ if every triangle of $\mathbf{T}'$ is contained in a triangle of $\mathbf{T}$.

Remark 6.2.7. Note that if $\mathbf{T}'$ is a refinement of $\mathbf{T}$, then every triangle of $\mathbf{T}$ is a union of triangles of $\mathbf{T}'$ (why?).

In general, two different triangulations do not admit a common refinement (see Remark 6.2.9). The next theorem (proved too in Section 6.5 of the supplementary material of this chapter) provides us with an alternative way, more complex but as valid, to compare different triangulations:

Theorem 6.2.8. *Let $\mathbf{T}_0$ and $\mathbf{T}_1$ be two topological (respectively, regular) triangulations of a regular region $R \subseteq S$. Then there always exist a topological* *triangulation* $\mathbf{T}^*$ *(respectively, topological with piecewise regular sides) and two topological (respectively, regular) triangulations* $\mathbf{T}_0^*$ *and* $\mathbf{T}_1^*$ *such that* $\mathbf{T}_0^*$ *is a refinement both of* $\mathbf{T}_0$ *and of* $\mathbf{T}^*$, *and* $\mathbf{T}_1^*$ *is a refinement both of* $\mathbf{T}_1$ *and of* $\mathbf{T}^*$.

Remark 6.2.9. If a side of $\mathbf{T}_1$ intersects in infinitely many isolated points a side of $\mathbf{T}_0$, a situation that may well happen, then $\mathbf{T}_0$ and $\mathbf{T}_1$ cannot have a common refinement. This is the reason why it is in general necessary to go through the intermediate triangulation $\mathbf{T}^*$.

As a first example of application of Theorem 6.2.8, let us show how to use triangulations to define a fundamental topological invariant of regular regions (which is the invariant promised at the end of the previous section).

Theorem 6.2.10. *Let* $\mathbf{T}_0$ *and* $\mathbf{T}_1$ *be two topological triangulations of a regular region* $R \subseteq S$. *Then:*

$$f(\mathbf{T}_0) - l(\mathbf{T}_0) + v(\mathbf{T}_0) = f(\mathbf{T}_1) - l(\mathbf{T}_1) + v(\mathbf{T}_1) . \tag{6.2}$$

Proof. We begin by showing that (6.2) holds if $\mathbf{T}_1$ is a refinement of $\mathbf{T}_0$. Starting from $\mathbf{T}_1$ we may get back $\mathbf{T}_0$ by a finite sequence of the two following operations (see Fig. 6.6):

(1) remove a side connecting two vertices each incident with at least two other sides;
(2) remove a vertex incident with exactly two sides.

Obviously, in intermediate steps, the faces we obtain (in the sense of connected components of the complement in R of the union of the sides) are not necessarily triangles (in the sense of having exactly three vertices in their boundary); but they are still connected open sets, and will be again triangles at the end of the operations.

Now, operation (1) lowers by 1 both the number of sides and the number of faces, while operation (2) lowers by 1 both the number of sides and the number of vertices. In both cases, the quantity we obtain by subtracting the number of sides from the sum of the number of faces and the number of vertices does not change; so we have proved (6.2) for refinements.

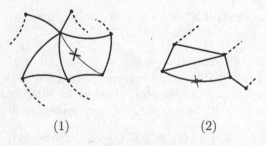

(1)   (2)

Fig. 6.6.

Let now $\mathbf{T}^*$, $\mathbf{T}_0^*$ and $\mathbf{T}_1^*$ be the triangulations provided by Theorem 6.2.8. The above implies

$$
\begin{aligned}
f(\mathbf{T}_0) - l(\mathbf{T}_0) + v(\mathbf{T}_0) &= f(\mathbf{T}_0^*) - l(\mathbf{T}_0^*) + v(\mathbf{T}_0^*) \\
&= f(\mathbf{T}^*) - l(\mathbf{T}^*) + v(\mathbf{T}^*) \\
&= f(\mathbf{T}_1^*) - l(\mathbf{T}_1^*) + v(\mathbf{T}_1^*) = f(\mathbf{T}_1) - l(\mathbf{T}_1) + v(\mathbf{T}_1) \,,
\end{aligned}
$$

and we are done. $\qquad\qquad\qquad\qquad\qquad\qquad\qquad\qquad\qquad\qquad\qquad\quad$ $\square$

Definition 6.2.11. The *Euler-Poincaré characteristic* $\chi(R)$ of a regular region R contained in a surface S is defined by

$$
\chi(R) = f(\mathbf{T}) - l(\mathbf{T}) + v(\mathbf{T}) \in \mathbb{Z} \,,
$$

where $\mathbf{T}$ is an arbitrary topological triangulation of R.

Theorem 6.2.10 guarantees that the Euler-Poincaré characteristic is well defined, that is, it does not depend on the triangulation used to compute it. In particular, it is a topological invariant:

Proposition 6.2.12. (i) *If $R \subseteq S$ and $R' \subseteq S'$ are two homeomorphic regular regions then $\chi(R) = \chi(R')$;*
(ii) *the Euler-Poincaré characteristic of a simple regular region is 1;*
(iii) *the Euler-Poincaré characteristic of a sphere is 2;*
(iv) *the Euler-Poincaré characteristic of a torus is 0.*

Proof. (i) If $\psi \colon R \to R'$ is a homeomorphism and $\mathbf{T} = \{T_1, \ldots, T_r\}$ is a topological triangulation of R, then $\psi(\mathbf{T}) = \{\psi(T_1), \ldots, \psi(T_r)\}$ is a topological triangulation of R' with the same number of faces, sides and vertices, and so $\chi(R) = \chi(R')$.

(ii) A simple regular region is homeomorphic to a closed disk, which in turn is homeomorphic to the standard triangle, whose Euler-Poincaré characteristic is clearly 1.

(iii) Modelling the sphere with a tetrahedron (which is homeomorphic to a sphere), we easily get a triangulation of a sphere S with 4 face, 6 sides and 4 vertices, so $\chi(S) = 2$; see Fig. 6.7.(a).

(iv) Using as our model a square with opposite sides identified, we get a triangulation of a torus with 8 faces, 12 sides and 4 vertices, and so the Euler-Poincaré characteristic of the torus is zero; see Fig. 6.7.(b). $\qquad\quad$ $\square$

We conclude this section by stating the famous *classification theorem for orientable compact surfaces* which, while being beyond the scope of this book (and not being needed in the proof of the global Gauss-Bonnet theorem), is useful to put in the right context some of the results we shall see, and to give an idea of the importance of the Euler-Poincaré characteristic.

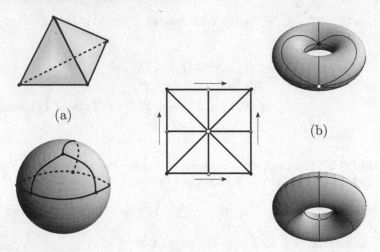

Fig. 6.7.

Definition 6.2.13. A *handle* in a surface S is a regular region $H \subset S$ homeomorphic to a closed (finite) circular cylinder and such that $S \setminus H$ is connected. Given $g \in \mathbb{N}$, a *sphere with g handles* is a surface S containing g disjoint handles $H_1, \ldots, H_g$ in such a way that $S \setminus (H_1 \cup \cdots \cup H_g)$ is a sphere with $2g$ disjoint closed geodesic balls removed; see Fig. 6.8.

Theorem 6.2.14 (classification of orientable compact surfaces). *Every orientable compact surface is homeomorphic to a sphere with $g \geq 0$ handles, that has Euler-Poincaré characteristic $2 - 2g$. In particular:*

(i) *two orientable compact surfaces are homeomorphic if and only if the have the same Euler-Poincaré characteristic;*

(ii) *the sphere is the only orientable compact surface with positive Euler-Poincaré characteristic;*

Fig. 6.8. A surface with 2 handles

(iii) *the torus is the only orientable compact surface with Euler-Poincaré characteristic equal to zero.*

You can find a proof of this result in [23].

In particular, all orientable compact surfaces have Euler-Poincaré characteristic even and less than or equal to 2, and it is not difficult to construct regular regions with Euler-Poincaré characteristic odd and less than or equal to 1. (Exercise 6.4).

Remark 6.2.15. In case you are wondering, in $\mathbb{R}^3$ there are no non-orientable compact surfaces, by Theorem 4.7.15 proved in the supplementary material of Chapter 4.

6.3 The global Gauss-Bonnet theorem

We are almost ready for the proof of the global Gauss-Bonnet theorem which, as mentioned, will express the value of the left-hand side of (6.1) for any regular region in topological terms; however, we have still to clarify some details.

Firstly, (6.1) contains the integral of the Gaussian curvature, but up to now we have only defined the integral of functions defined on a regular region contained in the image of a local parametrization. Triangulations allow to easily overcome this obstacle:

Lemma 6.3.1. *Let $R \subseteq S$ be a regular region, and choose two regular triangulations $\mathbf{T}_0 = \{T_{01}, \ldots, T_{0r}\}$ and $\mathbf{T}_1 = \{T_{11}, \ldots, T_{1s}\}$ of R such that each of their triangles is contained in the image of a local parametrization. Then*

$$\sum_{h=1}^{r} \int_{T_{0h}} f \, d\nu - \sum_{k=1}^{s} \int_{T_{1k}} f \, d\nu \qquad (6.3)$$

for every continuous function $f \colon R \to \mathbb{R}$.

Proof. Let

$$\mathbf{T}^* = \{T_1^*, \ldots, T_u^*\}, \qquad \mathbf{T}_0^* = \{T_{01}^*, \ldots, T_{0v}^*\}, \qquad \mathbf{T}_1^* = \{T_{11}^*, \ldots, T_{1w}^*\}$$

be the triangulations given by Theorem 6.2.8. Since every triangle of $\mathbf{T}_0$ is (by Remark 6.2.7) a union of finitely many triangles of $\mathbf{T}_0^*$ and, conversely, every triangle of $\mathbf{T}_0^*$ is contained in a single triangle of $\mathbf{T}_0$ (and the sides have no effect on integrals), we find

$$\sum_{h=1}^{r} \int_{T_{0h}} f \, d\nu = \sum_{j=1}^{v} \int_{T_{0j}^*} f \, d\nu .$$

Repeating this argument with $\mathbf{T}^*$ and $\mathbf{T}_0^*$, and then with $\mathbf{T}^*$ and $\mathbf{T}_1^*$, and finally with $\mathbf{T}_1$ and $\mathbf{T}_1^*$ we obtain the assertion.   □

Hence we may use (6.3) to define the integral of an arbitrary continuous function f on an arbitrary regular region $R \subseteq S$:

Definition 6.3.2. Let $R \subseteq S$ be a regular region of a surface S, and $f \colon R \to \mathbb{R}$ a continuous function. Then the *integral* of f on R is

$$\int_R f \, \mathrm{d}\nu = \sum_{j=1}^{r} \int_{T_r} f \, \mathrm{d}\nu \, , \qquad (6.4)$$

where $\mathbf{T} = \{T_1, \ldots, T_r\}$ is an arbitrary regular triangulation of R whose triangles are contained in the images of local parametrizations; such a triangulation exists by Theorem 6.2.5, and $\int_R f \, \mathrm{d}\nu$ does not depend on the choice of $\mathbf{T}$ because of the previous lemma.

The second obstacle to overcome for the computation the left-hand side of (6.1) for arbitrary regular regions is related to the appearance of the geodesic curvature. Firstly, the geodesic curvature is only defined for orientable surfaces; so we shall have to assume that our regular region R is contained in an orientable surface. But this is not enough: the geodesic curvature changes sign if we invert the orientation of a curve, and so we have to find a way to fix once and for all the orientation of the curvilinear polygons making up the boundary of R.

In the case of small curvilinear polygons, we managed by using the rotation index; the general case requires instead a somewhat different approach (which however will be compatible with the previous one). We need some definitions and a lemma.

Definition 6.3.3. Let $\sigma \colon I \to S$ be a curve parametrized by arc length in an oriented surface S, and $N \colon S \to S^2$ the Gauss map of S. The *intrinsic normal versor* of σ is the map $\hat{\mathbf{n}} \in \mathcal{T}(\sigma)$ given by

$$\hat{\mathbf{n}} = (N \circ \sigma) \wedge \dot{\sigma} \, .$$

Note that $\hat{\mathbf{n}}(s)$ is the unique versor in $T_{\sigma(s)}S$ such that $\{\dot{\sigma}(s), \hat{\mathbf{n}}(s)\}$ is a positive orthonormal basis of $T_{\sigma(s)}S$, for all $s \in I$.

Definition 6.3.4. Let $R \subset S$ be a regular region of a surface S. We shall say that a regular curve $\tau \colon (-\varepsilon, \varepsilon) \to S$ *enters* R if $\tau(0) \in \partial R$, $\tau(t) \in \mathring{R}$ for all $t > 0$, and $\tau(t) \notin R$ for all $t < 0$. Moreover, if $\tau(0)$ is not a vertex of ∂R, $\tau'(0)$ will be required not to be tangent to ∂R.

Definition 6.3.5. Let $R \subset S$ be a regular region of an oriented surface S, and $\sigma \colon [a, b] \to S$ a curvilinear polygon that parametrizes one of the components of the boundary of R. We shall say that σ is *positively oriented with respect to R* if the intrinsic normal versor $\hat{\mathbf{n}}$ of σ (defined out of the vertices) points toward the interior of R, in the sense that, for all $s_0 \in [a, b]$ such that $\sigma(s_0)$

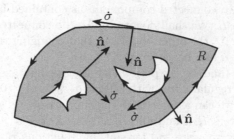

Fig. 6.9.

is not a vertex and for every regular curve $\tau\colon(-\varepsilon,\varepsilon)\to S$ with $\tau(0)=\sigma(s_0)$ that enters R, we have $\langle\tau'(0),\hat{\mathbf{n}}(s_0)\rangle > 0$; see Fig. 6.9. Finally, we shall say that the boundary of the regular region R is *positively oriented* if each of its components is so with respect to R.

For this to be a good definition we have to verify that the boundary of a regular region can always be positively oriented. To prove this we need a version of the notion of tubular neighborhood for curves on surfaces.

Lemma 6.3.6. *Let $\sigma\colon[a,b]\to S$ be a curve parametrized by arc length in an oriented surface S. Then there exists $\delta>0$ such that the map $\varphi\colon(a,b)\times(-\delta,\delta)\to S$ given by*

$$\varphi(s,t)=\exp_{\sigma(s)}\big(t\hat{\mathbf{n}}(s)\big)$$

is a local parametrization of S inducing the given orientation.

Proof. First of all, as $[a,b]$ is compact, Proposition 5.2.5 guarantees (why?) that for $\delta>0$ small enough φ is well defined and of class C^∞. Now, for all $s_0\in[a,b]$ we have

$$\frac{\partial\varphi}{\partial s}(s_0,0)=\dot\sigma(s_0)\quad\text{and}\quad\frac{\partial\varphi}{\partial t}(s_0,0)=\frac{\mathrm{d}}{\mathrm{d}t}\exp_{\sigma(s_0)}\big(t\hat{\mathbf{n}}(s_0)\big)\bigg|_{t=0}=\hat{\mathbf{n}}(s_0)\,;$$

then, by Corollary 3.4.28, for all $s_0\in[a,b]$ we may find $\delta_{s_0}>0$ such that $\varphi|_{(s_0-\delta_{s_0},s_0+\delta_{s_0})\times(-\delta_{s_0},\delta_{s_0})}$ is a local parametrization, inducing the given orientation because $\{\dot\sigma(s_0),\hat{\mathbf{n}}(s_0)\}$ is a positive basis. Finally, arguing as in the proof of Theorem 2.2.5 and recalling Proposition 3.1.31, we find a $\delta>0$ as required. □

Let then $\sigma\colon[a,b]\to S$ be a curvilinear polygon that is a component of the boundary of a regular region R in an oriented surface S, and suppose first that σ has no vertices. Let $\varphi\colon U\to S$, where $U=(a,b)\times(-\delta,\delta)$, be the local parametrization given by the previous lemma; then the complement in $\varphi(U)$ of the support of σ has two connected components, $\Sigma^+=\varphi\big((a,b)\times(0,\delta)\big)$ and $\Sigma^-=\varphi\big((a,b)\times(-\delta,0)\big)$; up to decreasing δ if necessary we can assume that one of these two connected components is contained in $\mathring{R}$ and the other one is disjoint from R. We shall denote by Σ^{i} the connected component contained in $\mathring{R}$, and by Σ^{e} the other one. Note that changing the orientation of σ changes the sign of $\hat{\mathbf{n}}$ and hence swap Σ^{+} with Σ^{-}.

Given $s_0 \in [a, b]$ (and, by the periodicity of σ, we may assume $s_0 \neq a, b$), let $\tau\colon (-\varepsilon, \varepsilon) \to S$ be a regular curve with $\tau(0) = \sigma(s_0)$ and $\tau'(0) \neq \pm\dot{\sigma}(s_0)$. Up to taking a smaller ε, we can assume that the support of τ is contained in $\varphi(U)$; in particular, τ enters R if and only if $\tau(t) \in \Sigma^{\mathrm{i}}$ for $t > 0$ and $\tau(t) \in \Sigma^{\mathrm{e}}$ for $t < 0$. Write now $\tau = \varphi(\tau_1, \tau_2)$ for suitable functions $\tau_1, \tau_2\colon (-\varepsilon, \varepsilon) \to \mathbb{R}$ of class C^∞, so that $\tau(t) \in \Sigma^{\pm}$ if and only if $\pm\tau_2(t) > 0$. We have

$$\langle \tau'(0), \hat{\mathbf{n}}(s_0) \rangle = \tau_2'(0) \ ,$$

which is different from zero because $\tau'(0) \neq \pm\dot{\sigma}(s_0)$. Hence $\hat{\mathbf{n}}(s_0)$ points toward the interior of R if and only if τ_2 is non-decreasing in 0. Since $\tau_2(0) = 0$, this means that we must have $\tau_2(t) > 0$ for small values of $t > 0$, and $\tau_2(t) < 0$ for small values of $t < 0$ (it cannot be zero because $\tau(t)$ is not in ∂R for $t \neq 0$). It then follows that $\hat{\mathbf{n}}(s_0)$ points toward the interior of R if and only if $\Sigma^{+} = \Sigma^{\mathrm{i}}$. Since changing the orientation of σ swaps Σ^{+} with Σ^{-} but does not change Σ^{i} or Σ^{e}, we can always orient σ positively (in exactly one way) with respect to R; moreover, as soon as $\hat{\mathbf{n}}(s_0)$ points toward the interior of R for some $s_0 \in [a, b]$, the curve σ is positively oriented with respect to R.

If, on the other hand, σ has vertices, by using the technique introduced at the end of the proof of Theorem 2.4.7 we may modify σ and R close to the vertices so to get a sequence $\sigma_k\colon [a, b] \to S$ of curvilinear polygons without vertices, coinciding on larger and larger subintervals, approaching σ and parametrizing suitable components of the boundary of regular regions R_k. The previous argument shows that as soon as we positively orient a σ_k with respect to R_k the other ones are too positively oriented (with respect to their regular region); so, taking the limit, we obtain exactly one orientation of σ that makes it positively oriented with respect to R.

Remark 6.3.7. In particular, the above argument shows that if R and R' are regular regions with disjoint interiors of the same oriented surface such that $\partial R \cap \partial R'$ consists in a segment of a curvilinear polygon σ then σ is positively oriented with respect to R if and only if it is negatively oriented (with the obvious meaning) with respect to R'.

Remark 6.3.8. We have still to verify that if $R \subset S$ is a simple regular region contained in the image of an orthogonal local parametrization $\varphi\colon U \to S$ then the two Definitions 6.1.4 and 6.3.5 of positively oriented boundary coincide. As usual, arguing as in the proof of Theorem 2.4.7 we may without loss of generality assume that the boundary $\sigma\colon [a, b] \to S$ of R has no vertices. Moreover, we have already remarked (Remark 2.4.9) that the two definitions coincide for curves in the plane.

Set $\sigma_o = \varphi^{-1} \circ \sigma = (\sigma_1, \sigma_2)$: we know (Proposition 6.1.3) that $\rho(\sigma) = +1$ if and only if $\rho(\sigma_o) = +1$. Moreover, $\rho(\sigma_o) = +1$ if and only if (Remark 2.4.9) the normal versor of σ_o points toward the interior of $\varphi^{-1}(R)$. Now we may write

$$\dot{\sigma} = \dot{\sigma}_1 \partial_1 + \dot{\sigma}_2 \partial_2 = \sqrt{E}\dot{\sigma}_1 \frac{\partial_1}{\sqrt{E}} + \sqrt{G}\dot{\sigma}_2 \frac{\partial_2}{\sqrt{G}} \; ;$$

so the intrinsic normal versor of σ is given by

$$\hat{\mathbf{n}} = -\sqrt{\frac{G}{E}}\dot{\sigma}_2 \partial_1 + \sqrt{\frac{E}{G}}\dot{\sigma}_1 \partial_2 \; .$$

The oriented normal versor $\tilde{\mathbf{n}}_o$ of σ_o is instead $\tilde{\mathbf{n}}_o = (-\dot{\sigma}_2, \dot{\sigma}_1)/\|\dot{\sigma}_o\|$.

Fix now $s_0 \in [a, b]$. A curve $\tau\colon (-\varepsilon, \varepsilon) \to S$ enters R at $\sigma(s_0)$ if and only if $\tau_o = \varphi^{-1} \circ \tau = (\tau_1, \tau_2)$ enters $\varphi^{-1}(R)$ at $\sigma_o(s_0)$. Since $\tau' = \tau_1'\partial_1 + \tau_2'\partial_2$, we have

$$\langle \tau'(0), \hat{\mathbf{n}}(s_0)\rangle = \sqrt{EG}\big[-\tau_1'(0)\dot{\sigma}_2(s_0) + \tau_2'(0)\dot{\sigma}_1(s_0)\big] \; ,$$

$$\langle \tau_o'(0), \tilde{\mathbf{n}}_o(s_0)\rangle = \frac{1}{\|\dot{\sigma}_o(s_0)\|}\big[-\tau_1'(0)\dot{\sigma}_2(s_0) + \tau_2'(0)\dot{\sigma}_1(s_0)\big] \; ;$$

so $\tilde{\mathbf{n}}_o$ points toward the interior of $\varphi^{-1}(R)$ if and only if $\hat{\mathbf{n}}$ points toward the interior of R, and we are done.

At last, we are ready for:

Theorem 6.3.9 (global Gauss-Bonnet). *Let $R \subseteq S$ be a regular region (if S is compact $R = S$ is allowed) of an oriented surface S, with ∂R positively oriented. Let $C_1, \ldots, C_s$ be the connected components of the boundary of R parametrized, for $j = 1, \ldots, s$, by curves $\sigma_j\colon [a_j, b_j] \to S$ with geodesic curvature κ_g^j (and if $R = S$ then $\partial R = \varnothing$). Denote further by $\{\varepsilon_1, \ldots, \varepsilon_p\}$ the set of all external angles of the curves $\sigma_1, \ldots, \sigma_s$. Then*

$$\int_R K \, d\nu + \sum_{j=1}^{s} \int_{a_j}^{b_j} \kappa_g^j \, ds + \sum_{h=1}^{p} \varepsilon_h = 2\pi\chi(R) \; . \qquad (6.5)$$

Proof. Let $\mathbf{T}$ be a triangulation of R such that every triangle of $\mathbf{T}$ is contained in the image of an orthogonal local parametrization compatible with the orientation (such a triangulation exists by Theorem 6.2.5). Give a positive orientation to the boundary of each triangle; by the above discussion, this orientation is compatible with that of the boundary of R.

Apply the local Gauss-Bonnet theorem to each triangle, and sum. Since (by Remarks 6.2.4 and 6.3.7) the integrals of the geodesic curvature on the internal sides of the triangulation cancel pairwise, we get

$$\int_R K \, d\nu + \sum_{j=1}^{s} \int_{a_j}^{b_j} \kappa_g^j \, ds + \sum_{i=1}^{f(\mathbf{T})} \sum_{j=1}^{3} \varepsilon_{ij} = 2\pi f(\mathbf{T}) \; , \qquad (6.6)$$

where ε_{i1}, ε_{i2} and ε_{i3} are the external angles of the triangle $T_i \in \mathbf{T}$. If we denote by $\phi_{ij} = \pi - \varepsilon_{ij}$ the *internal* angles of the triangle T_i, we have

$$\sum_{i=1}^{f(\mathbf{T})} \sum_{j=1}^{3} \varepsilon_{ij} = 3\pi f(\mathbf{T}) - \sum_{i=1}^{f(\mathbf{T})} \sum_{j=1}^{3} \phi_{ij} .$$

Denote by l_{i} (respectively, l_{b}) the number of sides of the triangulation in the interior of R (respectively, on the boundary of R), and by v_{i} (respectively, v_{b}) the number of vertices of the triangulation in the interior of R (respectively, on the boundary of R); clearly, $l_{\mathrm{i}} + l_{\mathrm{b}} = l(\mathbf{T})$ and $v_{\mathrm{i}} + v_{\mathrm{b}} = v(\mathbf{T})$. Since the boundary of R consists of curvilinear polygons, $l_{\mathrm{b}} = v_{\mathrm{b}}$. Moreover, every face has three sides, every internal side is a side of two faces, and every side on the boundary is a side of a single face (Remark 6.2.4); so

$$3f(\mathbf{T}) = 2l_{\mathrm{i}} + l_{\mathrm{b}} .$$

Hence,

$$\sum_{i=1}^{f(\mathbf{T})} \sum_{j=1}^{3} \varepsilon_{ij} = 2\pi l_{\mathrm{i}} + \pi l_{\mathrm{b}} - \sum_{i=1}^{f(\mathbf{T})} \sum_{j=1}^{3} \phi_{ij} .$$

Write $v_{\mathrm{b}} = v_{\mathrm{bc}} + v_{\mathrm{bt}}$, where $v_{\mathrm{bc}} = p$ is the number of vertices of the curvilinear polygons making up the boundary of R, and v_{bt} is the number of the other vertices of the triangulation on the boundary. Now, the sum of the internal angles around every internal vertex is 2π; the sum of the internal angles on every external vertex that is not a vertex of ∂R is π; and the sum of the internal angles on a vertex of ∂R is π minus the corresponding external angle. So we obtain

$$\sum_{i=1}^{f(\mathbf{T})} \sum_{j=1}^{3} \varepsilon_{ij} = 2\pi l_{\mathrm{i}} + \pi l_{\mathrm{b}} - 2\pi v_{\mathrm{i}} - \pi v_{\mathrm{bt}} - \sum_{h=1}^{p} (\pi - \varepsilon_h)$$

$$= 2\pi l_{\mathrm{i}} + 2\pi l_{\mathrm{b}} - \pi v_{\mathrm{b}} - 2\pi v_{\mathrm{i}} - \pi v_{\mathrm{bt}} - \pi v_{\mathrm{bc}} + \sum_{h=1}^{p} \varepsilon_h$$

$$= 2\pi l(\mathbf{T}) - 2\pi v(\mathbf{T}) + \sum_{h=1}^{p} \varepsilon_h ,$$

and by (6.6) we are done.   □

It is impossible to overrate the importance of the Gauss-Bonnet theorem in differential geometry. Perhaps the easiest way to realize the power of this result is to consider the simplest situation, that is, the case $R = S$:

Corollary 6.3.10. *Let S be an orientable compact surface. Then*

$$\int_S K \, \mathrm{d}\nu = 2\pi\chi(S) . \tag{6.7}$$

Fig. 6.10. A deformation of the sphere

Proof. This immediately follows from the Gauss-Bonnet theorem, since S is a regular region without boundary.   □

Remark 6.3.11. In the supplementary material of Chapter 4 we have proved (Theorem 4.7.15) that every closed surface (and so, in particular, every compact surface) in $\mathbb{R}^3$ is orientable, so the previous corollary actually applies to all compact surfaces.

We thus have a powerful and unexpected link between a purely local object, depending on the differential structure (the Gaussian curvature) and a purely global object, depending only on the topology (the Euler-Poincaré characteristic). For instance, in whichever way we deform a sphere (see, e.g., Fig. 6.10), the integral of the Gaussian curvature cannot change: it stays equal to 4π, no matter what we do.

As a consequence, recalling Theorem 6.2.14, the sign of the Gaussian curvature may force the topological kind of a surface:

Corollary 6.3.12. *An orientable compact surface whose Gaussian curvature is $K \geq 0$, and positive in at least one point, is homeomorphic to a sphere.*

Proof. It follows from the last corollary and from the fact that the sphere is the only orientable compact surface with positive Euler-Poincaré characteristic (Theorem 6.2.14).   □

Remark 6.3.13. There are deep links between the sign of the Gaussian curvature and the topology of surfaces. For instance, in the supplementary material of the next chapter we shall see that every orientable compact surface with strictly positive Gaussian curvature is *diffeomorphic* to the sphere; the diffeomorphism is the Gauss map $N\colon S \to S^2$ (Theorem 7.4.9). Conversely, we shall also see that every simply connected closed surface in $\mathbb{R}^3$ with Gaussian curvature $K \leq 0$ is diffeomorphic to a plane; the diffeomorphism is the exponential map $\exp_p\colon T_pS \to S$ (Theorem 7.6.4).

Corollary 6.3.10 was an application of the Gauss-Bonnet theorem to regions without boundary, in which two of the three summands in the left-hand side of (6.5) are zero. We find equally interesting consequences for regular regions with a boundary consisting of geodesic polygons:

Definition 6.3.14. A *geodesic polygon* in a surface S is a curvilinear polygon whose regular segments are all geodesics. A geodesic polygon is *simple* if it is the boundary of a simple regular region.

If the boundary of a regular region R consists of geodesic polygons we can exploit the fact that the middle summand of the left-hand side of (6.5) is zero. For instance, Gauss was especially proud of the following:

Corollary 6.3.15. *Let $T \subset S$ be a geodesic triangle, that is, a triangle whose three sides are geodesics, and denote by $\phi_j = \pi - \varepsilon_j$, for $j = 1, 2, 3$, the three internal angles (where $\varepsilon_1, \varepsilon_2, \varepsilon_3$ are the external angles). Then*

$$\phi_1 + \phi_2 + \phi_3 = \pi + \int_T K \, d\nu \,.$$

In particular, if the Gaussian curvature K is constant then the difference between the sum of the internal angles of a geodesic triangle and π equals K times the area of the triangle.

Proof. It immediately follows from (6.5). □

Remark 6.3.16. It is well known from elementary geometry that one of the conditions equivalent to Euclid's parallel postulate is the fact that the sum of the internal angles of a triangle is exactly π. Hence, if we consider geodesics as the natural equivalent of line segments on an arbitrary surface, this corollary gives a measure of how much Euclid's postulate is not valid on surfaces different from the plane.

Another interesting application of the Gauss-Bonnet theorem to geodesics is the following:

Proposition 6.3.17. *Let $S \subset \mathbb{R}^3$ be an orientable surface diffeomorphic to a circular cylinder, with $K < 0$ everywhere. Then on S there is (up to reparametrizations) at most one simple closed geodesic, and this geodesic cannot be the boundary of a simple regular region.*

Remark 6.3.18. A surface of revolution S having an open Jordan arc C as its generatrix is always diffeomorphic to a circular cylinder (Exercise 3.49). Moreover, if C has a parametrization by arc length of the form $s \mapsto (\alpha(s), 0, \beta(s))$ with α positive everywhere as usual, then S has Gaussian curvature negative everywhere if and only if $\ddot{\alpha} > 0$ (Example 4.5.22). In particular, α has at most one critical point, so (Example 5.1.17) at most one parallel of S is a (simple closed) geodesic. Finally, the central parallel of the one-sheeted hyperboloid is a geodesic, and so one can exist.

Proof (of Proposition 6.3.17). Let $C \subset S$ be the support of a simple closed geodesic $\sigma : \mathbb{R} \to S$. Suppose by contradiction that C is the boundary of a simple regular region $R \subset S$. Then the Gauss-Bonnet theorem and Proposition 6.2.12.(ii) imply

$$2\pi = 2\pi\chi(R) = \int_R K \, d\nu < 0 \, ,$$

impossible.

Since S is diffeomorphic to a circular cylinder, which in turn is diffeomorphic to the plane with a point removed (Exercise 3.53), there has to exist a diffeomorphism $\Phi : S \to \mathbb{R}^2 \setminus \{O\}$ between S and the plane minus the origin. So the curve $\Phi \circ \sigma$ is a Jordan curve in the plane, with support $\Phi(C)$. The Jordan curve Theorem 2.3.6 tells us that $\mathbb{R}^2 \setminus \Phi(C)$ has two connected components, with exactly one bounded, and both having as their boundary $\Phi(C)$. If the origin were in the unbounded component, then $\Phi(C)$ would be the boundary of a simple regular region, its interior Ω (Remark 2.3.7), and so C would be the boundary of the simple regular region $\Phi^{-1}(\Omega)$, contradicting what we have just seen. Consequently, the origin has to be in Ω.

Suppose now, by contradiction, that $\tilde{\sigma} : \mathbb{R} \to S$ is another simple closed geodesic with support $\tilde{C}$ different from C; note that, by the previous argument, the origin is in the interior of $\Phi(\tilde{C})$ too. If $C \cap \tilde{C} = \varnothing$, then $\Phi(C)$ and $\Phi(\tilde{C})$ are within one another, and thus they form the boundary of a regular region $R \subset \mathbb{R}^2 \setminus \{O\}$. It is easy (exercise) to construct a triangulation of R as in Fig. 6.11.(a) with 4 faces, 8 sides, and 4 vertices, so that $\chi(R) = 0$. But in this case C and $\tilde{C}$ would make up the boundary of the regular region $\Phi^{-1}(R) \subset S$, and we would have

$$0 = 2\pi\chi(\Phi^{-1}(R)) = \int_{\Phi^{-1}(R)} K \, d\nu < 0 \, ,$$

impossible.

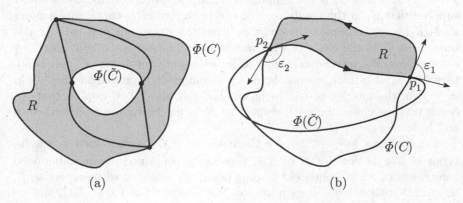

(a)   (b)

Fig. 6.11.

So C and $\tilde{C}$ have to intersect. Since the uniqueness of the geodesic through a point tangent to a given direction implies that C and $\tilde{C}$ cannot be tangent in an intersection point, they necessarily intersect in at least two points. So we may find (do this as an exercise, using as usual the diffeomorphism Φ) two consecutive intersection points p_1 and p_2, an arc in C from p_1 to p_2 and an arc in $\tilde{C}$ from p_2 to p_1 that form the boundary of a simple regular region R with two vertices having external angles ε_1 and ε_2; see Fig. 6.11.(b). As a consequence,

$$2\pi = 2\pi\chi(R) = \int_R K\,d\nu + \varepsilon_1 + \varepsilon_2 < 2\pi\,,$$

since $|\varepsilon_1| + |\varepsilon_2| < 2\pi$, and we have once more reached a contradiction. $\square$

The Gauss-Bonnet theorem also provides us with a simple way to compute the Euler-Poincaré characteristic of regular regions of the plane:

Corollary 6.3.19. *Let $R \subset \mathbb{R}^2$ be a regular region of the plane, and suppose that the boundary of R has $r \geq 1$ connected components. Then*

$$\chi(R) = 2 - r.$$

Proof. Let $\sigma_1, \ldots, \sigma_r$ be the curvilinear polygons making up the boundary of R, positively oriented with respect to R. Firstly, Proposition 5.1.28 tells us that the geodesic curvature of a curve in the plane coincides with the derivative of the rotation angle; recalling the definition of rotation index and the Gauss-Bonnet theorem we obtain

$$\sum_{j=1}^r \rho(\sigma_j) = \chi(R)\,.$$

By the turning tangents Theorem 2.4.7, we know that $\rho(\sigma_j) = \pm 1$, where the sign only depends on the orientation of σ_j. Now, since the interior $\mathring{R}$ of R is connected, it is contained in one of the two components of $\mathbb{R}^2 \setminus C_j$ for every $j = 1, \ldots, r$, where C_j is the support of σ_j. Up to reordering indices, we may suppose that C_1 is the component of the boundary of R containing a point $p_0 \in R$ having maximum distance d_0 from the origin. In particular, C_1 is tangent at p_0 to the circle C with radius d_0 and center the origin. Since, by construction, the interior of R has to lie within the interior of C and is adherent to p_0, it cannot be contained in the unbounded component of $\mathbb{R}^2 \setminus C_1$; so it is in the interior of C_1. In particular, Remark 2.3.10 implies that σ_1, being positively oriented with respect to R, is positively oriented *tout-court* and hence $\rho(\sigma_1) = +1$.

Finally, since for $j = 2, \ldots, r$ the component C_j is contained in the interior of C_1, the interior $\mathring{R}$ of R is necessarily contained in the unbounded component of $\mathbb{R}^2 \setminus C_j$; hence σ_j, being positively oriented with respect to R, has rotation index -1. So we have $\rho(\sigma_1) + \cdots + \rho(\sigma_r) = 1 - (r - 1)$, and we are done. $\square$

In the exercises of this chapter you will find further applications of the Gauss-Bonnet theorem in the spirit of those we have just discussed; the next section contains instead a completely different, and quite important, application.

6.4 The Poincaré-Hopf theorem

The last application of Gauss-Bonnet theorem we describe is concerned with vector fields or, more precisely, with their zeros.

Definition 6.4.1. Let $X \in \mathcal{T}(S)$ be a vector field on a surface S. A point $p \in S$ is *singular* (or a *zero*) for X if $X(p) = O$. The set of singular points of X will be denoted by $\mathrm{Sing}(X)$. A non-singular point will be called *regular*.

If $p \in S$ is a regular point of the vector field $X \in \mathcal{T}(S)$, by using the techniques seen in Section 5.3 it is not hard to prove (Exercise 5.25) that there exists a local parametrization $\varphi : U \to S$ centered at p such that $X = \partial_1$ in $\varphi(U)$. So the behavior of X in a neighborhood of a non-singular point is completely determined; in particular, its integral curves are locally coordinate curves of a local parametrization.

If, on the other hand, $p \in S$ is a *singular* point of the vector field X, the behavior of X and of its integral curves in a neighborhood of p can be very diverse and complicated; see for instance Figs. 5.3 and 6.12.

We want to introduce now a way to associate with each isolated singular point of a vector field an integer number that, in a sense, summarizes the qualitative behavior of the field near the point.

Definition 6.4.2. Let $X \in \mathcal{T}(S)$ be a vector field on a surface S, and $p \in S$ either a regular or an *isolated* singular point of X. Take a local parametrization

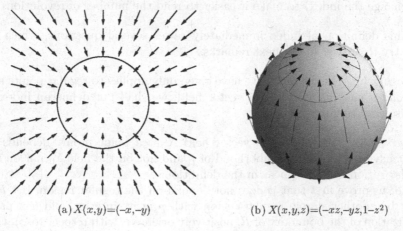

(a) $X(x,y)=(-x,-y)$ (b) $X(x,y,z)=(-xz,-yz,1-z^2)$

Fig. 6.12.

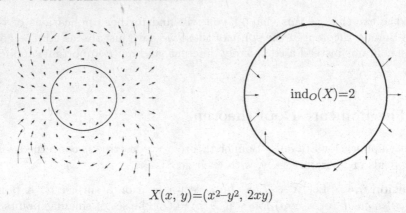

$$X(x, y)=(x^2-y^2,\ 2xy)$$

Fig. 6.13. Index at an isolated singular point

$\varphi : U \to S$ centered at p, with U homeomorphic to an open disk, and suppose that $\varphi(U) \cap \mathrm{Sing}(X) \subseteq \{p\}$, that is, that $\varphi(U)$ does not contain singular points of X except at most p. Fix next a vector field $Y_0 \in \mathcal{T}\big(\varphi(U)\big)$ nowhere zero; for instance, $Y_0 = \partial_1$. Let finally $R \subset \varphi(U)$ be a simple regular region with p in its interior, and $\sigma : [a, b] \to \partial R \subset \varphi(U)$ a parametrization of the boundary of R, positively oriented (with respect to φ). Denote by $\theta : [a, b] \to \mathbb{R}$ a continuous determination of the angle between $Y_0 \circ \sigma$ and $X \circ \sigma$. Then the *index* of X at p is given by

$$\mathrm{ind}_p(X) = \frac{1}{2\pi}\big(\theta(b) - \theta(a)\big) \in \mathbb{Z}\,.$$

Roughly speaking, $\mathrm{ind}_p(X)$ measures the number of revolutions of X along a curve that surrounds p; see Fig. 6.13 where, for the sake of simplicity, in the image on the right we have modified the length of X (an operation that does not change the index) to make it easier to read the number of revolutions.

This definition of index immediately raises several questions, which we shall try to answer in the next remarks.

Remark 6.4.3. The definition we have given only applies to regular points and to *isolated* singular points of a vector field; we shall not define an index in non-isolated singular points.

Remark 6.4.4. The index of a vector field X at a point p only depends on the behavior of X in a neighborhood of p and not on the field Y_0 nor on the regular region R that appear in the definition.

Let us prove first that it does not depend on the regular region. Let $\tilde{R} \subset \varphi(U)$ be another simple regular region with p in its interior, and $\tilde{\sigma}$ a parametrization of the boundary of $\tilde{R}$, positively oriented (with respect to φ), that without loss of generality we may assume to be also defined on $[a, b]$. Since U

is homeomorphic to an open disk and R and $\tilde{R}$ are simple regions, it is not difficult (see Exercise 2.13) to prove that there is a homotopy $\Phi : [0,1] \times [a,b] \to \varphi(U) \setminus \{p\}$ between σ and $\tilde{\sigma}$ as closed curves. Since X is nowhere zero in $\varphi(U) \setminus \{p\}$, we may define (using Corollary 2.1.14) a continuous determination $\Theta : [0,1] \times [a,b] \to \mathbb{R}$ of the angle between $Y_0 \circ \Phi$ and $X \circ \Phi$. So the function $s \mapsto [\Theta(s,b) - \Theta(s,a)]/2\pi$ is an integer-valued continuous function defined on a connected set; hence it is constant, and the index of X at p computed using R is equal to the index computed using $\tilde{R}$.

The index does not depend on the field Y_0 either. If $Y_1 \in \mathcal{T}\big(\varphi(U)\big)$ is another nowhere zero vector field, since R is homeomorphic to a closed disk we may again use Corollary 2.1.14 (how?) to define a continuous determination $\psi : R \to \mathbb{R}$ of the angle from Y_0 to Y_1. Then, if $\theta : [a,b] \to \mathbb{R}$ is a continuous determination of the angle from $Y_0 \circ \sigma$ to $X \circ \sigma$ we have that $\theta - \psi \circ \sigma$ is a continuous determination of the angle from $Y_1 \circ \sigma$ to $X \circ \sigma$; since

$$\big[\theta(b) - \psi\big(\sigma(b)\big)\big] - \big[\theta(a) - \psi\big(\sigma(a)\big)\big] = \theta(b) - \theta(a) \,,$$

it follows that the index of X at p computed using Y_1 coincides with the index computed using Y_0.

Finally, the index does not depend on the orientation given on $\varphi(U)$ either. Indeed, if we invert the orientation we change the sign of the determination of the angle from $Y_0 \circ \sigma$ to $X \circ \sigma$ but we also invert the orientation of σ, and so the value of the index does not change.

Hence the index of a vector field at a (regular or) isolated singular point is well defined. The next remark suggests a technique for computing it:

Remark 6.4.5. If X is a vector field defined on an open set $U \subset \mathbb{R}^2$ of the plane and $p \in U$, then the index of X at p is simply (why?) the degree of the map $\phi : [a,b] \to S^1$ defined by

$$\phi(t) = \frac{X\big(\sigma(t)\big)}{\|X\big(\sigma(t)\big)\|} \,,$$

where $\sigma : [a,b] \to U$ is an arbitrary piecewise regular, positively oriented Jordan curve with p in its interior.

The general case may be reduced to this one by using a local parametrization, as usual. Let $X \in \mathcal{T}(S)$, take $p \in S$ and a local parametrization $\varphi : U \to S$ centered at p, and write $X \circ \varphi = X_1 \partial_1 + X_2 \partial_2$. Then $X_o = (X_1, X_2) = \mathrm{d}\varphi^{-1}(X)$ is a vector field on U, and arguing as in the proof of Proposition 6.1.3 (and taking $Y_0 = \partial_1$) it is easy to see that the index of X at $p = \varphi(O)$ is exactly equal to the index of X_o at O.

Example 6.4.6. Let $X \in \mathcal{T}(\mathbb{R}^2)$ be given by $X(x_1, x_2) = (-x_1, -x_2)$; see Fig. 6.12.(a). The origin is an isolated singular point of X. If we take as Jordan curve $\sigma : [0, 2\pi] \to \mathbb{R}^2$ the circle $\sigma(t) = (\cos t, \sin t)$, we find $\phi(t) = (-\cos t, -\sin t)$, and so, using for instance Corollary 2.1.18, we obtain $\mathrm{ind}_O(X) = 1$.

Example 6.4.7. Let $X \in \mathcal{T}(S^2)$ be the vector field defined by $X(p) = \pi_p(\mathbf{e}_3)$, where $\pi_p \colon \mathbb{R}^3 \to T_p S^2$ is the orthogonal projection, and $\mathbf{e}_3 = (0,0,1)$; see Fig. 6.12.(b). Then X has exactly two singular points, the north pole $N = \mathbf{e}_3$ and the south pole $S = -N$. To compute the index of X at N, choose as local parametrization $\varphi \colon U \to \mathbb{R}^2$ the first one in Example 3.1.15, that is, $\varphi(x,y) = (x, y, \sqrt{1 - x^2 - y^2})$, and as simple regular region the one bounded by the curve $\sigma \colon [a, b] \to S^2$ given by

$$\sigma(t) = \varphi\left(\tfrac{1}{2}\cos t, \tfrac{1}{2}\sin t\right) = \left(\tfrac{1}{2}\cos t, \tfrac{1}{2}\sin t, \tfrac{\sqrt{3}}{2}\right).$$

It is easy to verify (check it!) that $X(p) = (-p_1 p_3, -p_2 p_3, 1 - p_3^2)$ for all $p = (p_1, p_2, p_3) \in S^2$, and so

$$X\big(\sigma(t)\big) = \left(-\tfrac{\sqrt{3}}{4}\cos t, -\tfrac{\sqrt{3}}{4}\sin t, \tfrac{1}{4}\right).$$

Now, $\partial_1|_\sigma = (1, 0, -\tfrac{1}{\sqrt{3}}\cos t)$ and $\partial_2|_\sigma = (0, 1, -\tfrac{1}{\sqrt{3}}\sin t)$; so, if we write $X \circ \sigma = X_1 \partial_1|_\sigma + \partial_2|_\sigma$, we find $(X_1, X_2) = \left(-\tfrac{\sqrt{3}}{4}\cos t, -\tfrac{\sqrt{3}}{4}\sin t\right)$, and hence $\phi(t) = (-\cos t, -\sin t)$, and $\mathrm{ind}_N(X) = 1$. Analogously (exercise), we find $\mathrm{ind}_S(X) = 1$.

Remark 6.4.8. The index is only interesting at singular points; *the index of a vector field at a regular point is always 0*. Indeed, let $p \in S$ be a regular point of the vector field $X \in \mathcal{T}(S)$. Then we may find a local parametrization $\varphi \colon U \to S$ centered at p, with U homeomorphic to an open disk, such that X is nowhere zero in $\varphi(U)$. Hence (up to shrinking slightly U) we may define a continuous determination $\Theta \colon \varphi(U) \to \mathbb{R}$ of the angle from ∂_1 to X on all $\varphi(U)$, by the usual Corollary 2.1.14. Now, if $\sigma \colon [a, b] \to \varphi(U)$ is a parametrization of the boundary of any simple regular region containing p in its interior, we have that $\Theta \circ \sigma$ is a continuous determination of the angle from $\partial_1|_\sigma$ to $X \circ \sigma$, and $\mathrm{ind}_p(X) = \big[\Theta\big(\sigma(b)\big) - \Theta\big(\sigma(a)\big)\big]/2\pi = 0$, as claimed.

The following proposition shows an alternative way to compute the index of a vector field at a point, with a formula suggesting that there might be a relation with the Gauss-Bonnet theorem:

Proposition 6.4.9. *Let $X \in \mathcal{T}(S)$ be a vector field on a surface S, and $p \in S$ a regular or isolated singular point of X. Let $R \subset \varphi(U)$ be a simple regular region containing p in its interior, contained in the image of an orthogonal local parametrization $\varphi \colon U \to S$, and such that $R \cap \mathrm{Sing}(X) \subseteq \{p\}$. Let now $\sigma \colon [a, b] \to \partial R \subset \varphi(U)$ be a parametrization by arc length of the boundary of R, positively oriented with respect to φ, and choose an arbitrary parallel non-zero vector field $\eta \in \mathcal{T}(\sigma)$ along σ. Then:*

$$\mathrm{ind}_p(X) = \frac{1}{2\pi} \int_R K \, d\nu - \frac{1}{2\pi}\big[\phi(b) - \phi(a)\big], \tag{6.8}$$

where $\phi \colon [a, b] \to \mathbb{R}$ is a continuous determination of the angle from $X \circ \sigma$ to η.

Proof. Denote by $\psi\colon [a,b] \to \mathbb{R}$ a continuous determination of the angle from $\partial_1|_\sigma$ to η. As a first thing, we want to prove that

$$\frac{\mathrm{d}\psi}{\mathrm{d}s} = -\frac{1}{2\sqrt{EG}} \left[\dot{\sigma}_2 \frac{\partial G}{\partial x_1} - \dot{\sigma}_1 \frac{\partial E}{\partial x_2} \right] , \qquad (6.9)$$

where, as usual, we have written $\sigma = \varphi(\sigma_1, \sigma_2)$. Set $\xi_j = \partial_j|_\sigma / \|\partial_j|_\sigma\|$ for $j = 1, 2$; in particular, $\eta/\|\eta\| = (\cos\psi)\xi_1 + (\sin\psi)\xi_2$. Lemma 5.1.24 tells us that to obtain (6.9) it suffices to prove that

$$\frac{\mathrm{d}\psi}{\mathrm{d}s} + \langle D\xi_1, \xi_2\rangle_\sigma \equiv 0 .$$

Now, $\{\xi_1, \xi_2\}$ is an orthonormal basis; so $\langle D\xi_1, \xi_1\rangle_\sigma \equiv \langle D\xi_2, \xi_2\rangle_\sigma \equiv 0$ and $\langle D\xi_1, \xi_2\rangle_\sigma = -\langle \xi_1, D\xi_2\rangle_\sigma$. Hence from $D\xi_j = \langle D\xi_j, \xi_1\rangle_\sigma \xi_1 + \langle D\xi_j, \xi_2\rangle_\sigma \xi_2$ we obtain

$$D\xi_1 = \langle D\xi_1, \xi_2\rangle_\sigma \xi_2 \quad \text{and} \quad D\xi_2 = -\langle D\xi_1, \xi_2\rangle_\sigma \xi_1 .$$

Keeping in mind that η is parallel (and in particular $\|\eta\|$ is constant), we deduce

$$O = D(\eta/\|\eta\|) = -\dot{\psi}(\sin\psi)\xi_1 + (\cos\psi)D\xi_1 + \dot{\psi}(\cos\psi)\xi_2 + (\sin\psi)D\xi_2$$

$$= \left(\frac{\mathrm{d}\psi}{\mathrm{d}s} + \langle D\xi_1, \xi_2\rangle_\sigma \right) \left[-(\sin\psi)\xi_1 + (\cos\psi)\xi_2 \right] .$$

So $\mathrm{d}\psi/\mathrm{d}s + \langle D\xi_1, \xi_2\rangle_\sigma \equiv 0$, and (6.9) is proved.

Integrating (6.9) between a and b we then find

$$\psi(b) - \psi(a) = \int_a^b \frac{\mathrm{d}\psi}{\mathrm{d}s}\, \mathrm{d}s = -\int_a^b \frac{1}{2\sqrt{EG}} \left[\dot{\sigma}_2 \frac{\partial G}{\partial x_1} - \dot{\sigma}_1 \frac{\partial E}{\partial x_2} \right] \mathrm{d}s = \int_R K\, \mathrm{d}\nu ,$$

where the last equality follows from the computation already seen in the proof of Theorem 6.1.7.

Now, if $\theta\colon [a,b] \to \mathbb{R}$ is a continuous determination of the angle from $\partial_1|_\sigma$ to $X \circ \sigma$ then $\phi = \psi - \theta$ is a determination of the angle from $X \circ \sigma$ to η. So

$$\mathrm{ind}_p(X) = \frac{1}{2\pi}[\theta(b) - \theta(a)] = \frac{1}{2\pi}[\psi(b) - \psi(a)] - \frac{1}{2\pi}[\phi(b) - \phi(a)]$$

$$= \frac{1}{2\pi}\int_R K\, \mathrm{d}\nu - \frac{1}{2\pi}[\phi(b) - \phi(a)] ,$$

and we are done. $\qquad\qquad\square$

We are now able to prove the promised application of the Gauss-Bonnet theorem to vector fields:

Theorem 6.4.10 (Poincaré-Hopf). *Let* $X \in \mathcal{T}(S)$ *be a vector field whose singular points are all isolated on a orientable compact surface* S. *Then:*

$$\sum_{p \in S} \mathrm{ind}_p(X) = \chi(S) . \qquad (6.10)$$

In particular, if S is not homeomorphic to a torus then every vector field on S has necessarily singular points.

Remark 6.4.11. Remember again that in the supplementary material to Chapter 4 we proved (Theorem 4.7.15) that every compact surface is orientable, and so the Poincaré-Hopf theorem applies to all compact surfaces.

Proof (of Theorem 6.4.10). First of all, note that the set of singular points of X (being a discrete subset of a compact surface) is finite; so Remark 6.4.8 assures us that the sum in the left-hand side of (6.10) is actually a finite sum.

Let $\mathbf{T} = \{T_1, \ldots, T_r\}$ be a triangulation of S whose faces are contained in the image of orthogonal local parametrizations compatible with the given orientation. We may also assume (why?) that every face contains in its interior at most one singular point of X, and that no singular point belongs to a side of the triangulation. Finally, orient positively the boundary of each face, and for $i = 1, \ldots, r$ choose a parallel vector field η_i along the boundary of T_i; let ϕ_i be a continuous determination of the angle from X to η_i.

If $\ell_{i,j}$ is a side of T_i with vertices v_1 and v_0, then the number

$$\phi_{i,j} = \phi_i(v_1) - \phi_i(v_0)$$

does not depend on the particular determination of the angle we have chosen. Moreover, $\phi_{i,j}$ does not depend on the particular parallel field either. Indeed, if we denote by ψ a continuous determination of the angle from ∂_1 to η along $\ell_{i,j}$ and by θ a continuous determination of the angle from ∂_1 to X along $\ell_{i,j}$, where ∂_1 is induced by an orthogonal local parametrization containing $\ell_{i,j}$ in its image, then (6.9) implies

$$\phi_{i,j} = -\int_{s_0}^{s_1} \frac{1}{2\sqrt{EG}} \left[\dot{\sigma}_2 \frac{\partial G}{\partial x_1} - \dot{\sigma}_1 \frac{\partial E}{\partial x_2} \right] \mathrm{d}s - \left[\theta(v_1) - \theta(v_0) \right] ,$$

where σ is a parametrization by arc length of the boundary of T_i with $\sigma(s_j) = v_j$ for $j = 0, 1$. In particular, if $\ell_{i,j}$ is a side of the face $T_{i'}$ too, say $\ell_{i,j} = \ell_{i',j'}$, we get

$$\phi_{i,j} = -\phi_{i',j'} , \tag{6.11}$$

by Remark 6.3.7.

Now, the previous proposition applied to T_i tells us that

$$\int_{T_i} K \, \mathrm{d}\nu - \sum_{j=1}^{3} \phi_{i,j} = \begin{cases} 0 & \text{if } T_i \cap \mathrm{Sing}(X) = \varnothing , \\ 2\pi \, \mathrm{ind}_p(X) & \text{if } T_i \cap \mathrm{Sing}(X) = \{p\} . \end{cases}$$

So, summing on all triangles we get:

$$\int_S K \, \mathrm{d}\nu - 2\pi \sum_{p \in S} \mathrm{ind}_p(X) = \sum_{i=1}^{f(\mathbf{T})} \sum_{j=1}^{3} \phi_{i,j} . \tag{6.12}$$

But every side of the triangulation is a side of exactly two distinct faces (Remark 6.2.4); so (6.11) implies that the right-hand side of (6.12) is zero, and the assertion follows from Corollary 6.3.10.   □

The Poincaré-Hopf theorem is a good example of a result that unexpectedly combines local and global properties, analytic and geometric objects. The left-hand side of (6.10) depends on the *local* differential properties of an analytic object, the vector field; the right-hand side only depends instead on the *global* topological properties of a geometric object, the surface, and does *not* depend on the particular vector field we are considering.

To appreciate the power of this theorem, consider the fact that if $p \in S$ is a singular point of a vector field $X \in \mathcal{T}(S)$ on an oriented compact surface, it is very easy to modify X in an arbitrarily small neighborhood of the point p so to obtain a vector field X_1 that is not zero at p. But in doing so, we *necessarily* create new singular points (or change the index of old ones) with a kind of action at a distance, since the value of the left-hand sum in (6.10) cannot change.

An especially eloquent consequence of the Poincaré-Hopf theorem is the so-called *hairy ball theorem*:

Corollary 6.4.12. *Let $S \subset \mathbb{R}^3$ be an orientable compact surface with non-zero Euler-Poincaré characteristic. Then every vector field on S has at least a singular point.*

Proof. Let $X \in \mathcal{T}(S)$. If X had no singular points, the sum in the left-hand side of (6.10) would be equal to zero, whereas $\chi(S) \neq 0$ by assumption.   □

Remark 6.4.13. The reason behind the curious name given to this corollary is that we may imagine a vector field as if it were a hairstyle (a tidy one: no mohawks here!) on the surface S. Then Corollary 6.4.12 tells us that if $\chi(S) \neq 0$ then any hairdresser, no matter how good, will be forced to leave on the surface a small tonsure (an isolated singular point) or a parting (consisting of non-isolated singular points).

Remark 6.4.14. If S is an orientable compact surface with $\chi(S) = 0$, then vector fields on S without singular points may exist; see Example 5.3.3 and Fig. 5.2.

Guided problems

Problem 6.1. *Prove that every compact surface $S \subset \mathbb{R}^3$ contains an open set consisting of elliptic points.*

Solution. Since S is compact, it has a point p_0 having maximum distance from the origin; we want to show that p_0 is elliptic.

Denote by $S_0 \subset \mathbb{R}^3$ the sphere with center in the origin and radius $\|p_0\|$. Clearly, $p_0 \in S_0 \cap S$, and S is contained in the closed ball with boundary S_0. In particular, S is tangent to S_0 at p_0 (Exercise 3.38); hence p_0 is orthogonal to $T_{p_0}S$.

Since we are only interested in what happens in a neighborhood of p_0, we may assume without loss of generality S to be orientable, and choose a Gauss map $N: S \to S^2$ such that $N(p_0) = p_0/\|p_0\|$. Let now $\sigma: (-\varepsilon, \varepsilon) \to S$ be a curve parametrized by arc length such that $\sigma(0) = p_0$. The function $s \mapsto \|\sigma(s)\|^2$ has an absolute maximum for $s = 0$; so its second derivative is not positive at 0. By differentiating we obtain

$$\kappa_n(0) \le -\frac{1}{\|p_0\|} \, ,$$

where κ_n is the normal curvature of σ.

Since σ was an arbitrary curve, it follows that all normal curvatures of S at p_0 have the same sign, and thus p_0 is elliptic. Finally, $K(p_0) > 0$ implies that K is positive in a neighborhood of p_0, and so S contains an open set consisting of elliptic points. $\qquad \square$

Problem 6.2. *Let S be an orientable compact surface in $\mathbb{R}^3$, not homeomorphic to a sphere. Show that S contains elliptic points, hyperbolic points and points with Gaussian curvature equal to zero.*

Solution. Since by assumption S is not homeomorphic to a sphere, its Euler-Poincaré characteristic $\chi(S)$ is non-positive. From the global Gauss-Bonnet theorem we get

$$\int_S K \, d\nu \le 0 \, .$$

We know that S contains an open set consisting of elliptic points, by Problem 6.1. So we may consider a triangulation $\{T_1, \ldots, T_r\}$ of S such that T_1 is contained in an open set consisting of elliptic points. The definition of integral tells us that $\int_S K \, d\nu = \sum_{i=1}^n \int_{T_i} K \, d\nu \le 0$. Since $\int_{T_1} K \, d\nu > 0$, we must have $\sum_{i=2}^r \int_{T_i} K \, d\nu < 0$. In particular, K is strictly negative somewhere on S and consequently, by the continuity of K and the connectedness of S, it mush also vanish somewhere else. $\qquad \square$

Problem 6.3. *Let S be an orientable surface in $\mathbb{R}^3$ with $K \le 0$ everywhere. Let $R \subset S$ be a simple regular region whose boundary is a geodesic polygon. Show that the polygon has at least three vertices. Conclude that in S neither a closed simple geodesic nor a geodesic self-intersecting once nor two geodesics intersecting in two points can be the boundary of a simple region.*

Solution. Since R is simple, it has Euler-Poicaré characteristic $\chi(R) = 1$. The Gauss-Bonnet theorem yields

$$\int_R K \, d\nu = 2\pi - \sum_{h=1}^p \varepsilon_h \, ,$$

where $\varepsilon_1, \ldots, \varepsilon_h$ are the external angles, as usual. Since $\int_R K \, d\nu \leq 0$, we have $\sum_{h=1}^p \varepsilon_h \geq 2\pi$. On the other hand, $\varepsilon_h \in (-\pi, \pi)$; hence $p > 2$, as required. $\square$

Exercises

THE GAUSS-BONNET THEOREM

6.1. Show that the claim in Problem 6.1 is false without the hypothesis that S is compact by finding a regular surface in $\mathbb{R}^3$ without elliptic points.

6.2. Let $H \colon S_1 \to S_2$ be an isometry between two regular surfaces S_1 and S_2 in $\mathbb{R}^3$. If $R \subset S_1$ is a regular region and $f \colon R \to \mathbb{R}$ is continuous show that $H(R)$ is a regular region of S_2 and that

$$\int_R f \, d\nu = \int_{H(R)} (f \circ H^{-1}) \, d\nu \, .$$

6.3. Show that two closed simple geodesics in a compact surface $S \subset \mathbb{R}^3$ entirely consisting of elliptic points always intersect. Show that the claim is false if S is not compact.

6.4. For which values of $d \in \mathbb{Z}$ is there a regular region R contained in a surface S with $\chi(R) = d$?

6.5. Construct on S^2 simple geodesic polygons with exactly 0 and 2 vertices, and prove that on S^2 there are no simple geodesic polygons with exactly 1 vertex.

6.6. The goal of this exercise is to construct a simple geodesic polygon with exactly 1 vertex. Let $S \subset \mathbb{R}^3$ be the ellipsoid of revolution

$$S = \left\{ (x, y, z) \in \mathbb{R}^3 \ \middle| \ x^2 + y^2 + \frac{z^2}{100} = 1 \right\}.$$

If $p_0 = (\frac{1}{2}, 0, 5\sqrt{3}) \in S$ and $v_0 = (0, -1, 0) \in T_{p_0}S$, let $\sigma(s) = \big(x(s), y(s), z(s)\big)$ be the geodesic parametrized by arc length with $\sigma(0) = p_0$ and $\dot{\sigma}(0) = v_0$. Set further $r(s) = \sqrt{x(s)^2 + y(s)^2}$, and write

$$\sigma(s) = \big(r(s) \cos \phi(s), r(s) \sin \phi(s), z(s)\big) \, ,$$

where $\phi(s)$ is the determination of the angle between $(1, 0)$ and $\big(x(s), y(s)\big)$ with initial value $\phi(0) = 0$.

Keeping in mind the properties of geodesics and in particular Clairaut's relation prove that:

(i) σ is not the parallel passing through p_0;

(ii) we have $1 \geq r(s) \geq 1/2$ for all $s \in \mathbb{R}$; in particular $(x(s), y(s)) \neq (0,0)$ everywhere, and so $\phi(s)$ is well defined for every value of s;

(iii) we have $x\dot{y} - \dot{x}y \equiv 1/2 \equiv r^2\dot{\phi}$; in particular, ϕ is monotonically increasing, $2 \geq \dot{\phi} \geq 1/2$, and $2s \geq \phi(s) \geq s/2$;

(iv) we have $r\dot{r} + (z\dot{z}/100) \equiv 0$ and $\dot{r}^2 + \dot{\phi}/2 + |\dot{z}|^2 \equiv 1$; hence $|\dot{z}| \leq \sqrt{3}/2$ and $\dot{z}(s) = 0$ if and only if $r(s) = 1/2$;

(v)   if $s_0 > 0$ is such that $\phi(s_0) = \pi$, then $\pi/2 \leq s_0 \leq 2\pi$, the coordinate z is monotonically decreasing in $[0, s_0]$, we have $z(s_0) \geq (5 - \pi)\sqrt{3} > 0$ and $r(s_0) < 1$;

(vi) we have $\sigma(-s_0) = \sigma(s_0)$; (*Hint:* use the symmetry of S with respect to the xz-plane.)

(vii) we have $\dot{\sigma}(s_0) \neq \pm\dot{\sigma}(-s_0)$, and so $\sigma|_{[-s_0, s_0]}$ is a simple geodesic polygon with only one vertex.

6.7. Let $S \subset \mathbb{R}^3$ be the ellipsoid of equation $x^2 + y^2 + 4z^2 = 1$.

(i) Parametrize the sides of the geodesic polygon on S having as vertices the points
$$\begin{cases} p_1 = (1, 0, 0), \\ p_2 = (\cos\theta_0, \sin\theta_0, 0), \quad \text{for a fixed } \theta_0 \in [0, \pi/2], \\ p_3 = (0, 0, 1/2). \end{cases}$$

(ii) Compute $\int_T K \, d\nu$ both directly and by using the Gauss-Bonnet theorem.

6.8. Let $S = S^2$ be the unit sphere. Given $\theta_0 \in [0, \pi/2]$, let $\sigma_1 : [0, \theta_0] \to S^2$, $\sigma_2, \sigma_3 : [0, \pi/2] \to S^2$ be the geodesics

$$\sigma_1(s) = (\cos s, \sin s, 0),$$
$$\sigma_2(s) = (\cos\theta_0 \cos s, \sin\theta_0 \cos s, \sin s),$$
$$\sigma_3(s) = (\sin s, 0, \cos s),$$

and denote by $T \subset S^2$ the geodesic triangle with sides σ_1, σ_2 and σ_3.

(i) Compute the area of T when $\theta_0 = \pi/2$.

(ii) Compute the area of T for any $\theta_0 \in [0, \pi/2]$.

6.9. Let $f_t : \mathbb{R}^3 \to \mathbb{R}$ be the function

$$f_t(x, y, z) = \frac{1}{t^2}x^2 + y^2 + z^2 - 1,$$

where $t > 0$ is a positive real parameter, and set

$$S_t = \{(x, y, z) \in \mathbb{R}^3 \mid f_t(x, y, z) = 0\}.$$

(i) Prove that $S_t \subset \mathbb{R}^3$ is a regular surface for all $t > 0$.

(ii) Compute the first and the second fundamental form of S_t at the point $(t, 0, 0)$.

(iii) Having set $D_t = S_t \cap \{(x, y, z) \in \mathbb{R}^3 \mid y \geq 0\}$, compute the integral of the Gaussian curvature over D_t.

6.10. Let $S \subset \mathbb{R}^3$ be the ellipsoid

$$S = \left\{ (x, y, z) \in \mathbb{R}^3 \;\middle|\; x^2 + y^2 + \frac{1}{2} z^2 = 1 \right\} .$$

(i) Find the principal, Gaussian and mean curvature of S at $p = (1, 0, 0)$.
(ii) Let $T \subset S$ be the regular region given by

$$T = \{ p = (x, y, z) \in S \mid x \geq 0, y \geq 0, z \geq 0 \} .$$

Compute the integral over T of the Gaussian curvature K of S.

6.11. On the sphere $S^2 \subset \mathbb{R}^3$ consider a geodesic triangle bounded by arcs, each of which is a quarter of a great circle, and let $\sigma \colon [0, \ell] \to S^2$ be a parametrization by arc length of this geodesic triangle.

(i)   Prove that, even if σ is just piecewise regular, for all $v \in T_p S^2$ there is a unique parallel vector field $\xi_v \in \mathcal{T}(\sigma)$ along σ such that $\xi_v(0) = v$.
(ii) Prove that the map $L \colon T_p S^2 \to T_p S^2$ given by $L(v) = \xi_v(\ell)$ is linear.
(iii) Determine explicitly L.

6.12. Determine the Gaussian curvature K of the surface $S \subset \mathbb{R}^3$ of equation $x^2 + y^2 = (\cosh z)^2$, and compute

$$\int_S K \, d\nu .$$

6.13. Consider a closed regular plane curve σ in $\mathbb{R}^3$, parametrized by arc length. If $\mathbf{v}$ is a versor orthogonal to the plane that contains the support of σ, and $a > 0$ is small enough, determine the integral (over the whole surface) of the Gaussian curvature of the surface S parametrized by

$$\varphi(s, v) = \sigma(s) + a \cos v \, \mathbf{n}(s) + a \sin v \, \mathbf{v} .$$

THE POINCARÉ-HOPF THEOREM

6.14. Let $X \colon S^2 \to \mathbb{R}^3$ be given by $X(x, y, z) = \left(-xy, 1 - y^2 - z, y(1 - z) \right)$. Prove that X is a vector field on S^2, find its singular points and compute its indices.

6.15. Let $S \subset \mathbb{R}^3$ be the surface given by the equation

$$x^4 + y^4 + z^4 - 1 = 0 ,$$

and let $X \in T(S)$ be the vector field given by

$$X(p) = \pi_p \left(\frac{\partial}{\partial x} \right)$$

for all $p \in S$, where $\pi_p \colon \mathbb{R}^3 \to T_pS$ is the orthogonal projection on the tangent plane to S at p.

(i)   Find the singular points of X.
(ii) Prove that the map $F \colon S \to S^2$ given by $F(p) = p/\|p\|$ is a diffeomorphism between S and the unit sphere S^2.
(iii) Compute the index of the singular points of X.

6.16. Let $S = S^2$ be the unit sphere, and for all $p \in S^2$ let $\pi_p \colon \mathbb{R}^3 \to T_pS^2$ be the orthogonal projection on the tangent plane to S^2 at p. Let $X \colon S^2 \to TS^2$ be the vector field on S given by

$$X(p) = \pi_p \left(\frac{\partial}{\partial z} \right) .$$

(i)   Prove that X has exactly two singular points, p_1 and p_2.
(ii) Compute the sum of the indices of the singular points of X on S^2.
(iii) Show that $\mathrm{ind}_{p_1}(X) = \mathrm{ind}_{p_2}(X) = 1$ using the symmetries of the problem.
(iv) Compute explicitly the index of X at the point p_2 using the local parametrization $\varphi \colon \mathbb{R}^2 \to S^2$ given by

$$\varphi(u, v) = \left(\frac{2u}{u^2 + v^2 + 1}, \frac{2v}{u^2 + v^2 + 1}, \frac{u^2 + v^2 - 1}{u^2 + v^2 + 1} \right) ,$$

that is, the inverse of the stereographic projection.

6.17. Let S be the torus obtained by rotating the circle $(x - 2)^2 + z^2 = 1$ around the z-axis. For all $p \in S$, denote by $\pi_p \colon \mathbb{R}^3 \to T_pS$ the orthogonal projection, and let $X \in T(S)$ be given by $X(p) = \pi_p(\partial/\partial y)$.

(i)   Determine the singular points of X.
(ii) Compute the sum of the indices of the singular points of X on S.
(iii) Compute the index of X at $(0, 3, 0)$.

6.18. Let $F \colon S_1 \to S_2$ be a local diffeomorphism between surfaces, and $X_1 \in T(S_1)$ a vector field with a singular point $p_1 \in S_1$. Prove that there is a neighbourhood $U \subseteq S_1$ of p_1 such that the map $X_2 \colon F(U) \to \mathbb{R}^3$ given by $X_2(q) = \mathrm{d}F_{F^{-1}(q)} X(F^{-1}(q))$ is a well-defined vector field on $\varphi(U)$, show that $F(p_1)$ is a singular point of X_2 and that $\mathrm{ind}_{p_1}(X_1) = \mathrm{ind}_{F(p_1)}(X_2)$. In other words, the index is invariant under local diffeomorphisms.

6.19. Let $X \in T(S)$ be a vector field on a surface $S \subset \mathbb{R}^3$, and $p \in S$ a singular point of X.

(i)   Prove che $\mathrm{d}X_p$ induces an endomorphism of T_pS. If $\det \mathrm{d}X_p \neq 0$ we say that p is a *non-degenerate* singular point.

(ii)   Prove that if p is non-degenerate then it is an isolated singular point, and that in this case $\mathrm{ind}_p(X)$ is given by the sign of $\det \mathrm{d}X_p$.

(iii) For $a \in S^2$ let $X_a: S^2 \to \mathbb{R}^3$ be given by $X_a(p) = a - \langle a, p \rangle p$. Prove that $X_a \in T(S^2)$, determine its singular points, and compute the indices of X_a at the singular points.

(iv)   Find a vector field on S^2 with exactly one singular point.

6.20. Let $S \subset \mathbb{R}^3$ be a surface, and fix $f \in C^\infty(S)$.

(i)   Prove that there exists a unique vector field $\nabla f \in T(S)$, the *gradient* of f, such that

$$\langle \nabla f(p), v \rangle_p = \mathrm{d}f_p(v)$$

for all $p \in S$ and $v \in T_pS$.

(ii)   Show that $p \in S$ is a singular point of ∇f if and only if it is a critical point for f.

(iii) Let $p \in S$ be a critical point of f. Prove that setting

$$\forall v \in T_pS \qquad \mathrm{Hess}_p f(v) = \frac{\mathrm{d}^2}{\mathrm{d}t^2}(f \circ \sigma)\Big|_{t=0},$$

where $\sigma: (-\varepsilon, \varepsilon) \to S$ is a curve with $\sigma(0) = p$ and $\sigma'(0) = v$, we get a well-defined quadratic form on T_pS, the *Hessian* of f in p.

(iv)   Let $p \in S$ be a singular point of ∇f. Show that p is non-degenerate (in the sense of Exercise 6.19) if and only if $\mathrm{Hess}_p f$ is a non-degenerate quadratic form.

(v)   Let $p \in S$ be a non-degenerate singular point of ∇f. Prove that the following assertions are equivalent:
(a) $\mathrm{Hess}_p f$ is positive or negative definite;
(b) $\mathrm{ind}_p(\nabla f) = +1$;
(c) p is a local minimum or a local maximum for f.

(vi)   Let $p \in S$ be a non-degenerate singular point of ∇f. Prove that the following assertions are equivalent:
(a) $\mathrm{Hess}_p f$ is indefinite;
(b) $\mathrm{ind}_p(\nabla f) = -1$;
(c) p is a saddle point for f.

(vii) Assume that all singular points of ∇f are non-degenerate, and that S is compact and orientable. Denote by $m(f)$ the number of local minima or maxima of f, and by $s(f)$ the number of saddle points of f. Prove that $m(f) - s(f) = \chi(S)$.

(viii) Let S be compact and oriented by a Gauss map $N: S \to S^2$, and let $a \in S^2$ be a regular value for both N and $-N$. Prove that $I_a = N^{-1}(\{a, -a\})$ is a finite set, and that the number of elliptic points in I_a minus the number of hyperbolic points in I_a is equal to the Euler-Poincaré characteristic of S. (*Hint:* use the function $h_a \in C^\infty(S)$ given by $h_a(p) = \langle p, a \rangle$.)

Supplementary material

6.5 Existence of triangulations

The goal of this section is to prove the theorems about triangulations (Theorems 6.2.5 and 6.2.8) stated in Section 6.2. Let us begin with a couple of preliminary definitions.

Definition 6.5.1. In this section, a *Jordan arc (respectively, curve)* in a surface S will be the support C of a non-closed (respectively, closed) simple continuous curve $\sigma: [a, b] \to S$. If σ is furthermore a regular curve, we shall say that C is a *regular* Jordan arc; if σ is a geodesic, C will be a *geodesic arc*.

Definition 6.5.2. A *topological region* in a surface S is a compact subset $R \subseteq S$ obtained as the closure of its interior and whose boundary consists of finitely many disjoint Jordan curves. A *simple topological region* is a topological region homeomorphic to a closed disk.

Definition 6.5.3. Let $R \subset S$ be a simple topological region of a surface S. A *cut* of R is a Jordan arc $C \subset R$ that intersects ∂R in its endpoints only.

The proof of existence of triangulations will rely on two fundamental lemmas, the first of which depends on Schönflies' Theorem 2.8.29 (see Remark 2.3.7), and generalizes Lemmas 2.8.5, 2.8.16, and 2.8.24:

Lemma 6.5.4. *Let $R \subset S$ be a simple topological (respectively, regular) region, and $C \subset R$ a (respectively, regular) cut. Then $R \setminus C$ has exactly two connected components, which are simple topological (respectively, regular) regions whose boundary consists of C and an arc of the boundary of R.*

Proof. Since this is a topological statement (the regular case is an immediate consequence), we may assume that R is the unit disk in the plane.

Let $p_1, p_2 \in \partial R$ be the two intersection points of C with ∂R, and denote by C_1, C_2 the two arcs of a circle cut by p_1 and p_2 on the boundary of R. Then $C \cup C_j$ is the support of a Jordan curve, and so it divides the plane in two connected components, a bounded one Ω_1 (which, by Schönflies' theorem, is a simple topological region) and an unbounded one. The unbounded component of $\mathbb{R}^2 \setminus \partial R$ intersects the unbounded component of $\mathbb{R}^2 \setminus (C \cup C_j)$, and so it lies within it; hence, Ω_j is contained in the unit disk, and has $C \cup C_j$ as its boundary.

If $\Omega_1 \cap \Omega_2$ were not empty, they would coincide, whereas they have different boundaries. Hence $\Omega_1 \cap \Omega_2 = \varnothing$; to get the assertion we only have to prove that $R \setminus C$ has no further components. Indeed, if we identify the boundary of R with a point, we get a topological space X homeomorphic to a sphere. In X, the cut C becomes a Jordan curve; and so $X \setminus C$ has two components, which are necessarily Ω_1 and Ω_2. $\qquad \square$

The second fundamental lemma is the tool that will allow us to construct the sides of the triangulations. We need a definition.

Definition 6.5.5. Let $R \subseteq S$ be a topological region of a surface S. We shall say that R is *in good position* with respect to a finite set $\mathcal{J}$ of Jordan arcs and curves contained in R if every component of ∂R intersects $J = \cup \mathcal{J}$ in at most finitely many points or subarcs.

Then:

Lemma 6.5.6. *Let $\mathcal{J}$ be a finite set of Jordan arcs and curves contained in a topological region $R \subseteq S$ of a surface S. Assume that R is in good position with respect to $\mathcal{J}$, and set $J = \cup \mathcal{J}$. Then every pair of points of R can be joined by a Jordan arc C contained in the interior of R except at most its endpoints and such that $C \cap J$ consists of finitely many points or subarcs of C and so that $C \setminus J$ is a union of finitely many geodesic arcs.*

Proof. We begin by proving the claim for the points in the interior of R. The relation stating that two points of the interior of R are equivalent if and only if they can be joined by a Jordan arc as in the statement is an equivalence relation (why?); so it suffices to prove that the equivalence classes are open.

Since $\mathcal{J}$ is finite, if $p \notin J$ is in the interior of R we may find an $\varepsilon > 0$ such that the geodesic ball $B_\varepsilon(p)$ is contained in $R \setminus J$. Then every point of $B_\varepsilon(p)$ can be joined to p by a geodesic not intersecting J, and so all points of $B_\varepsilon(p)$ are equivalent to p.

Let now $p \in J$. Since $\mathcal{J}$ is finite (again), we may find an $\varepsilon > 0$ such that the geodesic ball $B_\varepsilon(p) \subset R$ only intersects the elements of $\mathcal{J}$ containing p. Take $q \in B_\varepsilon(p)$. Follow the radial geodesic from q to the first point (which might well be p or q) belonging to an element of $\mathcal{J}$, and then follow that element up to p. So in this case too all points of $B_\varepsilon(p)$ are equivalent to p.

Let now $p_0 \in \partial R$. If $p_0 \notin J$ or, more in general, if there exists a neighborhood U of p_0 such that $J \cap \partial R \cap U \subset \partial R$, we may find a radial geodesic through p_0 with support C_0 contained (except for p_0) in the interior of R and disjoint from J. Let p be the second endpoint of C_0. Then the above applied to $\mathcal{J}' = \mathcal{J} \cup \{C_0\}$ tells us that for any point q in the interior of R we may find a Jordan arc from q to p contained in the interior of R that intersects the elements of $\mathcal{J}'$ in finitely many points or subarcs. Follow this arc from q up to the first intersection point with C_0, and then follow C_0 up to p_0; in this way we get an arc from q to p_0 as required.

If, on the other hand, $p_0 \in \partial R \cap J$ and there exists an element J_0 of $\mathcal{J}$ containing p and that does not lie in ∂R in a neighborhood of p_0, since R is in good position with respect to $\mathcal{J}$ there is a subarc C_0 of J_0 starting at p_0 and contained in the interior of R (except for p_0). We may then repeat the previous construction, joining p_0 to any point in the interior of R by an arc with the required properties.

Finally, let q_0 be another point of ∂R. Again, we may find an arc C_0 starting from q_0, contained in the interior of R except for q_0, and disjoint from

$J \setminus \{q_0\}$ or contained in an element of $\mathcal{J}$. Let q be the other endpoint of C_0. Applying what we have seen to $\mathcal{J} \cup \{C_0\}$ we obtain a Jordan arc as required from p_0 to q. Following this arc from p_0 up to the first intersection with C_0, and then following C_0 up to q_0, we have joined p_0 to q_0 too, as required. $\square$

Remark 6.5.7. It is clear from this proof that if the elements of $\mathcal{J}$ and the region R are regular then the arcs we have constructed are piecewise regular. Moreover, it is easy to see (exercise) that if $\mathcal{J} = \varnothing$ then the arcs we have constructed can be taken regular everywhere.

To be able to deduce Theorem 6.2.8 about refinements we shall need a more general version of Theorem 6.2.5 about the existence of triangulations: to state it, we need two more definitions.

Definition 6.5.8. Let $\mathbf{T}$ be a topological triangulation of a topological region R in a surface S. We shall say that $\mathbf{T}$ is *subordinate* to an open cover $\mathfrak{U}$ of R if for all $T \in \mathbf{T}$ there is a $U \in \mathfrak{U}$ such that $T \subset U$.

Definition 6.5.9. Let $\mathcal{J}$ be a finite set of Jordan arcs and curves in a topological region R. We shall say that a topological triangulation $\mathbf{T}$ of R is *in good position* with respect to $\mathcal{J}$ if every side of $\mathbf{T}$ intersects $J = \cup \mathcal{J}$ in at most finitely many points or subarcs.

We may now prove the existence of triangulations:

Theorem 6.5.10. *Given a topological region $R \subseteq S$ of a surface S, let $\mathfrak{U}$ be an open cover of R and $\mathcal{J}$ a finite set of Jordan arcs and curves in R. Suppose that R is in good position with respect to $\mathcal{J}$. Then there exists a topological triangulation $\mathbf{T}$ of R subordinate to $\mathcal{U}$ and in good position with respect to $\mathcal{J}$.*

Proof. Let $\delta > 0$ be the Lebesgue number of the cover $\mathfrak{U}$ (computed with respect to the intrinsic distance of S introduced in Section 5.4). For all $p \in R$, choose $0 < \varepsilon_p < \varepsilon'_p < \mathrm{inj}\, \mathrm{rad}(p)$ such that $\varepsilon'_p < \delta$. We may further suppose that $B_{\varepsilon'_p}(p)$ only intersects the elements of $\mathcal{J}'$ that contain p, where $\mathcal{J}'$ is obtained from $\mathcal{J}$ by adding the components of the boundary of R (so, in particular, if p is in the interior of R then $B_{\varepsilon'_p}(p)$ is in the interior of R); and if $p \in \partial R$ we may also suppose that both $B_{\varepsilon_p}(p) \cap \partial R$ and $B_{\varepsilon'_p}(p) \cap \partial R$ are Jordan arcs. Note finally that, since every geodesic ball is a ball with respect to the intrinsic distance (Corollary 5.4.4), there is automatically a $U \in \mathfrak{U}$ such that $B_{\varepsilon'_p}(p) \subset U$.

The region R is compact; so we may find $p_1, \ldots, p_k \in R$ such that $\{B_{\varepsilon_1}(p_1), \ldots, B_{\varepsilon_k}(p_k)\}$ is an open cover of R, where we have set $\varepsilon_j = \varepsilon_{p_j}$. Our first goal is to find k simple regular (or topological) regions contained in R whose interiors (in R) form an open cover of R, and such that $\mathcal{J} \cup \{\partial R_1, \ldots, \partial R_k\}$ is in good position with respect to all of them.

Start with p_1. Suppose first that p_1 belongs to the interior of R, so the whole geodesic ball $B_{\varepsilon'_1}(p_1)$ is contained in the interior of R. Choose

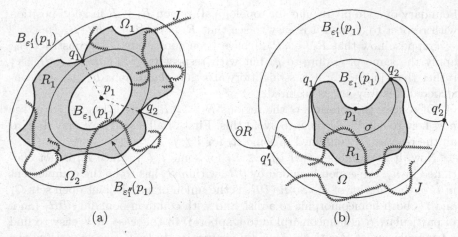

Fig. 6.14.

$\varepsilon_1 < \varepsilon_1'' < \varepsilon_1'$ and two points q_1, $q_2 \in \partial B_{\varepsilon_1''}(p_1)$. Then the segments of radial geodesic from p_1 to q_1 and q_2 taken between the geodesic circles with radii ε_1'' and ε_1' divide the annulus $B_{\varepsilon_1'}(p_1) \setminus B_{\varepsilon_1''}(p_1)$ in two simple regular regions Ω_1 and Ω_2; see Fig. 6.14.(a).

Now, Lemma 6.5.6 yields in each Ω_j a (piecewise regular if the elements of $\mathcal{J}$ are regular) Jordan arc that joins q_1 to q_2 intersecting $J = \cup\mathcal{J}$ in finitely many points or subarcs. These two arcs joins to form the boundary of a simple regular (or topological) region R_1, in good position with respect to $\mathcal{J}$. Moreover, the interior of R_1 contains $B_{\varepsilon_1}(p_1)$. Indeed, since $\partial R_1 \cap \partial B_{\varepsilon_1}(p_1) = \varnothing$, either $B_{\varepsilon_1}(p_1)$ lies in the interior of R_1, or it lies in the interior of $B_{\varepsilon_1'}(p_1) \setminus R_1$. But in this last case we would have $\partial\Omega_1 \cap R_1 = \{q_1, q_3\} \in \partial R_1$, and so the interior of Ω_1 would be disjoint from R_1, whereas one of the curves that make up the boundary of R_1 is contained in the interior of Ω_1.

Suppose now, on the other hand, that $p_1 \in \partial R$. By hypothesis, $\partial R \cap B_{\varepsilon_1}(p_1)$ is a Jordan arc, that intersects $\partial B_{\varepsilon_1}(p_1)$ in two points, q_1 and q_2. Follow the boundary of R beyond q_1 and q_2 up to the first (and only) intersection points with $\partial B_{\varepsilon_1'}(p_1)$, which we denote respectively by q_1' and q_2'. Since $\partial R \cap B_{\varepsilon_1}(p_1)$ is a cut of $B_{\varepsilon_1}(p_1)$, it divides $B_{\varepsilon_1}(p_1)$ in two connected components, one contained in R and the other one external to R. In particular, just one of the two arcs of geodesic circle from q_1 to q_2 is contained in R; call it σ. Denote by Ω the simple regular (or topological) region whose boundary consists of σ, the two boundary arcs from q_j to q_j', and the unique arc of geodesic circle from q_1' to q_2' for which $B_{\varepsilon_1}(p_1)$ is external to Ω; see Fig. 6.14.(b). In particular, $\Omega \subset R$. Lemma 6.5.6 (or, more precisely, its proof, which shows how in this case the way in which $\partial B_{\varepsilon_1'}(p_1)$ intersects J is irrelevant) provides us with a (piecewise regular if the elements of $\mathcal{J}$ are regular) Jordan arc in Ω that joins q_1 to q_2 intersecting J in finitely many points or subarcs, and contained in the interior of R but for the endpoints. Then this arc together with $\partial R \cap B_{\varepsilon_1}(p_1)$ forms the boundary of a simple regular (or topological) region $R_1 \subset R$ in good position with respect to $\mathcal{J}$, and it is easy to see that $R_1 \cap B_{\varepsilon_1}(p_1) = R \cap B_{\varepsilon_1}(p_1)$.

Suppose now that $R_1, \ldots, R_{j-1}$ have been constructed. To construct R_j, apply the same procedure to p_j but with respect to $\mathcal{J} \cup \{\partial R_1, \ldots, \partial R_{j-1}\}$ rather than just to $\mathcal{J}$; it is straightforward that the regions $R_1, \ldots, R_k$ constructed this way are as required.

Of course, the interiors of the regions $R_1, \ldots, R_k$ may intersect; let us see now how to subdivide them to avoid this. First of all, clearly we may discard all the regions R_j contained in some R_i for $i \neq j$.

Now, it may happen that $\partial R_j \subset R_i$ for some $i \neq j$. Since R_j is not contained in R_i, the region bounded by ∂R_j within R_i has to be the complement in R of the interior of R_j. So $R \setminus \partial R_j$ is the union of two disjoint open sets Ω_1 and Ω_2 both homeomorphic to a disk and with common boundary ∂R_j (and, in particular, R is homeomorphic to a sphere). In this case, it is easy to find a triangulation of R with the required properties. Indeed, take as vertices two points of ∂R_j. Join these two vertices with an arc in Ω_1 constructed using Lemma 6.5.6 as usual, and choose a point of the arc in the interior of Ω_1 as a new vertex (and join it to the other vertices of ∂R_j if necessary). Repeat this procedure in Ω_2, and we have the required triangulation.

Suppose next that $\partial R_j \not\subset R_i$ for all $i \neq j$, and take some R_j. By construction, ∂R_1 intersects R_j in finitely many (possibly zero) arcs of ∂R_j and cuts. The first cut divides R_j in two simple regions, by Lemma 6.5.4. The second cut divides one of these regions in two simple subregions, and so on. Hence, ∂R_1 divides R_j in finitely many simple regions. Each of those intersects ∂R_2 in finitely many (possibly zero) arcs of the boundary and cuts, and so it is subdivided by ∂R_2 in finitely many simple subregions. Proceeding like this we see that in the end R_j is divided in finitely many simple regions.

Denote by R_{jh} the regions in which R_j is divided. Clearly, two regions R_{jh} and R_{ik} either coincide or have disjoint interiors, since the boundary of either does not intersect the interior of the other one.

Analogously, every ∂R_j is divided into subarcs ℓ_{jr} by the intersection points with the other ∂R_i's or by the endpoints of the subarcs it shares with the other ∂R_j's. Moreover, the boundary of each R_{jh} is a union of arcs ℓ_{ir}.

We are done. Take first as vertices the endpoints of all ℓ_{jr}'s. If some R_{jh} happens to have a boundary with less than two vertices, we add any two vertices. For each R_{jh}, join two points of its boundary according to Lemma 6.5.6, add as a further vertex a point of the arc in the interior of R_{jh}, and join it (again using Lemma 6.5.6) to the other vertices of the boundary; in this way we have constructed the required triangulation. $\square$

Remark 6.5.11. Note that if R is a regular region and the elements of $\mathcal{J}$ are piecewise regular then the sides of the topological triangulation constructed in the previous theorem are piecewise regular.

Theorem 6.2.5 then follows from this remark, from Theorem 6.5.10 and from the following:

Lemma 6.5.12. *Let* **T** *be a topological triangulation of a regular region* $R \subseteq S$ *with piecewise regular sides. Then there exists a refinement of* **T** *that is a regular triangulation of* R.

Proof. Connect every vertex of a side of **T** (as a curve) that is not already a vertex of the triangulation with the opposite vertex (as a triangulation) of the same face, using Lemma 6.5.6 and Remark 6.5.7 to be sure that the new side is a regular Jordan arc; in this way, we obtain the required regular triangulation. □

To prove Theorem 6.2.8 we still have to work a little. A last definition:

Definition 6.5.13. Two triangulations are *in good position* if each of them is in good position with respect to the family of sides of the other one.

An immediate corollary of Theorem 6.5.10 is:

Corollary 6.5.14. *Let* $R \subseteq S$ *be a regular region,* $\mathfrak{U}$ *an open cover of* R, *and* $\mathbf{T}_0$, $\mathbf{T}_1$ *two topological (repsectively, regular) triangulations of* R. *Then there exists a topological triangulation (respectively, with piecewise regular sides)* $\mathbf{T}^*$ *subordinate to* $\mathfrak{U}$ *that is in good position both with* $\mathbf{T}_0$ *and with* $\mathbf{T}_1$.

Proof. Denote by $\mathcal{J}$ the family consisting of the sides of $\mathbf{T}_0$ and of $\mathbf{T}_1$; clearly, R is in good positions with respect to $\mathcal{J}$. So it suffices to apply Theorem 6.5.10 and Remark 6.5.11 to obtain a topological triangulation $\mathbf{T}^*$ (with piecewise regular sides if $\mathbf{T}_0$ and $\mathbf{T}_1$ are regular) in good position with respect to both $\mathbf{T}_0$ and $\mathbf{T}_1$. □

Theorem 6.2.8 follows now from this corollary and the following proposition:

Proposition 6.5.15. *Let* $R \subseteq S$ *be a regular region, and let* $\mathbf{T}_0$ *and* $\mathbf{T}_1$ *be two topological triangulations of* R *(respectively, with piecewise regular sides). Suppose that* $\mathbf{T}_0$ *and* $\mathbf{T}_1$ *are in good position. Then there exists a topological (respectively, regular) triangulation* $\mathbf{T}^*$ *that is a refinement both of* $\mathbf{T}_0$ *and* $\mathbf{T}_1$.

Proof. It suffices to prove that the union of the sides of $\mathbf{T}_0$ and $\mathbf{T}_1$ divides R into a finite union of simple topological regions (with piecewise regular boundary). Indeed, once we have proved this, we may give a triangulation of each region by joining (by a regular Jordan arc, thanks to Lemma 6.5.6 and Remark 6.5.7) an internal point to the vertices on the boundary of the region. In this way we get a topological triangulation (with piecewise regular sides) that refines both $\mathbf{T}_0$ and $\mathbf{T}_1$, and Lemma 6.5.12 tells us that, if necessary, we may find a further refinement that is a regular triangulation, as required.

Let T_0 be a face of $\mathbf{T}_0$; we have then to verify that the sides of $\mathbf{T}_1$ subdivide it into finitely many simple topological regions (with piecewise regular boundary).

Since $\mathbf{T}_0$ and $\mathbf{T}_1$ are in good position, every side of $\mathbf{T}_1$ is either completely contained in the interior of T_0 or intersects the boundary of T_0 in finitely many points or subarcs. First of all, Lemma 6.5.4 used as in the second part of the proof of Theorem 6.5.10 shows that the sides of $\mathbf{T}_1$ without vertices in the interior of T_0 divide T_0 in finitely many simple topological regions (with piecewise regular boundary).

After this, suppose that there exists a side ℓ of $\mathbf{T}_1$ with a vertex in the interior of one of these simple regular regions R_j, but not completely contained in the interior of R_j. Since the faces of $\mathbf{T}_1$ completely cover R_j, we may construct a (piecewise regular) Jordan arc that starts from the internal vertex of ℓ and reaches the boundary of R_j going first, if necessary, along sides of $\mathbf{T}_1$ that are completely internal to R_j, and then along a subarc of another side of $\mathbf{T}_1$ with a vertex in the interior of R_j. Joining this arc with the subarc of ℓ that goes from the internal vertex to the boundary of R_j, we get a cut, which consequently divides (again by Lemma 6.5.4) R_j in two (piecewise regular) simple regions.

By repeating this procedure we deal with all the sides of $\mathbf{T}_1$, since clearly each side of $\mathbf{T}_1$, if it is completely internal to a region R_j, can be joined to the boundary of R_j using an arc consisting of sides of $\mathbf{T}_1$; hence, the existence of such a side necessarily implies the existence of a side with an internal vertex and intersecting the boundary. Since $\mathbf{T}_1$ has finitely many sides, we complete the construction in finitely many steps. $\square$

Global theory of surfaces

The Gauss-Bonnet theorem is just the first (though one of the most important) of many theorems about the global theory of surfaces. It is a theory so wide-reaching and rich in results that t cannot be reasonably presented in a single chapter (or in a single book, for that matter). So we shall confine ourselves to introduce some significant theorems to give at least an idea of the techniques used and of the kind of results one can obtain.

The main goal of this chapter is to obtain the complete classification of closed surfaces in $\mathbb{R}^3$ having constant Gaussian curvature. In Section 7.1 we shall prove that the only closed surfaces with constant positive Gaussian curvature are spheres (actually, in this section we shall prove this for compact surfaces only; to get the complete statement we shall need the material we shall introduce in Section 7.4 of the supplementary material of this chapter). In Section 7.2 we shall prove that the only closed surfaces having Gaussian curvature everywhere zero are planes and surfaces, and in Section 7.3 we prove that there are no closed surfaces with constant negative Gaussian curvature. As you will see, the proofs of these results are rather complex, and use crucially the properties of asymptotic curves and Codazzi-Mainardi equations (4.28).

In the supplementary material we shall go further, introducing quite sophisticated techniques to study closed surfaces whose Gaussian curvature has constant sign. In Section 7.4 we shall prove that every compact surface with positive Gaussian curvature is diffeomorphic to a sphere (Hadamard's theorem), while every non-compact closed surface with positive Gaussian curvature is diffeomorphic to a convex open subset of the plane (Stoker's theorem). In Section 7.5 we shall introduce the notions of covering map and of simply connected surface, and in Section 7.6 we shall prove another theorem by Hadamard: every simply connected closed surface having non-positive Gaussian curvature is diffeomorphic to a plane.

Abate M., Tovena F.: Curves and Surfaces.
DOI 10.1007/978-88-470-1941-6_7, © Springer-Verlag Italia 2012

7.1 Surfaces with constant positive curvature

The goal of this section is to prove that the only compact surfaces with constant Gaussian curvature are spheres (Liebmann's theorem). This kind of results, where assumptions on local objects implies that the surface belongs to a very specific class, are sometimes called *rigidity theorems*.

We begin by remarking that we may actually assume that the surface has positive constant curvature, because Problem 6.1 tells us that every compact surface has elliptic points.

We have seen in Problem 4.3 that spheres are characterized by the fact that they consist entirely of umbilical points (and are not planes). Problem 6.1 implies, in particular, that a compact surface cannot be contained in a plane (why?); so this suggest that we try and prove that a compact surface with constant Gaussian curvature consists entirely of umbilical points.

To get there, we shall use the following criterion to obtain umbilical points, originally due to Hilbert:

Proposition 7.1.1. *Let $S \subset \mathbb{R}^3$ be an oriented surface, and k_1, $k_2 \colon S \to \mathbb{R}$ the corresponding principal curvatures, with $k_1 \leq k_2$ as usual. Let $p \in S$ be a point that is simultaneously a local minimum for k_1 and a local maximum for k_2; suppose further that $K(p) > 0$. Then p is an umbilical point.*

Proof. Let $N \colon S \to S^2$ be the Gauss map of S. Without loss of generality, we may suppose that p is the origin $O \in \mathbb{R}^3$, and that $N(p) = \mathbf{e}_3$, the third vector of the canonical basis. Moreover, by Proposition 3.1.29, the surface S is a graph in a neighborhood of p; having set $N(p) = \mathbf{e}_3$, it has to be a graph with respect to the $x_1 x_2$-plane, and so we have a local parametrization $\varphi \colon U \to S$ centered at p of the form $\varphi(x_1, x_2) = \bigl(x_1, x_2, h(x_1, x_2)\bigr)$ for a suitable function $h \colon U \to \mathbb{R}$ such that $h(O) = 0$ and $\nabla h(O) = (0, 0)$. Finally, up to a rotation around the z-axis, we may also suppose (Proposition 4.5.2) that $\mathbf{e}_j = \partial_j|_p$ is a principal direction with respect to $k_j(p)$ for $j = 1, 2$.

Let now $\sigma_1, \sigma_2 \colon (-\varepsilon, \varepsilon) \to S$ be given by $\sigma_1(t) = \varphi(0, t)$ and $\sigma_2(t) = \varphi(t, 0)$, and define the functions $\ell_1, \ell_2 \colon (-\varepsilon, \varepsilon) \to \mathbb{R}$ by setting

$$\ell_j(t) = Q_{\sigma_j(t)}(\partial_j / \|\partial_j\|) \, ;$$

in other words, $\ell_1 = (e/E) \circ \sigma_1$ and $\ell_2 = (g/G) \circ \sigma_2$, where E, F and G are the metric coefficients and e, f and g the form coefficients of S with respect to φ.

The hypotheses on p imply that for $|t|$ small enough we have

$$\ell_1(0) = Q_p(\mathbf{e}_1) = k_1(p) \leq k_1\bigl(\sigma_1(t)\bigr) \leq Q_{\sigma_1(t)}(\partial_1 / \|\partial_1\|) = \ell_1(t) \, ,$$
$$\ell_2(0) = Q_p(\mathbf{e}_2) = k_2(p) \geq k_2\bigl(\sigma_2(t)\bigr) \geq Q_{\sigma_2(t)}(\partial_2 / \|\partial_2\|) = \ell_2(t) \, ;$$

so 0 is a point of local minimum for ℓ_1 and of local maximum for ℓ_2. In particular, $\ell_2''(0) \leq 0 \leq \ell_1''(0)$.

Using Examples 4.1.8 and 4.5.19, which contain the formulas for the metric and form coefficients of a graph, we obtain the following expression for ℓ_1 and ℓ_2:

$$\ell_1(t) = \left(\frac{1}{(1 + |\partial h/\partial x_1|^2)\sqrt{1 + \|\nabla h\|^2}} \frac{\partial^2 h}{\partial x_1^2} \right)(0, t) \,,$$

$$\ell_2(t) = \left(\frac{1}{(1 + |\partial h/\partial x_2|^2)\sqrt{1 + \|\nabla h\|^2}} \frac{\partial^2 h}{\partial x_2^2} \right)(t, 0) \,;$$

in particular,

$$\frac{\partial^2 h}{\partial x_j^2}(O) = \ell_j(0) = k_j(p) \,.$$

Moreover, as $\partial_1|_p$ and $\partial_2|_p$ are principal directions, we also have $f(p) = 0$, and so $\partial^2 h/\partial x_1 \partial x_2(O) = 0$.

Differentiating twice the expressions we have found for ℓ_j and evaluating the result at 0 we obtain

$$\frac{\partial^4 h}{\partial x_1^2 \partial x_2^2}(O) - \left[\frac{\partial^2 h}{\partial x_1^2}(O) \right]^2 \frac{\partial^2 h}{\partial x_2^2}(O) = \ell_2''(0)$$

$$\leq \ell_1''(0) = \frac{\partial^4 h}{\partial x_1^2 \partial x_2^2}(O) - \left[\frac{\partial^2 h}{\partial x_2^2}(O) \right]^2 \frac{\partial^2 h}{\partial x_1^2}(O) \,,$$

or

$$k_1(p)k_2(p)\big[k_2(p) - k_1(p)\big] \leq 0 \,.$$

Since we know that $k_1(p)k_2(p) = K(p) > 0$, we get $k_2(p) \leq k_1(p)$. This is possible only if $k_2(p) = k_1(p)$, and so p is an umbilical point. $\qquad\square$

We are now able to prove the promised *Liebmann's theorem*:

Theorem 7.1.2 (Liebmann). *The only compact surfaces with constant Gaussian curvature are spheres.*

Proof. Let $S \subset \mathbb{R}^3$ be a compact surface with constant Gaussian curvature $K \equiv K_0$. Firstly, Problem 6.1 tells us that $K_0 > 0$; so S cannot be contained in a plane. Next, since the Gaussian curvature is always positive, the absolute value of the mean curvature cannot be zero anywhere; hence by Problem 4.16 we know that S is orientable.

Fix an orientation of S, and let $k_1, k_2 \colon S \to \mathbb{R}$ be the principal curvatures of S, with $k_1 \leq k_2$ as usual. Since k_1 and k_2 are continuous (a fact that follows, for instance, from Proposition 4.5.14) and S is compact, there exists a point $p \in S$ where k_2 has a maximum. Since the product $k_1 k_2$ is constant, the same point p has to be a minimum point for $k_1 = K_0/k_2$. Hence we may apply Proposition 7.1.1, and deduce that p is an umbilical point, that is, $k_1(p) = k_2(p)$. Then for all $q \in S$ we have

$$k_2(q) \leq k_2(p) = k_1(p) \leq k_1(q) \leq k_2(q) \,,$$

and so all points of S are umbilical.

Now, Problem 4.3 implies that S is contained in a sphere S_0; we have to prove that $S = S_0$. Since S is compact, it is closed in S_0. Since it is a surface, it is a union of images of local parametrizations, so it is open in S_0 by Proposition 3.1.31.(i). But S_0 is connected; hence, $S = S_0$, and we are done.   □

Example 7.1.3. The assumption that the surface is compact (or, at least, closed in $\mathbb{R}^3$; see next remark) in Liebmann's theorem is essential: there are non-closed surfaces in $\mathbb{R}^3$ with positive constant Gaussian curvature that are not contained in a sphere. For example, choose a constant $C > 1$, and let $S \subset \mathbb{R}^3$ be the surface of revolution obtained by rotating the curve $\sigma: \big(- \arcsin(1/C),$ $\arcsin(1/C)\big) \to \mathbb{R}^3$ given by

$$\sigma(s) = \left(C \cos s, 0, \int_0^s \sqrt{1 - C^2 \sin^2 t}\, \mathrm{d}t \right) \; ;$$

see Fig. 7.1.

Since the curve σ is parametrized by arc length, Example 4.5.22 provides us with the following values for the principal curvatures of S:

$$k_1 = \frac{C \cos s}{\sqrt{1 - C^2 \sin^2 s}} \, , \qquad k_2 = \frac{\sqrt{1 - C^2 \sin^2 s}}{C \cos s} \; ;$$

hence $K = k_1 k_2 \equiv 1$. Moreover, $C > 1$ implies that S has no umbilical points, so it is not contained in a sphere; and, clearly, S is not compact (nor closed in $\mathbb{R}^3$).

Remark 7.1.4. In the supplementary material of this chapter we shall prove (Theorem 7.4.14) that every surface S closed in $\mathbb{R}^3$ whose Gaussian curvature is bounded from below by a positive constant is necessarily compact. So Liebmann's theorem implies that *the only closed surfaces in $\mathbb{R}^3$ with positive constant Gaussian curvature are spheres.*

Fig. 7.1.

7.2 Surfaces with constant zero curvature

The aim of this section is to classify the closed surfaces in $\mathbb{R}^3$ with Gaussian curvature identically zero. In addition to planes (Example 4.5.16), we have already met surfaces with this property: right circular cylinders (Example 4.5.17) and, more in general, all cylinders (Exercise 4.27). The theorem by Hartman and Nirenberg we shall shortly prove says that there are no other such surfaces: every closed surface in $\mathbb{R}^3$ with Gaussian curvature identically zero is a plane or a cylinder.

Let $S \subset \mathbb{R}^3$ be a surface S with Gaussian curvature identically zero. The points of S have to be either parabolic or planar. We shall denote by $\mathrm{Pl}(S) \subseteq S$ the set of planar points of S, and by $\mathrm{Pa}(S) \subseteq S$ the set of parabolic points of S. Since $\mathrm{Pl}(S) = \{p \in S \mid |H(p)| = 0\}$, where $|H|$ is the absolute value of the mean curvature of S, the set of planar points is closed in S; as a consequence, the set $\mathrm{Pa}(S) = S \setminus \mathrm{Pl}(S)$ of parabolic points is open in S.

Our first goal is to show that, given any parabolic point of S, there is actually a straight line passing through it. Remember (see Definition 4.6.16) that an *asymptotic direction* at $p \in S$ is a $v \in T_p S$ of length one such that $Q_p(v) = 0$; clearly, this notion does not depend on the local orientation chosen to define the second fundamental form, and so it is well defined for non-orientable surfaces too. Now, if $p \in S$ is a parabolic point, (4.10) tells us that, up to a sign, there is a unique asymptotic direction at p, and it necessarily is a principal direction. So we may define a vector field $X^{\mathrm{Pa}} \in \mathcal{T}(\mathrm{Pa}(S))$ such that $X^{\mathrm{Pa}}(p)$ is an asymptotic direction of S at p (convince yourself that X^{Pa} is smooth because of he uniqueness of the asymptotic direction); and, since S is connected, X^{Pa} is uniquely defined up to a sign (why?). So Theorem 5.3.7 yields for all $p \in \mathrm{Pa}(S)$ a unique maximal integral curve of X^{Pa} passing through p; we shall denote by $R(p) \subset \mathrm{Pa}(S)$ its support, which clearly is the support of the unique asymptotic curve passing through p. We have found the line we were looking for:

Proposition 7.2.1. *Let $S \subset \mathbb{R}^3$ be a surface with Gaussian curvature $K \equiv 0$, and $p \in \mathrm{Pa}(S)$. Then $R(p) \subset \mathrm{Pa}(S)$ is the unique (open) line segment passing through p and contained in S.*

Proof. Since p is not an umbilical point, Corollary 5.3.24.(ii) tells us that there exists a local parametrization $\varphi: U \to S$ centered at p whose coordinate curves are lines of curvature; clearly, we may assume that $V = \varphi(U) \subseteq \mathrm{Pa}(S)$. Moreover, one of the two coordinate curves passing through a point of V has to be tangent to the asymptotic direction through that point, and so its support is the intersection of the support of the corresponding maximal asymptotic curve with V. Without loss of generality, we may suppose that the asymptotic coordinate curves are the curves $\{x_2 = \text{cost.}\}$.

Let $N: V \to S^2$ be the Gauss map induced by φ. By construction, we have $\partial(N \circ \varphi)/\partial x_1 = dN(\partial_1) \equiv O$ in V; in particular, N is constant along the curves $\{x_2 = \text{cost.}\}$. Moreover, we also have $\partial^2(N \circ \varphi)/\partial x_1 \partial x_2 \equiv O$ in V;

hence $\partial(N \circ \varphi)/\partial x_2 = dN(\partial_2)$ is also constant (and nowhere zero, since we are working in the set of parabolic points) along the curves $\{x_2 = \text{cost.}\}$. Next,

$$\frac{\partial}{\partial x_1} \langle \varphi, N \circ \varphi \rangle = \langle \partial_1, N \circ \varphi \rangle + \left\langle \varphi, \frac{\partial(N \circ \varphi)}{\partial x_1} \right\rangle \equiv 0 .$$

As a consequence, the function $\langle \varphi, N \circ \varphi \rangle$ is constant along the curves $\{x_2 = \text{cost.}\}$, which means that every curve $\{x_2 = \text{cost.}\}$ lies within a plane orthogonal to the constant value assumed by N along that curve. Analogously,

$$\frac{\partial}{\partial x_1} \left\langle \varphi, \frac{\partial(N \circ \varphi)}{\partial x_2} \right\rangle = \left\langle \partial_1, \frac{\partial(N \circ \varphi)}{\partial x_2} \right\rangle + \left\langle \varphi, \frac{\partial^2(N \circ \varphi)}{\partial x_1 \partial x_2} \right\rangle \equiv 0 ,$$

since $\partial(N \circ \varphi)/\partial x_2 = dN(\partial_2)$ is a multiple of ∂_2, and so (Proposition 4.5.2) it is orthogonal to ∂_1. Hence the function $\langle \varphi, \partial(N \circ \varphi)/\partial x_2 \rangle$ is also constant along the curves $\{x_2 = \text{cost.}\}$, which means that every curve $\{x_2 = \text{cost.}\}$ lies within a plane orthogonal to the constant value assumed by $\partial(N \circ \varphi)/\partial x_2$ along that curve. So every curve $\{x_2 = \text{cost.}\}$ is contained in the intersection of two (distinct, since N is always orthogonal to $\partial(N \circ \varphi)/\partial x_2$) planes, and thus it is a line segment.

We have still to check the uniqueness. But every line segment R passing through p is necessarily (why?) an asymptotic curve; since $R(p)$ is the only asymptotic curve passing through p, we are done.   □

We shall shortly see that if S is closed in $\mathbb{R}^3$ then every $R(p)$ actually is a whole straight line. To prove this, we shall need the fact that the closure of $R(p)$ does not intersect the set $\text{Pl}(S)$ of planar points of S. To verify this claim we shall use the following consequence of the Codazzi-Mainardi equations:

Lemma 7.2.2. Let $S \subset \mathbb{R}^3$ be a surface with Gaussian curvature identically zero, $p \in \text{Pa}(S)$, and $\sigma: I \to \text{Pa}(S)$ a parametrization by arc length of $R(p)$. Then

$$\frac{d^2}{ds^2} \left(\frac{1}{|H| \circ \sigma} \right) \equiv 0 ,$$

where $|H|$ is the absolute value of the mean curvature of S.

Proof. Note first that $|H|$ is nowhere zero on $\text{Pa}(S)$, so the claim makes sense.

Let $\varphi: U \to S$ be the local parametrization centered at p already used in the previous proof. In particular, since ∂_1 and ∂_2 are principal directions, we have $F \equiv f \equiv 0$ in $V = \varphi(U)$. Moreover, as $dN(\partial_1) \equiv O$ we also have $e \equiv 0$, and so the absolute value of the mean curvature is given by $|H| = |g|/2G$. Finally, g has constant sign on $R(p)$, since it does not vanish in $\text{Pa}(S)$. Hence

$$|H| = \pm \frac{g}{2G}$$

in $R(p)$, and we have to compute the second derivative of $(G/g) \circ \sigma$.

To this end, we resort to the Codazzi-Mainardi equations (4.28). Since $F \equiv 0$, the Christoffel symbols are given by (4.23); so, recalling that $h_{11} = e \equiv 0$, $h_{12} = h_{21} = f \equiv 0$ and $h_{22} = g$, the Codazzi-Mainardi equations become

$$\frac{g}{2G}\frac{\partial E}{\partial x_2} = 0, \qquad \frac{g}{2G}\frac{\partial G}{\partial x_1} = \frac{\partial g}{\partial x_1}. \tag{7.1}$$

The first equation implies that $\partial E/\partial x_2 \equiv 0$; so E only depends on x_1. Hence the map $\chi: U \to \mathbb{R}^2$ given by

$$\chi(x) = \left(\int_0^{x_1} \sqrt{E(t, x_2)}\, dt, x_2 \right)$$

is a diffeomorphism with its image, and the map $\hat{\varphi} = \varphi \circ \chi^{-1}$ is still a local parametrization with all the properties of φ, since the only thing we did was to change the parametrization of the curves $\{x_2 = \text{cost.}\}$ so that now they are parametrized by arc length.

So, up to changing the parametrization as explained, we may also assume that $E \equiv 1$. In particular, Lemma 4.6.14 tells us that

$$-\frac{1}{\sqrt{G}}\frac{\partial^2 \sqrt{G}}{\partial x_1^2} = -\frac{1}{2\sqrt{G}}\frac{\partial}{\partial x_1}\left(\frac{1}{\sqrt{G}}\frac{\partial G}{\partial x_1} \right) = K \equiv 0;$$

so it is possible to write

$$\sqrt{G(x)} = a_1(x_2)x_1 + a_2(x_2) \tag{7.2}$$

for suitable functions a_1, a_2 that only depend on x_2.

Now, the second equation in (7.1) can be written as

$$\frac{\partial \log|g|}{\partial x_1} = \frac{1}{g}\frac{\partial g}{\partial x_1} = \frac{1}{2G}\frac{\partial G}{\partial x_1} = \frac{1}{\sqrt{G}}\frac{\partial \sqrt{G}}{\partial x_1} = \frac{\partial \log \sqrt{G}}{\partial x_1};$$

so we have $g(x) = a_3(x_2)\sqrt{G(x)}$ for a suitable function a_3 that only depends on x_2. Summing up,

$$\pm\frac{1}{2|H|} = \frac{G}{g} = \sqrt{G}\frac{\sqrt{G}}{g} = \frac{a_1(x_2)x_1 + a_2(x_2)}{a_3(x_2)}. \tag{7.3}$$

Now, in this local parametrization the curve σ is exactly the curve $s \mapsto \varphi(s, 0)$; therefore

$$\frac{\partial^2}{\partial s^2}\left(\frac{1}{|H \circ \sigma|} \right) = \pm 2\frac{\partial^2}{\partial x_1^2}\frac{G}{g}\bigg|_{x_2=0} \equiv 0,$$

as required.   □

Corollary 7.2.3. Let $S \subset \mathbb{R}^3$ be a surface having Gaussian curvature identically zero, and $p \in \text{Pa}(S)$. Then $\overline{R(p)} \cap \text{Pl}(S) = \varnothing$. In particular, if S is closed in $\mathbb{R}^3$ then $R(p)$ is a straight line, and all the lines $R(p)$, for every p in a connected component of $\text{Pa}(S)$, are parallel.

Proof. Let $\sigma\colon(a,b)\rightarrow S$ be a parametrization by arc length of $R(p)$. If $\overline{R(p)}\cap\mathrm{Pl}(S)$ were not empty, up to changing the orientation of σ we would have $\lim_{s\to b}\sigma(s)=q_0\in\mathrm{Pl}(S)$. Now, the previous lemma tell us that there exist constants a_1, $a_2\in\mathbb{R}$ such that

$$\left|H\big(\sigma(s)\big)\right|=\frac{1}{a_1 s+a_2}\,;\qquad(7.4)$$

so, since the absolute value of the mean curvature is zero in all planar points, we should have

$$0=|H(q_0)|=\lim_{s\to b}\left|H\big(\sigma(s)\big)\right|=\lim_{s\to b}\frac{1}{a_1 s+a_2}=\frac{1}{a_1 b+a_2}\neq 0\,,$$

a contradiction.

So we have proved that $\overline{R(p)}\cap\mathrm{Pl}(S)=\varnothing$. Since $R(p)$, being the support of an integral curve, is closed in $\mathrm{Pa}(S)$, it has to be closed in S. If S is closed in $\mathbb{R}^3$, it follows that $R(p)$ is closed in $\mathbb{R}^3$, and so it has to be a whole line.

Finally, we have to prove that all the lines $R(p)$, for each p in a component of $\mathrm{Pa}(S)$, are parallel. Given $p\in\mathrm{Pa}(S)$, let $\sigma\colon\mathbb{R}\rightarrow S$ be a parametrization by arc length of $R(p)$. Clearly, $\left|H\big(\sigma(s)\big)\right|$ is defined (and nowhere zero) for all values of s; but (7.4) tells us that this is only possible if $|H|\circ\sigma$ is constant.

Let then $\varphi\colon U\rightarrow S$ be the local parametrization centered at p constructed in the proof of the previous lemma. The fact that $|H|\circ\sigma$ is constant corresponds to requiring that $|H|\circ\varphi$ does not depend on x_1; hence, recalling (7.3) and (7.2), to requiring that G does not depend on x_1. In particular, $\partial G/\partial x_1\equiv 0$, and we already remarked that $E\equiv 1$ and $F\equiv e\equiv f\equiv 0$. But then (4.18) and (4.23) imply

$$\frac{\partial^2\varphi}{\partial x_1^2}=\Gamma_{11}^1\partial_1+\Gamma_{11}^2\partial_2+eN\equiv O\,,\qquad\frac{\partial^2\varphi}{\partial x_2\partial x_1}=\Gamma_{21}^1\partial_1+\Gamma_{21}^2\partial_2+fN\equiv O\,;$$

hence $\partial_1=\partial\varphi/\partial x_1$ is constant, that is, the tangent direction to the lines $R(q)$ for $q\in\varphi(U)$ does not depend on q, and thus all $R(q)$'s are parallel.

So we have proved that all lines $R(q)$ for q in a neighborhood of p are parallel; hence the set of points $q\in\mathrm{Pa}(S)$ such that $R(q)$ is parallel to a given line is open, and this immediately implies (why?) the assertion.   □

So if S is closed in $\mathbb{R}^3$ the set of parabolic points is a disjoint union of straight lines; let us see now what happens for $\partial_S\mathrm{Pa}(S)$, the boundary in S of the set of parabolic points. We shall need the following:

Lemma 7.2.4. *Let $S\subset\mathbb{R}^3$ be a surface with Gaussian curvature identically zero, and take $p\in\partial_S\mathrm{Pa}(S)$. Suppose we have a sequence $\{q_n\}\subset\mathrm{Pa}(S)$ approaching p such that the line segments $R(q_n)$ approach a segment $C\subset S$ passing through p of positive length. Then C is the only segment through p contained in S.*

Proof. Suppose that there is another line segment $C' \subset S$ passing through p. Then for n large enough the segment $R(q_n)$ has to intersect (and be distinct from) C'. But a segment always has zero normal curvature, so it always is an asymptotic curve; so the point $q \in C' \cap R(q_n)$ should simultaneously be parabolic and have two distinct asymptotic directions, a contradiction. $\square$

Then:

Proposition 7.2.5. *Let $S \subset \mathbb{R}^3$ be a surface with Gaussian curvature identically zero. Then exactly one open line segment $C(p) \subset S$ passes through each $p \in \partial_S \mathrm{Pa}(S)$. Moreover, $C(p) \subset \partial_S \mathrm{Pa}(S)$, and if S is closed in $\mathbb{R}^3$ then $C(p)$ is a straight line.*

Proof. Given $p \in \partial_S \mathrm{Pa}(S)$, we want to show that if $q \in \mathrm{Pa}(S)$ approaches p then the line segment $R(q)$ tends to a segment $C(p)$ passing through p contained in S. Note first that

$$\liminf_{\substack{q \to p \\ q \in \mathrm{Pa}(S)}} |R(q)| > 0 , \tag{7.5}$$

where $|R(q)|$ is the length of the segment $R(q)$, since otherwise p would be an accumulation point of endpoints not belonging to S (by Corollary 7.2.3) of segments $R(q)$, contradicting the fact that p is in the interior of S. Let us show now that the asymptotic direction in $q \in \mathrm{Pa}(S)$ admits a limit for q tending to p. If it were not so, we could find two sequences $\{q_n\}$ and $\{q'_n\}$ tending to p such that the corresponding segments $R(q_n)$ and $R(q'_n)$ tend to distinct segments C and C' passing through p, both of positive length by (7.5), and this would contradict Lemma 7.2.4.

Hence the segments $R(q)$ tend to a segment $C(p)$ contained in S of positive length, which is unique by Lemma 7.2.4, and is an open segment since otherwise its endpoints (in S) would be limit of points not in S, impossible. We want to show now that $C(p) \subset \partial_S \mathrm{Pa}(S)$. Clearly, $C(p) \subset \overline{\mathrm{Pa}(S)} = \mathrm{Pa}(S) \cup \partial_S \mathrm{Pa}(S)$. If a $q \in \mathrm{Pa}(S) \cap C(p)$ existed, then $C(p) \cap \mathrm{Pa}(S)$ should be contained (being a limit of asymptotic curves) in $R(q)$; but in this case we would have $\overline{R(q)} \cap \mathrm{Pl}(S) \neq \varnothing$, contradicting Corollary 7.2.3.

Finally, if S is closed in $\mathbb{R}^3$, since $C(p)$ is closed in $\partial_S \mathrm{Pa}(S)$, it follows that $C(p)$ is closed in $\mathbb{R}^3$, and so it is a straight line. $\square$

So $\mathrm{Pa}(S) \cup \partial_S \mathrm{Pa}(S)$ is a disjoint union of segments (of lines when S is closed). We are finally able to prove the promised *Hartman-Nirenberg's theorem*:

Theorem 7.2.6 (Hartman-Nirenberg). *Let $S \subset \mathbb{R}^3$ be a closed surface with Gaussian curvature identically zero. Then S is either a plane or a cylinder.*

Proof. Suppose that S is not a plane. Then Problem 4.3 tells us that S contains parabolic points. What we have proved above implies that $\mathrm{Pa}(S) \cup$

$\partial_S \mathrm{Pa}(S)$ is a disjoint union of straight lines. Now the connected components of $\mathrm{Pl}(S) \setminus \partial_S \mathrm{Pa}(S)$ are open sets consisting of planar points; thus (Problem 4.3) they are portions of planes, each with a boundary consisting of straight lines, since it is contained in $\partial_S \mathrm{Pa}(S)$. This lines have to be parallel, since they would otherwise intersect, contradicting the fact that a unique line belonging to S may pass through each point of $\partial_S \mathrm{Pa}(S)$. Hence exactly one line passes through each point of $\mathrm{Pl}(S) \setminus \partial_S \mathrm{Pa}(S)$ as well, and all the lines belonging to the same component are parallel.

On the other hand, we have seen that all the lines belonging to a component K of $\mathrm{Pa}(S)$ are parallel too, and this implies (also recalling the proof of Proposition 7.2.5) that the lines making up the boundary of K are also parallel.

Let now p_0 and p_1 be two arbitrary points of S, and $\sigma: [0,1] \to S$ a curve joining them (it exists because S is connected). By compactness, we may find a partition $0 = t_0 < \cdots < t_k = 1$ of $[0,1]$ such that each $\sigma([t_{j-1}, t_j])$ is contained in the closure of a connected component of $S \setminus \partial_S \mathrm{Pa}(S)$. Then for all $j = 1, \ldots, k$ all the lines passing through each $\sigma(t)$ (with $t \in [t_{j-1}, t_j]$) are parallel. But t_j lies both in $[t_{j-1}, t_j]$ and in $[t_j, t_{j+1}]$; so all the lines passing through points of the support of σ are parallel. In particular, the lines through p_0 and p_1 are parallel. But p_0 and p_1 are generic points, and thus S is a cylinder, as claimed. $\square$

Remark 7.2.7. We may replace the assumption of being closed in $\mathbb{R}^3$ by the assumption of being complete (see Definition 5.4.9; note that a closed surface is always complete, by Corollary 5.4.11). Indeed, every line segment in a surface has to be a geodesic, and every geodesic in a complete surface has to be defined for all times (Theorem 5.4.8); so all line segments contained in S are lines, and we may repeat the previous arguments.

Example 7.2.8. Once more, the closure (or completeness) hypothesis is essential: there are non-closed surfaces with Gaussian curvature identically zero not contained in a plane or in a cylinder. A very simple example is the upper sheet of a circular cone

$$S = \{(x,y,z) \in \mathbb{R}^3 \mid z^2 = x^2 + y^2, \ z > 0\},$$

the surface of revolution of the curve $\sigma: \mathbb{R}^+ \to \mathbb{R}$ given by $\sigma(t) = (t,0,t)$. By Example 4.5.22 we know that $K \equiv 0$ and $H(x,y,z) = 1/z$. In particular, S only consists of parabolic points, and so it is not contained in a plane. It is not contained in a cylinder either: indeed, through each parabolic point of S passes a unique segment contained in S (Proposition 7.2.1), and thus it should be a generator (see Definition 4.P.4) of the cone. Hence if S were contained in a cylinder the generators would be parallel, but they are not.

7.3 Surfaces with constant negative curvature

In the previous sections we have classified the closed surfaces with constant positive or zero Gaussian curvature; we are left with the closed surfaces having constant negative Gaussian curvature. Our goal is to prove a theorem due to Hilbert which states that there is nothing to classify: there are no closed surfaces with constant negative Gaussian curvature.

If $S \subset \mathbb{R}^3$ is a surface with negative Gaussian curvature, then every point of S is hyperbolic, and so it has two distinct asymptotic directions. By Corollary 5.3.24.(ii) we know that there is a local parametrization centered at p in which the coordinate curves are asymptotic curves. As we shall see, when the Gaussian curvature of S is constant (negative), we can say much more about this local parametrization, finding the key to the proof of Hilbert's theorem. But let us begin with a definition.

Definition 7.3.1. A local parametrization $\varphi \colon U \to S$ of a surface S is called *Chebyshev* if $E \equiv G \equiv 1$, that is, if the coordinate curves are parametrized by arc length. Furthermore, a local parametrization is *asymptotic* if the coordinate curves are asymptotic curves.

To show the existence of Chebyshev asymptotic local parametrizations we need a preliminary computation.

Lemma 7.3.2. *Let $\varphi \colon U \to S$ be an asymptotic local parametrization of a surface S with form coefficients e, f, $g \colon U \to \mathbb{R}$. Then*

$$\frac{\partial f}{\partial x_1} = \frac{1}{EG - F^2} \left[\frac{1}{2} \frac{\partial(EG - F^2)}{\partial x_1} + F \frac{\partial E}{\partial x_2} - E \frac{\partial G}{\partial x_1} \right] f , \qquad (7.6)$$

$$\frac{\partial f}{\partial x_2} = \frac{1}{EG - F^2} \left[\frac{1}{2} \frac{\partial(EG - F^2)}{\partial x_2} + F \frac{\partial G}{\partial x_1} - G \frac{\partial E}{\partial x_2} \right] f . \qquad (7.7)$$

Proof. Recalling that in this local parametrization we have $e \equiv g \equiv 0$, the Codazzi-Mainardi equations (4.28) give

$$\frac{\partial f}{\partial x_1} = (\Gamma_{11}^1 - \Gamma_{21}^2) f , \qquad \frac{\partial f}{\partial x_2} = (\Gamma_{22}^2 - \Gamma_{12}^1) f .$$

Since

$$\Gamma_{11}^1 - \Gamma_{21}^2 = \frac{1}{2(EG - F^2)} \left[G \frac{\partial E}{\partial x_1} - 2F \frac{\partial F}{\partial x_1} + 2F \frac{\partial E}{\partial x_2} - E \frac{\partial G}{\partial x_1} \right] ,$$

$$\Gamma_{22}^2 - \Gamma_{12}^1 = \frac{1}{2(EG - F^2)} \left[E \frac{\partial G}{\partial x_2} - 2F \frac{\partial F}{\partial x_2} + 2F \frac{\partial G}{\partial x_1} - G \frac{\partial E}{\partial x_2} \right] ,$$

we obtain the assertion. $\qquad \square$

Then:

Proposition 7.3.3. *Let $S \subset \mathbb{R}^3$ be a surface with negative constant Gaussian curvature $K \equiv -K_0 < 0$. Then every point $p \in S$ has a Chebyshev asymptotic local parametrization centered at p.*

Proof. Let $\varphi \colon U \to S$ be an asymptotic local parametrization centered at p; it exists by Corollary 5.3.24.(ii). Up to two parameter changes (performed independently on the two variables), we may assume that the two coordinate curves passing through $p = \varphi(0,0)$ are parametrized by arc length. So,

$$E(\cdot, 0) \equiv G(0, \cdot) \equiv 1 \; ;$$

we want to prove that negative constant Gaussian curvature forces φ to be Chebyshev.

Note first that $K \equiv -K_0$ implies $f^2 = K_0(EG - F^2)$. Substituting this in (7.6) multiplied by $2f$, we get

$$F\frac{\partial E}{\partial x_2} - E\frac{\partial G}{\partial x_1} \equiv 0 \; .$$

Proceeding analogously in (7.7), we get

$$-G\frac{\partial E}{\partial x_2} + F\frac{\partial G}{\partial x_1} \equiv 0 \; .$$

The system formed by these two equations has determinant $F^2 - EG$, which is nowhere zero; hence, $\partial E/\partial x_2 \equiv \partial G/\partial x_1 \equiv 0$. But this means that E does not depend on x_2 and G does not depend on x_1; so $E(x_1, x_2) = E(x_1, 0) \equiv 1$ and $G(x_1, x_2) = G(0, x_2) \equiv 1$, as required. $\square$

Let $\varphi \colon U \to S$ be a Chebyshev local parametrization, with U homeomorphic to a square. Since ∂_1 and ∂_2 have norm 1, we have $N = \partial_1 \wedge \partial_2$, and $\{\partial_1, N \wedge \partial_2\}$ always is an orthonormal basis of the tangent plane. Up to restricting U, we may use Corollary 2.1.14 to define a continuous determination $\theta \colon U \to \mathbb{R}$ of the angle from ∂_1 to ∂_2 to write

$$\partial_2 = (\cos\theta)\partial_1 + (\sin\theta)N \wedge \partial_1 \; .$$

Moreover, since ∂_1 and ∂_2 are always linearly dependent, θ cannot be equal to 0 or to π; so, by choosing a suitable initial value, we may obtain $\theta(U) \subseteq (0, \pi)$. In particular, as

$$\cos\theta = F \quad \text{and} \quad \sin\theta = \sqrt{1 - F^2} = \sqrt{EG - F^2} \; ,$$

we get $\theta = \arccos F$; in particular, θ is of class C^∞. The function θ satisfies an important differential equation:

Lemma 7.3.4. *Let* $\varphi: U \to S$ *be a Chebyshev local parametrization, with* U *homeomorphic to a square, of a surface* S, *and let* $\theta: U \to (0, \pi)$ *be the continuous determination of the angle from* ∂_1 *to* ∂_2 *as described above. Then*

$$\frac{\partial^2 \theta}{\partial x_1 \partial x_2} = (-K) \sin \theta . \tag{7.8}$$

Proof. Keeping in mind that $\theta = \arccos F$ and $\sin \theta = \sqrt{1 - F^2}$, we easily obtain

$$-\frac{1}{\sin \theta} \frac{\partial^2 \theta}{\partial x_1 \partial x_2} = \frac{1}{1 - F^2} \frac{\partial^2 F}{\partial x_1 \partial x_2} + \frac{F}{(1 - F^2)^2} \frac{\partial F}{\partial x_1} \frac{\partial F}{\partial x_2} .$$

Now, setting $E \equiv G \equiv 1$ in the definition of the Christoffel symbols we find

$$\Gamma_{11}^1 = \frac{-F}{1 - F^2} \frac{\partial F}{\partial x_1} , \quad \Gamma_{12}^1 = \Gamma_{21}^1 \equiv 0 , \quad \Gamma_{22}^1 = \frac{1}{1 - F^2} \frac{\partial F}{\partial x_2} ,$$

$$\Gamma_{11}^2 = \frac{1}{1 - F^2} \frac{\partial F}{\partial x_1} , \quad \Gamma_{12}^2 = \Gamma_{21}^2 \equiv 0 , \quad \Gamma_{22}^2 = \frac{-F}{1 - F^2} \frac{\partial F}{\partial x_2} .$$

So (4.27) implies

$$K = \frac{\partial}{\partial x_1} \left(\frac{1}{1 - F^2} \frac{\partial F}{\partial x_2} \right) - \frac{F}{(1 - F^2)^2} \frac{\partial F}{\partial x_1} \frac{\partial F}{\partial x_2}$$

$$= \frac{1}{1 - F^2} \frac{\partial^2 F}{\partial x_1 \partial x_2} + \frac{F}{(1 - F^2)^2} \frac{\partial F}{\partial x_1} \frac{\partial F}{\partial x_2} ,$$

and we are done. $\square$

The main idea in the proof of the non-existence of closed surfaces with negative constant Gaussian curvature consists in using Chebyshev asymptotic local parametrizations to define a function $\theta: \mathbb{R}^2 \to (0, \pi)$ that satisfies (7.8), and then proving that such a function cannot possibly exist.

We need a further lemma, in whose proof the fact that we are working with closed surfaces is used in a crucial way:

Lemma 7.3.5. *Let* $S \subset \mathbb{R}^3$ *be a closed surface with Gaussian curvature everywhere negative, and* $\sigma: I \to S$ *a parametrization by arc length of a maximal asymptotic curve of* S. *Then* $I = \mathbb{R}$.

Proof. Suppose by contradiction that $I \neq \mathbb{R}$, and let $s_0 \in \mathbb{R}$ be a (finite) endpoint of I; without loss of generality, we may assume that it is the supremum of I. Since σ is parametrized by arc length, we have

$$\|\sigma(s) - \sigma(s')\| \leq L(\sigma|_{[s,s']}) = |s - s'| \tag{7.9}$$

for all s, $s' \in I$ with $s < s'$. So, for every sequence $\{s_n\} \subset I$ approaching s_0 the sequence $\{\sigma(s_n)\}$ is Cauchy in $\mathbb{R}^3$, and thus it tends to a point $p_0 \in \mathbb{R}^3$.

But S is closed in $\mathbb{R}^3$; hence $p_0 \in S$. It is straightforward to verify that p_0 does not depend on the sequence chosen, so by setting $\sigma(s_0) = p_0$ we have extended continuously σ to s_0.

Now, Theorem 5.3.7.(ii) provides us with a neighborhood $V \subseteq S$ of p_0 and an $\varepsilon > 0$ such that for all $p \in V$ the asymptotic curves parametrized by arc length passing through p are defined on the interval $(-\varepsilon, \varepsilon)$. So, if we take $s_1 \in I$ such that $\sigma(s_1) \in V$ and $|s_0 - s_1| < \varepsilon$ then the curve σ, which is an asymptotic curve passing through $\sigma(s_1)$, has to be defined at least up to $s_1 + \varepsilon > s_0$, against the choice of s_0.   □

We are now ready to prove *Hilbert's theorem*:

Theorem 7.3.6 (Hilbert). *There are no closed surfaces in $\mathbb{R}^3$ with negative constant Gaussian curvature.*

Proof. Suppose by contradiction that $S \subset \mathbb{R}^3$ is a closed surface with negative constant Gaussian curvature $K \equiv -K_0 < 0$. Fix a point $p_0 \in S$, and let $\sigma: \mathbb{R} \to S$ be an asymptotic curve parametrized by arc length with $\sigma(0) = p_0$; the existence of σ is assured by the previous lemma.

Let $\xi_0 \in T_{p_0}S$ be an asymptotic direction of length 1 at p_0, with $\xi_0 \neq \pm\dot\sigma(0)$; then there is (why?) a unique vector field $\xi \in \mathcal{T}(\sigma)$ along σ_0 with $\xi(0) = \xi_0$ such that $\xi(s) \in T_{\sigma(s)}S$ is an asymptotic direction of length 1 at $\sigma(s)$, different from $\pm\dot\sigma(s)$. Now define $\Phi: \mathbb{R}^2 \to S$ by setting

$$\Phi(x_1, x_2) = \sigma_{x_1}(x_2) \,,$$

where $\sigma_{x_1}: \mathbb{R} \to S$ is the only asymptotic curve parametrized by arc length such that $\sigma_{x_1}(0) = \sigma(x_1)$ and $\dot\sigma_{x_1}(0) = \xi(x_1)$; see Fig. 7.2.

We want to prove that for all $x^o = (x_1^o, x_2^o) \in \mathbb{R}^2$ there is an $\varepsilon > 0$ such that Φ restricted to $(x_1^o - \varepsilon, x_1^o + \varepsilon) \times (x_2^o - \varepsilon, x_2^o + \varepsilon)$ is a Chebyshev asymptotic local parametrization at $\Phi(x^o)$.

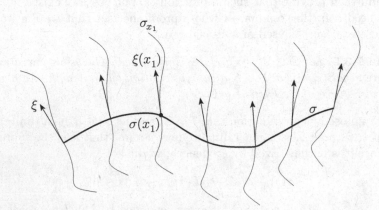

Fig. 7.2.

To this end, fix $x_1^o \in \mathbb{R}$. We know that the curves $s \mapsto \Phi(s,0)$ and $s \mapsto \Phi(x_1^o,s)$ are asymptotic curves parametrized by arc length, with tangent versors linearly independent in the intersection point $p = \Phi(x_1^o,0)$. A first remark is that, by Proposition 7.3.3, there exists a Chebyshev asymptotic local parametrization $\varphi\colon (x_1^o - \varepsilon, x_1^o + \varepsilon) \times (-\varepsilon, \varepsilon) \to S$ with $\varphi(x_1^o,0) = p$, $\partial_1|_p = \partial\Phi/\partial x_1(x_1^o,0)$ and $\partial_2|_p = \partial\Phi/\partial x_2(x_1^o,0)$. But then $s \mapsto \varphi(x_1,s)$ is the asymptotic curve passing through $\sigma(x_1)$ tangent to $\xi(x_1)$ parametrized by arc length, that is $\varphi(x_1,s) = \Phi(x_1,s)$ for all $x_1 \in (x_1^o - \varepsilon, x_1^o + \varepsilon)$ and $s \in (-\varepsilon, \varepsilon)$. In other words, Φ restricted to $(x_1^o - \varepsilon, x_1^o + \varepsilon) \times (-\varepsilon, \varepsilon)$ is a Chebyshev asymptotic local parametrization.

Suppose now that $x_2 \in \mathbb{R}$ is such that $s \mapsto \Phi(s,x_2)$ is an asymptotic curve in a neighborhood of x_1^o. Note that, for fixed s and t near x_2, the point $\Phi(s,t)$ is found by following σ_s from $\sigma_s(x_2)$ to $\sigma_s(t)$, which is an asymptotic curve passing through $\sigma_s(x_2)$. A second remark then is that it is possible to repeat the previous argument and find an $\varepsilon > 0$ such that Φ restricted to $(x_1^o - \varepsilon, x_1^o + \varepsilon) \times (x_2 - \varepsilon, x_2 + \varepsilon)$ is a Chebyshev asymptotic local parametrization.

Since $[0, x_2^o]$ is compact, we may cover the support of $\sigma_{x_1^o}|_{[0,x_2^o]}$ with the images of finitely many Chebyshev asymptotic local parametrizations $\varphi_1, \ldots, \varphi_n$ in such a way that each image intersects the next one as in Fig. 7.3. So the first remark above implies that Φ coincides with φ_1 (up to changing the orientation of φ_1), and the second remark implies that, one step after another, we get that Φ is a Chebyshev asymptotic local parametrization in a neighborhood of (x_1^o, x_2^o).

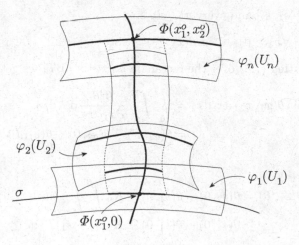

Fig. 7.3.

Define now $\theta \colon \mathbb{R}^2 \to (0, \pi)$ by

$$\theta = \arccos \left\langle \frac{\partial \Phi}{\partial x_1}, \frac{\partial \Phi}{\partial x_2} \right\rangle .$$

Lemma 7.3.4 tells that θ satisfies (7.8) on all $\mathbb{R}^2$, with $-K \equiv K_0 > 0$; we want to show that such a function cannot possibly exist.

First of all, $\partial^2 \theta / \partial x_1 \partial x_2 > 0$ everywhere; so $\partial \theta / \partial x_1$ is an increasing function of x_2 and, in particular,

$$\frac{\partial \theta}{\partial x_1}(x_1, x_2) > \frac{\partial \theta}{\partial x_1}(x_1, 0) \tag{7.10}$$

for all $x_2 > 0$. Integrating with respect to x_1 this relation over the interval $[a, b]$ we get

$$\theta(b, x_2) - \theta(a, x_2) > \theta(b, 0) - \theta(a, 0) \tag{7.11}$$

for every $x_2 > 0$ and every $a < b$. Now, since $\partial \theta / \partial x_1$ is not identically zero, up to a translation we may assume that $\partial \theta / \partial x_1(0, 0) \neq 0$. Since the function $(x_1, x_2) \mapsto \theta(-x_1, -x_2)$ also satisfies (7.8), we may also assume that $\partial \theta / \partial x_1(0, 0) > 0$. Let $0 < s_1 < s_2 < s_3$ with $\partial \theta / \partial x_1(\cdot, 0) > 0$ on $[0, s_3]$, and set

$$\delta = \min\{\theta(s_3, 0) - \theta(s_2, 0), \theta(s_1, 0) - \theta(0, 0)\} \in (0, \pi) .$$

So, recalling (7.10) and (7.11), we find that for all $x_2 > 0$ the function $s \mapsto \theta(s, x_2)$ is increasing on $[0, s_3]$ and

$$\min\{\theta(s_3, x_2) - \theta(s_2, x_2), \theta(s_1, x_2) - \theta(0, x_2)\} > \delta .$$

Keeping in mind that the image of θ is contained in $(0, \pi)$ all of the above implies that

$$\delta \leq \theta(x_1, x_2) \leq \pi - \delta$$

for every $x_1 \in [s_1, s_2]$ and every $x_2 \geq 0$, so

$$\forall x_1 \in [s_1, s_2]\ \forall x_2 \geq 0 \qquad \sin \theta(x_1, x_2) \geq \sin \delta .$$

But by integrating (7.8) over the rectangle $[s_1, s_2] \times [0, T]$ we get

$$K_0 \int_0^T \int_{s_1}^{s_2} \sin \theta(x_1, x_2)\, \mathrm{d}x_1 \mathrm{d}x_2 = \int_0^T \int_{s_1}^{s_2} \frac{\partial^2 \theta}{\partial x_1 \partial x_2}\, \mathrm{d}x_1 \mathrm{d}x_2$$
$$= \theta(s_2, T) - \theta(s_2, 0) - \theta(s_1, T) + \theta(s_1, 0) .$$

Hence,

$$\theta(s_2, T) - \theta(s_1, T) = \theta(s_2, 0) - \theta(s_1, 0) + K_0 \int_0^T \int_{s_1}^{s_2} \sin \theta(x_1, x_2)\, \mathrm{d}x_1 \mathrm{d}x_2$$
$$\geq \theta(s_2, 0) - \theta(s_1, 0) + K_0(s_2 - s_1)T \sin \delta$$

for all $T > 0$, and this is impossible since the left-hand side is less than π. $\square$

Remark 7.3.7. As usual, Hilbert's theorem also holds for complete (in the sense of Definition 5.4.9) surfaces. Indeed, the only step where we have used the closure of S was in the proof of Lemma 7.3.5, and the same proof holds for complete surfaces replacing (7.9) by

$$d_S\big(\sigma(s),\sigma(s')\big) \leq L(\sigma|_{[s,s']}) = |s - s'|\,,$$

where d_S is the intrinsic distance of S introduced in Definition 5.4.1.

Remark 7.3.8. As usual, again, the closure (or completeness) assumption is essential: we have seen that the pseudosphere (Example 4.5.23) is a non-closed (nor complete) surface with negative constant Gaussian curvature. On the other hand, there exist closed surfaces with Gaussian curvature everywhere negative (but not constant): for instance, the helicoid (Example 4.5.20). However, note that in that case the supremum of the values assumed by the curvature is zero. This is not a coincidence: *Efimov's theorem* states that there are no closed surfaces in $\mathbb{R}^3$ with $K \leq -C < 0$ for any constant $C > 0$. You may find a proof of this theorem in [14].

Guided problems

We recall that a parametrization in ruled form (see Exercise 4.69) of a ruled surface S is a local parametrization $\varphi\colon I \times \mathbb{R}^* \to \mathbb{R}^3$ of S of the form

$$\varphi(t,v) = \sigma(t) + v\,\mathbf{v}(t)\,,$$

where $\sigma\colon I \to \mathbb{R}^3$ is a regular curve (the *base curve* of the parametrization), and $\mathbf{v}\colon I \to S^2$ is a smooth map.

Definition 7.P.1. A ruled surface S parametrized in ruled form is *non-cylindrical* if $\mathbf{v}'(t) \neq 0$ for all $t \in I$. A ruled surface S is *developable* if its tangent planes are constant along the points of every generator.

Problem 7.1. *Let S be a ruled surface, with local parametrization in ruled form $\varphi(t,v) = \sigma(t) + v\,\mathbf{v}(t)$, with the basis curve σ is parametrized by arc length. Show that S is developable if and only if*

$$\det |\dot\sigma \quad \mathbf{v} \quad \mathbf{v}'| \equiv 0\,. \tag{7.12}$$

Solution. The tangent plane to S at $p_0 = \varphi(t_0, v_0)$ is generated by

$$\varphi_t = \dot\sigma(t_0) + v_0\mathbf{v}'(t_0) \qquad \text{and} \qquad \varphi_v = \mathbf{v}(t_0)\,.$$

Note that $\|\mathbf{v}\| \equiv 1$ implies $\mathbf{v}'(t) \perp \mathbf{v}(t)$ for all $t \in I$; in particular, $\mathbf{v}'(t) \wedge \mathbf{v}(t) = O$ if and only if $\mathbf{v}'(t) = O$.

If $\mathbf{v}'(t_0) = O$ or $\dot{\sigma}(t_0) \wedge \mathbf{v}(t_0) = O$ then the determinant (7.12) vanishes automatically in t_0 and the tangent planes along the generator passing through p_0 are automatically constant, and so there is nothing to prove.

Assume then $\mathbf{v}'(t_0), \dot{\sigma}(t_0) \wedge \mathbf{v}(t_0) \neq O$. If the determinant in (7.12) vanishes in t_0 then $\dot{\sigma}(t_0)$ can be written as a linear combination of $\mathbf{v}(t_0)$ and $\mathbf{v}'(t_0)$, and thus the the tangent planes along the generator passing through p_0 are constant. Conversely, if the tangent planes along the generator passing through p_0 are constant then the direction of the vector $(\dot{\sigma}(t_0) + v\mathbf{v}'(t_0)) \wedge \mathbf{v}(t_0)$ is independent of v. Letting v tend to 0 we see that this direction is the direction of $\dot{\sigma}(t_0) \wedge \mathbf{v}(t_0)$; hence $\{\dot{\sigma}(t_0), \mathbf{v}(t_0)\}$ is a basis of all these tangent planes. But this can happen only if $\mathbf{v}'(t_0)$ can be expressed as linear combination of $\dot{\sigma}(t_0)$ and $\mathbf{v}(t_0)$, and we are done.   □

Problem 7.2. *Let $\sigma : (a, b) \to \mathbb{R}^3$ be a regular curve of class C^∞, parametrized by arc length. Let $S \subset \mathbb{R}^3$ be a ruled regular surface, admitting a parametrization $\varphi : (a, b) \times \mathbb{R} \to S$ in ruled form with base curve σ.*

(i) *Show that if $\mathbf{v}(t)$ and $\mathbf{v}'(t)$ are linearly dependent for all $t \in (a, b)$ then the generators of S are parallel and S is contained in a plane or in a cylinder.*

(ii) *Prove that if S is developable and non-cylindrical then $\dot{\sigma}$ is always a linear combination of $\mathbf{v}$ and $\mathbf{v}'$.*

Solution. (i) The vectors $\mathbf{v}$ and $\mathbf{v}'$ are orthogonal, since $\|\mathbf{v}\| \equiv 1$; so the only way they can be everywhere linearly dependent is $\mathbf{v}' \equiv O$. Thus $\mathbf{v}$ is constant, and this means that the generators are parallel.

(ii) As in (i), S non-cylindrical implies that $\mathbf{v}(t)$ and $\mathbf{v}'(t)$ are always linearly independent. The assertion immediately follows because, by Problem 7.1, if S is developable then the vectors $\dot{\sigma}$, $\mathbf{v}$ and $\mathbf{v}'$ are always linearly dependent.   □

Problem 7.3. *Let $S \subset \mathbb{R}^3$ be a compact surface with Gaussian curvature K everywhere positive, and assume that the absolute value $|H|$ of the mean curvature is constant. Prove that S is a sphere.*

Solution. Since the Gaussian curvature is always positive the absolute value of the mean curvature is different from zero everywhere; by Problem 4.16 it follows that S is orientable. Choose then an orientation so that $H \equiv H_0 > 0$, and denote by $k_1 \leq k_2$ the principal curvatures of S. Being S compact and k_1, k_2 continuous, we can find a maximum point $p \in S$ for k_2. From $k_1 + k_2 \equiv 2H_0$ we deduce that p is a minimum point for k_1; hence p is an umbilical point, by Proposition 7.1.1. Then

$$\forall q \in S \qquad k_2(q) \leq k_2(p) = k_1(p) \leq k_1(q) \leq k_2(q) ,$$

and thus all points of S are umbilical. Problem 4.3 shows then that S is contained in a sphere; and the argument already used at the end of the proof of Theorem 7.1.2 implies that S is a sphere.   □

Exercises

RIGIDITY

7.1. Let $S \subset \mathbb{R}^3$ be a compact surface with Gaussian curvature K everywhere positive, and assume it to be oriented (as we may by Exercise 4.33). Prove that if:

(i) the quotient H/K between the mean curvature and the Gaussian curvature is constant; or
(ii) a principal curvature (k_1 or k_2) is constant; or
(iii) there is a decreasing function $h \colon \mathbb{R} \to \mathbb{R}$ such that $k_2 = h(k_1)$,

then S is a sphere.

7.2. Let S be a regular surface in $\mathbb{R}^3$ closed in $\mathbb{R}^3$, with constant Gaussian and mean curvatures and with no umbilical points. Show that S is a right circular cylinder.

7.3. Let S be a orientable closed regular surface in $\mathbb{R}^3$. Prove that it is possible to give S an orientation such that $Q_p = -I_p$ for all $p \in S$ if and only if S is a sphere with radius 1.

THIRD FUNDAMENTAL FORM

Definition 7.E.1. Let $S \subset \mathbb{R}^3$ be an oriented surface in $\mathbb{R}^3$ with Gauss map $N \colon S \to S^2$. The *third fundamental form* of S is the map $III_p \colon T_p S \times T_p S \to \mathbb{R}$ given by

$$III_p(v, w) = \langle dN_p(v), dN_p(w) \rangle .$$

7.4. Let $S \subset \mathbb{R}^3$ be an oriented surface $\mathbb{R}^3$ having Gauss map $N \colon S \to S^2$.

(i) Prove that III_p is a symmetric bilinear form on $T_p S$ independent on the choice of N.
(ii) Prove that

$$III_p - 2H(p)Q_p + K(p)I_p \equiv O .$$

7.5. Let S be a closed surface in $\mathbb{R}^3$. Prove that the first and third fundamental forms of S are equal if and only if S is a sphere with radius 1.

7.6. Let S be an orientable closed surface in $\mathbb{R}^3$. Prove that if it is possible to give S an orientation such that the second and third fundamental form of S coincide then S is a sphere with radius 1, or a plane, or a right circular cylinder with radius 1.

DEVELOPABLE SURFACES

7.7. Prove that a cylinder is a developable ruled surface.

7.8. Show that the regular part of the tangent surface to a curve of class C^∞ is developable (see also Problem 4.20).

7.9. Consider a parametrization $\varphi\colon U \to S$ of a surface S of the form

$$\varphi(x,y) = a(y) + xb(y) \,,$$

where a and b are suitable maps with values in $\mathbb{R}^3$ such that $b'(y) = h(y)a'(y)$ for some real-valued function h. Prove that the Gaussian curvature of $\varphi(U)$ is zero everywhere. Which conditions have a and b to satisfy for $\varphi(U)$ not to consist only of planar points?

7.10. Determine a parametrization in ruled form (see Exercise 4.69) of the ruled surface S having as base curve the curve $\sigma\colon(0,+\infty) \to \mathbb{R}^3$ parametrized by $\sigma(t) = (t, t^2, t^3)$ and generators parallel to $\mathbf{v} = (\cos u, \sin u, 0)$. Is the surface S developable?

7.11. Show that the normal surface (see Exercise 4.73) to a curve σ is developable if and only if the curve is plane.

7.12. Consider a curve σ of class C^∞ in a regular surface S in $\mathbb{R}^3$, and let S_1 be the surface described by the union of the affine normal lines to S passing through the points of the support of σ. Show that S_1 is developable if and only if σ is a line of curvature of S.

Supplementary material

7.4 Positive Gaussian curvature

In this section, we are going to study in detail the closed surfaces in $\mathbb{R}^3$ having positive Gaussian curvature. In particular, we shall prove that they are always diffeomorphic to either a sphere or a convex open subset of the plane (Hadamard's and Stoker's theorems).

Let us begin with some general remarks. Let $S \subset \mathbb{R}^3$ be a closed surface. In the supplementary material of Chapter 4 we proved that S is orientable (Theorem 4.7.15) and that $\mathbb{R}^3 \setminus S$ consists exactly of two connected components (Theorem 4.8.4). If we fix a Gauss map $N\colon S \to S^2$, Theorem 4.8.2 tells us further that there exists a unique component of $\mathbb{R}^3 \setminus S$ that contains points of the form $p + tN(p)$, with $p \in S$ and $0 < t < \varepsilon(p)$, where $\varepsilon\colon S \to \mathbb{R}^+$ is a continuous function such that $N_S(\varepsilon)$ is a tubular neighborhood of S. So we may give the following:

Definition 7.4.1. Let $S \subset \mathbb{R}^3$ be an oriented closed surface with Gauss map $N \colon S \to S^2$. Then the *interior* of S is the only component $\mathbb{R}^3 \setminus S$ containing points of the form $p + tN(p)$ with $p \in S$ and $0 < t < \varepsilon(p)$, where $\varepsilon \colon S \to \mathbb{R}^+$ is a continuous function such that $N_S(\varepsilon)$ is a tubular neighborhood of S.

Suppose now that the Gaussian curvature of S is positive everywhere. In particular, the principal curvatures of S are nowhere zero; as a consequence, there is a Gauss map $N \colon S \to S^2$ with respect to which the principal curvatures of S are positive everywhere. From now on, *we shall always assume that the surfaces with Gaussian curvature positive everywhere are oriented with the Gauss map that makes all the principal curvatures positive.*

The first goal of this section is to show that the hypothesis $K > 0$ implies the convexity of the interior of S and (in a sense to be made precise) of S itself. But let us first introduce the convexity notions we are interested in.

Definition 7.4.2. Given two points x, $y \in \mathbb{R}^n$, we shall denote by $[x, y] \subset \mathbb{R}^n$ the closed line segment from x to y, and by $]x, y[\subset \mathbb{R}^n$ the open line segment from x to y. A subset $K \subseteq \mathbb{R}^n$ is *convex* if $[x, y] \subseteq K$ for all x, $y \in K$; *strictly convex* if $]x, y[$ is contained in the interior of K for all x, $y \in K$.

Remark 7.4.3. Clearly, every strictly convex set is convex, and its interior is convex. Moreover, every convex open set is trivially strictly convex, so the notion of strict convexity is only interesting for non-open sets. Finally, it is easy to verify (exercise) that the closure of a convex set is convex.

It is apparent that a surface cannot be convex in this sense unless it is contained in a plane. For this reason we introduce another notion of convexity for surfaces, which will be useful in this section.

Definition 7.4.4. Let $S \subset \mathbb{R}^3$ be a surface, and $p \in S$. The *affine tangent plane* to S at p is the plane $H_pS = p + T_pS$ parallel to T_pS and passing through p. If S is oriented and $N \colon S \to S^2$ is the corresponding Gauss map, we shall denote by $H_p^{\pm}S \subset \mathbb{R}^3$ the open half-space determined by H_pS and containing $p \pm N(p)$.

Definition 7.4.5. A surface $S \subset \mathbb{R}^3$ is *convex* if for all $p \in S$ it is contained in one of the two closed half-spaces determined by H_pS; and *strictly convex* if $S \setminus \{p\}$ is contained in one of the two open half-spaces determined by H_pS for any $p \in S$.

The result that will relate the positivity of the Gaussian curvature with the various notions of convexity is the following:

Lemma 7.4.6. *Let $S \subset \mathbb{R}^3$ be a closed surface, and $p \in S$ with $K(p) > 0$. Let $N \colon S \to S^2$ be the Gauss map of S such that the principal curvatures are positive in a neighborhood of p, and $\Omega \subset \mathbb{R}^3$ the interior of S. Let $h_p \colon \mathbb{R}^3 \to \mathbb{R}$ be the function*

$$h_p(x) = \langle x - p, N(p) \rangle \,,$$

so that $H_p S = h_p^{-1}(0)$. Then there is a neighborhood $V_p \subset \mathbb{R}^3$ of p in $\mathbb{R}^3$ such that h_p is strictly positive in $(\overline{\Omega} \setminus \{p\}) \cap V_p$; moreover, no line segment $[x, y] \subset V_p \cap \overline{\Omega}$ is such that $p \in]x, y[$.

Proof. For every curve $\sigma \colon (-\varepsilon, \varepsilon) \to S$ with $\sigma(0) = p$ and $\sigma'(0) = v \in T_p S$ we have

$$\mathrm{d}(h_p)_p(v) = \langle \sigma'(0), N(p) \rangle = 0 \quad \text{and} \quad (h_p \circ \sigma)''(0) = \langle \sigma''(0), N(p) \rangle = Q_p(v) \,,$$

by (4.8). Since Q_p is strictly positive definite by the choice of N, it follows that p is a strict local minimum for $h_p|_S$, and so there exists a neighborhood $V_p \subset \mathbb{R}^3$ of p such that $h_p(q) > 0$ for all $q \in (S \setminus \{p\}) \cap V_p$.

Let $N_S(\varepsilon) \subset \mathbb{R}^3$ be a tubular neighborhood of S, and $\pi \colon N_S(\varepsilon) \to S$ and $h \colon N_S(\varepsilon) \to \mathbb{R}$ the maps defined in Theorem 4.8.2; in particular, $h(x) > 0$ for all $x \in \Omega \cap N_S(\varepsilon)$. Up to restricting V_p, we may assume that $V_p = \pi^{-1}(V_p \cap S) \subset N_S(\varepsilon)$, and that $\langle N(q), N(p) \rangle > 0$ for all $q \in V_p \cap S$. So for all $x \in V_p \cap \Omega$ we have

$$h_p(x) = h_p\big(\pi(x)\big) + h(x)\langle N\big(\pi(x)\big), N(p) \rangle > 0 \,;$$

hence we have $h_p(q) > 0$ for all $q \in (\overline{\Omega} \setminus \{p\}) \cap V_p$.

Finally, suppose there exists a line segment $[x, y] \subset V_p \cap \overline{\Omega}$ with $p \in]x, y[$. Let $\sigma \colon [0, 1] \to \mathbb{R}^3$ be its parametrization given by $\sigma(t) = x + t(y - x)$, and let $t_0 \in (0, 1)$ be such that $\sigma(t_0) = p$; in particular,

$$h_p\big(\sigma(t)\big) = (t - t_0)\langle y - x, N(p) \rangle \,.$$

So either $h_p \circ \sigma$ is zero everywhere or it changes sign in $[0, 1]$; but, from the fact that $[x, y] \subset \overline{\Omega} \cap V_p$ we deduce that $h_p\big(\sigma(t)\big) > 0$ for all $t \neq t_0$, a contradiction. $\qquad\Box$

Then:

Proposition 7.4.7. *Let $S \subset \mathbb{R}^3$ be a closed surface with Gaussian curvature positive everywhere, and let $\Omega \subset \mathbb{R}^3$ be its interior. Then:*

(i) $\overline{\Omega}$ *is strictly convex;*

(ii) S *is strictly convex;*

(iii) $\overline{\Omega} = \bigcap_{p \in S} \overline{H_p^+ S}$ *and* $\Omega = \bigcap_{p \in S} H_p^+ S$.

Proof. We begin by proving that Ω is convex, that is, that the subset $A = \{(x, y) \in \Omega \times \Omega \mid [x, y] \subset \Omega\}$ of $\Omega \times \Omega$ coincides with the whole $\Omega \times \Omega$. Since Ω is connected, if A were not equal to $\Omega \times \Omega$ we could find $(x_0, y_0) \in \partial A$. So there must exists two sequences $\{x_n\}, \{y_n\} \subset \Omega$ such that $x_n \to x_0$, $y_n \to y_0$, and $[x_n, y_n] \subset \Omega$ for all $n \geq 1$ but $[x_0, y_0] \not\subset \Omega$. Clearly, we have $[x_0, y_0] \subset \overline{\Omega} = \Omega \cup S$; choose a point $p \in]x_0, y_0[\cap S$, and let $V_p \subset \mathbb{R}^3$ be the neighborhood of p given by the previous lemma. Then by intersecting $[x_0, y_0]$ with V_p we would find a segment contained in $\overline{\Omega} \cap V_p$ and containing p in its interior, against Lemma 7.4.6.

So Ω is convex, and hence $\overline{\Omega}$ is too. If $\overline{\Omega}$ were not strictly convex, there should exist two points $x_0,\, y_0 \in S$ such that $]x_0, y_0[\, \cap\, S \neq \varnothing$, and we would again contradict Lemma 7.4.6.

Fix now $p \in S$, and suppose by contradiction that there exists $q \in H_p S \cap \overline{\Omega}$ different from p. Then the whole segment $]p, q[$ should be contained in $H_p S \cap \Omega$. But Lemma 7.4.6 implies that the function h_p is strictly positive in $V_p \cap \Omega$, while it is zero everywhere on $H_p S$, a contradiction.

Hence $\overline{\Omega} \setminus \{p\}$ is connected and does not intersect $H_p S = h_p^{-1}(0)$; so h_p cannot change sign in $\overline{\Omega} \setminus \{p\}$. Since it is positive near p, it is positive everywhere, and thus $\overline{\Omega} \setminus \{p\} \subset H_p^+ S$. Since this holds for every $p \in S$, we have proved, in particular, that S is strictly convex, that $\overline{\Omega} \subseteq \bigcap_{p \in S} \overline{H_p^+ S}$ and that $\Omega \subseteq \bigcap_{p \in S} H_p^+ S$.

Suppose by contradiction that there exists $q_0 \in \bigcap_{p \in S} \overline{H_p^+ S} \setminus \overline{\Omega}$, and choose a point $p_0 \in \overline{\Omega}$ nearest to q_0; it exists because $\overline{\Omega}$ is closed. Clearly, $p_0 \in \partial\overline{\Omega} = S$. Let $\sigma\colon (-\varepsilon, \varepsilon) \to S$ be a curve with $\sigma(0) = p$, and set $f(t) = \|\sigma(t) - q_0\|^2$. Then we have

$$2\langle \sigma'(0), p_0 - q_0 \rangle = f'(0) = 0 \; ;$$

since this holds for every curve σ in S passing through p, it follows that $p_0 - q_0$ is orthogonal to $T_{p_0} S$, that is, $q_0 - p_0 = \lambda N(p_0)$ for some $\lambda \in \mathbb{R}^*$. Now, since $q_0 \in \overline{H_{p_0}^+ S}$, we have

$$0 \leq h_{p_0}(q_0) = \lambda \, ,$$

and so $h_{p_0}\big(p_0 + t(q_0 - p_0)\big) = t\lambda \geq 0$ for all $t > 0$. On the other hand, $p_0 + t(q_0 - p_0) \notin \overline{\Omega}$ for all $t \in (0, 1]$, as p_0 is the point of $\overline{\Omega}$ closest to q_0; hence $p_0 + t(q_0 - p_0) = p_0 - sN(p_0)$ with $s > 0$ when $t > 0$ is small enough, and so $h_{p_0}\big(p_0 + t(q_0 - p_0)\big) < 0$, contradiction.

So $\overline{\Omega} = \bigcap_{p \in S} \overline{H_p^+ S} \supset \bigcap_{p \in S} H_p^+ S$. But $\partial\Omega \cap \bigcap_{p \in S} H_p^+ S = \varnothing$; so $\Omega \supseteq \bigcap_{p \in S} H_p^+ S$, and we are done. $\qquad\square$

A consequence of this proposition is that the Gauss map of a surface with positive Gaussian curvature behaves particularly well:

Proposition 7.4.8. *Let $S \subset \mathbb{R}^3$ be a closed surface with Gaussian curvature positive everywhere. Then the Gauss map $N\colon S \to S^2$ is a globally injective local diffeomorphism.*

Proof. Since the Gaussian curvature is exactly the determinant of the differential of N, the inverse function theorem for surfaces (Corollary 3.4.28) tells us that the Gauss map is a local diffeomorphism.

Suppose by contradiction that there exist two points $p,\, q \in S$ such that $N(p) = N(q)$; in particular, $H_p S$ e $H_q S$ are parallel. By the previous proposition, we know that $S \subset \overline{H_p^+ S} \cap \overline{H_q^+ S}$; since $H_p S$ and $H_q S$ are parallel, this intersection has to coincide with one of the two half-spaces, say $\overline{H_p^+ S}$. But from this follows that $q \in H_q S \subset \overline{H_p^- S}$; so $q \in \overline{H_p^+ S} \cap \overline{H_p^- S} = H_p S$, and hence $H_p S = H_q S$. From the strict convexity of S we deduce that

$$\{p\} = S \cap H_p S = S \cap H_q S = \{q\} ,$$

that is $p = q$, and N is injective.   □

So we have obtained *Hadamard's theorem*:

Corollary 7.4.9 (Hadamard). *Let $S \subset \mathbb{R}^3$ be a compact surface with Gaussian curvature positive everywhere. Then the Gauss map $N: S \to S^2$ is a diffeomorphism.*

Proof. By the previous proposition, N is a globally injective local diffeomorphism; in particular, $N(S)$ is open in S^2. But S is compact; so $N(S)$ is closed in S^2 too. As S^2 is connected, this implies that $N(S) = S^2$. So N is a bijective local diffeomorphism, that is, a diffeomorphism.   □

So every compact surface with Gaussian curvature positive everywhere is diffeomorphic to a sphere. To find the geometry of the non-compact ones, we have still some work to do. Let us begin with:

Lemma 7.4.10. *Let $S \subset \mathbb{R}^3$ be a closed surface with Gaussian curvature positive everywhere, and let $\Omega \subset \mathbb{R}^3$ be its interior. Then:*

(i) *if $\overline{\Omega}$ contains a closed half-line ℓ then Ω contains all the open half-lines parallel to ℓ passing through points of $\overline{\Omega}$;*
(ii) *$\overline{\Omega}$ does not contain whole straight lines.*

Proof. (i) Suppose that there exist $x \in \overline{\Omega}$ and a non-zero $v \in \mathbb{R}^3$ such that $x + tv \in \overline{\Omega}$ for all $t \geq 0$. If $y \in \overline{\Omega}$ strict convexity implies that $]y, x + tv[\subset \Omega$ for all $t \geq 0$. Passing to the limit as $t \to +\infty$ we find that the open half-line through y parallel to v lies in $\overline{\Omega}$, and cannot intersect $\partial\Omega = S$ without contradicting Lemma 7.4.6.

(ii) Suppose that $\overline{\Omega}$ contains a line ℓ. Then part (i) implies that the line through any $p \in S$ parallel to v is contained in $\overline{\Omega}$, and this again contradicts Lemma 7.4.6.   □

We are now ready to prove *Stoker's theorem*:

Theorem 7.4.11 (Stoker). *Every non-compact closed surface $S \subset \mathbb{R}^3$ with Gaussian curvature positive everywhere is a graph over a convex open subset of a plane. In particular, S is diffeomorphic to a convex plane open set.*

Proof. Since S is not compact, it is not bounded. So, for every point $q \in S$ we may find a sequence $\{q_n\} \subset S$ such that $\|q_n - q\| \to +\infty$; up to extracting a subsequence, we may also assume that $(q_n - q)/\|q_n - q\| \to v \in S^2$. Let $\Omega \subset \mathbb{R}^3$ be the interior of S. Since $\overline{\Omega}$ is convex, we have $[q, q_n] \subset \overline{\Omega}$ for all n; by letting n approach infinity, we see that $\overline{\Omega}$ has to contain the open half-line ℓ_q^+ passing through q and parallel to v. So the previous lemma tells us that Ω contains the open half-line ℓ_p^+ passing through p and parallel to v for all $p \in S$.

Denote by ℓ_p the line passing through p parallel to v. First of all, if ℓ_p intersects S in a point p' different from p we would have $p' \in \ell_p^+$ or $p \in \ell_{p'}^+$, against Lemma 7.4.6. So $\ell_p \cap S = \{p\}$ for all $p \in S$. Moreover, ℓ_p cannot be tangent to S; if it were so, necessarily at p, we would have $\ell_p \subset H_p S$ and $\ell_p^+ \subset \Omega$, against Proposition 7.4.7.(iii). So $v \notin T_p S$ for all $p \in S$.

Let $H \subset \mathbb{R}^3$ be the plane passing through the origin and orthogonal to v. What we have seen so far tells us that every line orthogonal to H intersects S in at most one point; so the orthogonal projection $\pi \colon \mathbb{R}^3 \to H$ restricted to S is injective. In particular, S is a graph over the image $\pi(S) \subseteq H$. Now, for every $p \in S$ and every non-zero $w \in T_p S$ we have

$$\mathrm{d}\pi_p(w) = w - \langle w, v \rangle v \neq O ,$$

since $v \notin T_p S$. Therefore $\pi|_S \colon S \to H$ is a globally injective local diffeomorphism, that is, a diffeomorphism between S and its image $\pi(S)$, which has to be open in H. Finally, $\pi(S)$ is convex. Indeed, let $\pi(p), \pi(q) \in \pi(S)$. We know that the segment $]p, q[$ lies in Ω, and that for all $x \in]p, q[$ the open half-line through x parallel to v lies in Ω. But we also know that $\overline{\Omega}$ does not contain lines; hence the line through x parallel to v has to intersect S, and so the whole segment $[\pi(p), \pi(q)]$ is contained in $\pi(S)$. □

In particular, *every non-compact closed surface with positive Gaussian curvature, being diffeomorphic to a convex open subset of the plane, is homeomorphic to a plane.*

In Remark 7.1.4 we claimed that closed surfaces with Gaussian curvature bounded from below by a positive constant are necessarily compact. To prove this we need two more lemmas. The first one is a general one:

Lemma 7.4.12. *Let $S \subset \mathbb{R}^3$ be a surface that is a graph over an open subset Ω of a plane $H \subset \mathbb{R}^3$ passing through the origin, and denote by $\pi \colon \mathbb{R}^3 \to H$ the orthogonal projection. Then for every regular region $R \subset S$ we have*

$$\mathrm{Area}(R) \geq \mathrm{Area}\big(\pi(R)\big) .$$

Proof. Up to a rotation in $\mathbb{R}^3$ we may assume that H is the xy-plane, and that S is the graph of a function $h \colon \Omega \to \mathbb{R}$. Then Theorem 4.2.6 and Example 4.1.8 imply

$$\mathrm{Area}(R) = \int_{\pi(R)} \sqrt{1 + \|\nabla h\|^2}\, \mathrm{d}x\, \mathrm{d}y \geq \int_{\pi(R)} \mathrm{d}x\, \mathrm{d}y = \mathrm{Area}\big(\pi(R)\big) ,$$

as required. □

The second lemma deals instead with surfaces having positive curvature:

Lemma 7.4.13. *Let* $S \subset \mathbb{R}^3$ *be a closed surface with positive Gaussian curvature, let* $\Omega \subset \mathbb{R}^3$ *be its interior, and let* $H \subset \mathbb{R}^3$ *be a plane such that* $H \cap \Omega \neq \varnothing$. *Then there exists an open subset* $V \subseteq S$ *of* S *that is a graph over* $H \cap \Omega$.

Proof. Let $U = H \cap \Omega$, and choose a versor $v \in S^2$ orthogonal to H. Suppose that there exists $x \in U$ such that the half-line ℓ_x^+ through x parallel to v does not intersect S. Then ℓ_x^+ must be contained in one of the two connected components of $\mathbb{R}^3 \setminus S$; since $x \in \Omega$, we find $\ell_x^+ \subset \Omega$. Hence Lemma 7.4.10 implies that $\ell_y^+ \subset \Omega$ for all $y \in U$, where ℓ_y^+ is the half-line through y parallel to v. But in this case, since Ω cannot contain whole straight lines, the opposite half-line ℓ_y^- has to intersect S for all $y \in U$.

So, up to swapping v with $-v$, we may suppose that all half-lines ℓ_x^+ through points $x \in U$ intersect S; then the assertion follows arguing as in the final part of the proof of Theorem 7.4.11. $\square$

The promised result about compactness of closed surfaces with Gaussian curvature bounded below by a positive constant is then a consequence of *Bonnet's theorem,* that we can now prove:

Theorem 7.4.14 (Bonnet). *Let* $S \subset \mathbb{R}^3$ *be a closed surface such that*

$$K_0 = \inf_{p \in S} K(p) > 0 \,.$$

Then S *is compact, diffeomorphic to a sphere, and moreover*

$$\text{Area}(S) \leq \frac{4\pi}{K_0} \,. \tag{7.13}$$

Proof. Proposition 7.4.8 tells us that the Gauss map $N \colon S \to S^2$ is a diffeomorphism with an open subset of S^2. In particular, for every regular region $R \subseteq S$ contained in the image of a local parametrization we have

$$K_0 \, \text{Area}(R) = \int_R K_0 \, d\nu \leq \int_R K \, d\nu = \int_R |\det dN| \, d\nu = \text{Area}\big(N(R)\big) \leq 4\pi \,,$$

where we have used Proposition 4.2.10. So we get

$$\text{Area}(R) \leq \frac{4\pi}{K_0} \tag{7.14}$$

for every regular region $R \subseteq S$ contained in the image of a local parametrization.

Suppose now by contradiction that S is not compact. Then Theorem 7.4.11 tells us that S is a graph over a convex subset of a plane $H_0 \subset \mathbb{R}^3$, orthogonal to a versor $v \in S^2$. In particular, if $\Omega \subset \mathbb{R}^3$ is the interior of S, we may choose

v so that for all $x \in \Omega$ the half-line ℓ_x^+ through x parallel to v is contained in Ω. Thus if $[x_0, y_0] \subset \Omega$ and $n \in \mathbb{N}^*$ then the set

$$Q_n = \{x + tv \mid x \in [x_0, y_0],\, 0 \le t \le n\} \subset \Omega$$

is a rectangle in H containing $]x_0, y_0[$ and parallel to v. Lemma 7.4.13 tells us that there exists a regular region $R_n \subset S$ that is a graph over Q_n. Hence Lemma 7.4.12 implies

$$\frac{4\pi}{K_0} \ge \text{Area}(R_n) \ge \text{Area}(Q_n) = n\|y_0 - x_0\|$$

for all $n \in \mathbb{N}^*$, impossible.

So S is compact, and hence (Corollary 7.4.9) diffeomorphic to a sphere. In particular, S is an increasing union of regular regions contained in the image of a local parametrization, and (7.13) follows from (7.14). $\qquad\square$

Remark 7.4.15. Inequality (7.13) is the best possible: indeed, for the unit sphere S^2 we have $\text{Area}(S^2) = 4\pi$ and $K \equiv 1$.

Remark 7.4.16. There is a completely different proof of Bonnet's theorem (see [4, p. 352]), based on a study of the behavior of geodesics, giving an estimate of the diameter (with respect to the intrinsic distance) of the surface: $\text{diam}(S) \le \pi/\sqrt{K_0}$.

Remark 7.4.17. In case you were wondering, there are non-compact (but still closed in $\mathbb{R}^3$) surfaces with positive Gaussian curvature. A typical example is the elliptic paraboloid, which is the graph of the function $h\colon \mathbb{R}^2 \to \mathbb{R}$ given by $h(x_1, x_2) = x_1^2 + x_2^2$. Indeed, Example 4.5.19 implies

$$K = \frac{4}{\left(1 + 4(x_1^2 + x_2^2)\right)^2} > 0\,.$$

7.5 Covering maps

To study surfaces with non-positive Gaussian curvature we need some preliminary topological notions which, for the sake of simplicity, we shall directly describe for surfaces, only proving the results we shall need later on. You may find a more complete exposition of the theory of covering maps in, e. g., [11].

Let $F\colon \tilde{S} \to S$ be a local diffeomorphism between surfaces. In particular, this means that every $\tilde{p} \in \tilde{S}$ has an open neighborhood $\tilde{U} \subseteq \tilde{S}$ such that $F|_{\tilde{U}}$ is a diffeomorphism with its image. As already remarked before, this does not say a lot about the structure of S. Basically, two things may happen. The first, more evident, is that F might not be surjective, and thus outside the image of F there is no relation between S and $\tilde{S}$. The second one is that, in a sense, there might exist a connected component $\hat{U}$ of the inverse image of an open subset U of $F(S)$, albeit small, lacking some points, in the sense that $F(\hat{U}) \ne U$.

Example 7.5.1. Let $S = \{(x, y, z) \in \mathbb{R}^3 \mid x^2 + y^2 = 1\} \subset \mathbb{R}^3$ be the right cylinder of radius 1, and $F: \mathbb{R}^2 \to S$ the local diffeomorphism given by

$$F(x_1, x_2) = (\cos x_1, \sin x_1, x_2) \ .$$

If $\tilde{S} = (0, 3\pi) \times (-1, 1)$, then $F|_{\tilde{S}}$ is a non-surjective local diffeomorphism. If, on the other hand, $\tilde{S}_1 = (0, 3\pi) \times \mathbb{R}$, then $F|_{\tilde{S}_1}$ is a surjective local diffeomorphism, but the second kind of problem appears. Indeed, given $\varepsilon > 0$, set $U_\varepsilon = F\big((\pi - \varepsilon, \pi + \varepsilon) \times (-\varepsilon, \varepsilon)\big)$. Then

$$F|_{\tilde{S}_1}^{-1}(U_\varepsilon) = (\pi - \varepsilon, \pi + \varepsilon) \times (-\varepsilon, \varepsilon) \cup (3\pi - \varepsilon, 3\pi) \times (-\varepsilon, \varepsilon)$$

but $F\big((3\pi - \varepsilon, 3\pi) \times (-\varepsilon, \varepsilon)\big) \neq U_\varepsilon$.

The following definition introduces a particular kind of local diffeomorphism for which these problems do not arise.

Definition 7.5.2. A map $F: \tilde{S} \to S$ of class C^∞ between surfaces is a *smooth covering map* if every $p \in S$ has a connected open neighborhood $U_p \subseteq S$ such that the connected components $\{\tilde{U}_\alpha\}$ of $F^{-1}(U_p)$ are such that $F|_{\tilde{U}_\alpha}$ is a diffeomorphism between $\tilde{U}_\alpha$ and U for every α. The neighborhoods U_p are said to be *evenly covered*, the connected components $\tilde{U}_\alpha$ are called *sheets* over U, and the set $F^{-1}(p)$ is the *fiber* of p.

Remark 7.5.3. If we replace S and $\tilde{S}$ by two topological spaces, F by a continuous map, and we ask that $F|_{\tilde{U}_\alpha}$ is a homeomorphism between $\tilde{U}_\alpha$ and U_p, we get the general topological definition of a covering map. For instance, it is easy to see (e.g., recalling the beginning of the proof of Proposition 2.1.4) that the map $\pi: \mathbb{R} \to S^1$ given by $\pi(x) = (\cos x, \sin x)$ is a topological covering map. Indeed, several of the results we are going to discuss in this section are generalizations of facts we have seen in Section 2.1.

Example 7.5.4. The map $F: \mathbb{R}^2 \to S$ introduced in Example 7.5.1 is a smooth covering map. Indeed, for all $p = F(x_1, x_2)$ set $U_p = F\big((x_1 - \pi, x_1 + \pi) \times \mathbb{R}\big)$. Then

$$F^{-1}(U_p) = \bigcup_{k \in \mathbb{Z}} (x_1 + (2k - 1)\pi, x_1 + (2k + 1)\pi) \times \mathbb{R} \ ,$$

and it is easy to verify that $F|_{(x_1 + (2k-1)\pi, x_1 + (2k+1)\pi) \times \mathbb{R}}$ is a diffeomorphism between $(x_1 + (2k - 1)\pi, x_1 + (2k + 1)\pi) \times \mathbb{R}$ and U_p for all $k \in \mathbb{Z}$.

A smooth covering map is clearly a surjective local diffeomorphism, and the evenly covered neighborhoods are diffeomorphic to each component of their inverse image; so we are justified in saying that a smooth covering map F from the surface $\tilde{S}$ over the surface S provides us with a global way of comparing the local structures of $\tilde{S}$ and S.

The main goal of this section is to obtain a characterization of those local diffeomorphisms that are smooth covering maps. As we shall see, what is important is the possibility of lifting curves.

Definition 7.5.5. Let $F: \tilde{S} \to S$ be a smooth map between surfaces, and X a topological space. A *lift* of a continuous map $\psi: X \to S$ is a continuous map $\tilde{\psi}: X \to \tilde{S}$ such that $\psi = F \circ \tilde{\psi}$. We shall say that F has *the continuous (respectively, C^1) lifting property* if every continuous (respectively, piecewise C^1) curve in S has a continuous (respectively, piecewise C^1) lift.

Remark 7.5.6. In the application to surfaces with non-positive Gaussian curvature we have in mind (Proposition 7.6.3) we shall be interested in lifting piecewise C^1 curves only; for this reason, we have decided to introduce the C^1 lifting property (instead of limiting ourselves to the more standard continuous lifting property). However, at the end of this section we shall be able to prove that a local diffeomorphism with the C^1 lifting property also has the continuous lifting property.

Remark 7.5.7. Every lift of a piecewise C^1 curve with respect to a local diffeomorphism is necessarily piecewise C^1 (why?); so a local diffeomorphism with the continuous lifting property has the C^1 lifting property too.

Remark 7.5.8. A map $F: \tilde{S} \to S$ with the C^1 lifting property is necessarily surjective. Indeed, take arbitrarily $p \in F(\tilde{S})$ and $q \in S$. Choose a smooth curve (Corollary 4.7.11) $\sigma: [a, b] \to S$ joining p to q, and let $\tilde{\sigma}: [a, b] \to \tilde{S}$ be a lift of σ; then $q = \sigma(b) = F\big(\tilde{\sigma}(b)\big) \in F(\tilde{S})$, and F is surjective.

An important remark is that lifts (with respect to a local diffeomorphism), if they exist, are essentially unique:

Lemma 7.5.9. *Let $F: \tilde{S} \to S$ be a local diffeomorphism, X a connected Hausdorff topological space, and $\psi: X \to S$ a continuous map. Suppose we have two lifts $\tilde{\psi}, \hat{\psi}: X \to \tilde{S}$ of ψ such that $\tilde{\psi}(x_0) = \hat{\psi}(x_0)$ for some $x_0 \in X$. Then $\tilde{\psi} \equiv \hat{\psi}$.*

Proof. Let $C = \{x \in X \mid \tilde{\psi}(x) = \hat{\psi}(x)\}$. We know that C is a non-empty closed subset of X; to conclude we have just to prove that it is open too.

Take $x_1 \in C$. Since F is a local diffeomorphism, there is an open neighborhood $\tilde{U}$ of $\tilde{\psi}(x_1) = \hat{\psi}(x_1)$ such that $F|_{\tilde{U}}$ is a diffeomorphism between $\tilde{U}$ and $U = F(\tilde{U})$; in particular, U is an open neighborhood of $\psi(x_1) = F\big(\tilde{\psi}(x_1)\big)$. By continuity, there is an open neighborhood $V \subseteq X$ of x_1 such that $\tilde{\psi}(V) \cup \hat{\psi}(V) \subset \tilde{U}$. But then from $F \circ \tilde{\psi}|_V \equiv \psi|_V \equiv F \circ \hat{\psi}|_V$ we deduce that $\tilde{\psi}|_V \equiv F|_{\tilde{U}}^{-1} \circ \psi|_V \equiv \hat{\psi}|_V$; thus $V \subseteq C$, and we are done. □

In the last theorem of this section it will be proved that smooth covering maps can be characterized as the local diffeomorphisms having the C^1 lifting property; to get there, we must first show that smooth covering maps have the C^1 lifting property.

Proposition 7.5.10. *Every smooth covering map between surfaces has the continuous lifting property (and hence the C^1 lifting property too).*

Proof. Let $F\colon \tilde{S} \to S$ be a smooth covering map, and $\sigma\colon [a,b] \to S$ a continuous curve; we have to construct a lift of σ.

By continuity, for all $t \in [a,b]$ there is an open interval $I_t \subseteq [a,b]$ containing t such that $\sigma(I_t)$ lies within an evenly covered neighborhood $U_{\sigma(t)}$ of $\sigma(t)$. Choose a point $\tilde{p}_0 \in F^{-1}\big(\sigma(a)\big)$, and choose the component $\tilde{U}_{\sigma(a)}$ of $F^{-1}(U_{\sigma(a)})$ containing $\tilde{p}_0$. We may then define a lift $\tilde{\sigma}$ of σ over I_a by setting $\tilde{\sigma} = F|_{\tilde{U}_{\sigma(a)}}^{-1} \circ \sigma$.

If $I_a = [a,b]$, we are done. Otherwise, let $[a,t_0) \subset [a,b]$ be the largest interval where a lift $\tilde{\sigma}\colon [a,t_0) \to \tilde{S}$ with $\tilde{\sigma}(a) = \tilde{p}_0$ is defined. Choose $t \in [a,t_0)$ and $\varepsilon > 0$ such that $\sigma([t,t_0+\varepsilon]) \subset U_{\sigma(t_0)}$, and let $\tilde{U}$ be the connected component of $F^{-1}(U_{\sigma(t_0)})$ containing $\tilde{\sigma}(t)$. Then we can define a lift $\hat{\sigma}\colon [t,t_0+\varepsilon] \to \tilde{S}$ of $\sigma|_{[t,t_0+\varepsilon]}$ by setting $\hat{\sigma} = F|_{\tilde{U}}^{-1} \circ \sigma$. By construction, $\hat{\sigma}(t) = \tilde{\sigma}(t)$; therefore Lemma 7.5.9 yields $\hat{\sigma}|_{[t,t_0)} \equiv \tilde{\sigma}|_{[t,t_0)}$. But this means that using $\hat{\sigma}$ we can extend $\tilde{\sigma}$ beyond t_0, contradiction.

The last assertion follows from Remark 7.5.7. $\qquad\square$

In Section 2.1 we have introduced the notion of homotopy between curves (Definition 2.1.9). In the present context we need some more terminology:

Definition 7.5.11. Let σ_0, $\sigma_1\colon [a,b] \to S$ be two continuous curves in a surface S, and $\Psi\colon [0,1] \times [a,b] \to S$ a homotopy between σ_0 and σ_1. If there is a $p \in S$ such that $\Psi(\cdot,a) \equiv p$ (so, in particular, $\sigma_0(a) = p = \sigma_1(a)$), we shall say that Ψ has *origin* p. If, moreover, $\Psi(\cdot,b)$ is constant, we shall say that Ψ has *fixed endpoints*. Finally, we shall say that Ψ is *piecewise* C^1 if all curves $t \mapsto \Psi(s_0,t)$ and $s \mapsto \Psi(s,t_0)$ are piecewise C^1.

It is not difficult to see that homotopies lift too:

Proposition 7.5.12. *Let $F\colon \tilde{S} \to S$ be a local diffeomorphism between surfaces with the continuous (C^1) lift property, and $\Phi\colon [0,1] \times [a,b] \to S$ a (piecewise C^1) homotopy with origin $p_0 \in S$. Then for all $\tilde{p}_0$ in the fiber of p_0 there is a unique lift $\tilde{\Phi}\colon [0,1] \times [a,b] \to \tilde{S}$ of Φ with origin $\tilde{p}_0$.*

Proof. We shall prove the statement for the continuous case; the piecewise C^1 case is completely analogous.

Uniqueness immediately follows from Lemma 7.5.9. To prove existence, let $\tilde{\Psi}\colon [0,1] \times [a,b] \to \tilde{S}$ be given by $\tilde{\Psi}(s,t) = \tilde{\sigma}_s(t)$, where $\tilde{\sigma}_s\colon [a,b] \to \tilde{S}$ is the unique lift of the curve $\sigma_s = \Psi(s,\cdot)$ such that $\tilde{\sigma}_s(a) = \tilde{p}_0$. Clearly, $\tilde{\Psi}$ is a lift of Ψ and $\tilde{\Psi}(\cdot,a) \equiv \tilde{p}_0$; to finish the proof we only need to show that $\tilde{\Psi}$ is continuous.

Take $(s_0,t_0) \in [0,1] \times [a,b]$. Since F is a local diffeomorphism, there is an open neighborhood $\tilde{U}$ of $\tilde{\Psi}(s_0,t_0)$ such that $F|_{\tilde{U}}$ is a diffeomorphism with its image $U = F(\tilde{U})$, an open neighborhood of $\Psi(s_0,t_0)$ in S. Let $Q_0 \subset [0,1] \times [a,b]$ be an open square of side 2ε centered at (s_0,t_0) such that $\Psi(Q_0) \subset U$; if we show that $\tilde{\Psi}|_{Q_0} \equiv F|_{\tilde{U}}^{-1} \circ \Psi$, we have proved that $\tilde{\Psi}$ is continuous at (s_0,t_0), and we are done.

Note that $s \mapsto F|_{\tilde{U}_0}^{-1}\big(\Psi(s, t_0)\big)$ defined in $(s_0 - \varepsilon, s_0 + \varepsilon)$ is a lift of the curve $s \mapsto \Psi(s, t_0)$ starting at $\tilde{\Psi}(s_0, t_0)$; by the uniqueness of the lift we get $\tilde{\Psi}(s, t_0) = F|_{\tilde{U}_0}^{-1}\big(\Psi(s, t_0)\big)$ for all $s \in (s_0 - \varepsilon, s_0 + \varepsilon)$. Analogously, for all $s \in (s_0 - \varepsilon, s_0 + \varepsilon)$ the curve $t \mapsto F|_{\tilde{U}_0}^{-1}\big(\Psi(s, t)\big)$ defined in $(t_0 - \varepsilon, t_0 + \varepsilon)$ is a lift of the curve $t \mapsto \Psi(s, t)$ starting at $\tilde{\Psi}(s, t_0)$; by the uniqueness of the lift we get $\tilde{\Psi}(s, t) = F|_{\tilde{U}_0}^{-1}\big(\Psi(s, t)\big)$ for all $(s, t) \in Q_0$, as required. $\qquad\square$

An important consequence of this result is that the lifts of homotopic curves are homotopic:

Corollary 7.5.13. *Let $F: \tilde{S} \to S$ be a local diffeomorphism between surfaces with the continuous (C^1) lift property, and σ_0, $\sigma_1: [a, b] \to \mathbb{R}$ two (piecewise C^1) curves such that $\sigma_0(a) = p = \sigma_1(a)$ and $\sigma_0(b) = q = \sigma_1(b)$. Having chosen $\tilde{p} \in F^{-1}(p)$, let $\tilde{\sigma}_0$, $\tilde{\sigma}_1: [a, b] \to \tilde{S}$ be the lifts of σ_0 and σ_1 issuing from $\tilde{p}$. Then there exists a (piecewise C^1) homotopy with fixed endpoints between σ_0 and σ_1 if and only if there exists a (piecewise C^1) homotopy with fixed endpoints between $\tilde{\sigma}_0$ and $\tilde{\sigma}_1$.*

Proof. If $\tilde{\Psi}: [0, 1] \times [a, b] \to \tilde{S}$ is a (piecewise C^1) homotopy with fixed endpoints between $\tilde{\sigma}_0$ and $\tilde{\sigma}_1$, then $F \circ \Psi$ is a (piecewise C^1) homotopy with fixed endpoints between σ_0 and σ_1.

Conversely, suppose that $\Psi: [0, 1] \times [a, b] \to S$ is a (piecewise C^1) regular homotopy with fixed endpoints between σ_0 and σ_1, and let $\tilde{\Psi}: [0, 1] \times [a, b] \to \tilde{S}$ be the lift with origin $\tilde{p}$ given by the previous proposition. Since F is a local diffeomorphism, $\tilde{\Psi}$ is piecewise C^1 if Ψ is. By the uniqueness of lift of curves we know that $\tilde{\Psi}$ is a homotopy between $\tilde{\sigma}_0$ and $\tilde{\sigma}_1$. Moreover, $s \mapsto \tilde{\Psi}(s, b)$ has to be a lift of the constant curve $s \mapsto \Psi(s, b) \equiv q$; hence we must have $\tilde{\Psi}(\cdot, b) \equiv \tilde{\sigma}_0(b)$, and so $\tilde{\Psi}$ has fixed endpoints. $\qquad\square$

Homotopies allow us to identify a particular class of surfaces whose covering maps are necessarily trivial.

Definition 7.5.14. *A surface $S \subset \mathbb{R}^3$ is* simply connected *if for every closed curve $\sigma_0: [a, b] \to S$ there is a homotopy with fixed endpoints between σ_0 and the constant curve $\sigma_1 \equiv \sigma_0(a)$.*

Example 7.5.15. Every convex open subset of the plane is simply connected. Indeed, let $U \subseteq \mathbb{R}^2$ be a convex open set, and $\sigma: [a, b] \to U$ a closed curve. Then the map $\Psi: [0, 1] \times [a, b] \to U$ given by

$$\Psi(s, t) = s\sigma(a) + (1 - s)\sigma(t) \tag{7.15}$$

is a homotopy between σ and the constant curve $\sigma_1 \equiv \sigma(a)$. In particular, every surface homeomorphic to a convex open subset of the plane is simply connected.

Remark 7.5.16. Note that if σ is a piecewise C^1 closed curve then the homotopy defined in (7.15) is piecewise C^1 too. In particular, every piecewise C^1 closed curve in a surface *diffeomorphic* to a convex open subset of the plane is *piecewise C^1* homotopic to a constant curve.

So we may finally reap the benefits of our work, and prove that every local diffeomorphism with the C^1 lifting property is a smooth covering map:

Theorem 7.5.17. *Let $F\colon \tilde{S} \to S$ be a local diffeomorphism between surfaces with the C^1 lift property. Then F is a smooth covering map, and so in particular it has the continuous lifting property.*

Proof. Given $p_0 \in S$, let $U \subseteq S$ be the image of a local parametrization at p_0 whose domain is an open disk in the plane; in particular, U is a simply connected, connected, open neighborhood of p_0. We want to prove that U is evenly covered.

Let $F^{-1}(U) = \bigcup \tilde{U}_\alpha$ be the decomposition of $F^{-1}(U)$ into connected components. If we prove that $F|_{\tilde{U}_\alpha}$ is a diffeomorphism between $\tilde{U}_\alpha$ and U we have shown that F is a smooth covering map.

We begin by proving that $F(\tilde{U}_\alpha) = U$. Chosen $p \in U$, let $\sigma\colon [a,b] \to U$ be a piecewise C^1 curve from a point $q \in F(\tilde{U}_\alpha)$ to p. Since F has the C^1 lifting property, there is a lift $\tilde{\sigma}$ of σ issuing from a point $\tilde{q} \in \tilde{U}_\alpha$. Clearly, $\tilde{\sigma}([a,b]) \subset F^{-1}(U)$; being $\tilde{U}_\alpha$ a connected component of $F^{-1}(U)$, the whole support of $\tilde{\sigma}$ is contained in $\tilde{U}_\alpha$. In particular, $p = F\big(\tilde{\sigma}(b)\big) \in F(\tilde{U}_\alpha)$, and $F|_{\tilde{U}_\alpha}$ is surjective.

Thus $\tilde{F}|_{U_\alpha}\colon U_\alpha \to U$ is a local diffeomorphism with the C^1 lifting property; to conclude, we only have to prove that it is injective.

Let $\tilde{p}_1, \tilde{p}_2 \in \tilde{U}_\alpha$ be such that $F(\tilde{p}_1) = F(\tilde{p}_2) = p$. Since $\tilde{U}_\alpha$ is connected, Corollary 4.7.11 yields a piecewise C^1 curve $\tilde{\sigma}_0\colon [a,b] \to \tilde{U}_\alpha$ from $\tilde{p}_1$ to $\tilde{p}_2$. Then the curve $\sigma_0 = F \circ \tilde{\sigma}_0$ is a piecewise C^1 closed curve. Since U is diffeomorphic to a disk in the plane, Remark 7.5.16 tells us that there is a piecewise C^1 homotopy $\Psi\colon [0,1] \times [a,b] \to U$ with fixed endpoints between σ_0 and the constant curve $\sigma_1 \equiv p$. Proposition 7.5.13 provides us then with a piecewise C^1 homotopy $\tilde{\Psi}\colon [0,1] \times [a,b] \to \tilde{U}_\alpha$ with fixed endpoints between the curve $\tilde{\sigma}_0$ and the lift $\tilde{\sigma}_1 \equiv \tilde{p}_1$ of σ_1. In particular, we have $\tilde{p}_2 = \tilde{\sigma}_0(b) = \tilde{\sigma}_1(b) = \tilde{p}_1$, and so F is injective. $\quad\square$

The argument given in the last part of this proof allows us to conclude that simply connected surfaces do not admit non-trivial covering maps:

Proposition 7.5.18. *Let $F\colon \tilde{S} \to S$ be a smooth covering map between surfaces. If S is simply connected then F is a diffeomorphism.*

Proof. We have to prove that F is injective. Let $\tilde{p}_1, \tilde{p}_2 \in \tilde{S}$ be such that $F(\tilde{p}_1) = F(\tilde{p}_2) = p$. Since $\tilde{S}$ is connected, there is a curve $\tilde{\sigma}_0\colon [a,b] \to \tilde{S}$ from $\tilde{p}_1$ to $\tilde{p}_2$. Then the curve $\sigma_0 = F \circ \tilde{\sigma}_0$ is a closed curve; since S is simply connected, there is a homotopy $\Psi\colon [0,1] \times [a,b] \to S$ with fixed endpoints between σ_0 and the constant curve $\sigma_1 \equiv p$. Hence Proposition 7.5.13 provides us with a homotopy $\tilde{\Psi}\colon [0,1] \times [a,b] \to \tilde{S}$ with fixed endpoints between the curve $\tilde{\sigma}_0$ and the lift $\tilde{\sigma}_1 \equiv \tilde{p}_1$ of σ_1. In particular, we have $\tilde{p}_2 = \tilde{\sigma}_0(b) = \tilde{\sigma}_1(b) = \tilde{p}_1$, and so F is injective. $\square$

Example 7.5.19. A consequence of the previous proposition is that no surface of revolution S is simply connected. Indeed, it is straightforward to verify that the map $\varphi\colon \mathbb{R}^2 \to S$ defined in Example 3.1.18 is a smooth covering map but is not a diffeomorphism.

7.6 Non-positive Gaussian curvature

We have reached the last section of this book, devoted to surfaces with non-positive Gaussian curvature. We know, by Problem 6.1, that they cannot be compact; however, we shall be able to say a lot about their geometric structure. Indeed, our goal is to prove that every complete surface with non-positive Gaussian curvature is covered by the plane (Cartan-Hadamard theorem). In particular, every simply connected complete surface with non-positive Gaussian curvature is diffeomorphic to a plane.

Let $S \subset \mathbb{R}^3$ be a complete surface (see Definition 5.4.9); in particular, for all $p \in S$ the exponential map $\exp_p$ is defined on the whole tangent plane T_pS (Theorem 5.4.8). The crucial property of surfaces with non-positive Gaussian curvature is that the exponential map increases the length of tangent vectors.

Proposition 7.6.1. *Let $S \subset \mathbb{R}^3$ be a complete surface with non-positive Gaussian curvature K, and $p \in S$. Then:*

$$\forall v, w \in T_pS \qquad \|d(\exp_p)_v(w)\| \geq \|w\| \,. \qquad (7.16)$$

In particular, $\exp_p$ is a local diffeomorphism.

Proof. If $v = O$, inequality (7.16) is trivial. If $v \neq O$, Lemma 5.2.7 tells us that $\|d(\exp_p)_v(v)\| = \|v\|$, and that if w is orthogonal to v then $d(\exp_p)_v(w)$ is orthogonal to $d(\exp_p)_v(v)$; hence (why?) we may restrict ourselves to prove (7.16) for the vectors w orthogonal to v.

Since S is complete, the map $\Sigma\colon \mathbb{R} \times [0,1] \to S$ given by

$$\Sigma(s,t) = \exp_p\big(t(v + sw)\big)$$

is well defined and of class C^∞. Define now $J\colon [0,1] \to \mathbb{R}^3$ by setting

$$J(t) = \frac{\partial \Sigma}{\partial s}(0,t) = d(\exp_p)_{tv}(tw) \in T_{\sigma_v(t)}S \,,$$

where $\sigma_v\colon \mathbb{R} \to S$ is the geodesic $\sigma_v(t) = \exp_p(tv)$ passing through p and tangent to v. Lemma 5.5.4 tells us that the vector field $J \in \mathcal{T}(\sigma_v)$ is such that

$J(0) = O$, $DJ(0) = w$ and $\langle J, \sigma_v' \rangle \equiv 0$. Moreover, $J(1) = \mathrm{d}(\exp_p)_v(w)$; so our goal has become proving that $\|J(1)\| \geq \|w\|$.

First of all, by Lemma 5.5.4.(iv), we have

$$\frac{\mathrm{d}}{\mathrm{d}t}\langle J, DJ \rangle = \|DJ\|^2 + \langle J, D^2 J \rangle = \|DJ\|^2 - \|v\|^2 (K \circ \sigma_v)\|J\|^2 \geq 0 \,,$$

since $K \leq 0$ everywhere. As $J(0) = O$, this implies that $\langle J, DJ \rangle \geq 0$; so

$$\frac{\mathrm{d}}{\mathrm{d}t}\|DJ\|^2 = 2\langle D^2 J, DJ \rangle = -2\|v\|^2 (K \circ \sigma_v)\langle J, DJ \rangle \geq 0 \,,$$

again by the assumption on the sign of the Gaussian curvature. In particular,

$$\|DJ(t)\|^2 \geq \|DJ(0)\|^2 = \|w\|^2$$

for all $t \in [0,1]$. Hence it follows that

$$\frac{\mathrm{d}^2}{\mathrm{d}t^2}\|J\|^2 = 2\|DJ\|^2 + 2\langle J, D^2 J \rangle \geq 2\|w\|^2 - 2\|v\|^2 (K \circ \sigma_v)\|J\|^2 \geq 2\|w\|^2 \,.$$

By integrating with respect to t from 0 to 1, we find

$$\frac{\mathrm{d}}{\mathrm{d}t}\|J\|^2 \geq 2\|w\|^2 t + \frac{\mathrm{d}\|J\|^2}{\mathrm{d}t}(0) = 2\|w\|^2 t + 2\langle DJ(0), J(0) \rangle = 2\|w\|^2 t \,.$$

Integrating again we get

$$\|J\|^2 \geq \|w\|^2 t^2 + \|J(0)\|^2 = \|w\|^2 t^2 \,,$$

and setting $t = 1$ we obtain $\|J(1)\|^2 \geq \|w\|^2$, as required.

So we have proved (7.16). In particular, it implies that the differential $\mathrm{d}(\exp_p)_v$ of $\exp_p$ is injective for all $v \in T_p S$, and so $\exp_p$ is a local diffeomorphism (Corollary 3.4.28).   □

For the sake of simplicity of exposition, let us introduce the following:

Definition 7.6.2. A map $F: \tilde{S} \to S$ of class C^∞ between surfaces is *expansive* if $\|\mathrm{d}F_p(v)\| \geq \|v\|$ for all $p \in \tilde{S}$ and every $v \in T_p\tilde{S}$.

Clearly, every expansive map is a local diffeomorphism. But if $\tilde{S}$ is complete we may be more precise:

Proposition 7.6.3. *If $\tilde{S}$ is a complete surface, then every expansive map $F: \tilde{S} \to S$ between surfaces is a smooth covering map.*

Proof. Since we know that F is a local diffeomorphism, by Theorem 7.5.17 it suffices to prove that F has the C^1 lifting property.

We begin by showing that we may lift piecewise C^1 curves issuing from a point of the image of F. Let $\sigma: [0, \ell] \to S$ be a piecewise C^1 curve with

$\sigma(0) = p = F(\tilde{p})$. Since F is a local diffeomorphism, there is a neighborhood $\tilde{U} \subseteq \tilde{S}$ of $\tilde{p}$ such that $F|_{\tilde{U}}$ is a diffeomorphism; in particular, $F(\tilde{U})$ is open in S. By continuity, there exists an $\varepsilon > 0$ such that $\sigma([0, \varepsilon)) \subset F(\tilde{U})$; so, by setting $\tilde{\sigma} = (F|_{\tilde{U}})^{-1} \circ \sigma \colon [0, \varepsilon) \to \tilde{S}$ we have found a lift of σ over $[0, \varepsilon)$.

Clearly, we may repeat this argument starting from any point of the support of σ contained in the image of F. Thus the set

$$A = \{t \in [0, \ell] \mid \text{there is a lift } \tilde{\sigma} \colon [0, t] \to \tilde{S} \text{ of } \sigma \text{ with } \tilde{\sigma}(0) = \tilde{p}\}$$

is open in $[0, \ell]$. Let $t_0 = \sup A$; if we prove that $t_0 \in A$, then $t_0 = \ell$ necessarily, and we have lifted σ over all $[0, \ell]$, as required.

Choose a sequence $\{t_n\} \subset A$ approaching t_0; we want first to prove that $\{\tilde{\sigma}(t_n)\}$ has an accumulation point in $\tilde{S}$. Suppose by contradiction that it does not. Since $\tilde{S}$ is complete, the Hopf-Rinow Theorem 5.4.8 implies that the intrinsic distance $d_{\tilde{S}}(\tilde{p}, \tilde{\sigma}(t_n))$ has to diverge for $n \to +\infty$. So we have $L(\tilde{\sigma}|_{[0, t_n]}) \to +\infty$ too. But F is expansive; hence, $L(\sigma|_{[0, t_n]}) \geq L(\tilde{\sigma}|_{[0, t_n]})$, and so $L(\sigma|_{[0, t_n]})$ should diverge too, whereas it actually converges to $L(\sigma|_{[0, t_0]})$.

So, up to taking a subsequence, we may suppose that $\tilde{\sigma}(t_n)$ converges to a point $\tilde{q} \in \tilde{S}$. By continuity, $F(\tilde{q}) = \sigma(t_0)$; in particular, $\sigma(t_0)$ is in the image of F. Let $\tilde{U} \subseteq \tilde{S}$ be an open neighborhood of $\tilde{q}$ over which F is a diffeomorphism; in particular, $F(\tilde{U})$ is an open neighborhood in S of $\sigma(t_0)$. Since $\tilde{q}$ is an accumulation point of $\{\tilde{\sigma}(t_n)\}$, we may find a $n_0 \in \mathbb{N}$ such that $\tilde{\sigma}(t_{n_0}) \in \tilde{U}$. Moreover, there exists an open neighborhood $I \subseteq [0, \ell]$ of t_0 such that $\sigma(I) \subset F(\tilde{U})$. So we may define a lift of σ over I with $(F|_{\tilde{U}})^{-1} \circ \sigma$. Since $(F|_{\tilde{U}})^{-1} \circ \sigma(t_{n_0}) = \tilde{\sigma}(t_{n_0})$, this lift has to coincide with $\tilde{\sigma}$ on $I \cap [0, t_0)$, and thus we have found a lift of σ on $[0, t_0]$. In particular, $t_0 \in A$, as required.

So we have proved that we are able to lift all piecewise C^1 curves issuing from a point of the image of F. Now, if $q \in S$ is an arbitrary point, we may always find a piecewise C^1 curve $\sigma \colon [0, \ell] \to S$ from a point $p \in F(\tilde{S})$ to q. Hence, if $\tilde{\sigma}$ is a lift of σ we have $F(\tilde{\sigma}(\ell)) = \sigma(\ell) = q$, and so $q \in F(\tilde{S})$. Thus F is surjective, and we are done. $\square$

Putting together Propositions 7.6.1 and 7.6.3, we get the promised *Cartan-Hadamard theorem*:

Theorem 7.6.4 (Cartan-Hadamard). *Let $S \subset \mathbb{R}^3$ be a complete surface with non-positive Gaussian curvature. Then for all $p \in S$ the exponential map $\exp_p \colon T_p S \to S$ is a smooth covering map. In particular, if S is simply connected then it is diffeomorphic to a plane.*

Proof. Indeed, Proposition 7.6.1 tells us that $\exp_p$ is expansive, and so it is a covering map by Proposition 7.6.3. Finally, the last claim immediately follows from Proposition 7.5.18. $\square$

Remark 7.6.5. There are non simply connected, complete surfaces with Gaussian curvature negative everywhere; an example is the catenoid (see Examples 4.5.21 and 7.5.19).

Remark 7.6.6. In contrast to what happens with non-compact, closed surfaces with positive Gaussian curvature, a simply connected, non-compact, closed surface with negative Gaussian curvature might well not be a graph over any plane: an example is given by the helicoid S (see Problem 3.2). Indeed, it has Gaussian curvature negative everywhere (Example 4.5.20), and is simply connected, since it admits a global parametrization $\varphi \colon \mathbb{R}^2 \to S$ given by $\varphi(x_1, x_2) = (x_2 \cos x_1, x_2 \sin x_1, a x_1)$.

To prove that it is not a graph with respect to any plane, it suffices (why?) to show that for all $\mathbf{v} = (u_0, v_0, w_0) \in S^2$ there exists a line parallel to $\mathbf{v}$ that intersects S in at least two points. If $w_0 \neq 0, \pm 1$, let $x_1 \in \mathbb{R}^*$ be such that

$$
(\cos x_1, \sin x_1) = \left(\operatorname{sgn}\left(\frac{u_0}{w_0}\right) \sqrt{\frac{u_0^2}{u_0^2 + v_0^2}}, \operatorname{sgn}\left(\frac{v_0}{w_0}\right) \sqrt{\frac{v_0^2}{u_0^2 + v_0^2}} \right) ,
$$

and set $x_2 = a x_1 \sqrt{(u_0^2 + v_0^2)/w_0^2}$; note that we have infinitely many distinct choices for x_1. Then

$$
\varphi(x_1, x_2) = \frac{a x_1}{w_0} \mathbf{v} ,
$$

and so the line through the origin parallel to $\mathbf{v}$ intersects S in infinitely many points.

Since the x- and z-axis are contained in S, the lines passing through the origin and parallel to $(0, 0, \pm 1)$ or to $(\pm 1, 0, 0)$ also intersect S in infinitely many points.

We are left with the lines parallel to $(u_0, v_0, 0)$ with $v_0 \neq 0$. Let $\psi \in \mathbb{R}^*$ be such that $(u_0, v_0) = (\cos \psi, \sin \psi)$, and let $p_0 = (0, 0, a\psi) = \varphi(\psi, 0) \in S$. But then

$$
\varphi(\psi, x_2) = (x_2 u_0, x_2 v_0, a\psi) = p_0 + x_2 \mathbf{v} ;
$$

hence the whole line parallel to $\mathbf{v}$ passing through p_0 is contained in S, and so we have proved that S is not a graph.

References

1. Berger, M.: Geometry I, II. Second corrected printing. Springer-Verlag, Berlin (1996).
2. Blåsjö, V.: The isoperimetric problem. Amer. Math. Monthly, **112**, 526–566 (2005).
3. Chern, S.-S.: An elementary proof of the existence of isothermal parameters on a surface. Proc. Amer. Math. Soc., **6**, 771–782 (1955).
4. do Carmo, M.P.: Differential geometry of curves and surfaces. Prentice-Hall, Englewood Cliffs (1976).
5. Fleming, W.: Functions of several variables. Second edition, Springer-Verlag, Berlin (1977).
6. Holm, P.: The theorem of Brown and Sard. Enseign. Math , **33**, 199–202 (1987).
7. Kelley, J.L.: General topology. GTM 27, Springer-Verlag, Berlin (1975).
8. Lang, S.: Differential and Riemannian manifolds. GTM 160, Springer-Verlag, Berlin (1995).
9. Lax, P.D.: A short path to the shortest path. Amer. Math. Monthly, **102**, 158–159 (1995).
10. Lee, J.M.: Riemannian manifolds. GTM 176, Springer-Verlag, Berlin (1997).
11. Lee, J.M.: Introduction to topological manifolds. GTM 202, Springer-Verlag, Berlin (2000).
12. Lee, J.M.: Introduction to smooth manifolds. GTM 218, Springer-Verlag, Berlin (2003).
13. Lipschutz, M.M.: Schaum's outline of differential geometry. McGraw-Hill, New York (1969).
14. Milnor, T.K.: Efimov's theorem about complete immersed surfaces of negative curvature. Adv. Math., **8**, 474–543 (1972).
15. Milnor, J., Munkres, J.: Differential topology. Princeton University Press, Princeton (1958).
16. Montiel, S., Ros, A.: Curves and surfaces. American Mathematical Society, Providence (2005).
17. Morgan, F.: Geometric measure theory. A beginner's guide. Academic Press, San Diego (2000).
18. Osserman, R.: The four-or-more vertex theorem. Amer. Math. Monthly, **92**, 332–337 (1985).

19. Pederson, R.N.: The Jordan curve theorem for piecewise smooth curves. Amer. Math. Monthly, **76**, 605–610 (1969).
20. Samelson, H.: Orientability of hypersurfaces in $\mathbb{R}^n$. Proc. Amer. Math. Soc., **22**, 301–302 (1969).
21. Sernesi, E.: Linear algebra. A geometric approach. Chapman and Hall/CRC, London (1993).
22. Spivak, M.: A comprehensive introduction to differential geometry (5 voll.). Second edition. Publish or Perish, Berkeley (1979).
23. Thomassen, C.: The Jordan-Schönflies theorem and the classification of surfaces. Amer. Math. Monthly, **99**, 116–131 (1992).
24. Walter, W.: Ordinary differential equations. GTM 182, Springer-Verlag, Berlin (1998).

List of symbols

Index

Collana Unitext – La Matematica per il 3+2

Series Editors:
A. Quarteroni (Editor-in-Chief)
L. Ambrosio
P. Biscari
C. Ciliberto
G. van der Geer
G. Rinaldi
W.J. Runggaldier

Editor at Springer:
F. Bonadei
francesca.bonadei@springer.com

As of 2004, the books published in the series have been given a volume number. Titles in grey indicate editions out of print.
As of 2011, the series also publishes books in English.

S. Margarita, E. Salinelli
MultiMath - Matematica Multimediale per l'Università
2004, XX+270 pp, ISBN 88-470-0228-1

A. Quarteroni, R. Sacco, F.Saleri
Matematica numerica (2a Ed.)
2000, XIV+448 pp, ISBN 88-470-0077-7
2002, 2004 ristampa riveduta e corretta
(1a edizione 1998, ISBN 88-470-0010-6)

13. A. Quarteroni, F. Saleri
    Introduzione al Calcolo Scientifico (2a Ed.)
    2004, X+262 pp, ISBN 88-470-0256-7
    (1a edizione 2002, ISBN 88-470-0149-8)

14. S. Salsa
    Equazioni a derivate parziali - Metodi, modelli e applicazioni
    2004, XII+426 pp, ISBN 88-470-0259-1

15. G. Riccardi
    Calcolo differenziale ed integrale
    2004, XII+314 pp, ISBN 88-470-0285-0

16. M. Impedovo
    Matematica generale con il calcolatore
    2005, X+526 pp, ISBN 88-470-0258-3

17. L. Formaggia, F. Saleri, A. Veneziani
    Applicazioni ed esercizi di modellistica numerica
    per problemi differenziali
    2005, VIII+396 pp, ISBN 88-470-0257-5

18. S. Salsa, G. Verzini
    Equazioni a derivate parziali – Complementi ed esercizi
    2005, VIII+406 pp, ISBN 88-470-0260-5
    2007, ristampa con modifiche

19. C. Canuto, A. Tabacco
    Analisi Matematica I (2a Ed.)
    2005, XII+448 pp, ISBN 88-470-0337-7
    (1a edizione, 2003, XII+376 pp, ISBN 88-470-0220-6)

20. F. Biagini, M. Campanino
    Elementi di Probabilità e Statistica
    2006, XII+236 pp, ISBN 88-470-0330-X

21. S. Leonesi, C. Toffalori
    Numeri e Crittografia
    2006, VIII+178 pp, ISBN 88-470-0331-8

22. A. Quarteroni, F. Saleri
    Introduzione al Calcolo Scientifico (3a Ed.)
    2006, X+306 pp, ISBN 88-470-0480-2

23. S. Leonesi, C. Toffalori
    Un invito all'Algebra
    2006, XVII+432 pp, ISBN 88-470-0313-X

24. W.M. Baldoni, C. Ciliberto, G.M. Piacentini Cattaneo
    Aritmetica, Crittografia e Codici
    2006, XVI+518 pp, ISBN 88-470-0455-1

25. A. Quarteroni
    Modellistica numerica per problemi differenziali (3a Ed.)
    2006, XIV+452 pp, ISBN 88-470-0493-4
    (1a edizione 2000, ISBN 88-470-0108-0)
    (2a edizione 2003, ISBN 88-470-0203-6)

26. M. Abate, F. Tovena
    Curve e superfici
    2006, ristampa con modifiche 2008, XIV+394 pp, ISBN 88-470-0535-3

27. L. Giuzzi
    Codici correttori
    2006, XVI+402 pp, ISBN 88-470-0539-6

28. L. Robbiano
    Algebra lineare
    2007, XVI+210 pp, ISBN 88-470-0446-2

29. E. Rosazza Gianin, C. Sgarra
    Esercizi di finanza matematica
    2007, X+184 pp, ISBN 978-88-470-0610-2

30. A. Machì
Gruppi - Una introduzione a idee e metodi della Teoria dei Gruppi
2007, XII+350 pp, ISBN 978-88-470-0622-5
2010, ristampa con modifiche

31. Y. Biollay, A. Chaabouni, J. Stubbe
Matematica si parte!
A cura di A. Quarteroni
2007, XII+196 pp, ISBN 978-88-470-0675-1

32. M. Manetti
Topologia
2008, XII+298 pp, ISBN 978-88-470-0756-7

33. A. Pascucci
Calcolo stocastico per la finanza
2008, XVI+518 pp, ISBN 978-88-470-0600-3

34. A. Quarteroni, R. Sacco, F. Saleri
Matematica numerica (3a Ed.)
2008, XVI+510 pp, ISBN 978-88-470-0782-6

35. P. Cannarsa, T. D'Aprile
Introduzione alla teoria della misura e all'analisi funzionale
2008, XII+268 pp, ISBN 978-88-470-0701-7

36. A. Quarteroni, F. Saleri
Calcolo scientifico (4a Ed.)
2008, XIV+358 pp, ISBN 978-88-470-0837-3

37. C. Canuto, A. Tabacco
Analisi Matematica I (3a Ed.)
2008, XIV+452 pp, ISBN 978-88-470-0871-3

38. S. Gabelli
Teoria delle Equazioni e Teoria di Galois
2008, XVI+410 pp, ISBN 978-88-470-0618-8

39. A. Quarteroni
Modellistica numerica per problemi differenziali (4a Ed.)
2008, XVI+560 pp, ISBN 978-88-470-0841-0

40. C. Canuto, A. Tabacco
Analisi Matematica II
2008, XVI+536 pp, ISBN 978-88-470-0873-1
2010, ristampa con modifiche

41. E. Salinelli, F. Tomarelli
Modelli Dinamici Discreti (2a Ed.)
2009, XIV+382 pp, ISBN 978-88-470-1075-8

42. S. Salsa, F.M.G. Vegni, A. Zaretti, P. Zunino
Invito alle equazioni a derivate parziali
2009, XIV+440 pp, ISBN 978-88-470-1179-3

43. S. Dulli, S. Furini, E. Peron
Data mining
2009, XIV+178 pp, ISBN 978-88-470-1162-5

44. A. Pascucci, W.J. Runggaldier
Finanza Matematica
2009, X+264 pp, ISBN 978-88-470-1441-1

45. S. Salsa
Equazioni a derivate parziali – Metodi, modelli e applicazioni (2a Ed.)
2010, XVI+614 pp, ISBN 978-88-470-1645-3

46. C. D'Angelo, A. Quarteroni
Matematica Numerica – Esercizi, Laboratori e Progetti
2010, VIII+374 pp, ISBN 978-88-470-1639-2

47. V. Moretti
Teoria Spettrale e Meccanica Quantistica – Operatori in spazi di Hilbert
2010, XVI+704 pp, ISBN 978-88-470-1610-1

48. C. Parenti, A. Parmeggiani
Algebra lineare ed equazioni differenziali ordinarie
2010, VIII+208 pp, ISBN 978-88-470-1787-0

49. B. Korte, J. Vygen
Ottimizzazione Combinatoria. Teoria e Algoritmi
2010, XVI+662 pp, ISBN 978-88-470-1522-7

50. D. Mundici
Logica: Metodo Breve
2011, XII+126 pp, ISBN 978-88-470-1883-9

51. E. Fortuna, R. Frigerio, R. Pardini
Geometria proiettiva. Problemi risolti e richiami di teoria
2011, VIII+274 pp, ISBN 978-88-470-1746-7

52. C. Presilla
Elementi di Analisi Complessa. Funzioni di una variabile
2011, XII+324 pp, ISBN 978-88-470-1829-7

53. L. Grippo, M. Sciandrone
Metodi di ottimizzazione non vincolata
2011, XIV+614 pp, ISBN 978-88-470-1793-1

54. M. Abate, F. Tovena
Geometria Differenziale
2011, XIV+466 pp, ISBN 978-88-470-1919-5

55. M. Abate, F. Tovena
Curves and Surfaces
2011, XIV+390 pp, ISBN 978-88-470-1940-9

The online version of the books published in this series is available at SpringerLink.
For further information, please visit the following link:
http://www.springer.com/series/5418